Reviews in Fluorescence 2005

Reviews in Fluorescence 2005

Edited by

CHRIS D. GEDDES

Institute of Fluorescence and Center for Fluorescence Spectroscopy
University of Maryland Biotechnology Institute
Baltimore, Maryland

and

JOSEPH R. LAKOWICZ

Center for Fluorescence Spectroscopy
University of Maryland School of Medicine
Baltimore, Maryland

 Springer

Chris D. Geddes
Institute of Fluorescence and Center for Fluorescence Spectroscopy
University of Maryland Biotechnology Institute
Baltimore, Maryland

and

Joseph R. Lakowicz
Center for Fluorescence Spectroscopy
University of Maryland School of Medicine
Baltimore, Maryland

[Cover illustration:]

A C.I.P. Catalogue record for this book is available
from the Library of Congress.

ISBN 0-387-23628-7   e-ISBN 0-387-23690-2   Printed on acid-free paper.

Printed in the United States of America.   (HAM)

9 8 7 6 5 4 3 2 1

springeronline.com

CONTRIBUTORS AND BIOGRAPHIES

Ramachandram Badugu, Center for Fluorescence Spectroscopy, University of Maryland School of Medicine, Baltimore, MD, USA.

Dr. Badugu is a Research Associate at the Center for Fluorescence Spectroscopy, University of Maryland School of Medicine, Baltimore, MD, USA. He has a B.Sc. in Biological and Chemical Sciences and a M.Sc. in Chemistry from Osmania University, Hyderabad, India, in 1992 and 1994, respectively. Subsequently he moved to University of Hyderabad, Hyderabad, India, and obtained his M.Phil. and Ph.D. degrees in Chemistry in 1995 and 1999, respectively, while working with Professor Anunay Samanta. In early 2000, he has been selected for the Japanese Society for the Promotion of Science (JSPS) fellowship under which he worked for two years with Professor Mitsuo Kira at Department of Chemistry, Tohoku University, Sendai, Japan. From April 2002 he worked as a Frontier Research Scientist with Professor Kenkichi Sakamoto for about a year at Photodynamics Research Center, The Institute of Physical and Chemical Research (RIKEN), Sendai, Japan. He has been involved in the design and development of Chemical Sensors/Fluorosensors for the detection of various analytes including transition metal ions, glucose and other anionic analytes such as cyanide, fluoride, etc. Along with sensing he is also interested in the photophysics of small molecules and organosilicon compounds.

Karsten Bahlmann, Massachusetts Institute of Technology, Department of Mechanical Engineering and Division of Biological Engineering, 77 Massachusetts Avenue, Cambridge, MA 02139

Wolfgang Becker, Becker & Hickl GmbH, Berlin, Germany.

Dr. Becker is a specialist of optical short-time measurement techniques and obtained his PhD 1979 in Berlin, Germany. Since 1993 he is the head of Becker & Hickl GmbH in Berlin. His field of interest is development and application of time-correlated single photon counting techniques. He likes cats, skiing and beach volleyball.

Aleš Benda, J. Heyrovský Institute, Academy of Sciences of Czech Republic and Center for Complex Molecular Systems and Biomolecules, Dolejškova 3, Cz-18223 Prague 8, Czech Republic.

Axel Bergmann, Becker & Hickl GmbH, Berlin, Germany.

Dr. Bergmann received his PhD in physics from the Technical University of Berlin in 1999. In 2000 he joined Becker & Hickl in Berlin as a senior designer of hard- and software products for photon counting instrumentation. His field of interest are biomedical applications of time-correlated single-photon counting, in including FLIM and FRET microscopy. He likes horses and currently spends his spare time re-building his home.

Kankan Bhattacharyya, Department of Physical Chemistry, Indian Association for the Cultivation of Science, Jadavpur, Kolkata 700 032, India.

Dr. Bhattacharyya obtained Ph. D. degree from Calcutta University in 1984 (supervisor: M. Chowdhury). He was a post-doctoral research associate at the Radiation Laboratory, University of Notre Dame (with P. K. Das) and at Columbia University (with K. B. Eisenthal). He joined the faculty of Indian Association for the Cultivation of Science in 1987 and became a Professor in 1998. His major research interest includes ultrafast laser spectroscopy and organized assemblies.

Amitabha Chattopadhyay, Centre for Cellular and Molecular Biology, Hyderabad 500 007, India.

Dr. Chattopadhyay obtained his Ph.D. from the State University of New York (SUNY) at Stony Brook and worked as a Postdoctoral Fellow at the University of California, Davis. He subsequently joined the Centre for Cellular and Molecular Biology (CCMB) in Hyderabad, India and is now a Group Leader there. His research is focussed on the organization and dynamics of biological membranes utilizing fluorescence approaches. His work has contributed significantly to the understanding of membrane organization and domains, and the interplay between membrane lipids and proteins, especially for neuronal membranes. He is currently on the editorial boards of Journal of Fluorescence, Cellular and Molecular Neurobiology, Molecular Membrane Biology, Bioscience Reports, and Journal of Biosciences. He was awarded the prestigious Shanti Swarup Bhatnagar Prize in 2001 for his research accomplishments. He is also a Fellow of the Indian Academy of Sciences.

Ben DeGraff, Chemistry Department, James Madison University, Harrisonburg, VA 22807.

Dr. DeGraff is a professor of chemistry at James Madison University in Virginia. His research interests lie in physical inorganic chemistry with a focus on luminescent materials and their characterization. Outside the office he enjoys skiing and hiking.

James Demas, Chemistry Department, University of Virginia, Charlottesville, VA 22904.

Dr. Demas is a professor of chemistry at the University of Virginia. His research interests include luminescence, especially of inorganic complexes, their applications, and data analysis. His hobbies include photography, skiing, hiking, karate, and biking.

Pinghau Ge, Loomis Laboratory of Physics, University of Illinois, 1110 W. Green St., Urbana, IL 61801.

Chris D. Geddes, Institute of Fluorescence, University of Maryland Biotechnology Institute, 725 W. Lombard St., Baltimore, MD, 21201. USA.

Dr. Geddes is an Associate Professor and Director of the Institute of Fluorescence at the University of Maryland Biotechnology Institute in Baltimore. The Institute of

Fluorescence features both the Laboratory for Advanced Medical Plasmonics (LAMP) and the Laboratory for Advanced Fluorescence Spectroscopy (LAFS). Dr Geddes also has an adjunct appointment (Assistant Professor) in the Center for Fluorescence Spectroscopy, University of Maryland School of Medicine. He has a B.Sc. in chemistry from Lancaster University in England and a Ph.D. in physical chemistry (fluorescence spectroscopy) from the University of Wales Swansea. He is the editor-in-chief of the Journal of Fluorescence, and both editor-in-chief and founding editor of the Who's Who in Fluorescence and Annual Reviews in Fluorescence volumes. He is also executive director of the Society of Fluorescence, and co-series editor of the Topics in Fluorescence Spectroscopy series.

Michelle Gee, School of Chemistry, University of Melbourne, Parkville, 3010, Victoria, Australia.

Dr. Gee returned to the University of Melbourne after post-doctoral terms at UC Santa Barbara and Princeton University. She has worked extensively in the area of surface and colloid science, and has applied variable angle of incidence evanescent wave-based techniques to the study of polymers and proteins.

Martin Hof, J. Heyrovský Institute, Academy of Sciences of Czech Republic and Center for Complex Molecular Systems and Biomolecules, Dolejškova 3, Cz-18223 Prague 8, Czech Republic.

Dr. Hof was born in Friedberg/Germany in 1962. He finished his PhD work and his habilitation work at the University Wurzburg/Germany in the years 1990 and 1999, respectively. In 1997 he joined the J. Heyrovský Institute of Physical Chemistry, Academy of Sciences of the Czech Republic, where he is presently heading the biospectroscopy laboratory. Aleš Benda was born in Olomouc/Czech Republic in 1980. He is working on his PhD thesis in the biospectroscopy laboratory. Teresa Kral was born in Ostrowiec Sw/Poland in 1966. She obtained her PhD at Łódz University in 1999. Presently she is working on her habilitation at the Agr) University and in the above mentioned biospectroscopy lab. Marek Langner was born in Ostrow Wlkp/Poland in 1956. He finished his PhD work at the Medical Academy of Wrocław in 1986 and his habibilitation work at the Łódz University in 1999. In 1999 he joined the Institute of Physics at Wrocław Technical University, where he is presently heading the Laboratory for Biophysics of Macromolecular Aggregates.

Hayden Huang, Massachusetts Institute of Technology, Department of Mechanical Engineering and Division of Biological Engineering, 77 Massachusetts Avenue, Cambridge, MA 02139

Devaki A. Kelkar, Centre for Cellular and Molecular Biology, Hyderabad 500 007, India.
Teresa Kral, J. Heyrovský Institute, Academy of Sciences of Czech Republic and Center for Complex Molecular Systems and Biomolecules, Dolejškova 3, Cz-18223 Prague 8, Czech Republic.

Ki Hean Kim, Massachusetts Institute of Technology, Department of Mechanical Engineering and Division of Biological Engineering, 77 Massachusetts Avenue, Cambridge, MA 02139

Hyuk-Sang Kown, Massachusetts Institute of Technology, Department of Mechanical Engineering and Division of Biological Engineering, 77 Massachusetts Avenue, Cambridge, MA 02139

Irina M Kuznetsova, Institute of Protein Biochemistry, CNR, Via Pietro Castellino, 111 80131 Naples, Italy.

Joseph R. Lakowicz, Center for Fluorescence Spectroscopy, University of Maryland School of Medicine, Baltimore, MD, USA.

Dr. Lakowicz is director of the Center for Fluorescence Spectroscopy at the University of Maryland School of Medicine in Baltimore. He has a BS in chemistry from LaSalle University and an MS and Ph.D. in Biochemistry from the University of Illinois at Urbana. He is also the founding editor of the Journal of Biomedical Optics and the Journal of Fluorescence. He is a co-founder and co-President of the Society of Fluorescence. He is also co-editor of the Who's Who in Fluorescence and Annual Reviews in Fluorescence.

Marek Langner, Wrocław University of Technology, Institute of Physics, Wyb. Wyspiańskiego 27, 50-370, Wrocław, Poland.

Richard T. Lee, TissueVision, Inc., 98 Line Street Suite 2, Somerville, MA 02143.

Raj Mutharasan, Department of Chemical Engineering, Drexel University Philadelphia, PA 19104.

Dr. Mutharasan received his B. S degree in chemical engineering from IIT Madras (India) and a Ph. D in Chemical Engineering from Drexel University. He has been at Drexel University since 1974 where he is the Frank A. Fletcher Professor of Chemical Engineering. His research interests are in biophysics, biophotonics and microcantilever for sensor development, and process biotechnology.

Manuel Prieto, Ph. D. (Group leader), **Luís Loura**, Ph. D. (Staff member) and **Rodrigo de Almeida**, Ph. D. (Pos-doc), received their Ph.D. degrees at the Instituto Superior Técnico, Lisbon, Portugal. Their scientific interests are on the application of fluorescence techniques on membrane biophysics, namely the study of lipid domains (rafts), lipid-peptide, lipid-protein interaction, DNA-lipid interaction, and polyene antibiotics. The derivation of models for energy transfer, taking into account the specific topologies for these systems and model fitting to time-resolved data, is one of the main group activities.

Timothy Ragan, Massachusetts Institute of Technology, Department of Mechanical Engineering and Division of Biological Engineering, 77 Massachusetts Avenue, Cambridge, MA 02139

H. Raghuraman, Centre for Cellular and Molecular Biology, Hyderabad 500 007, India.

Jeff G. Reifernberger, Loomis Laboratory of Physics, University of Illinois, 1110 W. Green St., Urbana, IL 61801

Alan G. Ryder,   Department of Chemistry, and National Centre for Biomedical Engineering Science, National University of Ireland – Galway, Galway, Ireland.

Dr. Ryder is a Science Foundation Ireland Investigator in the National Centre for Biomedical Engineering Science, and an adjunct lecturer in the Dept. of Chemistry, both at the National University of Ireland–Galway. He obtained a B.Sc. honours (Chemistry, 1989) and Ph.D. degrees (Inorganic chemistry, 1994) from NUI-Galway. After a stint as a postdoctoral researcher in University College Cork, he rejoined NUI-Galway to work on developing quantitative Raman spectroscopic methods for measuring illicit narcotic concentrations.   He leads the Nanoscale Biophotonics research group, which is an interdisciplinary team, developing novel analytical methods and instrumentation. Current research involves the use of fluorescence and Raman spectroscopies for the development of quantitative and qualitative analysis methods. Research projects include: fluorescence lifetime based pH sensors, quantitative Raman spectroscopy for forensic applications, time-resolved fluorescence instrumentation development, and the fluorescence analysis of petroleum fluids. More details are available on the group website:
http://www.nuigalway.ie/chem/AlanR/.

D'Auria Sabato, Institute of Protein Biochemistry, CNR, Via Pietro Castellino, 111 80131 Naples, Italy.

Dr. Sabato is senior scientist at the Institute of Protein Biochemistry Italian National Research Council, Naples. He is member of the editorial board of Chemical Biology, BioMed, UK. The research interests of D'Auria's lab are the use of fluorescence techniques for the characterization of enzymes and proteins also isolated from extremophilic organisms.

Herbert Schneckenburger, Hochschule Aalen, Institut für Angewandte Forschung, Beethovenstraße 1, 73430 Aalen, Germany.

Dr. Schneckenburger is a professor of Physics, Optics and Biophotonics at Hochschule Aalen and private lecturer at the Medical faculty of the University of Ulm. He passed his PhD in Physics at the University of Stuttgart in 1979 and his habilitation in Biomedical Optics at the University of Ulm in 1992. From 19791986 he worked as a scientist at the GSF Research Center for Environment and Health in Munich. His work is concentrated on in vitro diagnostics and biomedical screening using time-resolving spectroscopic, microscopic and laser techniques.

Colin Scholes is currently a PhD student developing total internal reflection-base spectroscopic techniques principally applied to protein adsorption at the School of Chemistry, University of Melbourne, Parkville, 3010, Victoria, Australia.

Paul Selvin, Loomis Laboratory of Physics, University of Illinois, 1110 W. Green St., Urbana, IL 61801.

Dr. Selvin undertook his undergraduate studies at the University of Michigan in Ann Arbor. He then did his Ph.D. in physics at the University of California, Berkeley, with Melvin Klein, where he studied the motion of DNA, using EPR and fluorescence. He fell in love with fluorescence and did a postdoc with John Hearst in Chemistry at UC Berkeley. He then became a staff scientist at the Lawrence Berkeley Laboratory, where he did single molecule fluorescence and Lanthanide luminescent. in 1997 he went to the University of Illinois, Urbana-Champaign, where he has studied Lanthanides applied to ion channels and muscles, and also single molecules associated with molecular motors.

P. M. Shankar, Department of Electrical and Computer Engineering, Drexel University Philadelphia, PA 19104.

Dr. Shankar received a Masters degree in Physics from Kerala University (India) and a Masters degree in Applied Optics and a Ph. D in Electrical Engineering, both from Indian Institute of Technology, Delhi. He has been at Drexel University since 1982 where he is the Allen Rothwarf Professor of Electrical and Computer Engineering. His research interests are in Fiberoptics, Biophotonics, Wireless Communications, and Medical Ultrasound.

Trevor Smith, School of Chemistry, University of Melbourne, Parkville, 3010, Victoria, Australia.

Following a post-doctoral period working in the Chemistry Department at Imperial College, London, where he was introduced to the field of evanescent wave-induced fluorescence spectroscopy, Mr. Smith returned to the School of Chemistry at University of Melbourne where he has held several research positions including an Australian Research Council QEII Fellowship, and is now on the academic staff. He has worked in various areas involving applications of time-resolved fluorescence techniques throughout his career.

Peter So, Massachusetts Institute of Technology, Department of Mechanical Engineering and Division of Biological Engineering, 77 Massachusetts Avenue, Cambridge, MA 02139.

Dr. So completed his Ph.D. in 1992 at Princeton University working on the morphological phase transition of lipid water systems. He subsequently worked as a postdoctoral research associate in the Laboratory for Fluorescence Dynamics at the University of Illinois developing fluorescence microscopy and spectroscopy techniques. He joined the faculty in Massachusetts Institute of Technology in 1996. He is currently an associate professor in both the Mechanical Engineering Department and the Biological Engineering Division. The research of his laboratory focuses on the developing of novel optical imaging techniques. The current projects include the development of two-photon endoscopy, high through 3D tissue image cytometry, spectral resolved fluorescence correlation spectroscopy, lifetime resolved fluorescence energy transfer, fluorescence laser tracking microrheometry, and high resolution total internal refraction imaging. These techniques are applied to biological and medical studies ranging from single molecules to tissues.

Maria Staiano, Institute of Protein Biochemistry, CNR, Via Pietro Castellino, 111 80131 Naples, Italy.

Richard Blair Thompson, University of Maryland School of Medicine, Baltimore, Maryland 21201.

Dr. Thompson was born in Ohio and raised north of Chicago, Illinois. He received a B.A. in Biology from Northwestern University; while there, he began biochemical studies with E. Margoliash. He received the Ph.D. in Biochemistry from the University of Illinois in Urbana-Champaign working under the direction of Thomas O. Baldwin. He trained as a postdoctoral fellow in the laboratory of Joseph Lakowicz at the University of Maryland at Baltimore before moving to the U.S. Naval Research Laboratory as a National Research Council Associate. At the Naval Research Laboratory he began work on fluorescence-based biosensors under Paul Schoen and subsequently became a Supervisory Research Chemist under the direction of Frances Ligler; he received a Navy Special Act Award for activity related to Operation Desert Storm. He joined the faculty of the University of Maryland School of Medicine in the Department of Biochemistry and Molecular Biology where he is now Associate Professor. He serves on the Editorial Boards of the Journal of Fluorescence and the Journal of Biomedical Optics, as well as panels for the National Research Council, National Institutes of Health, National Science Foundation, and other agencies.

Konstantin K.Turoverov, Institute of Protein Biochemistry, CNR, Via Pietro Castellino, 111   80131 Naples, Italy.

Michael Wagner, Hochschule Aalen, Institut für Angewandte Forschung, Beethovenstraße 1, 73430 Aalen, Germany.

Mr. Wagner passed his engineering diploma in Optoelectronics in 2001 and his Master in Photonics in 2003, both at Hochschule Aalen. He is presently preparing his PhD on "Membrane Dynamics of Living Cells" at the University of Ulm.

Gerald M. Wilson, University of Maryland School of Medicine, Baltimore, Maryland 21201.

Dr. Wilson is an Assistant Professor in the Department of Biochemistry and Molecular Biology at the University of Maryland, School of Medicine, and is a member of the Center for Fluorescence Spectroscopy and the Greenebaum Cancer Center. His principal research interests are the cellular mechanisms regulating messenger RNA decay kinetics, particularly those controlling the production of oncoproteins, cytokines, and inflammatory mediators in mammalian cells. These studies have included the application of fluorescence anisotropy-based systems to quantitatively assess the thermodynamics and kinetics of RNA-protein binding, and resonance energy transfer-based systems to monitor RNA folding events that regulate or result from protein binding.

PREFACE

Last year we launched Volume 1 of the Reviews in Fluorescence series. The volume was well-received by the fluorescence community, with many e-mails and letters providing valuable feedback, we subsequently thank you all for your continued support.

After the volume was published we were most pleased to learn that the volume is to be citable and indexed, appearing on the ISI database. Subsequently, as well as the series having an impact number in due course, individual chapters will appear on the database and be both citable and keyword searchable. We feel that this will be a powerful resource to both authors and readers, further disseminating leading-edge fluorescence based material.

Our intention with this new series is to both disseminate and archive the most recent developments in both past and emerging fluorescence based disciplines. While all chapters are invited, we welcome and indeed encourage the fluorescence community to suggest areas of interest that they feel need to be covered by the series.

In this new volume, Reviews in Fluorescence 2005, Volume 2, we have invited reviews in areas such as: Multi-dimensional Time-correlated Single Photon Counting; Fluorescence Correlation Spectroscopy; RNA folding; Lanthanide Probes and Fluorescent Biosensors to name but just a few. We hope you find this volume a useful resource and we look forward to receiving any suggestions you may have.

Finally we would like to thank the authors for their timely articles, Caroleann Aitken for the front cover design, Kadir Aslan for typesetting and Mary Rosenfeld for administrative support.

Chris D. Geddes
Joseph R. Lakowicz,

August 25th 2004.
Baltimore, Maryland, USA.

CONTENTS

5. SOME ASPECTS OF DNA CONDENSATION OBSERVED BY FLUORESCENCE CORRELATION SPECTROSCOPY

Teresa Kral, Aleš Benda, Martin Hof and Marek Langner

6. LUMINESCENCE-BASED OXYGEN SENSORS

B. A. DeGraff and J. N. Demas

Gerald M. Wilson

Trevor A. Smith, Michelle L. Gee and Colin A. Scholes

12. APPLICATION OF FLUORESCENCE TO UNDERSTAND THE INTERACTION OF PEPTIDES WITH BINARY LIPID MEMBRANES..... 271

Rodrigo F. M. de Almeida, Luís M. S. Loura, and Manuel Prieto

CONTENTS

ORGANIZED ASSEMBLIES PROBED BY FLUORESCENCE SPECTROSCOPY

Kankan Bhattacharyya[*]

1.1. INTRODUCTION

Self-organized assemblies originate in an aqueous solution because of the tendency of a biological macromolecule to expose its hydrophilic part to water and to keep the hydrophobic portion away from water.[1] Examples of such assemblies range from the unique biologically active structure (native form) of a protein and the DNA double helix to many supramolecules, guest-host complexes and aggregates of amphiphilic molecules (e.g. lipids, micelles). Structures of some organized assemblies are shown in Fig. 1.1. They play a key role in molecular recognition, bio-catalysis, targeted drug delivery,[2] and in many emerging areas such as, dynamic combinatorial chemistry,[3] and adaptive chemistry.[4]

The polarity and solvation dynamics in the hydration layer of the organized assemblies are crucial in many chemical processes. The very high polarity (dielectric constant) of water arises principally from the extended hydrogen bond network and mutual polarization of the water molecules in close proximity.[5] Because of polarization, the dipole moment of water in the liquid phase (2.6 D) is higher than that in vapor (1.85 D).[5] A water molecule bound to a macromolecule can not polarize another water molecule. This causes a marked reduction in the dielectric constant of the hydration layer of an organized assembly compared to bulk water. In recent years, time resolved fluorescence spectroscopy has yielded a wealth of information on the polarity and solvation dynamics in many organized assemblies.[6-7] The fluorescence data is supplemented by large scale computer simulations on dielectric constant and dynamics in organized assemblies.[8-10] In this chapter, we will summarize the most recent results in this area.

We will begin with fluorescence anisotropy decay in organized assemblies. A major portion of this chapter will be devoted to solvation dynamics of water (and a few other liquids) confined in many organized assemblies. The confined water displays a component of

*Department of Physical Chemistry, Indian Association for the Cultivation of Science, Jadavpur, Kolkata 700 032, India. e-mail: pckb@mahendra.iacs.res.in.

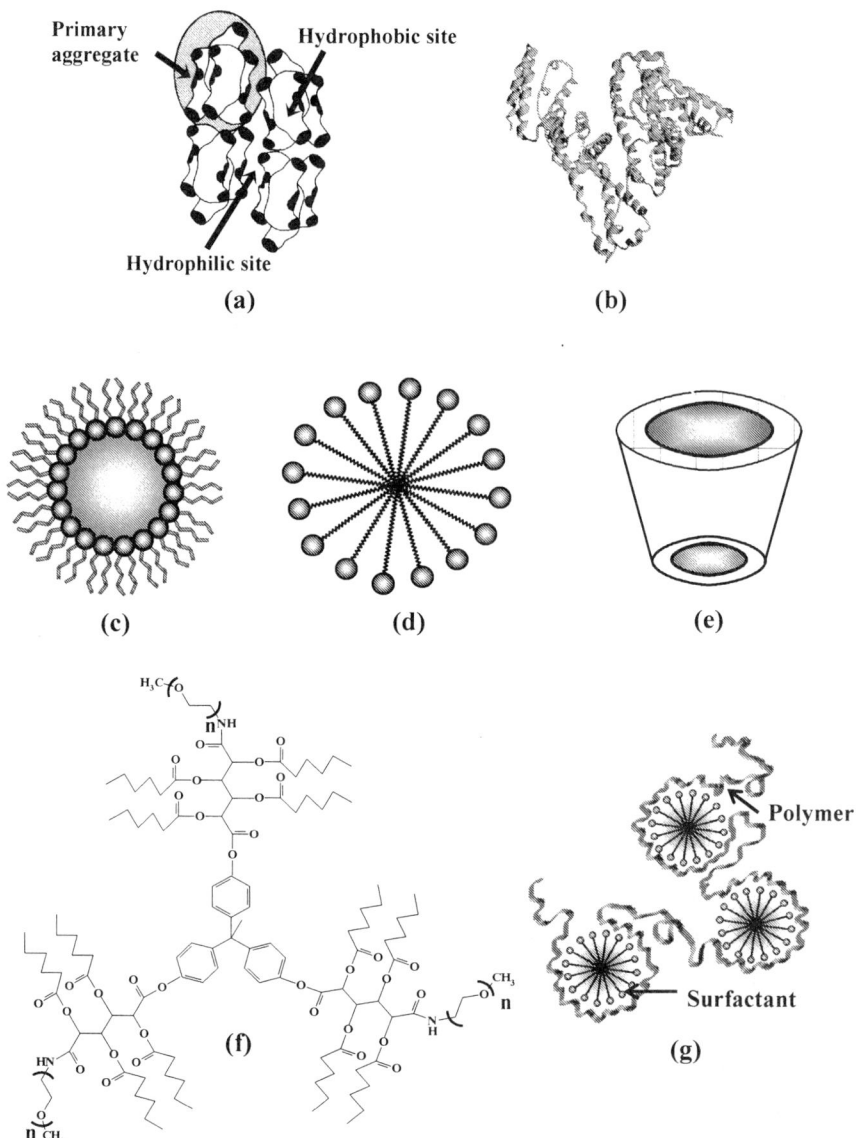

Fig. 1.1. Structure of some organized assemblies: (a) secondary aggregate of bile salt, (b) a protein, human serum albumin, (c) microemulsion, (d) micelle, (e) cyclodextrin, (f) amphiphilic star like macromolecule (ASM), and (g) polymer-surfactant aggregate.

solvation dynamics which is 100-1000 times slower compared to bulk water.[11-14] We will discuss several examples of ultraslow solvation dynamics in organized assemblies and the possible origins of the ultraslow component. The ultraslow component of solvation causes a marked retardation of many polar reactions such as electron and proton transfer. We will illustrate this with several examples.

1.2. FLUORESCENCE ANISOTROPY DECAY IN ORGANIZED ASSEMBLIES

Fluorescence anisotropy decay is monitored by the anisotropy function, $r(t)$ which is defined as,

$$r(t) = \frac{I_{\parallel}(t) - G\,I_{\perp}(t)}{I_{\parallel}(t) + 2\,G\,I_{\perp}(t)} \tag{1}$$

where $I_{\parallel}$ and $I_{\perp}$ denote emission intensities with polarization respectively, parallel and perpendicular to that of the exciting light and G, is the instrument response function. In a homogeneous liquid, the decay of $r(t)$ is governed by the rotational motion of the fluorescent probe. The time constant (τ_{rot}) of decay of $r(t)$ is given by $\tau_{rot}=(6D_R)^{-1}$ where D_R denotes the rotational diffusion coefficient of the probe. In an ordinary solution, D_R is inversely proportional to viscosity. In an organized assembly, the fluorescence anisotropy decay of a probe is complicated by the superposition of the motions of the macromolecular assembly. In

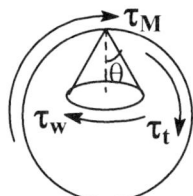

Fig. 1.2. Wobbling-in-cone-model.

certain cases, it is possible to extract the contributions of different independent motions to the anisotropy decay. For example, in the case of a nearly spherical micelle, many groups have used the "wobbling-in-cone" model.[15-18] According to this model, fluorescence depolarization in a micelle arises as a result of three independent motions, (a) wobbling motion, $r_w(t)$ of the probe in a cone of angle θ, (b) translational motion, $r_t(t)$ of the probe along the surface of the spherical micellar aggregates and (c) overall rotation or tumbling, $r_M(t)$ of the micelles (Fig. 1.2). As a result,

$$r(t) = r_w(t)\,r_t(t)\,r_M(t) \tag{2}$$

and the decay of $r(t)$ is biexponential,

$$r(t) = r_0\,[\beta\,\exp(-t/\tau_{slow}) + (1-\beta)\,\exp(-t/\tau_{fast})] \tag{3}$$

where, r_0 denotes the initial anisotropy (at $t = 0$). If τ_R, τ_D and τ_M, respectively denote the time constants for wobbling of the dye molecule, translation of the dye molecule and overall rotation of the spherical micelle,[15-18]

$$r(t) = r_0\,[S^2+(1-S^2)\,\exp(-t/\tau_R)]\,\exp\{-t(1/\tau_D + 1/\tau_M)\} \tag{4}$$

where S is the order parameter. Comparing equation (3) with equation (2) one immediately obtains

$$\beta = S^2 \tag{5}$$
$$1/\tau_{fast} = 1/\tau_R + 1/\tau_D + 1/\tau_M \tag{6}$$
$$1/\tau_{slow} = 1/\tau_D + 1/\tau_M \tag{7}$$

When the probe is attached to a spherical particle, τ_M is given by[15-18]

$$\tau_M = 4\pi\eta r_h^3 / 3KT \tag{8}$$

where, η = viscosity of water and r_h = hydrodynamic radius of the spherical micelle. In the case of sodium dodecyl sulfate (SDS) micelle, r_h = 20 Å and at 25°C, τ_M = 8.3 ns.[17-18] The translational diffusion coefficient(D_t) for translation of the fluorescent probe along the surface of a micelle is given by[17-18]

$$D_t = r_M^2/6\tau_D. \tag{9}$$

D_t of many probes in micelles and polymer-surfactant aggregates[15-18] ($\sim 10^{-10}$ m^2 s^{-1}) are very similar to that of an organic molecule in bulk water.[19] This is an important result and clearly demonstrates that the dramatic slowing down of many chemical processes in an organized assembly is not due to any change in translational diffusion of the probe. It is also important to note that because of translational diffusion a probe diffuses over a distance of $(6D_t\tau)^{1/2}$ within its life time, τ. From the magnitude of D_t (10^{-10} m^{2}s^{-1}) this corresponds to about 10 Å per nanosecond. In summary, fluorescence anisotropy decay provides information on the motion of the organized assembly and self-diffusion of the probe. However, anisotropy decay gives no information on the motion of confined water molecules around a biological macromolecule. In the following section, we will discuss how the dynamics of the water molecules or more precisely, solvation dynamics in an organized assembly may be studied using time dependent shift of the fluorescence maximum.

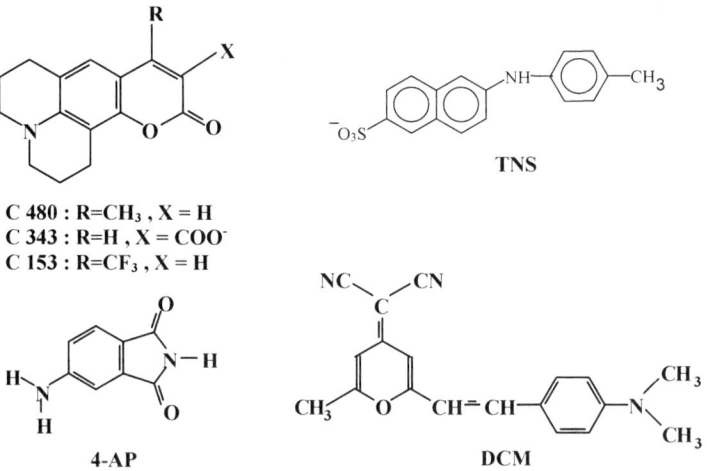

C 480 : R=CH$_3$, X = H
C 343 : R=H , X = COO$^-$
C 153 : R=CF$_3$, X = H

4-AP   DCM

Fig. 1.3. Structure of some solvation probes.

1.3. SLOW SOLVATION DYNAMICS IN ORGANIZED ASSEMBLIES

Solvation refers to the stabilization of a charged or dipolar solute by the collective reorientation of the solvent dipoles. In order to study solvation dynamics one chooses a fluorescent probe (Fig. 1.3) which is non-polar in the ground state (i.e. dipole moment $\mu \approx 0$) but possesses a very large dipole moment in the excited state. In a polar solvent, the solvent dipoles remain randomly arranged around the non-polar solute in its ground state. When the solute (fluorescent probe) is excited by an ultrashort light pulse, a dipole is created suddenly. Immediately after creation of the solute dipole, the solvent dipoles are randomly oriented and the energy of the system is high. With increase in time, as the solvent dipoles reorient the energy of the solute dipole decreases and its fluorescence maximum gradually shifts to lower energy i.e. towards longer wavelength. This is known as time dependent fluorescence Stokes shift (TDFSS) (Fig. 1.4). Evidently, at a short wavelength, the fluorescence corresponds to the unsolvated solute and exhibits a decay. At a long wavelength, the fluorescence originates from the solvated species and a rise precedes the decay. The rise at a long wavelength corresponds to the formation of the solvated species. Such a wavelength dependence of fluorescence decays is a clear manifestation of solvation dynamics. Fig. 1.5 shows the wavelength dependent fluorescence decays of a probe in an organized assembly.

The first step in studying solvation dynamics involves construction of time resolved emission spectra. Lakowicz has discussed in detail how one can construct the time resolved emission spectra using the steady state emission spectrum and the fluorescence decays.[20] Solvation dynamics is monitored by the decay of the time correlation function C(t) which is defined in terms of emission energies as,

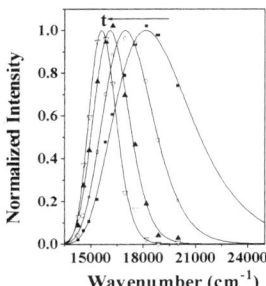

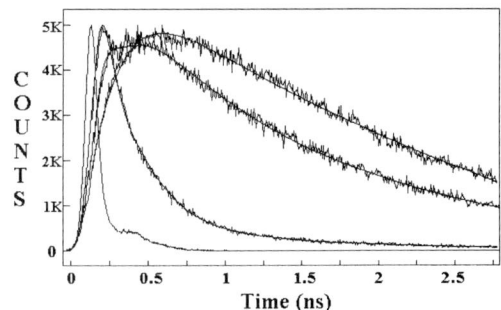

Fig. 1.4. Time dependent fluorescence Stokes shift (TDFSS) of a probe.[38]

Fig. 1.5. Wavelength dependent decays of DCM in a microemulsion.[38]

$$C(t) = \frac{E(t) - E(\infty)}{E(0) - E(\infty)} \qquad (10)$$

E(0), E(t) and E(∞) denote the observed emission energies of the system at time zero, t and infinity, respectively. Obviously, E(0)> E(t)> E(∞). At t=0, the value of C(t) is one and at t=∞ it is zero. If the decay of C(t) is single exponential, the time constant of the decay is known as

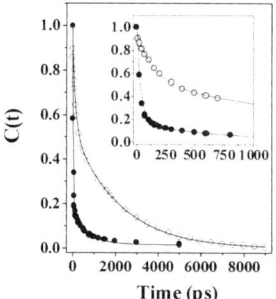

Fig. 1.6. Decay of C(t) in a protein, GlnRS, with a covalent probe (TAP, filled circle) and a non-covalent probe (bis-ANS, open circle.[69,78]

the longitudinal relaxation time or solvation time (τ_L or τ_s). If the decay of C(t) is mutiexponential e.g. $\Sigma a_i \exp(-t/\tau_i)$ one may consider an average solvation time, defined as $<\tau_s> = \Sigma a_i \tau_i$. Fig. 1.6 displays decay of C(t) in a protein. According to continuum model,[21] τ_s is given by

$$\tau_S = \frac{2\varepsilon_\infty + \varepsilon_C}{2\varepsilon_0 + \varepsilon_C}\tau_D \qquad (11)$$

In this equation, ε_∞ and ε_0 are respectively, the high frequency and static dielectric constants of the solvent. ε_∞ arises from the electronic polarization while ε_0 includes both electronic and orientational polarization. In earlier studies, it was assumed that ε_∞ is approximately equal to the square of refractive index of the medium (n^2). However, more recently many groups have measured dielectric constants at very high frequencies. For instance, for water $n^2 = 1.7$ while actual high frequency dielectric constant measured by Barthel et al. is about 5.[22] ε_C is the dielectric constant of the cavity around the fluorescent probe. For an organic fluorophore one may assume $\varepsilon_C \approx 1$ (i.e. same as that of a hydrocarbon). τ_D is the dielectric relaxation time and is the time taken for a system to reach equilibrium when an external electric field is switched on or off. For water, τ_D is 8.3 ps[23] while ε_∞ and ε_0, are respectively, about 5 and 78. Thus according to the continuum theory the solvation time of water should be about 0.5 ps. Actual experiments suggest that solvation dynamics in bulk water occurs in <1 ps time scale.[24-26] In bulk water, about 70% of solvation occurs in 0.03 ps (i.e. 30 fs) and 20% occurs in 0.2 ps time scale.[25] The longest component of solvation dynamics in bulk water is about 1 ps (10%). This gives rise to an average solvation time of 0.16 ps in bulk water.

    Solvation dynamics is almost entirely described by the solvent dipoles in the first solvation shell of the fluorophore. Thus solvation dynamics probes the immediate surrounding of the fluorophore and hence, offers excellent spatial and time resolution. Other relaxation techniques such as dielectric relaxation[27-28] and NMR[29-30] offer no such spatial resolution and also the temporal resolution of ultrafast fluorescence technique is far superior.

    In this chapter, we will give special emphasis on solvation dynamics in an organized assembly which directly probes the motion of the water molecules. According to many recent experiments the water molecules confined in many organized and biological assemblies exhibit a slow component of solvation dynamics in 100-1000 ps time scale.[11-14] The contribution of the slow component in most cases is 10-30%. We will now discuss several examples of slow solvation dynamics in various organized assemblies.

1.3.1. Reverse Micelles and Microemulsions

A microemulsion refers to a nanometer sized water droplet ("water pool") surrounded by a layer of surfactant molecules and dispersed in a non-polar medium (fig. 1.1c). The water pool in a microemulsion is an elegant model of confined water molecules. For the surfactant, AOT (sodium dioctyl sulfosuccinate) radius of the water pool is approximately $2w_0$ (Å) where w_0 denotes the water to surfactant molar ratio.[31] Different additives affects the size of a microemulsion.[32-33] In a water pool with $w_0 > 10$, solvation dynamics of water exhibits a component in 100-1000 ps time scale which is slower by three orders of magnitude compared to bulk water.[11-14,34-37]

In the water pool of a microemulsion the polarity and dynamics of water molecules depends strongly on the distance from the polar head group of the surfactants. Evidently, in the peripheral region, the water molecules are rigidly held by the head groups of the surfactants. As a result, in this region movement of the water molecule is highly constrained and also the polarity is quite low. In contrast, at the core of the water pool, the water molecules are relatively free and the polarity, higher.

In a microemulsion, an organic fluorophore having a low dipole moment is often located at the peripheral region near the water-surfactant interface. On electronic excitation the fluorophore becomes highly polar and as a result, may move towards a more polar region at the core of the water pool. Self-diffusion of a solvation probe from the interface to the core of the water pool has been proposed to be a source of the ultraslow component of solvation. In a microemulsion the full width at half maximum (FWHM, Γ) of the time resolved emission spectra (TRES) changes appreciably ($\sim$50%) with time (Fig. 1.7).[38] This is much larger compared to the change in Γ (10-20%) observed in bulk water (or other liquids).[39]

The time dependence of Γ of a solvation probe, DCM may be explained as follows.[38] DCM is insoluble in bulk water and hence, resides at the water-surfactant (AOT) interface

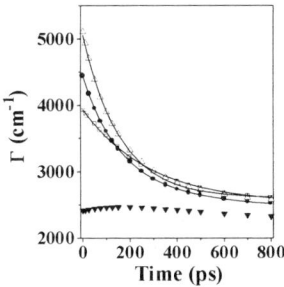

Fig. 1.7. Decay of Γ of DCM in AOT microemulsion without any additive ( ), and containing Na-cholate (Δ), and Na-salicylate (•). Decay of Γ in 100 mM Na-cholate in bulk water ( ) is also shown.[38]

where the medium is highly heterogeneous and changes from pure hydrocarbon to water over a small distance. Due to the superposition of the emission spectra of DCM in different environments at short times the spectral width is very large. On electronic excitation DCM becomes highly polar and presumably, more soluble in water. Thus, after excitation DCM may migrate from the AOT interface towards a more polar core of the water pool. Thus at a long time, the emission spectrum originates from the DCM molecules at the core of the water pool. This corresponds to a more uniform environment and hence, leads to a smaller spectral width. Thus the decay of Γ with time may be ascribed to self-diffusion of the probe inside the water pool. It seems the rate determining step in the solvation dynamics in the water pool is the diffusion of the probe from the interface to the core of the water pool where the water molecules are quite free and exhibit ultrafast dynamics. Thus the slow component (in 100-1000 ps time scale) in the decay of $\Gamma(t)$ and $C(t)$ of DCM in a water pool is ascribed to self-diffusion of the probe.[38]

In the case of aggregates of Na-cholate in bulk water Γ does not change much with time and the time dependence of Γ is similar to that in simple liquids (fig. 1.7).[38] Thus self-diffusion is not a universal mechanism of slow solvation dynamics. In the case of secondary aggregate of the bile salt, Na-cholate the central water filled channel is quite long ($\approx$50 Å, Fig. 1.1a).[40] It appears that in a bile salt aggregate, the probe (DCM) remains in a uniform environment within the channel during its entire life time.

The absorption maximum of a solvation probe depends on polarity of the medium and usually shifts to a longer wavelength with increase in polarity. Thus by variation of excitation wavelength it is possible to excite selectively probes in different regions. This is the basis of the red edge excitation effect (REES) in lipids.[41-42] Recently, Tominaga and co-workers found that solvation dynamics in the water pool of a microemulsion depends on the excitation wavelength. For excitation at the blue end the total dynamic spectral shift $\Delta v = v(0) - v(\infty)$ is found to be quite large while excitation at the red end results in a small Δv. The time constants of decay of $C(t)$ is however, almost independent of the excitation wavelength. It is proposed that in a water pool there are two regions of drastically different dynamic behavior. The fluorophores present in the core of the water pool display bulk water like ultrafast solvation dynamics in sub-picosecond time scale. The probes at the periphery however, exhibit slow solvation dynamics.

Several groups have applied large scale computer simulation to explain solvation dynamics in a microemulsion. Senapati and Chandra[44] showed that the dielectric constant and solvation time inside a microemulsion is lower than those in bulk water by less than one order of magnitude. Faeder and Ladanyi[45] carried out a simulation up to 10 ps and did not detect the slow dynamics in 100-1000 ps time scale. Senapati and Berkowitz,[46] studied dynamics and residence time of water molecules in the solvation shell of the head group of the surfactant layer by layer. In the first solvation shell ("region I") 25 water molecules stay for 1.8 ns while 18 water molecules stay in "region II" for 160 ps and 15 water molecules stay for 20 ps in "region III". The translational mobility of the water molecules in "region I" is lower by a factor of six compared to bulk water. The water molecules in "region II" are more mobile. The water molecules in "region III" are bulk-like. The simulations reveal that 50% of the water molecules in the first shell form two surfactant to water hydrogen bonds and 40% of them form single hydrogen bond to surfactant oxygen. The average reorientational correlation times are 396, 41, 3.8 and 3.2 ps for "region I", "region II", "region III" and bulk water, respectively.[46] The long component of the reorientational motion is 1700 ps for "region I". Thus, the slow component of the reorientational relaxation of the water molecules in the first solvation shell is slowed down by three orders of magnitude compared to bulk water.

1.3.2. Micelles

According to small angle neutron scattering studies,[47] ordinary micelles are spherical aggregates of surfactants with a "dry" hydrocarbon core and a polar peripheral shell which contains polar/ionic head groups, counter ions and the water molecules (Fig. 1.1d). Solvation dynamics of the water molecules hydrogen bonded to the polar head groups of a micelle exhibit a very slow component in 100-1000 ps time scale.[48-49] The bile salt micelles are structurally different from ordinary spherical micelles.[40] The bile salts are natural amphiphiles which exhibit two critical micellar concentrations at ~ 10 mM (CMC_1) and ~ 60 mM (CMC_2). Above CMC_1, bile salts form primary aggregates containing very few monomers with the hydrophilic groups pointing outwards.[40] Above CMC_2, secondary aggregates are formed which resemble an elongated rod with a central hydrophilic core filled with water and the ions (Fig. 1.1a).[40] The solvation dynamics of DCM in a secondary aggregate of sodium deoxycholate (NaDC) is found to be triexponential with components of 110 ps, 700 ps and 2750 ps.[50] These components are significantly slower than those in bulk water.

Sen et al.[51] showed that solvation dynamics in a micelle exhibits strong temperature dependence. For triton X-100 micelles, the average solvation times are 800, 400, and 110 ps at 283, 303 and 323 K respectively. The temperature dependence corresponds to an activation energy of 9±1 Kcal mol^{-1} for solvation dynamics in a micelle.

It is interesting to note that the experimental results on temperature dependence of solvation dynamics is very close to the results obtained in a simulation. Pal et al.[52] found that the interfacial water molecules form one or two hydrogen bonds with the polar head groups of the micelles and the hydrogen bond energy is 13 kcal mol^{-1}. This is nearly 8 kcal mol^{-1} higher than the hydrogen bond energy between two water molecules in bulk water. Thus it appears that the rate determining step in the micellar solvation is rupture of the hydrogen bond(s) between the water and the polar head groups. Thus according to simulations solvation in a micelle would involve an activation barrier of about 8 Kcal mol^{-1}.

Bagchi and Balasubramanian carried out a fully atomistic molecular dynamics simulations on solvation,[53] and hydrogen bond dynamics[54] at the surface of a micelle, cesium pentadecafluorooctanoate (CsPFO). The decay of C(t) at the micellar surface[53] obtained by them may be fitted to a tri-exponential function having components, 1.6 ps, 4.3 ps and 30 ps. The ultrafast components (1.6 ps and 4.3 ps) detected in this simulation are close to the components of solvation dynamics detected in a recent femtosecond up-conversion study.[55]

According to computer simulations, the translation diffusion coefficient of water at the surface of a micelle is not too different from bulk water but the reorientation and solvation dynamics are significantly slower. In the case of sodium dodecyl sulfate (SDS) micelle, the translational diffusion coefficient of the interfacial water molecules in the first shell, second shell and third shell are in the ratio 0.6:0.8:1.[56] One reason behind almost bulk like diffusion in the solvation shell of a micelle may be the fact that for outward diffusion normal to the micellar surface the water molecules become bulk like just after travelling one water layer. Diffusion along the surface of the micelle where the water molecule is enclosed in the first layer at all times may be much slower.

1.3.3. Cyclodextrins

A supramolecule consisting of a cyclodextrin cavity as a host and a solvation probe with several solvent molecules as guests (Fig. 1.1e) is perhaps the most well defined example of a confined liquid. Fleming et al. studied solvation dynamics of water confined in a γ-cyclodextrin cavity and detected three very slow components of 13 ps, 109 ps and 1200 ps.[57] They ascribed the very slow components of solvation dynamics to three processes. First, the restricted motion of the highly confined water molecules, second, the motion of the guest probe molecule in and out of the cavity, and third, the fluctuations of the γ-cyclodextrin ring. Sen et al. studied solvation dynamics of a non-aqueous solvent, dimethylformamide (DMF) in a β-cyclodextrin cavity.[58] The dynamics of confined DMF molecules is found to be described by two slow components of 400 ps and 8000 ps. This is substantially slower than the solvation (~ 1 ps) in bulk DMF.[21] Nandi and Bagchi[59] attributed the slow dynamics in a cyclodextrin cavity to complete suppression of the translational degrees of freedom of the solvent molecules.

1.3.4. Proteins

The biological function of a protein is largely controlled by the water molecules at its surface.[60-61] A water molecule hydrogen bonded to a protein differs markedly from that in a bulk with an extended hydrogen bond network. The water molecules in the hydration layer of a protein are loosely described as *biological or structured water*.[62] Study of the dynamics of biological water is of fundamental importance to understand the role of water in biology. In the following section, we will summarize recent studies on solvation dynamics in the native state of several proteins using an intrinsic fluorescent probe (tryptophan) and several extrinsic probes attached to the protein covalently or non-covalently. This will be followed by two sections on solvation dynamics in denatured protein and in protein folding intermediates.

1.3.4.1. Native State

1.3.4.1a. Intrinsic probe, Tryptophan. Tryptophan is one of the most popular intrinsic fluorescence probes to study a protein. The principle advantage of using tryptophan is one can study a protein in the native state without worrying about the change in structure or biological activity caused by an external probe.

The early studies on solvation studies in the native state of a protein were based on phase modulation of fluorescence. Marzola and Gratton[63] studied solvation dynamics of several single tryptophan proteins in the water pool of a microemulsion. More recently, several other groups used tryptophan to study solvent relaxation in a protein.[64-65] In bulk water, tryptophan exhibits a solvation time of 1.1 ps. However, for many single tryptophan proteins a much longer component of solvation dynamics is reported.[65]

Spectral relaxation using tryptophan is complicated by the existence of two rotamers of tryptophan.[66-67] Lifetimes of the two rotamers of tryptophan in water are 0.5 ns and 3.5 ns, respectively and the corresponding emission maxima are at 335 nm and 350 nm.[66-67] The existence of two rotamers having different lifetimes gives rise to an apparent dynamic spectral shift in a few hundred ps time scale. Because of this artifact tryptophan is unsuitable for detection of the ultraslow component of solvation dynamics in a protein in 100-1000 ps time scale. It is suggested that the spectral shift of tryptophan due to solvation is in most cases, over by 20 ps. Hence, in order to study solvation dynamics one should normalize the time dependent emission spectra with that of tryptophan at 20-50 ps.[65] The justification of this procedure may be debated. In summary, though tryptophan is an attractive solvation probe it may be used only for the ultrafast component of solvation dynamics upto ~50 ps and may not be suitable to detect the long component in 100-1000 ps time scale.

1.3.4.1b. Covalent probes. Many proteins can be labeled using a covalent probe. For instance, the cysteine residue may be labeled by an acrylodan group. Bright and co-workers[68] studied acrylodan labeled serum albumins in a water pool of a microemulsion using phase

Protein-SH + Br-CH$_2$CO-NH⟨imide⟩ $\xrightarrow{\text{- HBr}}$ Protein-S-CH$_2$CO-NH⟨imide⟩

Scheme 1. Labeling of a sulfhydryl group with TAP.

modulation and detected a nanosecond component. In order to study solvation dynamics at the surface of a protein in aqueous solution, Mandal et al. attached a solvation probe, thioacetyl aminophthalimide (TAP) covalently to a sulfhydryl group of the cysteine residue of glutaminyl RNA synthetase (GlnRS) (scheme 1).[69] In the case of GlnRS, the most reactive sulfhydryl group is located at a distance of about 7 Å below the protein surface. As a result, about half (4 Å) of the probe (TAP) is buried inside the protein. Solvation dynamics of GlnRS labeled with TAP (Fig. 1.6) displays a major component of 40 ps and a minor component of 580 ps giving rise to an average solvation time of 120 ps.[69]

1.3.4.1c. Non-covalent extrinsic probe. Many fluorescent probes bind to the hydrophobic region of a protein non-covalently. In a few cases, x-ray crystal structure of the protein bound to a fluorescent probe (e.g. ANS bound to α-chymotrypsin)[70] is known and thus, the location of the probe is known accurately.

    The location of the probe within a protein may also be ascertained from fluorescence resonance energy transfer (FRET) from the tryptophan residue. Though such approach is most suitable for a single tryptophan protein, in favorable cases this approach may be extended to a protein containing more than one tryptophan residues. For instance, lysozyme contains 6 tryptophan (Trp) residues. Imoto et al.[71] found that replacement of two of them, Trp 62 and Trp 108 quenches about 80% of lysozyme emission. This suggests that these two tryptophan residues account for nearly 80% of the total emission of lysozyme. The distance between Trp 62 and Trp 108 is about 13 Å. Obviously, if a fluorescent probe quenches the tryptophan emission of lysozyme by fluorescence energy transfer, it must be located close to Trp 62 and 108. Baugher et al.[72] used this strategy to ascertain the location of a probe, eosin inside lysozyme. Using energy transfer they concluded that eosin resides in a "hydrophobic box" near the two tryptophan residues.[72] The hydrophobic box is defined by Trp 28, Trp 111, Tyr 23 and Met 105 residues of lysozyme.[72]

    Several groups studied solvation dynamics of a protein using TNS as a non-covalent extrinsic probe.[73-74] They detected a very slow component of solvation dynamics in several nanosecond time scale. Since late nineties, several groups applied ultrafast laser spectroscopy to study directly solvation dynamics in a protein. Pal et al. studied solvation dynamics of a non-covalent probe (DCM) bound to human serum albumin (HSA).[75] They detected two components of 600 ps (25%) and 10,000 ps (75%).[76] This demonstrates that the water molecules in the immediate vicinity of the protein are highly constrained. Fleming et al. applied three photon echo peak shift to study dynamics of a non-covalent probe, eosin in the hydration layer of a protein (lysozyme).[76] They found that most of the dynamics occurs in sub-picosecond time scale while there is a slow component of 530 ps which constitutes about 8% of the total decay. The slow component is absent for free eosin in bulk water and hence, is ascribed to protein bound water.[76] Dutta et al.[77] studied solvation dynamics of a non-covalent probe coumarin 153 (C153) bound to lysozyme. In the native state of lysozyme the average solvation time for C153 is 330 ps. Sen et al. studied solvation dynamics in GlnRS using a non-covalent probe bis-ANS and detected a component of 1400 ps.[78] This component is 12 times slower than that detected using a covalent probe TAP (Fig. 1.6). The difference in solvation times for the two probes in the native state of the same protein is ascribed to different locations of the probes.[78] The covalent probe is constrained to reside at the surface of the protein and hence, exhibits very fast dynamics. The non-covalent probe however, goes deep inside the protein and reports a much slower dynamics at a buried site.

1.3.4.2. Denatured state

Most recently, several groups have reported on solvation dynamics in the non-native states of a protein. Dutta et al.[77] studied solvation dynamics of C153 in lysozyme denatured by a surfactant SDS and dithiothreitol (DTT). On addition of SDS to lysozyme when the noncovalent interactions are destroyed and the protein is denatured the average solvation time increases to 7250 ps. This is more than 20 times longer than the average solvation time (330 ps) in the native state of lysozyme.[77] When DTT is added to lysozyme denatured by SDS, the di-sulfide bonds are ruptured and the protein assumes an extended polymer like structure so that it easily passes through porous poly-acrylamide gel during gel electrophoresis. Under this condition the average solvation time is found to be 1100 ps.[77] It is proposed that on addition of SDS the protein becomes decorated by small SDS micelles. The water molecules squeezed between SDS micelle and protein are highly retarded. Zewail and co-workers reported that solvation dynamics in several proteins denatured by acid and by guanidine hydrochloride is slower than that in the native state.[65]

1.3.4.3. Protein folding intermediates

Under mildly denaturing conditions many proteins may be trapped in a state intermediate between the native (folded state) and the denatured state. The most common folding intermediate is the molten globule state. Sen et al. studied solvation dynamics in the molten globule state of a protein, glutaminyl-tRNA synthetase (GlnRS).[78] For this purpose, they used both a non-covalent probe (bis-ANS) and a covalent probe (TAP). In this case for both covalent and non-covalent probes, the solvation times (250 and 400 ps respectively) are close whereas the solvation time of the non-covalent probe in the native state (1400 ps) is 12 times longer.

1.3.4.4. Origin of slow solvation dynamics

The discovery of slow water molecules in biological systems has shed a new light on dynamics and biological function of a protein and motivated many theoretical studies. The ultraslow component of solvation dynamics suggests that the water molecules in the vicinity of a protein are highly ordered and organized through strong hydrogen bonds with the protein. In many recent theoretical models and simulations, pre-organization at the active site of an enzyme has been suggested to be responsible for the high rate enhancement and molecular recognition by an enzyme.[79]

In perhaps the first accurate simulation of a protein including the hydration layer, Levitt and Sharon[80] detected a dramatic accumulation of water molecules at the protein surface. The number of water molecules is nearly double of that expected from the accessible surface area. The clustering of water molecules at the protein surface leads to a density 1.25 g/cm^3 which is higher than that in bulk water. They divided the water molecules at the protein surface into 4 classes depending on the distance from the protein surface. The water molecules closest to the protein (region I) were found to be energetically stabler by 1.9 Kcal mol^{-1} compared to bulk water with a diffusion coefficient lower than that of bulk water by at most a factor of 4.[80] The energy and diffusion coefficient of the water molecules increase with distance from the protein. About half of the water molecules reside at a distance >10 Å (region IV) and their properties are almost identical to those of bulk water.[80] More recently, many authors investigated number of hydration sites and residence time of water at different sites. In a recent simulation, Makarov et al. detected 294 hydration sites in myoglobin not all of which are occupied simultaneously.[81] They found that the buried sites (inside cavities or grooves or concave surfaces) display a long residence time (>80 ps). In contrast the exposed or convex sites are characterized by short residence time (<10 ps). This clearly shows that the residence times at the hydration sites of a protein are significantly longer than those in bulk water (0.34 ps and 4.1 ps).[81] The relation between residence time and solvation time is yet to be understood.

To explain the slow component of dielectric relaxation and solvation dynamics, Nandi and Bagchi proposed a phenomenological model.[62] In this model, the water in the vicinity of a biological system is considered to be made up of two kinds of water molecules-bound and free. The bound water refers to those which are attached to the biomolecule by one or two hydrogen bonds and are essentially immobilized. The free water molecules retain the hydrogen bond network in bulk water and exhibits bulk-like fast dynamics. Nandi and Bagchi showed that the slow dielectric relaxation results from a dynamic equilibrium between bound and free water.[62] A water-biomolecule hydrogen bond is stronger than a water-water hydrogen bond. Thus, the bound water molecules are more stable than the free water molecules.[11-14,62] The slow component depends on the free energy difference (ΔG^0) between bound and free water molecules. In the limit of very high binding energy,

$$\tau_{slow} \approx k_{bf}^{-1} \qquad (12)$$

where the rate of bound-to-free interconversion (k_{bf}) is given by,

$$k_{bf} = (k_B T/h) \exp (-E_{a,bf}/RT) \qquad (13)$$

The activation energy $E_{a,bf}$ is a sum of the free energy difference of the bound and free water and the activation energy $E_{a,fb}$ for free-to-bound interconversion.

In order to explain the origin of the very long 10,000 ps component very recently a new mechanism has been proposed.[82] The overall tumbling time (τ_M) of a protein is given by $\eta V/k_B T$ and is usually very long. For HSA (8 x 8 x 3 nm), τ_M is 25 ns. When the membrane sensitive protein, HSA is encapsulated in a lipid (DMPC) vesicle the protein molecules can not rotate individually and instead rotate with the entire vesicle. The diameter of DMPC vesicle is 30 nm and the corresponding τ_M is so slow (3500 ns) that one may neglect it. It is observed that when the overall tumbling of the HSA molecule is suppressed by entrapping the protein in DMPC vesicle the 10,000 ps component vanishes. Thus the 10,000 ps component observed earlier in bulk water[75] is ascribed to overall tumbling of the protein.[82]

In the denatured state, a protein is expected to be more open with greater penetration of water. Thus one expects faster dynamics of water molecules or solvation dynamics in the denatured state. The slower dynamics in the denatured state is contrary to this expectation. To reconcile this, it is proposed that on addition of SDS the protein becomes decorated by small SDS micelles. The motion of the water molecules squeezed between SDS micelle and protein are highly restricted. This may give rise to the slower dynamics in SDS denatured state.[77]

Bagchi and co-workers[83] invoked the concept of Rouse chain dynamics of a homopolymer to explain the slow dynamics in the denatured state. They concluded that in the denatured state the protein resembles a polymer and displays Rouse chain dynamics. According to this model, the time correlation function for the chain's solvation energy fluctuations is a multiexponential function $\Sigma a_i \exp(-t/\tau_i)$ where $\tau_i (=\lambda_1^{-1})$ are given by

$$\lambda_1 = 3D_0 \left(\frac{l\pi}{Nb}\right)^2 \qquad (14)$$

where N is the number of monomers in the chain and $l= 1,3,5...N-1$, b^2 is the mean square bond length, D_0 is the translational diffusion coefficient of a monomer (7×10^{-10} m^2 s^{-1}). In the case of lysozyme, the number of amino acids (N) is 129. Thus for the lysozyme-SDS complex, the slowest component of relaxation is ~200 ns (for $l = 1$) and the fastest component is ~10 ps. The solvation dynamics of C153 in the lysozyme-SDS complex is observed to be described by two components, 500 ps (50%) and 14 ns (50%).[77] Obviously, in the case of lysozyme-SDS complex the probe experiences multiple time scales. The superposition of the various components of chain dynamics may be responsible for the observed bimodal decay with a very slow component of 14 ns and a relatively fast 500 ps component.

Before closing, it may be interesting to speculate on the implication of slow solvation dynamics in enzyme catalysis. The slow water molecules in the hydration layer of an enzyme do not change their orientation much when a polar substrate approaches or binds to it. Thus, the slow water molecules at the surface of a protein protect its native structure. If the water molecules at the surface respond too quickly, the structure of the hydration layer and hence, of the protein may be seriously perturbed when a polar substrate approaches.

1.3.5. DNA

Berg and coworkers[84] studied solvation dynamics inside DNA by attaching a solvation probe, coumarin 102 (C102) covalently to the deoxyribose moiety of DNA double-helix such that it replaces a normal base pair. They detected logarithmic relaxation times over three decades (40 ps – 40 ns) indicating a complex relaxation among large number of conformational substates.[84] They further studied the effect of sodium ion binding to DNA.[85] Binding of sodium ion does not cause a time dependent decrease in emission energy but causes spectral narrowing.[85] No such change in spectral shape is observed when tetrabutyl ammonium ion (TBA) binds to DNA. The exact cause of the specific effect caused by sodium ion is not yet clearly understood.[85] Zewail and co-workers studied solvation dynamics of water in DNA using 2-aminopurine as an intrinsic probe and a minor groove binding non-covalent probe, pentamidine.[65,86] They detected a ~10 ps component of solvation dynamics.

1.3.6. Polymer and Polymer-Surfactant Aggregates

Supramolecular assembly of polymer and surfactants in an aqueous solution gives rise to many interesting phenomena.[1] For instance, the highly insoluble protein zein readily dissolves in water in the presence of a surfactant, SDS.[87] It is proposed that the protein (or polymer)-surfactant aggregate resembles a "necklace" with the spherical micelles as "beads" and polymer chains as the "thread" (Fig. 1.1g).[87] Such an aggregate is formed at a surfactant concentration, called critical association concentration (CAC). The CAC is often significantly lower than the critical micellar concentration (CMC). In a polymer-surfactant aggregate, the polymer segments reduce two unfavorable interactions in a micelle viz., electrostatic repulsion among the head groups and residual contact of the first few peripheral carbon atoms of the hydrocarbon chains with water. Thus, fewer number of surfactant molecules are required for the formation of the micelles in the presence of a polymer.[88] According to fluorescence correlation spectroscopy, the hydrodynamic diameter of a polymer (PVP)-surfactant (SDS) aggregate is much larger than that of the polymer and the micelle.[89]

Solvation dynamics in a polymer-surfactant aggregate is found to be appreciably slower than that in a micelle or in a polymer.[90-92] Solvation dynamics of TNS in PVP-SDS aggregate is described by two components, 300 ps and 2500 ps.[90] In contrast, solvation dynamics of TNS occurs in <50 ps in SDS micelles while in an aqueous solution of PVP the solvation dynamics is described by a major (85%) component of 60 ps.[90] The slower solvation dynamics in PVP-SDS aggregate compared to the polymer (PVP) or SDS micelle indicates severe restrictions of motion of the water molecules squeezed in between the polymer and the micelles.[90]

Recently, a large number of polymers have been developed which consist of a hydrophobic core and a hydrophilic peripheral shell. A single molecule of such a polymer resembles a micelle and hence, they are known as amphiphilic star-like macromolecule (ASM, fig. 1.1f).[93] They encapsulate highly hydrophobic drug molecules and make them water soluble. This has potential applications in drug delivery.[1] Castner et al.[93] reported that solvation dynamics in an aqueous solution of an ASM is described by an ultrafast component of 0.95 ps (44%) and two very slow components of 361 ps (19%) and 3962 ps (37%).[93]

1.3.7. Sol-gel Glass

Dynamics of liquids confined in a nanoporous sol-gel glass is the subject of many recent studies. Optical Kerr effect studies on various liquids confined in a sol-gel glass reveal a major bulk-like component.[94] In a sol-gel glass with 10 Å pores, the average solvation time of trapped water molecules is found to be 220 ps.[95] This is about 200 times slower than that in bulk water. The average solvation time of ethanol in bulk is 12.5 ps while it is 18.6 ps in a sol-gel glass with 75 Å pores and 35.9 ps in 50 Å pores.[96]

Very recently, Halder et al. studied solvation dynamics in an ormosil (i.e. a sol-gel glass containing entrapped organic guests).[97] In the presence of dimyristoyl-phosphatidyl-choline (DMPC) in the sol-gel glass, solvation dynamics of coumarin 480 (C480) is about 2.3 times slower than that in its absence.[97] The solvation dynamics inside the DMPC entrapped sol-gel is about 14 times faster than that in DMPC vesicles in bulk solution.

1.4. EFFECT OF SLOW SOLVATION ON POLAR REACTIONS IN ORGANIZED ASSEMBLIES

A polar reaction is facilitated by solvation because of differential stabilization of the transition state (TS) and the product with respect to the reactant and consequent reduction in the activation barrier. The slow solvation dynamics in organized assemblies causes incomplete solvation of the TS and consequently, markedly retards many polar reactions e.g. twisted intramolecular charge transfer (TICT),[99-108] excited state proton transfer (ESPT)[109-115] and intermolecular electron transfer.[116-121]

1.4.1. TICT: Polarity Dependent Barrier and Retardation in Organized Assemblies

The TICT process involves intramolecular transfer of an electron from a donor to an acceptor and twist about the bond joining the donor and the acceptor.[99-101] The TICT state (product) and the TS leading to it is more polar than the reactant (i.e. the non-polar or locally excited state, LE). Eisenthal and co-workers[102] first proposed that activation barrier for the TICT process decreases linearly with increase in the polarity parameter $E_T(30)$ as,

$$E_b = E_b^0 - A[E_T(30)-30] \qquad (15)$$

$E_T(30)$, a solvent polarity parameter,[98] is defined as the absorption energy of a betaine dye in Kcal mol^{-1}. If TICT is the main non-radiative pathway in the excited state of a fluorescent probe the polarity i.e. $E_T(30)$ may be obtained by measuring the non-radiative rate

$$k_{NR} \approx k_{TICT} = (1-\phi_f)/\tau_f \qquad (16)$$

where ϕ_f and τ_f, respectively denote the fluorescence quantum yield and lifetime, respectively. According to equation (15) and (16) logarithm of k_{TICT} is linearly related to $E_T(30)$. Using this approach polarities of many organized assemblies such as zeolite[103] and microemulsion[104] have been determined.

TICT of p-dimethylamino-benzonitrile (DMABN) inside a cyclodextrin (CD) cavity depends markedly on the cavity size.[105-106] In the small α-CD, TICT is quite fast as a major part of the DMABN molecule sticks out. In bigger β- or γ-CD cavity the entire DMABN molecule is enclosed and TICT is severely inhibited because of reduced polarity, and restrictions on the twisting motion.[101,105-106] This results in a dramatic enhancement of the "non-polar" emission of DMABN. 1-(2-naphthyl)-2-ethenyl-(2-benzothazolium) iodide (1,2-NEB) exhibits a twisting motion along the central C-C double bond.[108] In the smaller β-CD cavity, *trans*-1,2-NEB does not undergo twisting and intramolecular charge-transfer (ICT) and exhibits a lifetime of a few picosecond. However, in the larger γ-CD, *trans*-1,2-NEB has enough room to undergo twisting and ICT and this results in the *cis* isomer with a ns lifetime. Thus, the size of the CD cavity controls whether the isomerization may occur or not.[107-108]

1.4.2. Excited State Proton Transfer

The essential condition for dissociation of an acid (HA), is adequate □alvation of the anion (A$^-$) and the proton. Many organic molecules which are a weak acid (or even base) in the ground state, behave as a strong acid in the excited state. Such a molecule is called a "photoacid." For instance, 1-naphthol is a weak base in the ground state (pK_a = 9.5). In the excited state, 1-naphthol behaves as a strong acid (pK_a^* ≈ 0.5) and readily donates a proton to bulk water in 35 ps.[109] Dynamics of excited state proton transfer (ESPT) of 1-naphthol is monitored by the decay of the emission from the neutral form (at 360 nm) and the rise of the anion emission (at 460 nm). In a supersonic jet, ESPT of 1-naphthol occurs in a water cluster only if there are at least 30 water molecules in the cluster.[91] Methanol can not solvate 1-naphthol and the anion because of steric hindrance. Consequently, ESPT of 1-naphthol does not occur in a cluster with methanol in a supersonic jet and in a liquid solution in methanol.[110] Thus both alvation and steric effect play an important role in ESPT of 1-naphthol.

In an organized assembly ESPT of 1- naphthol is markedly retarded because of slow and inadequate alvation and also of steric hindrance by the surfactant for the approach of water molecules to solvate the anion and the proton. As a result ESPT of 1-naphthol is dramatically retarded in a cyclodextrin cavity,[111] micelle,[112] in a polymer-surfactant aggregate,[113] and in the water pool of a microemulsion.[114] The retardation of ESPT is manifested in a long decay of the neutral fluorescence (360 nm) and a slow rise of the anion emission (460 nm) and a significant increase in the intensity of neutral fluorescence.

Organero and Douhal studied excited state intramolecular proton transfer (ESIPT) of 1-hydroxy-2-acetonaphtone (HAN).[115] They reported a significant retardation of the ESIPT process of HAN inside the α-CD cavity. This results in an enhancement of fluorescence intensity and lifetime of the enol emission of HAN.[115]

1.4.3. Photoinduced Intermolecular Electron Transfer

The time constant of photoinduced intermolecular electron transfer between a donor and an acceptor is determined from the lifetime of the donor in the absence (τ_0) and presence of an acceptor (τ). If k_Q denotes the quenching constant and [Q] the quencher concentration, $k_Q[Q]=(\tau^{-1}-\tau_0^{-1})$. In bulk solution, the donor and acceptor remain at a distance and diffusion of the donor and acceptor puts an upper limit to rate of electron transfer. This obscures observation of the Marcus inverted region.[116-118] Intermolecular electron transfer in many systems have been studied in micelles.[119-122] In this case, both the donor and acceptor reside in the polar peripheral shell of the micelle.[119] It is observed that the electron transfer process in a micelle is much slower than that in bulk water.[119] The retardation of the electron transfer process in a micelle has been attributed to an increase of the distance between the donor and the acceptor because of the intervening surfactant chains.[119]

In a micelle, the donor and acceptor are always confined within the micelle and hence, are constrained to remain close. This eliminates the effect of diffusion and hence, it is possible to observe the Marcus inverted region as observed earlier in covalently linked donor-acceptor systems.[118] Most recently, several groups reported Marcus inverted behavior in intermolecular electron transfer in a micelle.[121-122]

1.5. CONCLUSION AND FUTURE OUTLOOK

In 1963 in his *Lectures in Physics* Richard Feynman[123] talked about an *assumption* that "Everything that living things do can be understood in terms of *jigglings and wigglings of atoms*." The picture of *jigglings and wigglings* of atoms in a protein was first captured in

1977 in a computer simulation carried out by Karplus and coworkers. On the occasion of 25 years of computer simulations of biomolecules, Karplus[124] stated that "the long range goal of biology is to describe living systems in terms of chemistry and physics." As outlined in this chapter, recent application of ultrafast fluorescence spectroscopy has provided a molecular picture of dynamics in biological assemblies in unprecedented details and in unprecedented time resolution. The most important discovery is undoubtedly, the slow component of solvation dynamics. The effect of slow solvation dynamics on many polar reactions in organized assemblies has been delineated by fluorescence spectroscopy. This has significantly improved our understanding of chemistry in confined systems. The biological implication of slow solvation dynamics in molecular recognition and bio-catalysis is beginning to be unfathomed.

In 1999, Lakowicz very correctly predicted that "it seems clear that measurements of spectral relaxation (i.e. solvation dynamics) will find increased use in the studies of biopolymer dynamics."[125] Within 5 years of this prediction this field has grown enormously. Further studies in this area may ultimately explain biology in chemical terms and with this end in view, this will remain a frontier area of research for quite sometime.

1.6. ACKNOWLEDGEMENTS

Thanks are due to Council of Scientific and Industrial Research (CSIR) and Department of Science and Technology (DST), Government of India for generous research grants. I thank Professor B. Bagchi for many stimulating discussions. Finally, I thank all my students for their enthusiasm and untiring efforts. I, particularly, thank Mr. P. Sen for his help in preparing the manuscript.

1.7. REFERENCES

1.   L. Maibaum, A. R. Dinner, D. Chandler, Micelle formation and the hydrophobic effect, *J. Phys. Chem. B* **104**, (2004) (in press).
2.   A. Muller, D. F. O'Brien, Supramolecular materials via polymerization of mesophases of hydrated amphiphiles, *Chem. Rev.* **102**(3), 727-758 (2002).
3.   J. R. Nitschke, J. –M. Lehn, Self-organization by selection: Generation of a metallosupramolecular grid architecture by selection of components in a dynamic library of ligands, *Proc. Natl. Acad. Sci. USA* **100**(21), 11970-11974 (2003).
4.   S. Fernandez-Lopez, H.-S. Kim, E. C. Choi, M. Delgado, J. R. Granja, A. Khasanov, K. Kraehenbuehl, G. Long, D. A. Weinberger, K. M. Wilcoxen, M. R. Ghadiri, Antibacterial agents based on the cyclic D,L-α-peptide architecture, *Nature* **412**(6845), 452-456 (2001).
5.   S. W. Rick, S. J. Stuart, B. J. Berne, Dynamical fluctuating charge force fields: applications to liquid water, *J. Chem. Phys.* **101**(7), 6141-6156 (1994).
6.   J. R. Lakowicz, *Principles of Fluorescence Spectroscopy*, (Kluwer/Plenum, New York, 1999).
7.   K. Bhattacharyya, Study of organized media using time-resolved fluorescence spectroscopy, *J. Fluorescence* **11**(3), 167-176 (2001).
8.   T. Simonson, G. Archontis, M. Karplus, Free energy simulations come of age: Protein-ligand recognition, *Acc. Chem. Res.* **35**(6), 430-437 (2002).
9.   A. Warshel, Molecular dynamic simulations of biological reactions, *Acc. Chem. Res.* **35**(6), 385-395 (2002).
10. T. Simonson, Gaussian fluctuations and linear response in an electron transfer protein, *Proc. Natl. Acad. Sc. USA* **99**(10), 6544-6549 (2002).
11. B. Bagchi, Water solvation dynamics in the bulk and in the hydration layer of protein and self-assemblies, *Annu. Rep. Prog. Chem., Sect. C.* **99**, 127-175 (2003).
12. K. Bhattacharyya, Solvation dynamics and proton transfer in supramolecular assemblies, *Acc. Chem. Res.* **36**(2), 95-101 (2003).

13.   N. Nandi, K. Bhattacharyya, B.Bagchi, Dielectric relaxation and solvation dynamics of water in complex chemical and biological systems, *Chem. Rev.* **100**(6), 2013-2045 (2000).

14.   K. Bhattacharyya, B. Bagchi, Slow dynamics of constrained water in complex geometries, *J. Phys. Chem. A* **104**(46), 10603-10613 (2000).

15.   E. L. Quitevis, A. H. Marcus, M. D. Fayer, Dynamics of ionic lipophilic probes in micelles: picosecond fluorescence depolarization measurements, *J. Phys. Chem.* **97**(21), 5762-5769 (1993).

16.   N. W. Wittouck, R. M. Negri, F. C. De Schryver, AOT reversed micelles investigated by fluorescence anisotropy of cresyl violet, *J. Am. Chem. Soc.* **116**(23), 10601-10611 (1994).

17.   N. C. Maiti, M. M. G. Krishna, P. J. Britto, N. Periasamy, Fluorescence dynamics of dye probes in micelles, *J. Phys. Chem. B* **101**(51), 11051-11060 (1997).

18.   S. Sen, D. Sukul, P. Dutta, K. Bhattacharyya, Fluorescence anisotropy decay in-polymer-surfactant aggregates, *J. Phys. Chem. A* **105**(32), 7495-7500 (2001).

19.   For instance, D_t of amino benzoic acid is 8×10^{-10} m^2/s^{-1}; *Handbook of Chemistry and Physics*, (CRC Press, Boca Raton, Fl, 1990) p. 6-151.

20.   Ref. 6, p. 211-233.

21.   M. Maroncelli, The dynamics of solvation in polar liquids, *J. Mol. Liq.* **57**, 1-37 (1993).

22.   J. Barthel, K. Bachuber, R. Buchner, H. Hetzenauer, Dielectric spectra of some common solvents in the microwave region. Water and lower alcohols, *Chem. Phys. Lett.* **165**(4), 369-373 (1990).

23.   U. Kaatze, Dielectric relaxation of H_2O/D_2O mixtures, *Chem. Phys. Lett.* **203**(1), 1-4 (1993).

24.   W. Jarzeba, G. C. Walker, A. E. Johnson, M. A. Kahlow, P. F. Barbara, Femtosecond microscopic solvation dynamics in aqueous solution, *J. Phys. Chem.* **92**(25), 7039-7041 (1988).

25.   R. Jimenez, G. R. Fleming, P. V. Kumar, M. Maroncelli, Femtosecond solvation dynamics in water, *Nature* **369**(6480), 471-473 (1994).

26.   N. Nandi, S. Roy, B. Bagchi, Ultrafast solvation dynamics in water: Isotope effects and comparison with experimental results, *J. Chem. Phys.* **102**(3), 1390-1397 (1995).

27.   E. H. Grant, R. J. Sheppard, G. P. South, *Dielectric Behavior of Biological Molecules* (Clarendon, Oxford, 1978).

28.   *Protein-Solvent Interactions*, edited by R. B. Gregory (Marcel Dekker, New York, 1995).

29.   G. Otting, E. Liepinsh, K. Wüthrich, Protein hydration in aqueous solution, *Science* **254**(5034), 974-980 (1991).

30.   K. Modig, E. Liepinsh, G. Otting, B. Halle, Dynamics of protein and peptide hydration, *J. Am. Chem. Soc.* **126**(1), 102-114 (2004).

31.   A. Maitra, Determination of size parameters of water-Aerosol OT-oil reverse micelles from their nuclear magnetic resonance data, *J. Phys. Chem.* **88**(21), 5122-5125 (1984).

32.   S. P. Moulik, G. C De, B. B. Bhowmik, A. K. Panda, Physicochemical studies on microemulsions. 6. Phase behavior, dynamics of percolation, and energetics of droplet clustering in water/AOT/*n*-heptane system influenced by additives (sodium cholate and sodium salicylate), *J. Phys. Chem. B* **103**(34), 7122-7129 (1999).

33.   S. Sen, D. Sukul, P. Dutta, K. Bhattacharyya, Solvation dynamics in the water pool of aerosol sodium dioctylsulfosuccinate microemulsion. Effect of polymer, *J. Phys. Chem. A* **106**(25), 6017 (2002).

34.   J. Zhang, F.V. Bright, Nanosecond reorganization of water within the interior of reversed micelles revealed by frequency-domain fluorescence spectroscopy, *J. Phys. Chem.* **95**(20), 7900-7907 (1991).

35.   N. Sarkar, K. Das, S. Das, A. Datta, K. Bhattacharyya, Solvation dynamics of coumarin 480 in reverse micelles. Slow relaxation of water molecules, *J. Phys. Chem.* **100**(25), 10523-10527 (1996).

36.   R. E. Riter, D. M. Willard, N. E. Levinger, Water immobilization at surfactant interfaces in reverse micelles, *J. Phys. Chem. B* **102**(15), 2705-2714 (1998).

37.   K. Bhattacharyya, K. Hara, N. Kometani, Y. Uozu, O. Kajimoto, Solvation dynamics in a microemulsion in near-critical propane, *Chem. Phys. Lett.* **361**(1-2), 136-142 (2002).

38.   P. Dutta, P. Sen, S. Mukherjee, A. Halder, K. Bhattacharyya, Solvation dynamics in the water pool of an aerosol-OT microemulsion. Effect of sodium salicylate and sodium cholate, *J. Phys. Chem. B* **107**(39), 10815-10822 (2003).

39.   M.-L. Horng, J. A. Gardecki, M. Maroncelli, Rotational dynamics of coumarin 153: Time-dependent friction, dielectric friction, and other non-hydrodynamic effects, *J. Phys. Chem. A* **101**(6), 1030-1047 (1997).

40.   C. Ju, C. Bohnne, Dynamics of probe complexation to bile salt aggregates, *J. Phys. Chem.* **100**(9), 3847-3854 (1996).

41.   A. P. Demchenko, The red-edge effects: 30 years of exploration, *Luminescence* **17**(1), 19-42 (2002).

42.   H. Raghuraman, A. Chattopadhyay, Organization and dynamics of melittin in environments of graded hydration. A fluorescence approach, *Langmuir* **19**(24), 10332-10341 (2003).

43.   T. Satoh, H. Okuno, K. Tominaga, K. Bhattacharyya, Excitation wavelength dependence of solvation dynamics in a water pool of a reverse micelle, *Chemistry Letters* (submitted).

44.   S. Senapati, A. Chandra, Dielectric constant of water confined in a nanocavity, *J. Phys. Chem. B* **105**(22), 5106-5109 (2001).

45.   J. Faeder, M. V. Albert, B. M. Ladanyi, Molecular dynamics simulations of the interior of aqueous reverse micelles: A comparison between sodium and potassium counterions, *Langmuir* **19**(6), 2514-2520 (2003).

46.   S. Senapati, M. L. Berkowitz, Water structure and dynamics in phosphate fluorosurfactant based reverse micelle: A computer simulation study, *J. Chem. Phys.* **118**(4), 1937-1944 (2003).

47.   S. S. Berr, Solvent isotope effects on alkyltrimethylammonium bromide micelles as a function of alkyl chain length, *J. Phys. Chem.* **91**(18), 4760-4765 (1987).

48.   N. Sarkar, A. Datta, S. Das, K. Bhattacharyya, Solvation dynamics of coumarin 480 in micelles, *J. Phys. Chem.* **100**(38), 15483-15486 (1996).

49.   K. Hara, H. Kuwabara, O. Kajimoto, Pressure effect on solvation dynamics in micellar environment, *J. Phys. Chem. A*, **105**(30), 7174-7179 (2001).

50.   S. Sen, P. Dutta, S. Mukherjee, K. Bhattacharyya, Solvation dynamics in bile salt aggregates, *J. Phys. Chem. B* **106**(32), 7745-7750 (2002).

51.   P. Sen, S. Mukherjee, A. Halder, K. Bhattacharyya, Temperature dependence of solvation dynamics in a micelle. 4-Aminophtalimide in triton X-100, *Chem. Phys. Lett.* **385**(5-6), 357-361 (2004).

52.   S. Pal, S. Balasubramanian, B. Bagchi, Identity, energy, and environment of interfacial water molecules in a micellar solution, *J. Phys. Chem. B* **107**(22), 5194-5202 (2003).

53.   S. Balasubramanian, B. Bagchi, Slow solvation dynamics near an aqueous micellar surface, *J. Phys. Chem. B* **105**(50), 12529-12533 (2001).

54.   S. Balasubramanian, S. Pal, B. Bagchi, Hydrogen bond dynamics near a micellar surface: origin of the universal slow relaxation at complex aqueous interfaces, *Phys. Rev. Lett.* **89**(11), 115505-1 (2002).

55.   D. Mandal, S. Sen, T. Tahara, K. Bhattacharyya, Femtosecond study of solvation dynamics of DCM in micelles, *Chem. Phys. Lett.* **359**(1-2), 77-82 (2002).

56.   C. D. Bruce, S. Senapati, M. L. Berkowitz, L. Perera, M. D. E. Forbes, Molecular dynamics simulations of sodium dodecyl sulfate micelle in water: The behavior of water, *J. Phys. Chem. B* **106**(42), 10902-10907 (2002).

57.   S. Vajda, R. Jimenez, S. J. Rosenthal, V. Fidler, G. R. Fleming, E. W. Castner Jr., Femtosecond to nanosecond solvation dynamics in water and inside the γ-cyclodextrin cavity, *J. Chem. Soc. Faraday Trans.* **91**(5), 867-873 (1995).

58.   S. Sen, D. Sukul, P. Dutta, K. Bhattacharyya, Slow solvation dynamics of dimethylformamide in a nanocavity. 4-Aminophthalimide in β-cyclodextrin, *J. Phys. Chem. A* **105**(47), 10635-10639 (2001).

59.   N. Nandi, B. Bagchi, Ultrafast solvation dynamics of an ion in the γ-cyclodextrin cavity: Role of restricted environment, *J. Phys. Chem.* **100**(33), 13914-13919 (1996).

60.   *Hydration Processes in Biology: Theoretical and Experimental Approaches*, Edited by M.–C. Bellisent-Funnel (IOS Press, Amsterdam, 1999).

61.   Hydration Processes in Biological and Macromolecular Systems, *Faraday Discuss.* **103**(1), 1-394 (1996).

62.   N. Nandi, B. Bagchi, Dielectric relaxation of biological water, *J. Phys. Chem. B* **101**(50), 10954-10962 (1997).

63.   P. Marzola, E. Gratton, Hydration and protein dynamics: frequency domain fluorescence spectroscopy of proteins in reverse micelles, *J. Phys. Chem.* **95**(23), 9488-9495 (1991).

64.   D. Toptygin, R. S. Savichenko, N. D. Meadow, L. Brand, Homogeneous spectrally- and time-resolved fluorescence emission from single-tryptophan mutants of IIA^{Glc} protein, *J. Phys. Chem. B* **105**(10), 2043-2055 (2001).

65.   S. K. Pal, A. H. Zewail, Dynamics of water in molecular recognition, *Chem. Rev.* **104**(4), 2099-2124 (2004).

66.   L. P. McMahon, H. T. Yu, M. A. Vela, G. A. Morales, L. Shui, F. R. Fronczek, M. L. McLaughlin, M. D. Barkley, Conformer interconversion in the excited state of constrained tryptophan derivatives, *J. Phys. Chem. B* **99**(16), 3269-3280 (1997).

67.   A. G. Szabo, D. M. Rayner, Fluorescence decay of tryptophan conformers in aqueous solution, *J. Am. Chem. Soc.* **102**(2), 554-563 (1980).

68.   J. S. Lundgren, M. P. Heitz, F. V. Bright, Dynamics of acrylodan-labeled bovine and human serum albumin sequestered within aerosol-OT reverse micelles, *Anal. Chem.* **67**(20), 3775-3781 (1995).

69. D. Mandal, S. Sen, D. Sukul, K. Bhattacharyya, A. K. Mandal, R. Banerjee and S. Roy, Solvation dynamics of a probe covalently bound to a protein and in AOT microemulsion. 4 (N-bromoacetylamino)-phthalimide, *J. Phys. Chem. B* **106**(41), 10741-10747 (2002).

70. L. D. Weber, A. Tulinsky, J. D. Johnson, M. A. El-Bayoumi, Expression of functionality of α-chymotrypsin. The structure of a fluorescent probe-α-chymotrypsin complex and the nature of its pH dependence, *Biochemistry* **18**(7), 1297-1303 (1979).

71. T. Imoto, L. S. Forster, J. A. Rupley, F. Tanaka, Fluorescence of lysozyme: Emission from tryptophan residues 62 and 108 and energy migration, *Proc. Natl. Acad. Sci. USA* **69**(5), 1151-1155 (1971).

72. J. F. Baugher, L. I. Grossweiner, J. Lewis, Intramolecular energy transfer in lysozyme-eosin complex, *J. Chem. Soc. Faraday Trans. II* **70**, 1389-1398 (1974).

73. J. S. Bashkin, G. McLedon, S. Mukamel, J. Marohn, Influence of medium dynamics on solvation and charge separation reactions: comparison of a simple alcohol and a protein "solvent," *J. Phys. Chem.* **94**(12), 4757-4761 (1990).

74. D. W. Pierce, S. G. Boxer, Dielectric relaxation in a protein matrix, *J. Phys. Chem.* **96**(13), 5560-5566 (1992).

75. S. K. Pal, D. Mandal, D. Sukul, S. Sen, K. Bhattacharyya, Solvation dynamics of DCM in human serum albumin, *J. Phys. Chem. B* **105**(7), 1438-1441 (2001).

76. X. J. Jordanides, M. J. Lang, X. Song, G. R. Fleming, Solvation dynamics in protein environments studied by photon echo spectroscopy, *J. Phys. Chem. B* **103**(37), 7995-8005 (1999).

77. P. Dutta, P. Sen, A. Halder, S. Mukherjee, S. Sen, K. Bhattacharyya, Solvation dynamics in a protein-surfactant complex, *Chem. Phys. Lett.* **377**(1-2), 229-235 (2003).

78. P. Sen, S. Mukherjee, P. Dutta, A. Halder, D. Mandal, R. Banerjee, S. Roy, K. Bhattacharyya, Solvation dynamics in the molten globule state of a protein, *J. Phys. Chem.* **107**(51), 14563-14568 (2003).

79. A. Warshel, Computer simulations of enzyme catalysis: Methods, progress, and insights, *Annu Rev Biophys Biomol. Struct.* **32**, 425-443 (2003).

80. M. Levitt, R. Sharon, Accurate simulation of protein dynamics in solution, *Proc. Natl. Acad. Sci. USA* **85**(20), 7557-7561 (1988).

81. V. Makrov, B. M. Petit, Solvation and hydration of protein and nucleic acids: A theoretical view of simulation and experiment, *Acc. Chem. Res.* **35**(6), 376-384 (2002).

82. P. Dutta, P. Sen, S. Mukherjee, K. Bhattacharyya, Solvation dynamics in DMPC vesicle in the presence of a protein, *Chem. Phys. Lett.* **382**(3-4), 426-433 (2003).

83. S. K. Pal, J. Peon, B. Bagchi, A. H. Zewail, Biological water: Femtosecond dynamics of macromolecular hydration, *J. Phys. Chem.* **107**(48), 12376-12395 (2003).

84. E. B. Brauns, M. L. Madaras, R. S. Coleman, C. J. Murphy, M. A. Berg, Complex local dynamics in DNA on the picosecond and nanosecond time scales, *Phys. Rev. Lett.* **88**(15), 158101-1 (2002).

85. L. A. Gearheart, M. M. Somoza, W. E. Rivers, C. J. Murphy, R. S. Coleman, M. A. Berg, Sodium-ion binding to DNA: Detection by ultrafast time-resolved Stokes-shift spectroscopy, *J. Am. Chem. Soc.* **125**(39), 11812–11813 (2003).

86. S. K. Pal, L. Zhao, T. Xia, A. H. Zewail, Site- and sequence-selective ultrafast hydration of DNA, *Proc. Natl. Acad. Sci. USA* **100**(24), 13746-13751 (2003).

87. N. Deo, S. Jockusch, N. J. Turro, P. Somasundaran, Surfactant interactions with zein protein, *Langmuir* **19**(12), 5083-5088 (2003).

88. E. A. Lissi, E. Abuin, Aggregation numbers of sodium dodecyl sulfate micelles formed on poly(ethylene oxide) and poly(vinyl pyrrolidone) chains, *J. Coll. Inter. Sci.* **105**(1), 1-6 (1985).

89. R. Narenberg, J. Kliger, D. Horn, Study of the interaction between poly(vinyl pyrrolidone) and sodium dodecyl sulfate by fluorescence correlation spectroscopy, *Angew. Chem. Int. Ed. Engl.* **38**(11), 1626-1629 (1999).

90. S. Sen, D. Sukul, P.Dutta, K. Bhattacharyya, Solvation dynamics in aqueous polymer solution and in polymer-surfactant aggregate, *J. Phys. Chem. B* **106**(15), 3763-3769 (2002).

91. P. Dutta, S. Sen, S. Mukherjee, K. Bhattacharyya, Solvation dynamics of TNS in polymer (PEG)-surfactant (SDS) aggregate, *Chem. Phys. Lett.* **359**(1-2), 15-21 (2002).

92. P. Dutta, D. Sukul, S. Sen, K. Bhattacharyya, Solvation dynamics of 4-aminophthalimide in a polymer (PVP)-surfactant (SDS) aggregate, *Phys. Chem. Chem. Phys.* **5**(21), 4875-4879 (2003).

93. L. Frauchiger, H. Shirota, K. E. Uhrich, E. W. Castner Jr., Dynamic fluorescence probing of the local environments within amphiphilic starlike macromolecules, *J. Phys. Chem. B* **106**(30), 7463-7468 (2002).

94. R. A. Farrrer, J. T. Fourkas, Orientational dynamics of liquids confined in nanoporous sol-gel glasses studied by optical Kerr effect spectroscopy, *Acc. Chem. Res.* **36**(8), 605-612 (2003).

95.   S. K. Pal, D. Sukul, D. Mandal, S. Sen, K. Bhattacharyya, Solvation dynamics of coumarin 480 in sol-gel matrix, *J. Phys. Chem. B* **104**(12) 2613-2616 (2000).

96.   R. Bauman, C. Ferrante, F. W. Deeg, C. Brauchle, Solvation dynamics of nile blue in ethanol confined in porous sol-gel glasses, *J. Chem. Phys.* **114**(13), 5781-5791 (2001).

97.   A. Halder, S. Sen, A. Das Burman, A. Patra, K. Bhattacharyya, Solvation dynamics in dimyristoyl-phosphatidylcholine entrapped inside a sol-gel matrix, *J. Phys. Chem. B* **108**(7), 2309-2312 (2004).

98.   C. Reichardt, Solvatochromic dyes as solvent polarity indicators, *Chem. Rev.* **94**(8), 2319-2358 (1994).

99.   Z. R. Grabowski, K. Rotkiewicz, W. Rettig, Structural changes accompanying intramolecular electron transfer: Focus on twisted intramolecular charge-transfer states and structures, *Chem. Rev.* **103**(10), 3899-4032 (2003).

100.   S. Techert, F. Schotte, M. Wulff, Picosecond X-ray diffraction probed transient structural changes in organic solids, *Phys. Rev. Lett.* **86**(10), 2030-2033 (2001).

101.   K. Bhattacharyya, M. Chowdhury, Environmental and magnetic field effects on exciplex and twisted charge transfer emission, *Chem. Rev.* **93**(1), 507-535 (1993).

102.   J. M. Hicks, M. T. Vandersall, Z. Babarogic, K. B. Eisenthal, The dynamics of barrier crossings in solution: The effect of a solvent polarity-dependent barrier, *Chem. Phys. Lett.* **116**(1), 18-24 (1985).

103.   N. Sarkar, K. Das, D. Nath, K. Bhattacharyya, Twisted intramolecular charge transfer processes of nile red in homogeneous solutions and in zeolite, *Langmuir* **10**(1), 326-329 (1994).

104.   A. Datta, D. Mandal, S. K. Pal, K. Bhattacharyya, Intramolecular charge transfer in confined systems. Nile red in reverse micelles, *J. Phys. Chem. B* **101**(49),10221-10225 (1997).

105.   A. Nag, K. Bhattacharyya, Twisted intramolecular charge transfer emission of dimethyl-aminobenzonitrile in α-cyclodextrin cavities, *Chem. Phys. Lett.* **151**(4-5), 474-476 (1988).

106.   A. Nag, R. Dutta, N. Chattopadhyay, K. Bhattacharyya, Effect of size of cyclodextrin cavity on twisted intramolecular charge transfer emission: Dimethylamino benzonitrile in β-cyclodextrin, *Chem. Phys. Lett.* **157**(1-2), 83-86 (1989).

107.   A. Douhal, Ultrafast guest dynamics in cyclodextrin nanocavities, *Chem. Rev.* **104**(4), 1955-1976 (2004).

108.   T. A. Fayed, J. A. Organero, I. Garcia-Ochoa, L. Tormo, A. Douhal, Ultrafast twisting motions and intramolecular charge-transfer reaction in a cyanine dye trapped in molecular nanocavities, *Chem. Phys. Lett.* **364**(1-2), 108-114 (2002).

109.   L. M. Tolbert, K. M. Solntsev, Excited-state proton transfer: From constrained systems to "super" photoacids to superfast proton transfer, *Acc. Chem. Res.* **35**(1), 19-27 (2002).

110.   M. Saeki, S.–I. Ishiuchi, M. Sakai, M. Fuji, Structure of 1-naphthol:alcohol clusters studied by IR dip spectroscopy and ab-initio molecular orbital calculations, *J. Phys. Chem. A* **105**(44), 10045-10053 (2001).

111.   J. E. Hansen, E. Pines, G. R. Fleming, Excited state proton transfer in 1-aminopyrene complexed with β-cyclodextrin, *J. Phys. Chem.* **96**(17), 6904-6910 (1992).

112.   D. Mandal, S. K. Pal, K. Bhattacharyya, Excited state proton transfer of 1-naphthol in micelles, *J. Phys. Chem. A* **102**(48), 9710-9714 (1998).

113.   P. Dutta, A. Halder, S. Mukherjee, P. Sen, S. Sen, K. Bhattacharyya, Excited state proton transfer of 1-naphthol in a hydroxypropylcellulose/sodium dodecyl sulfate system, *Langmuir* **18**(21), 7867-7871 (2002).

114.   B. Cohen, D. Huppert, K. M. Solntsev, Y. Tsfadia, E. Nachliel, M. Gutman, Excited state proton transfer in reverse micelles, *J. Am. Chem. Soc.* **124**(25), 7539-7547 (2002).

115.   J. A. Organero, A. Douhal, Confinement effects on the photorelaxation of a proton-transfer phototautomer, *Chem. Phys. Lett.* **373**(3-4), 426-431 (2003).

116.   R. A. Marcus, Electron transfer reactions in chemistry: Theory and experiments (Nobel lecture), *Angew. Chem. Int. ed. Engl.* **32**(8), 1111-1222 (1993).

117.   G. J. Kavaranos, *Fundamentals of Photoinduced Electron transfer* (VCH, New York, 1993).

118.   G. L. Closs, L. T. Calcaterra, N. J. Green, K. W. Penfield, J. R. Miller, Distance, stereoelectronic effects, and the Marcus inverted region in intramolecular electron transfer in organic radical anions, *J. Phys. Chem.* **90** (16), 3673-3683 (1986).

119.   S. K. Pal, D. Mandal, D. Sukul, K. Bhattacharyya, Photoinduced electron transfer between dimethyl aniline and oxazine 1 in micelles, *Chem. Phys.* **249**(1), 63-71 (1999).

120.   H. L. Tavernier, F. Laine, M. D. Fayer, Photoinduced intermolecular electron transfer in micelles: Dielectric and structural properties of micelle headgroup regions, *J. Phys. Chem. A* **105**(39), 8944-8957 (2001).

121.   M. Kumbhakar, S. Nath, H. Pal, A. V. Sapre, T. Mukherjee, Photoinduced electron transfer from aromatic amines to coumarin dyes in sodium dodecyl sulfate micellar solutions, *J. Chem. Phys.* **119**(1), 388-399 (2003).

122.   D. Chakraborty, A. Chakrabarty, D. Seth, N. Sarkar, Photoinduced electron transfer between coumarin dyes and electron donating solvents in cetyltrimethyl ammonium bromide micelles: Evidence for Marcus inverted region, *Chem. Phys. Lett.* **382**(5-6), 508-517 (2003).
123.   R. P. Feynman, R. B. Leighton, M. Sands, *The Feynman Lectures in Physics* (Addison-Wesley, MA, 1963) Vol. 1, p. 3.6.
124.   M. Karplus, Molecular Dynamics Simulations of Biomolecules (guest editorial), *Acc. Chem Res.* **35**(6), 321-323 (2002).
125.   Ref. 6, p. 233.

THE COMBINED USE OF FLUORESCENCE SPECTROSCOPY AND X-RAY CRYSTALLOGRAPHY GREATLY CONTRIBUTES TO ELUCIDATING STRUCTURE AND DYNAMICS OF PROTEINS

Sabato D'Auria, Maria Staiano, Irina M Kuznetsova and Konstantin K.Turoverov[1]

2.1. INTRODUCTION

Intrinsic fluorescence is a powerful tool for investigating the structure, dynamics, and folding-unfolding of proteins (Lakowicz, 1999; Eftink, 1991, 1998; Demchenko, 1986; Burshtein, 1976). This is due to high sensitivity of various parameters of the fluorescence of tryptophan residues (spectrum position, quantum yield, anisotropy, etc.) to their microenvironment and to the peculiarities of their location in protein macromolecules. The dependence of intrinsic protein fluorescence on the unique location of tryptophan and tyrosine residues was elaborated by the study of intrinsic fluorescence of model compounds and proteins in different structural states. The combined analysis of the characteristics of protein intrinsic fluorescence and protein three-dimensional structure with the peculiarities of the location of tryptophan residues in protein structure was performed for the first time on azurin (Turoverov et al., 1985). The dependence of the recorded fluorescence characteristics of tryptophan residues on the peculiarities of their microenvironment has been further studied on other proteins (Turoverov and Kuznetsova, 1986; Agekyan et al., 1988; Kuznetsova et al., 1996, 1999, 2000, Stepanenko, 2004, Staiano et al., in press). Carrying out these studies on recombinant proteins with altered tryptophan residues or other amino acid residues in the vicinity of indol residues essentially increases the experimental basis of these works (Martensson et al., 1995; Gopalan et al., 1997; Atkins et al., 1991; Doyle et al., 2001). Since detailed

[1] Sabato D'Auria and Maria Staiano at Institute of Protein Biochemistry, CNR, Via P. Castellino, 111 80131, Naples, Italy. Irina M Kuznetsova and Konstantin K.Turoverov at Institute of Cytology Russian Academy of Science, Tikhoretsky av.,4 194064 St.Petersburg, Russia

spatial structures of many proteins (co-ordinates of separate atoms in protein structures) determined by X-ray analysis are available from Protein Data Bank (Bernstein et al., 1977) this prompts the idea to fulfill the combined analysis of the characteristics of protein intrinsic fluorescence and their 3-D structures in order to elucidate the peculiarities of protein microenvironments such as protein dynamics and protein folding-unfolding processes. The performance of such analysis could allow to verify, to refine, and to improve the existing knowledge on protein structure.

2.2. ANALYSIS OF PROTEIN 3-D STRUCTURE TO ELUCIDATE THE ESSENTIAL FACTORS FOR INTERPRETATION OF THE PROTEIN INTRINSIC FLUORESCENCE FEATURES

It is evident that characteristics of the intrinsic fluorescence of proteins are determined by their structure. Nonetheless the data given in the works on proteins 3-D structure are insufficient to obtain even a general insight in the fluorescence properties of proteins. It was necessary to elaborate special approaches and algorithms on protein 3-D structure analysis that could elucidate potential factors for determining fluorescence properties of separate tryptophan residues in the protein matrix. The main principles of such an approach were elaborated in the work aimed to elucidate the extraordinary fluorescence of azurin on the base of the analysis of its 3-D structure (Turoverov et al., 1985) and were developed in subsequent works (Turoverov & Kuznetsova, 1986; Agekian et al., 1988; Kuznetsova et al., 1996, 1999, 2000).

Figure 2.1 shows how to visualize the location of tryptophan residues within the structure of protein macromolecule. It is convenient to represent the protein structure in the so called "tryptophanocentric" coordinate system with the center in the C_β atom and plane ZOX in the plane of indole ring and to represent only backbone and aromatic residues, or some other residues of interest. It is generally accepted that tryptophan fluorescence is determined by atoms and groups of atoms that have immediate contact with the indole ring.

This is why special attention should be paid to the analysis of tryptophan residue microenvironment, which is determined as a set of atoms that are no greater than r_0 distant from the geometrical center of the indole ring. The largest distance from the geometrical center of the indole ring to its periphery is 4.2 Å (Pauling and Pauling, 1975). The van der Waals radius of carbon atoms with the attached hydrogen atoms is 2.2 Å. As a consequence, in order to take into the account all atoms that are potentially in contact with the indole ring, the value of r_0 is chosen to be equal to 7 Å (Turoverov et al., 1985; Kuznetsova and Turoverov, 1998). For analysis of the vicinity of tryptophan residues it is convenient (1) to select atoms of microenvironment; (2) to determine their polar co-ordinates relative to tryptophan residue; (3) to determine the nearest atoms of microenvironment to each atom of the indole ring; and (4) to estimate the distance between them. At the same time some groups can affect tryptophan fluorescence though they are distant from tryptophan residue (Turoverov et al., 1985).

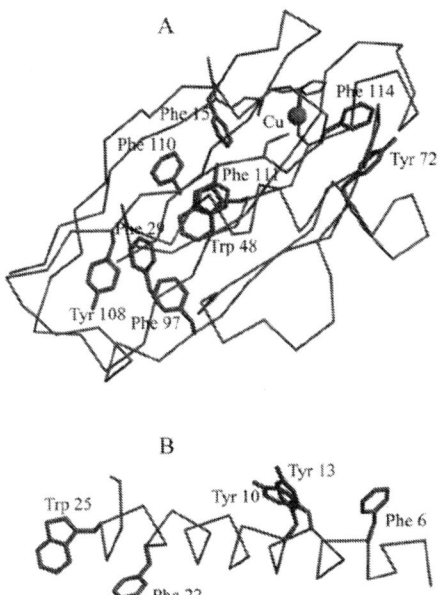

Figure 2.1. The structure of azurin (A) and glucagon (B) in tryptophanocentric coordinate system. Tryptophan, tyrosine and phenylalanine residues are shown. The structures of azurin and glucagon were taken from PDB (Berstein et al., 1977) files 1AZU.ent (Adman et al., 1980) and (Sasaki et al., 1975).

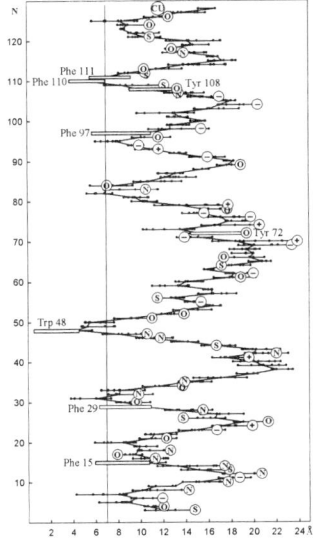

Figure 2.2. Distance of the polypeptide chain from the centre of Trp 48 of the indole ring in azurin. -, OD2 and OE2 atoms of the carboxylic groups of Asp and Glu residues; +, NZ and NH2 atoms of amino and guanidylic groups of Lys and Arg residues; N, ND2 and NE2 atoms of amide groups of Asn and Gln residues; O, OG and OG1 atoms of hydroxyl groups of Ser and Thr residues; S, SG and SD atoms of Cys and Met residues; O-Tyr, OH atoms of Tyr residues; N-His, ND1 atoms of His residues. File 1AZU.ent (Adman et al., 1980) was used.

As a consequence, all such groups must be examined throughout the protein macromolecule. The location of the polypeptide chain and amino acid residues that can affect tryptophan fluorescence can be readily illustrated by the diagrams of the distances of protein atoms from the geometrical center of indole ring of the test tryptophan residue. For small proteins like azurin, all atoms of protein can be depicted. Abscissa and ordinate show the distance between the atom and the geometrical center of indole ring of the tested tryptophan residue and between the atom and the number of residues, respectively. The broken line connecting CA atom shows the pathway of polypeptide chain around the tested tryptophan residue (Figure 2.2). However, it is important to state that for large proteins such a kind of depiction is impossible. In such cases the diagrams can show the pathway for the fragment of the polypeptide chain which are involved in the microenvironment of the tested tryptophan residues. This diagram shows N, CA and C atoms of the residue for which but one atom is involved in the microenvironment and N and C atoms of neighbor residues along the polypeptide chain. It has been shown that the conformation of the side chain of tryptophan residue can greatly influence the excitation lifetime (Szabo and Rayner, 1980) and other fluorescence characteristics of tryptophan residues. That is why the determination of the angles χ_1 and χ_2, which characterizes the conformation of the side chain of tryptophan residue, is useful for the analysis of the peculiarities of tryptophan residue location in protein.

The important characteristic of tryptophan residue location is the packing density of atoms in its microenvironment:

$$d_{micro} = \frac{\sum V_i}{V_0} \tag{1}$$

where V_0 is the volume of sphere with radius r_0 and V_i is the part of the $i's$ atom volume that is inside the sphere. The $i's$ atom volume is determined on the basis of its van der Waals radius. Of course, this estimation is not exact, because atoms are incorporated by chemical bonds and they occupy a smaller volume than the Van der Waals one. Nonetheless, this is not critical for the comparative estimation of packing density of microenvironment of different tryptophan residues.

It is known that parameters of intrinsic tryptophan fluorescence are sensitive to the accessibility of tryptophan residues to the solvent. The exposure of tryptophan residue to the solvent molecules depends obviously not only on the packing density of its microenvironment but also on its location in the protein macromolecule, i.e. whether it is located near the center of protein macromolecule or at the protein periphery. To estimate the accessibility of tryptophan residues to the solvent, the radial dependence of atoms packing density about the geometrical center of tryptophan residue is evaluated as:

$$d(r) = \frac{\sum V_i(r, r + \Delta r)}{V_0(r, r + \Delta r)} \tag{2}$$

where V_0 *(r, r+Δr)* is the volume of sphere layer, that is r distant from the geometrical center of the indole ring, Δr is layer thickness and V_i is the part of the $i's$ atom volume that is inside this sphere layer (Fig. 2.3).

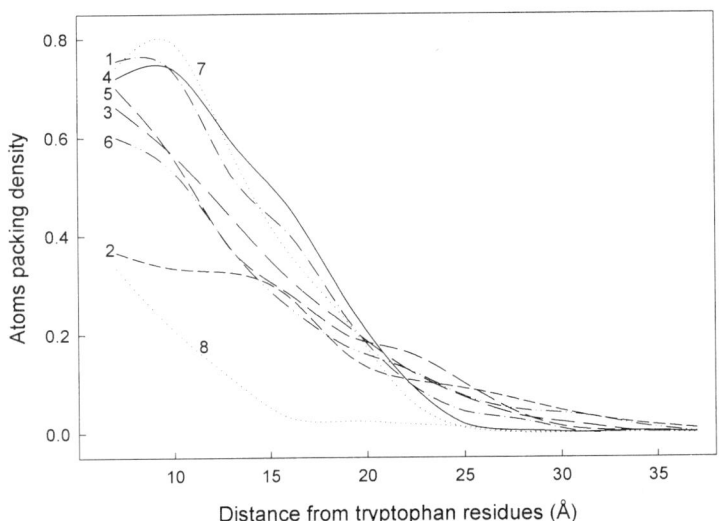

Figure 2.3. Radial dependence of macromolecule atoms packing density about geometrical centers of indole rings of tryptophan residues in lysozyme. Curves 1-6 are the dependencies for Trp 28, Trp 62, Trp 63, Trp 108, Trp111, and Trp 123, respectively. For comparison the dependencies for Trp 48 of azurin (internal tryptophan residue) and Trp 19 of melittin (external tryptophan residues) are given (curves 7 and 8).

Finally the evaluation of the efficiency of nonradiative energy transfer between tryptophan residues and from tyrosine to tryptophan is also important. According to Förster (1960), nonradiative energy transfer between can be evaluated as follows:

$$W = \frac{1}{1 + \frac{2/3}{k^2}\left(\frac{R}{R_0}\right)^6} \tag{3}$$

where R is the distance between donor and acceptor (the geometrical centers of indole and phenol rings), R_0 is the so-called critical Förster distance and k^2 is the orientation factor of dipole–dipole interaction:

$$k^2 = \left(\cos\theta - 3\cos\theta_A \cos\theta_D\right)^2 \tag{4}$$

where θ is the angle between the directions of the emission oscillator of donor and absorption oscillator of acceptor and θ_A and θ_D are the angles between the oscillators and the vector connecting the geometrical centers of donor and acceptor (Dale and Eisinger, 1974). The calculations are performed in the frame of the model of oscillators strictly oriented in space. Due to the uncertainty of donor quantum yield and the overlap integral donor fluorescence and acceptor absorption, it is preferably to estimate the range of the values of W corresponding to the values of $R_0 - 7.8$ and 8.7 Å and 13.1 and 15.0 Å for Trp – Trp and Tyr – Trp energy transfer, respectively (Eisinger et al., 1969; Steinberg, 1971). It is assumed that oscillators, responsible for the long-wavelength absorption band (^1L_a and ^1L_b), are located in the plane of the indole ring, oscillator ^1L_a being oriented at an angle of 60^0 to the C_β-C_γ bond (Yamane et al., 1977; Yamamoto and Tanaka, 1972; Umetskay and Turoverov, 1978) and ^1L_b oriented perpendicular to ^1L_a (Weber, 1960). It is also assumed that only oscillator ^1L_a is responsible for the fluorescence of tryptophan residues (Weber, 1960; Turoverov and Kuznetsova, 1986b). There are grounds for neglecting the ^1L_b oscillator when calculating Trp – Trp energy transfer in proteins (Turoverov and Kuznetsova, 1986a).

2.3. FACTORS DETERMINING THE CONTRIBUTION OF SEPARATE TRYPTOPHAN RESIDUES TO THE TOTAL PROTEIN FLUORESCENCE

The contribution of separate tryptophan to the protein fluorescence is determined by the value of its quantum yield that is primarily depending upon the existence of quenching groups of the amino acid side chains in its microenvironment. The most effective quenchers of tryptophan fluorescence are sulfur atoms of cysteine and methionine (especially sulfur atoms of cysteine incorporated in disulfide bond), imidazole rings of histidine, carboxyl groups of glutamic acid and aspartic acid etc. (Burshtein, 1976). Since the quenching effect of some of these groups depends upon their ionization the traditional and well developed method of the detection of such groups in the vicinity of tryptophan residues is the registration of pH dependencies of different fluorescence characteristics (see e.g. Shinitsky and Goldman, 1967, Shapova and Genov, 1983). The existence of such groups and their nature are detected by the change of registered characteristics and the value of pH at which this change appears. The polarity of tryptophan residue microenvironment and its accessibility to the solvent can also affect its quantum yield. Nonradiative energy transfer between tryptophan residues can also essentially affect the value of quantum yield of individual tryptophan residues as well as the total fluorescence characteristics of proteins. Nonradiative energy transfer is determined by the distance between the indole rings of tryptophan residues and their mutual orientation.

Since the majority of proteins contain more than one Trp, the importance of experimental data increases considerably if the contribution of an individual chromophore can be determined. To accomplish this, different approaches may be used, namely, selective quenching (Eftink and Ghiron, 1981; Eftink, 1991; Burstein, 1996), selective modification (Formoso and Forster, 1975; Imoto et al., 1971), selective excitation (Kuramitsu, 1974; 1978; Rao et al.,1981), separation of the integrated spectrum into components, assuming that there are limited numbers of the discrete forms of tryptophan residues (Burstein et al., 1976; 2001; Reshetnyak and Burstein, 2001; Reshetnyak et al., 2001) and determination of decay associated-spectra (Knutson et al., 1982; Philips et al., 1989; Beechem et al., 1991). The most direct determination of the contribution of individual tryptophan residues to bulk fluorescence is the recording of the fluorescence characteristics of mutant proteins, where tryptophan residues are either deleted or replaced (Martensson, et al., 1995; Gopalan et al., 1997). However, each of these approaches has some limitations.

At the same time the concept on the dependence of fluorescence characteristics of separate tryptophan residues upon their microenvironment allows the prediction of fluorescence characteristics if microenvironment of tryptophan residues is known. This can be done for protein structure that have been determined by X-ray analysis up to the co-ordinate of separate atoms.

The determination of actin structure (Kabsch, 1990) allowed a detailed analysis of microenvironment of each tryptophan residue in order to determine the contribution of each of them to the total actin fluorescence. Actin contains four tryptophan residues. All four tryptophan residues of actin are located in the subdomain 1 (Kabsch et al., 1990). Tryptophan residues Trp 79, Trp 86 and Trp 340 are incorporated in α-helix formed by Trp 79 – Asn 92 and Ser 338 – Ser 348. Tryptophan residue Trp 356 is situated in the unstructured region between α-helixes Ser 350 – Met 355 and Lys 359 – Ala 365 (Figure 2.4).

The analysis of the actin structure reveals that microenvironments of both Trp 79 and Trp 86 contain sulfur atoms (Fig. 2.5 A), which are known to be effective fluorescence quenchers (Burstein, 1976) atoms of Met 119 in the vicinity of Trp 79 and sulfur atoms of Met 82, Met 119, Met 123 and Cys 10 in the environment of Trp 86. The analysis of the dependence of fluorescent quantum yield upon the peculiar features of microenvironments of tryptophan residues for a number of proteins revealed that the efficiency of quenching depends not only upon the proximity of quenching group to the indole ring but also on the location of this group relative to the position of the indole rings of tryptophan residues (Kuznetsova and Turoverov 1998).

A number of sulfur atoms in the vicinity of the indole ring of tryptophan residue Trp 86 and especially the immediate neighborhood of SG of Cys 10 to NE1 of the indole ring of Trp 86 (Fig. 2.5 A) allow to consider that this tryptophan residue is completely quenched. The determination of the distances between the geometrical centers of the indole rings of tryptophan residues and their mutual orientation reveals the effective

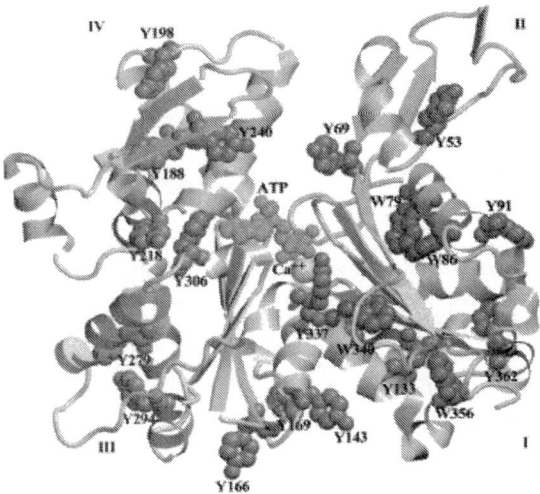

Figure 2.4. Spatial pattern of actin macromolecule. Backbone of the molecule is represented as cartoon diagram; tryptophan residues (red), tyrosine residues (blue), ATP and Ca^{++} are shown as spheres. Subdomains are indicated by roman numbers. This figure is constructed on the base of Protein Data Bank (Bernstein et al., 1977); file 1ATN.ent (Kabsch et al., 1990). The drawing was generated by the graphic programs VMD (Humphrey et al., 1996) and Raster 3D (Merritt, and Bacon, 1977).

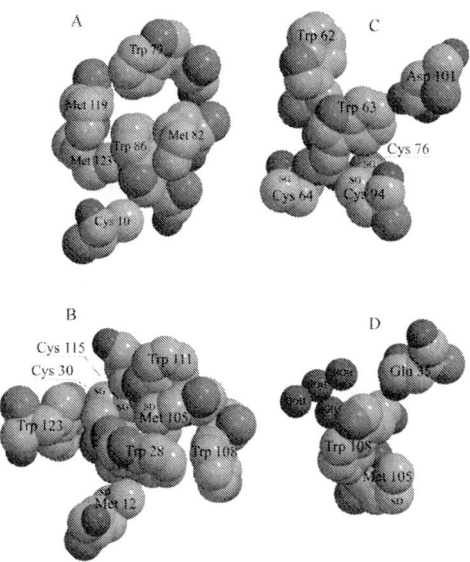

Figure 2.5. Localization of sulfur atoms of Met and Cys in the vicinity of tryptophan residues Trp 79 and Trp 86 of actin (A), and tryptophan residues of lysozyme, Trp 28, Tpr 111 and Trp 123 (B), Trp 62 and Trp63 (C) and Trp 108 (D). This figure is constructed on the base of Protein Data Bank (Bernstein et al., 1977); file 2LYZ.ent (Diamond et al., 1974). The drawing was generated by the graphic programs Molscript (Kraulis, 1991) and Raster 3D (Merritt, and Bacon, 1977).

nonradiative energy transfer between tryptophan residues Trp79 and Trp86 (Fig. 2.6).Consequently, even if the tryptophan residue Trp 79 is not quenched by SD atom of Met 119, it must have a low quantum yield due to effective energy transfer to Trp 86. The calculations show that nonradiative energy transfer between other tryptophan residues is of low efficiency (Table 2.1). Hence the intrinsic UV-fluorescence of the native actin is mainly determined by tryptophan residues Trp 340 and Trp 356 that are inaccessible to the solute molecules. The microenvironments of these residues are formed mainly by nonpolar groups of the protein which are closely packed.

Lysozyme is one of a few multi-tryptophan proteins where the contribution of each of six tryptophan residues has been evaluated using experimental approaches such selective modification, selective quenching, selective excitation and registration of the fluorescence decay curves (Formoso et al., 1975; Imoto, et al., 1971, Lehrer and Fasman, 1967; Teichberg and Shinitzky, 1973). However, an explanation for the differences in the contribution of individual Trp residues to the total lysozyme fluorescence has not been determined. The analysis of the microenvironments of tryptophan residues using the known 3-D structure of lysozyme (Diamond, 1974) can answer this question.

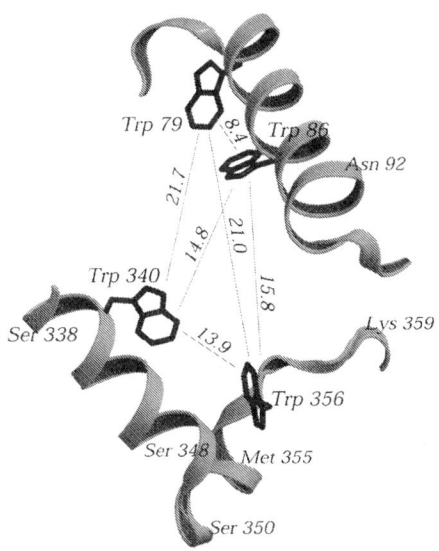

Figure 2.6. Location of tryptophan residues in actin macromolecule. The distances between the geometrical centers of indole rings of tryptophan residues are given in Å.

Table 2.1. Nonradiative energy transfer between tryptophan residues in actin*

Residues	Trp 79	Trp 86	Trp 340	Trp 356
Trp 79		0.64 – 0.77	0.01 – 0.02	0.01 – 0.02
Trp 86	1.6		0.11 – 0.19	0.00 – 0.01
Trp 340	2.7	3.4		0.05 – 0.09
Trp 356	2.8	0.1	1.0	

* The Table shows the values of the efficiency of nonradiative energy transfer W calculated with two values of $R_0 - 7.8$ and 8.7 Å (Eisinger et al., 1969; Steinberg, 1971) (upper right part of the table) and orientation factors k^2 (lower left part of the table). These values were calculated according to Eqs. 3 and 4. The distances between the geometrical centers of the indole rings of tryptophan residues, needed for the evaluation of nonradiative energy transfer, are given in the Figure 2.6.

The analysis of the microenvironments of tryptophan residues in lysozyme reveals that, in the vicinity of five Trp residues of lysozyme, there are sulfur atoms (Table 2.2) that are known to be effective quenchers of tryptophan fluorescence, especially those incorporated in S-S bonds (Burstein, 1976). In the microenvironment of Trp111 and Trp123 there are sulfur atoms incorporated in disulfide bond Cys30 – Cys115 (Table 2.2,

Fig. 2.5B). Sulfur atom SG Cys115 is 6.0 Å and 6.5 Å from the geometrical center of the indole ring of Trp111 and Trp123, respectively. Sulfur atom SG Cys30 is 4.5 Å from the geometrical center of the indole ring of Trp123, being located quite closely (3.3 Å) to NE1 atom of this residue. Moreover, as close as 3.9 Å to NE1 atom of the indole of Trp111, there is a sulfur atom SD Met105. The microenvironment of Trp28 contains sulfur atoms of Met12 and Met105 that are 4.2 Å and 6.2 Å distant from the geometrical center of indole ring, respectively (Table 2.2, Fig. 2.5B). Sulfur atoms incorporated in the disulfide bond Cys94–Cys76 were found close to Trp63 (Table 2.2, Fig. 2.5C). They are located at 4.6 Å and 3.7 Å from CG and CE3 of the indole ring, respectively.

There is no consensus about the contribution of Trp 63 to the total intrinsic fluorescence of lysozyme (Burstein et al., 1973; Turoverov et al., 1985). Due to the existence of a disulfide bond in the vicinity, Trp 63 could be assigned to be the Trp residues with low quantum yield. At the same time, the analysis of nonradiative energy transfer does not reveal this (see below). The microenvironment of Trp108 also contains sulfur atom of Met105 (Table 2.2, Fig. 2.5 D), but it was experimentally shown that this Trp gives the greatest contribution to the lysozyme fluorescence (Teichberg, and Sharon, 1970). More detailed analyses show that the location of S atoms nearby the indole rings of Trp28, Trp111 and Trp123 and that of Trp63 and Trp108 differ dramatically. In the first case, sulfur atoms are located near the indole ring plane, close to its NE1 atom. In the second case, the sulfur atoms are far away from this region of indole rings of Trp63 and Trp108 (Table 2.2). A summary of this analysis is presented in Table 2.2.

This table allows us extract factors determining the efficiency of Trp fluorescence quenching by S. The efficiency of sulfur quenching depends not only on the distance of sulfur atoms from the indole ring, but even more on their location relative to the indole ring.

Considering that quantum yield of Trp is determined not only by the presence of quenching groups in their vicinity, but also by the efficiency of nonradiative energy transfer from tyrosine to tryptophan and among the tryptophan residues, the distances between tryptophan residues and their mutual orientations were analyzed and the efficiency of energy transfer was calculated. As suggested earlier (Teichberg. and Shinitzky, 1973), an effective energy transfer between Trp62 and Trp63 was detected. Therefore, if the fluorescence of Trp63 would be quenched by sulfurs of the disulfide bond Cys76–Cys94, then the excitation energy of Trp62 would dissipate through Trp63. Nonetheless, it has been shown that Trp62 gives a significant contribution to the bulk intrinsic fluorescence of lysozyme (Imoto et al., 1975; Lehrer and Fasman, 1967). In other words, we assume that even though sulfur atoms in the disulfide bond Cys94–Cys76 are located near Trp63, this tryptophan residue is not quenched. This conclusion agrees with the theory of Lehrer & Fasman (Lehrer and Fasman, 1967). Results of our analyses also show that energy transfer from Trp108 to Trp62 through Trp63 (Formoso et al., 1975) does not occur, since the efficiency of energy transfer from Trp108 to Trp63 is negligible. The fluorescence quantum yield of Trp108 could be slightly diminished due to energy transfer from Trp108 to the quenched tryptophan residue Trp28.

Table 2.2. Characteristics of microenvironments and conformation of the side chains of tryptophan residues in lysozyme

Residue	N	d	χ_1 (deg)	χ_2 (deg)	Atom*	R_1 (Å)	R_2 (Å)
Trp 28	66	0.75	300	107	SD Met 105	4.2	4.2 (Cntr)
					SD Met 12	6.2	4.9 (CG)
Trp 62	34 (3)	0.42	326	89	NE Arg 61	5.0	4.6 (CZ2)
					NH2 Arg 61	6.6	5.8 (CZ2)
					NE Arg 73	6.7	5.0 (CZ3)
					NH1 Arg 73	5.7	3.6 (CZ3)
Trp 63	62 (4)	0.68	311	117	SG Cys 76	5.0	3.7 (CE3)
					SG Cys 94	5.4	4.6 (CG)
					OD2 Asp 101	4.8	2.6 (CH2)
Trp 108	62 (4)	0.70	281	286	SD Met 105	6.4	5.2 (CZ3)
					OE2 Glu 35	5.8	3.6 (CD1)
Trp 111	67	0.74	171	88	SD Met 105	4.4	3.9 (NE1)
					SG Cys 115	6.0	4.1 (CD1)
					NZ Lys 116	6.5	4.6 (CZ3)
Trp 123	53 (2)	0.61	286	113	SG Cys 30	4.5	3.3 (NE1)
					SG Cys 115	6.5	5.4 (NE1)
					NE Arg 5	6.8	5.3 (CE3)
					NH1 Arg 5	6.7	5.4 (CE3)
					NH1 Arg 125	6.6	6.2 (CD1)

N is number of atoms of microenvironment, in brackets are given the number of bounded water molecules;
d is packing density of atoms of microenvironment;
χ_1 and χ_2 are the angles that characterize the conformation of tryptophan residue side chain;
R_1 and R_2 are the distances between indicated atom and the center of the indole ring and the nearest atom of the indole ring (given in brackets);
Nitrogen, oxygen and sulfur atoms of amino acid side chains involved in the microenvironment of tryptophan residue.

In some cases the combined analysis of the characteristics of protein intrinsic fluorescence and peculiarities of tryptophan residues location can suggest a different interpretation of the experimental data. In connection with this the effect of "blue" copper center of azurin *Pseudomonas aeruginosa* on the fluorescence features of its single tryptophan residue Trp 48 is of particular interest. Fluorescence experiments show that the copper removal leads to a significant increase of the fluorescence intensity without any change of fluorescence spectrum position (Finazzi-Agro et al., 1973; Grinvald et al., 1975; Mallison et al., 1981; Szabo et al., 1983). Such essential quenching effect of copper in these works was explained by immediate contact of copper center to indole ring. However, the analysis of X-ray data (Turoverov et al., 1985) revealed that the distance from copper ion to geometric center of indole ring is 11.8 Å (Fig. 2.1, 2.2). The copper ligands are also significantly removed from the indole ring: SG of Cys112 and SD Met121 – at a distance of 10.6 and 11.0 Å; ND1 His 40 and 117 – at 11.8 and 13.9 Å; the geometrical centres of these histidine residues – at 12.1 and 15.0 Å. The assumption of immediate contact between the copper ion or one of its ligand with indole ring (Finazzi-Agro et al., 1973, 1973; Grinvald et al., 1975; Mallison et al., 1981; Szabo et al., 1983) is inconsistent with these data. In view of such localization of the tryptophan residue we can suggest that quenching action of copper ion is caused by the effect of electrostatic field of its positive charge on the indole ring. It seems plausible that the difference in the absorption spectra of apo- and holo forms of azurin registered in the works of Grinvald and Szabo (Grinvald et al., 1975; Szabo et al., 1983) is also determined by this effect.

There is an assumption that excitation of the indole ring leads to redistribution of charge that results in the accumulation of high electron density at one of the atoms of the ring. In some condition the electron detachment can occur, which entails in fluorescence quenching. On the basis of 3-D structure of azurin we could suggest that the electric field of copper ion provokes electron detachment from the indole ring (Turoverov et al., 1985) as was proved (Petrich, et al., 1987). This could arise the problem of long-range effects when analysing the effect of different protein groups on fluorescent characteristics of tryptophan residues.

2.4. FACTORS DETERMINING THE FLUORESCENCE SPECTRUM POSITION OF SEPARATE TRYPTOPHAN RESIDUES

Fluorescence spectrum of tryptophan residue is one of the most sensitive tool for investigating the protein structure. The position and the form of protein fluorescence spectrum are determined by superposition of fluorescence spectra of separate tryptophan residues – by their position in the wavelength scale and relative contribution to the bulk fluorescence. Maximum of fluorescence spectra varies from 308 nm for azurin from *Pseudomonas aeruginosa* (Finazzi-Agro et al., 1970) to 350-355 nm for such proteins as melittin, glucagon, mielin basic protein and all proteins in unfolded state. It is sufficient to look at the structures of azurin and glucagon (Fig. 2.1 A and B), to conclude that, in these two extreme cases, the notion that fluorescence spectrum position is dictated by the accessibility of tryptophan residue to the solvent is absolutely adequate. The existing knowledge of the dependence of fluorescence characteristics and properties of tryptophan residue microenvironment was elaborated as a result of investigations of model compounds (tryptophan and its derivatives) and proteins in different structural states. Usually it is assumed that the position of fluorescence spectrum of tryptophan residues is dictated by their accessibility to solvent molecules. Since the glucagon tryptophan residue Trp 25 is exposed to the solvent this protein must have a red shifted fluorescence spectrum, while tryptophan residue Trp 48 of azurin is located in the inner part of macromolecule and as a consequence this protein must display a blue fluorescence spectrum. At the same time the inaccessibility to the solvent is not the only condition of the appearance of blue fluorescence spectrum. For example, it is necessary to take into account that the polarity of microenvironment of tryptophan residues is determined not only by its exposure to the solvent but by intrinsic polar groups of amino acid side chains included in the microenvironment of this tryptophan residue. Thus, inactivated actin has a rather red shifted fluorescence spectrum ($\lambda_{\text{макс}}$ = 343 нм; A = 1.3; Kuznetsova et al., 1988) and very low efficiency of its fluorescence quenching by acrylamide (Kuznetsova et al., 1995).

2.5. GLUTAMINE-BINDING PROTEIN FROM *E. COLI*

Glutamine-binding protein (GlnBP) from *E. coli* is a monomer (25 kDa) that is responsible for the first step in the active transport of L-glutamine across the cytoplasmic membrane also has a rather red fluorescence spectrum (λ_{max} = 330 nm, A = 1.65; Fig. 2.7), suggesting that tryptophan residues of GlnBP are in a polar environment. This can be caused by the accessibility of tryptophan residues to the solvent molecules and/or by the existence of some polar groups of the protein in the vicinity of the Trp residues. Fluorescence spectra of GlnBP in the absence and in the presence of glutamine (Gln) are practically the same (Fig. 2.7).

GlnBP consists of two domains (termed large and small) linked by two antiparallel β-strands. The large domain is similar to the small domain but contains two additional α-helices and three more short antiparallel β-strands (Hsiao et al., 1996). The deep cleft formed between the two domains contains the ligand-binding site. The binding of L-glutamine leads to cleft closing and a significant structural change with the formation of the so-called "closed form" structure (Sun et al., 1998). There are no disulfide bonds in GlnBP. This protein contains two tryptophan residues (Trp 32 and Trp 220) and ten tyrosine residues (Fig. 2.8).

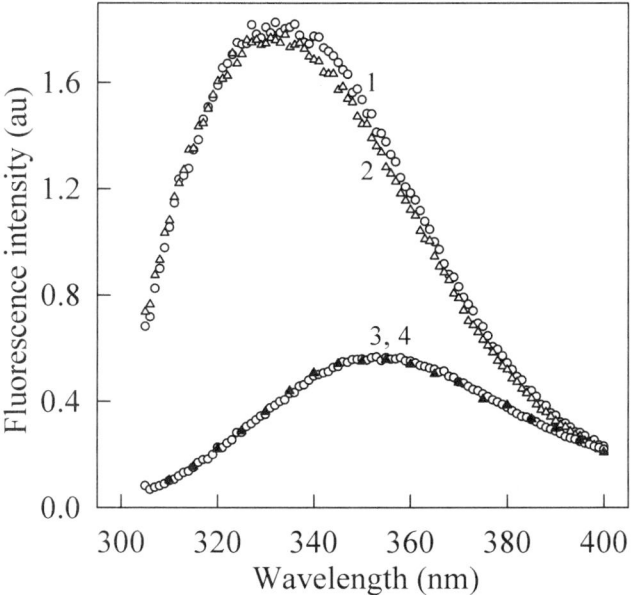

Figure 2.7. Fluorescence spectra of GlnBP and its complex with Gln in native (curves 1 and 2) and unfolded (curves 3 and 4) states. Denaturation was induced by 3.0 M of GdnHCl. The excitation wavelength was 297 nm. All values are reduced to fluorescence intensity of native GlnBP at 365 nm.

Both tryptophan residues are localized rather far from the ligand binding site. The fluorescence lifetimes of tryptophan residues of GlnBP in the absence and in the presence of Gln are also almost the same. Nonetheless the recording of the parameter A allowed us to conclude that the interaction of this protein with Gln leads to a slight blue shift in the tryptophan fluorescence spectra.

Analysis of the 3D structure of the ligand-free GlnBP (file 1GGG.ent) and ligand-bound GlnBP (file 1WND.ent) showed that both tryptophan residues of GlnBP are located in α-helices (Trp32 belongs to the first helix Phe 27 – Glu 38, and Trp 220 belongs to the final one Thr 212 – Phe 221). Their microenvironment density is not very high (there are 63 atoms and 66 atoms in the microenvironments of Trp 32 and Trp 220, respectively), and both tryptophan residues are partially accessible to solvent. At the same time the polarity of their microenvironments differs significantly. Densities of microenvironments of Trp 32 and Trp 220 change insignificantly upon interaction of the protein with Gln (there are 65 atoms in the microenvironments of both tryptophan residues), being Gln located far from the protein tryptophan residues. This explains why the ligand-binding does not influence the tryptophan fluorescence directly by itself.

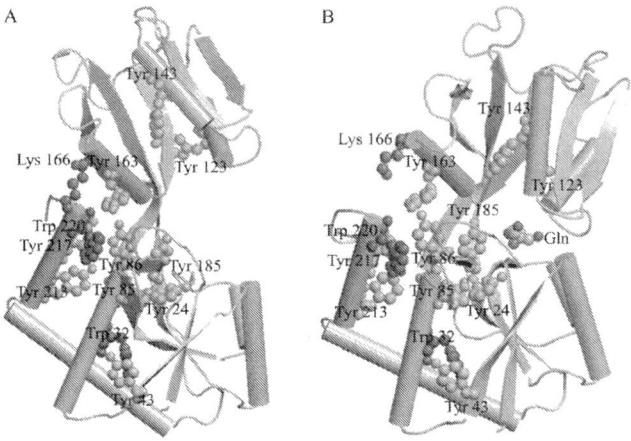

Figure 2.8. Spatial pattern of GlnBP in the absence (A) and in the presence (B) of Gln. Backbone of the molecule is represented as cartoon diagram; tryptophan residues (red), tyrosine residues (green), Lys 166 (blue) belonging to microenvironment of Trp 220 and Gln are shown as spheres. Complex formation leads to collapse of protein macromolecule around Gln resulting in cleft close. This figure is constructed based on Protein Data Bank (Bernstein et al., 1977), file 1GGG.ent (Hsiao et al., 1996) and 1WDN.ent (Sun et al., 1998). The drawing was generated by the graphic programs VMD (Humphrey et al., 1996) and Raster 3D (Merritt, and Bacon, 1977).

There is only one polar group of the amino acid side chain (OH group of Tyr 43) in the microenvironment of Trp 32. The position of this group changes slightly with the complex formation (GlnBP/Gln). At the same time there are many hydrophobic groups in the microenvironment of this tryptophan residue (Val 7, Ile 35, Ala 36, Leu 41, ring of Tyr 43, Leu 45, Leu 64, Leu 66, Ile 187 and Val 200) in ligand-free and ligand-bound protein. As a consequence, we can conclude that apparently Trp 32 must have a rather blue fluorescence spectrum. Furthermore the side chain of Trp 32 is t-conformer ($\chi_1 = 180^0$). This conformation is rarely detected in proteins. According to our observation this conformation has tryptophan residues located in a rather hydrophobic microenvironment with blue fluorescence spectrum position, e.g. Trp 48 of azurin, Trp 59 of ribonuclease T1, and Trp 340 of actin (Kuznetsova and Turoverov, 1998; Turoverov and Kuznetsova, 2003). Worthy of note is the work of Axelsen et al. (1991) where the fluorescence properties of mutant forms of GlnBP were examined with one of the tryptophan residues substituted by Phe or Tyr. Our conclusion that Trp 32 must have a rather blue fluorescence spectrum coincides with one of the conclusions of the work of Axelsen et al. (1991). At the same time Trp 32 is essentially quenched. Our analysis shows that there is only one quenching group, namely the OH group of Tyr 43 in the microenvironment of Trp 32. So, it is not clear why it could be quenched. At the same time we also observe a rather red fluorescence spectrum of GlnBP, proving that the contribution of Trp 32 to the bulk fluorescence spectrum is insignificant.

In the microenvironment of Trp 220 there are six polar groups of the side chains of amino acids and eleven O and N atoms of peptide bonds, which is more in comparison with Trp 32 (only one polar group of the side chains of amino acids and six O and N atoms of peptide bonds). This led us to conclude that the fluorescence spectrum of Trp 220 is more red shifted than that of Trp 32. Apparently, fluorescence spectrum of GlnBP is formed mainly by Trp 220.

Although the number of atoms in the microenvironment of Trp 220 residue is practically the same for ligand-free and ligand-bound GlnBP, its contents change. In fact, in the Trp 220 microenvironment of GlnBP, there are Pro 15, Phe 18, Phe 27, Phe 221, Tyr 163, while in the GlnBP/Gln there are only Pro 15, Phe 18, and Phe 27 remain in the microenvironment, while Phe 221 and Tyr 163 belonging to the other domain leave the microenvironment of Trp 220, when this domain turns at the complex formation. Interestingly, there are more polar groups in the protein in the absence of the ligand (six) than in the complex with Gln (four). At the complex formation, the OH groups of Tyr 163 and N2 Lys 166 leave the microenvironment, while other groups change only slightly both their orientation and distance from the tryptophan center (Fig. 2.8). Tryptophan residues Trp 32 and Trp 220 are located far from each other ($R = 16$ Å in the ligand-free protein and 15.3Å in ligand-bound protein) and the efficiency of energy transfer is negligible. It is necessary to take into account that the rigidity of microenvironment of its tryptophan residues can also influence the protein fluorescence spectrum position. Blue fluorescence spectra can be observed in case that tryptophan residues are located in hydrophobic microenvironment independently of their relaxation properties. In this case the fluorescence emission arises from the inequilibrium state that does not fit energy minimum of interconnection with media. There is also an opinion that for the formation of fluorescence spectrum the existence of the group which can form complexes with indole ring in the exited state is very important (exiplecs formation). That is why much more detailed analysis of azurin structure is need for explaining the anomalous blue structured fluorescence spectrum.

Diagram (Fig. 2.2) shows that, in the close vicinity of tryptophan residue Trp 48 of azurin, there is no polar groups of amino acid side chains. Only polar hydroxyl groups of threonin and serin residues Thr 17, Thr 30, Thr 84, Ser 94 and amide group of asparagine Asn 32 are located as far as 10 Å from the centre of indole ring. Tryptophan residue is shaded from all of these groups except hydroxyl group of threonin Thr 17 by the near nonpolar hydrocarbon atoms. Interestingly, all side chains that have polar groups are orientated from the indole ring situated in the centre of azurin to its periphery (Fig. 2.2).

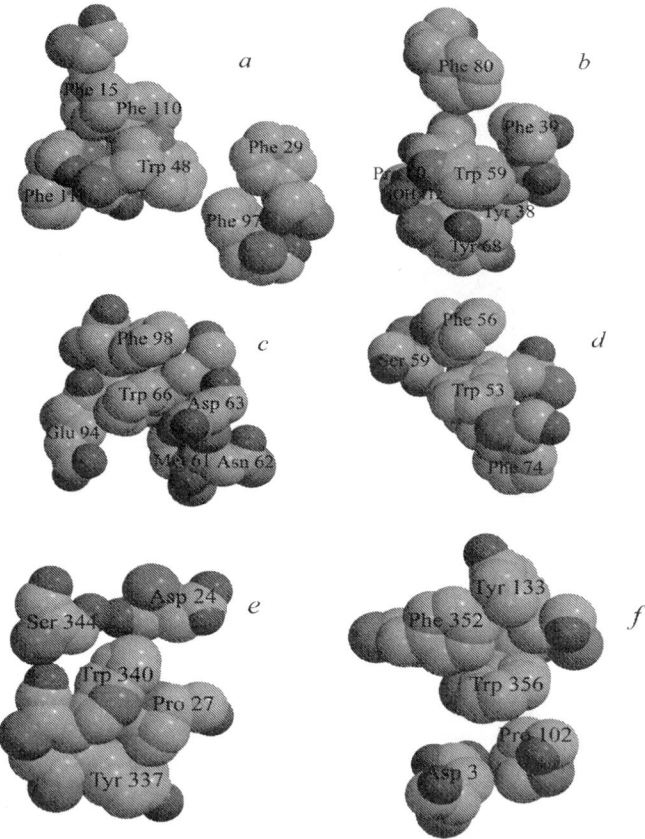

Figure 2.9. The location of Phe, Tyr and Pro residues in the vicinity of tryptophan residues of Trp 48 of azurin (*a*), Trp 59 of ribonuclease T1 (*b*), Trp 66 of L-asparaginase (*c*), Trp 53 of barstar (*d*), and Trp 340 (*e*) and Trp 356 (*f*) of actin. The drawing was generated by the graphic programs Molscript (Kraulis, 1991) and Raster 3D (Merritt, and Bacon, 1977).

Analysis of the properties of tryptophan residue Trp 48 of azurin and tryptophan residues of some other proteins with rather blue fluorescence spectrum position (tryptophan residue Trp 59 of ribonuclease T1 and tryptophan residue Trp 66 of L-asparaginase) allowed to conclude that for appearance of such blue fluorescent spectrum tryptophan residues must be inaccessible for the molecules of solution and have hydrophobic environment. At the same time in the vicinity of every tryptophan residues there are peptide bonds, which have essential dipole moment. Now it is not clear how the existing of polar peptide groups is combined with unique blue fluorescence spectra.

There is an opinion (Burstein, 1976) that all tryptophan residues in the exited state make complexes with the groups of their microenvironment. The only exclusion is a single tryptophan residue of azurin, which do not participate in any complex formation and which has a unique fluorescence spectrum position. The position of fluorescence spectra of all other molecules is determined be the exciplexes stoichiometry. For example it is believed that tryptophan residue Trp 59 of ribonuclease T1 in the excited state forms complex with bound water molecule located near NE1 atom of the indole ring (Fig. 2.9).

The idea that the indole ring of tryptophan residue in excited state forms exciplexes with some groups of its microenvironment was essentially proved in the work devoted to the examination of the correlation between fluorescence lifetime τ fluorescence quantum yield q of several proteins which have one tryptophan residue (Szabo and Faerman, 1992). It was found that the radiation lifetime τ_0 that was calculated on the base of these experiments

$$\tau_0 = \tau/q \qquad (5)$$

differ among proteins.

This can be due to the static fluorescence quenching or due to emission from exciplexes of indole ring with the neighbor groups but not from indole it self. Since different groups can form complexes with indole ring and differently orientated and distant from indole ring, the efficiency of the transition to the ground state for exciplexes $K_{f,excipl}$

$$K_{f,excipl} = \frac{1}{\tau_{0,excipl}} \qquad (6)$$

may be different, and just this was found experimentally (Szabo and Faerman, 1992).
According to the model of discrete forms of tryptophan residues (Burstein, 1976), exciplexes of indole ring with stoichiometry of 1:1 have fluorescence spectrum with maximum at 316 nm (ribonuclease T1, L-asparaginase, parvalbumine), exciplexes of indole ring with stoichiometry of 1:2 have fluorescence spectrum with maximum at 330 nm. If these assumptions are correct then one of the aims of tryptophan residues' microenvironment analysis must be the finding of the groups that could form exciplexes with the indole ring. At the same time microenvironments of different tryptophan residues vary essentially their content, distance and location of the group relative to the indole ring. So, we consider there are no grounds to divide tryptophan residues to discrete classes.

Analysis of a large number of tryptophan residues shows that very high density of microenvironment is insufficient for appearance of unique blue fluorescence spectrum. For example the density of microenvironment of tryptophan residue Trp 48 of azurin is lower than that of tryptophan residues whose fluorescence spectrum is much more red shifted (Kuznetsova and Turoverov, 1998).

It is not difficult that the conformation of tryptophan residue side chain could play an essential role in the formation of blue fluorescent spectrum. It turns out that tryptophan residues Trp 48 of azurin and Trp 59 of ribonuclease T1 has " t " conformation by angle χ_1 that is not very typical for tryptophan residues in protein. Interestingly, the same conformations of the side chain have tryptophan residues with very low density of microenvironment, accessible for solvent, e.g. tryptophan residue Trp 25 of glucagon.

The peculiar feature of microenvironment of Trp 48 of azurin is the availability of five phenylalanine residues (Phe 15, Phe 29, Phe 97, Phe 110 and Phe 111). Furthermore all atoms of phenol ring of Phe 110 are the most closely adjacent to the indole ring of tryptophan residue of Trp 48, and there is only a small angle between the planes of the rings of these residues. The analysis of other single-tryptophan containing proteins with blue fluorescence spectra (ribonuclease T1, L-asparaginase) revealed that in the close vicinity of their tryptophan residues there are clusters of aromatic rings (Fig. 2.9). Probably this microenvironment feature plays the main role in the formation of the blue fluorescence spectrum. The analysis of fluorescence properties of C5 protein, which is included in ribonuclease P and mutant form of barstar (Gopalan et al., 1997) gives strong evidence for this. The fluorescence spectra of tryptophan residue Trp 105 of C5 protein and tryptophan residue Trp 53 of barstar (mutant W38F/W44F) represent a superposition of two bands with maximum at 318 and 332 nm. The maximum at 318 nm disappears when the phenylalanine residues in the microenvironment of tryptophan residue Trp 53 of barstar and Trp 109 of C5 protein are replaced by other amino acids.

Another example with extraordinary blue fluorescence spectrum is G-actin (λ_{max} = 325 nm, A = 2.6 Kuznetsova et al., 1988). Among the known tryptophan-containing proteins only a few such as azurin from *P. aeruginosa* (Finazzi-Agro et al., 1970), RNasa T1 (Yamamoto and Tanaka, 1970), RNasa C2 (Agekian et al., 1988), and parvalbumin merlanga (Permyakov et al., 1980) have more blue fluorescence spectrum position. Analysis of actin 3D revealed that the packing density of microenvironment of separate tryptophan residue varies greatly. Thus, in the sphere of 7 Å in radius, whose center coincides with geometrical center of the indole ring of analyzed tryptophan residue, there are 50, 61, 78 and 69 atoms of the protein for tryptophan residues Trp 79, Trp 86, Trp 340 and Trp 356, respectively. For comparison, there are 71 atoms in the microenvironment of the inner tryptophan residue of azurin (Turoverov et al., 1985), which has the most unique blue fluorescence spectrum (λ_{max} = 308 nm (Finazzi-Agro et al., 1970)). Thus two tryptophan residues of actin – Trp 340 and Trp 356, has very high density of microenvironment (d = 0.84 and 0.76). And, though they are not located in the center protein macromolecule, but closer to its periphery (the value of d decreases rapidly with the increase of r_0; Fig. 2.10), they are apparently inaccessible to the solvent. The packing density of microenvironment of Trp 86 is lower than that of Trp 340 and Trp 356.

At the same time Fig. 2.10 shows that this tryptophan residue is located far from protein's periphery, and it is evidently inaccessible to the solvent. So, the only tryptophan residue that can be regarded as exposed to the solvent is Trp 79.

The averaged experimental characteristic of tryptophan residue exposure can be obtained by fluorescence quenching by the external quencher (Eftink and Ghiron, 1981). The experiments on tryptophan fluorescence quenching of actin reveal the low accessibility of tryptophan residues to the molecules of polar quencher acrylamide (Kuznetsova et al, 1999 a). The quenching constant determined from the initial slope of the curve (K = 1.7 M^{-1}) (Kuznetsova et al, 1999 a) is practically as low as that for the proteins like RNase T1 (K = 1.0 M^{-1}) (Eftink and Ghiron, 1977), which have blue, structural fluorescence spectra. The low value of quenching constant indicates that tryptophan residues of actin are inaccessible to the solvent. It correlates with the very blue fluorescence spectrum of this protein. The blue emission spectrum of actin fluorescence and the low efficiency of fluorescence quenching by acrylamide can be explained if the contribution of tryptophan residues Trp 79 (that is exposed to the solvent) to protein intrinsic fluorescence is low (see above).

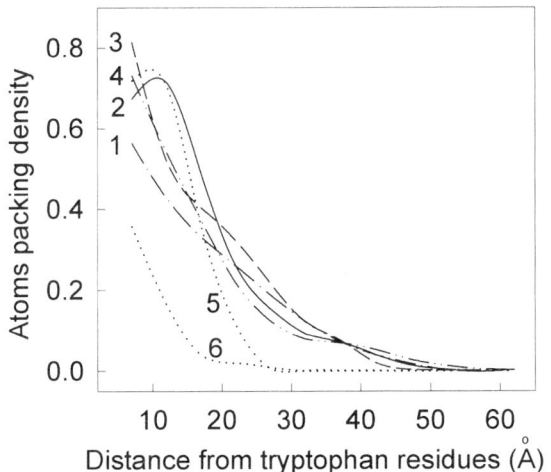

Figure 2.10. Radial dependence of packing density of atoms in macromolecule about geometrical centers of indole rings of tryptophan residues in actin. Curves 1 - 4 are the dependencies for Trp 79, Trp 86, Trp 340 and Trp 356 consequently. For comparison the dependence for Trp 48 of azurin (internal tryptophan residues) and Trp 19 of melittin (external tryptophan residues) are given (curves 5 and 6).

Hence the intrinsic UV-fluorescence of the native actin is mainly determined by practically inaccessible to the solute molecules tryptophan residues Trp 340 and Trp 356. The microenvironments of these residues are formed mainly by nonpolar groups of the protein and closely packed (Fig. 2.9 e, f). Though atoms OD2 of Asp 24, OG Ser 324 and OD1 and OD2 of Asp3 are incorporated in the microenvironment, they are distant rather away from the rings and it is unlikely that they can influence the fluorescent characteristics of these tryptophan residues. The distinctive feature of microenvironments of tryptophan residues Trp 340 and Trp 356 is the existence of aromatic rings of tyrosine and phenylalanine residues and rings of proline (Fig. 2.9 e, f). So, there is phenol ring of Tyr 337 and the ring of Pro 27 in the close vicinity of tryptophan residue Trp 340, and aromatic ring of Phe 352 and Tyr 133 and the ring of Pro 102 in vicinity of tryptophan residue Trp 356. Similar clusters of aromatic residues were found in the proteins with one tryptophan residue, which have blue fluorescent spectrum such as azurin, ribonuclease T1 and L-asparaginase (Fig 2.9 b). The high packing density of microenvironment of Trp 340 can be explained by the location of proline residue Pro 27, whose ring is placed almost in parallel with indole ring of Trp 340.

It is noteworthy that tryptophan residue Trp 340 is t-conformer ($\chi_1 = 190^0$, $\chi_2 = 89^0$) unlike other tryptophan residues of actin and tryptophan residues of many other proteins (Kuznetsova and Turoverov, 1998 Cytology). As mentioned above, the same conformation of the side chain was found for inner tryptophan residues with extremely blue fluorescence spectrum, like Trp 48 of azurin and Trp 59 of ribonuclease T1, and for tryptophan residue completely exposed to the solvent, like Trp 25 of glucagon and Trp 19 of melittin (Kuznetsova and Turoverov, 1998 Cytology). Thus this conformation is probably the most unstrained side chain conformation. At the same time the microenvironment of tryptophan residue Trp 340 is very closely packed. Its packing density is even higher than that of microenvironments of tryptophan residue Trp 48 of azurin (d = 0.75) and Trp 59 of ribonuclease T1 (d = 0.80). Furthermore the oscillations of tryptophan residue Trp 340 are restricted by proline residue Pro 27, whose ring is practically parallel to the indole ring (Fig. 2.9 *e*). It is not difficult to find that unstrained conformation of the side chain of tryptophan residue Trp 340, as well as the existence of aromatic rings of tyrosine and phenylalanine residues and proline residue, in the vicinity of tryptophan residue Trp340 and 356 are essential for the formation of blue fluorescence spectrum of actin.

2.6. TYROSINE FLUORESCENCE IN PROTEINS

Usually in tryptophan containing proteins the contribution of tyrosine is negligible. The contribution of tyrosine residues can be determined from the comparison of fluorescence spectra excited at 280 nm and 297 nm and normalized at 365 nm (where the contribution of tyrosine residues is negligible). Nonetheless the evaluation of the contribution of tyrosine residues to the bulk fluorescence is essential for understanding the fluorescence properties of proteins. And the registration of tyrosine fluorescence can be important for examination processes of folding-unfolding, because the number of tyrosine residues is greater than that of tryptophan and tyrosine residues can be localized all over the protein macromolecule. Thus in actin all four tryptophan residues are located in the subdomain I, while 16 tyrosine residues are distributed evenly over the actin macromolecule (Fig. 2.4). The results of the analysis of tyrosine residues location in some proteins are given below.

DsbC from *Escherichia coli* is a homodimeric molecule. Each subunit of DsbC molecule contains one tryptophan residue, Trp 140, and eight tyrosine residues (Fig. 2.11). With excitation at 280 nm DsbC showed a fluorescence spectrum with a maximum emission at 302 nm. Comparison of two fluorescence spectra excited at 280 and 297 nm, respectively, revealed that the parts in the red wavelength range are coincided but deviated significantly from the blue wavelength range (Stepanenko, 2004). The difference spectrum presents the contribution of tyrosine residues to DsbC fluorescence excited at 280 nm.

To determine the contribution of each tyrosine residue to the bulk protein fluorescence we examined their individual microenvironment, the energy transfer between tyrosine residues and from tyrosine residues to Trp 140. Microenvironment of tyrosine residues in DsbC molecule differs significantly (Table 2.3). Because there are only 40 atoms in the microenvironment of Tyr 100, it is considered as external and accessible to solvent. In compliance with it many molecules of bound water were found in its microenvironment.

In contrast, there are 77, 72, 79 and 79 atoms in the microenvironment of Tyr 38, Tyr 81, Tyr 111 and Tyr 120, respectively, and the microenvironment density of these residues are thus significantly higher. The fact that for all tyrosine residues the number of the atoms in the microenvironment of oxygen atom of hydroxyl groups (Table 2.3, in brackets) is less than that in the sphere with the geometrical center of phenol ring indicates that the hydroxyl groups of the tyrosine residues are all directed into the periphery of the molecule. It is particularly true for Tyr 52, Tyr 100 and Tyr 171, which are located near the surface of the molecule with low density of microenvironment. Except Tyr 100 and Tyr 171 there are a lot of potential quenching groups in the microenvironment of the other six tyrosine residues, such as SD atoms of Met 27, Met 51 and Met 66 and oxygen atom OD1 of Asn 61 in the microenvironment of Tyr 38; oxygen and nitrogen atoms of Gln 48 and Asn 61 in the microenvironment of Tyr 52. Nevertheless the contributions of Tyr 38 and Tyr 52 may not be completely ignored. As shown in Figure 2.12 there is efficient energy transfer from Tyr 81, Tyr 100, Tyr 111 and Tyr 120 to Trp 140 in the same C-terminal domain, but not from the tyrosine residues located in the N-terminal domain. Also, there is no energy transfer between the tyrosine residues located in the different domains. Tyr 196 contributes little fluorescence for two reasons. Firstly, there are many quenching groups in its microenvironment, such as Lys 102, Glu 201 and Gln 107, which NE2 atom is in immediate contact with the OH group of Tyr 196. Secondly, energy from Tyr 196 can be transferred with high efficiency of 0.99 and 0.91 to Tyr 100 and Tyr 111 respectively, and further to Trp 140. As a result, only Tyr 171 makes major contribution to the tyrosine fluorescence of the DsbC molecule.

The contribution of tyrosine residues to the bulk fluorescence of GlnBP (Fig. 2.13, Insert) and GlnBP/Gln (data not shown) is negligible. It increases significantly when the protein (or its complex) is unfolded in 3.0 M GdnHCl (Fig. 2.13, Insert). At first glance it is easy to explain. All tyrosine residues except Tyr 123 transfer their excitation energy to Trp 32 and/or Trp 220 (Table 2.1).

This could account for the insignificant contribution of tyrosine residues to the bulk fluorescence of native protein. The increased contribution of tyrosine residues to protein fluorescence when it is unfolded can be explained by an energy transfer condition failure. At the same time, if this were so, the intensity of tryptophan fluorescence ($\lambda_{em} = 365$ nm) should decrease on protein unfolding more significantly when excited at 280 nm than at 297 nm. Figure 2.4 shows that this is not so: the dependencies of I_{365} on GdnHCl concentration recorded at 280 nm and 297 nm are normalized to the value of completely unfolded protein (at 3.0 M GdnHCl) because under these conditions fluorescence emission at 365 nm is determined only by tryptophan residues. The result obtained uniquely suggests that the non-significant contribution of tyrosine residue in the bulk fluorescence of ligand-free GlnBP (and ligand-bound GlnBP) cannot be explained only by the energy transfer between Tyr residues and Trp residues. At the same time, since there are conditions for effective energy transfer between the Tyr residues and Trp residues in the ligand-free protein (see Table 2.4), the energy of tyrosine residues excitation dissipates even more effectively in some other ways, i.e., there are effective fluorescence quenchers, or excitation energy is transferred to other tyrosine residues which have fluorescence quenchers in the vicinity. Examination was carried out on the microenvironment of tyrosine residues uniformly distributed along the protein macromolecule (see Fig. 2.8), and the evaluation of energy transfer Tyr-Tyr. The results of this analysis are given in Table 2.5 and Table 2.6. Table 2.5 shows polar groups or oxygen atoms of bounded water, which are the nearest to OH groups of tyrosine residues and thus could be potential quenchers of tyrosine fluorescence. It is obvious that many tyrosine residues could be quenched. Table 2.6 shows that there is effective energy transfer among many tyrosine residues. Thus it is evident that most of the tyrosine residue could be quenched not only by energy transfer to tryptophan residues, but also by quenching groups in their vicinity or by effective energy transfer to tyrosine residues which are quenched.

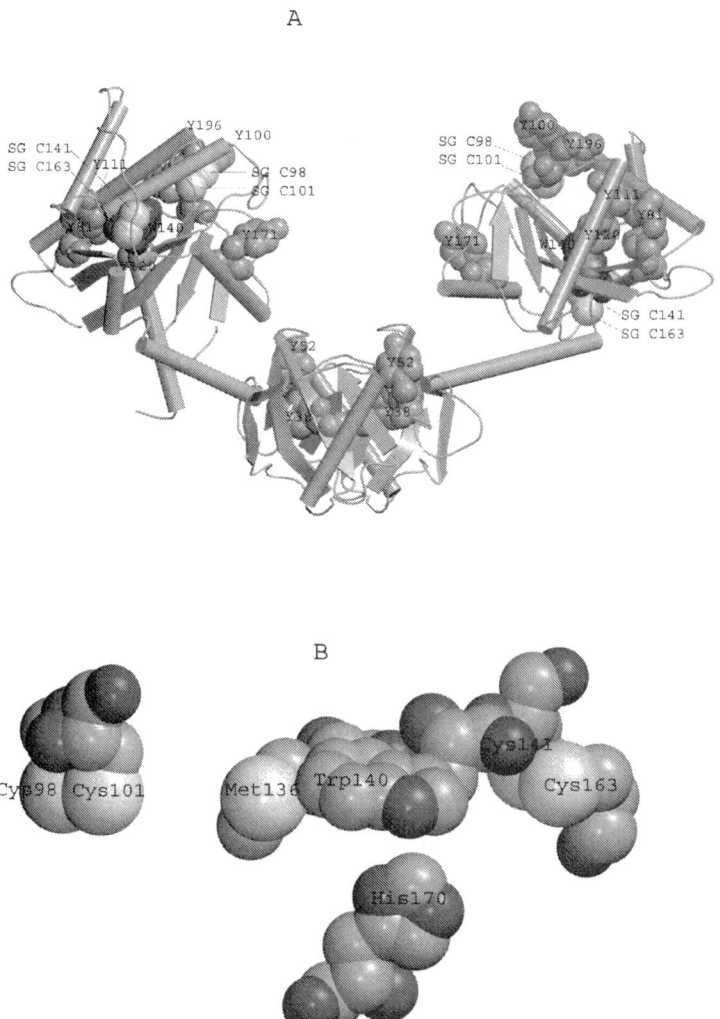

Figure 2.11. 3D structure of DsbC (file 1EEJ.ent from PDB(McCarthy et al, 2000, Berstein et al., 1977). A. Localization of tryptophan and tyrosine residues in the structure of DsbC. Backbone of the molecule is represented as cartoon diagram; tryptophan (red) and tyrosine (blue) residues, and sulfur atoms of disulfide bonds (yellow) are shown as solid van der Waal spheres. B. Localization of the residues of Cys, Met and His (potential quenchers of tryptophan fluorescence) in the vicinity of Trp 140. Graphic programs VMD (Humphrey et al., 1996) and Raster 3D (Merritt, and Bacon, 1977) were used.

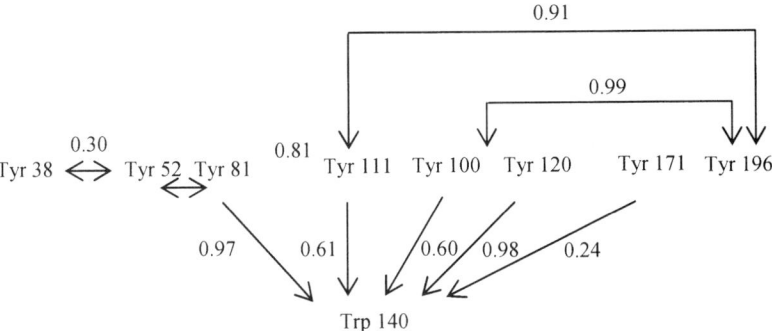

Figure 2.12. Efficiency of nonradiative energy transfer from tyrosine to tryptophan and between tyrosine residues in DsbC (1EEJ.ent (McCarthy et al, 2000), chain A.

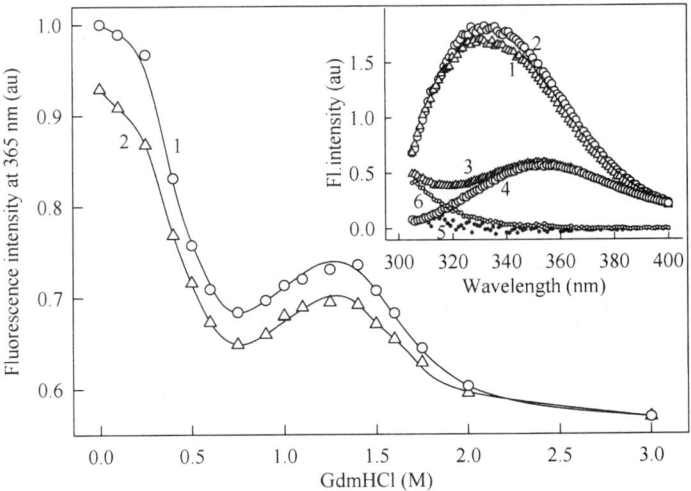

Figure 2.13. The change of the intensity of tryptophan fluorescence (λ_{em} =365 nm) of GlnBP in the process of protein unfolding induced by GdnCHl. Curve 1 (circles) represent the case when only tryptophan residues (λ_{ex} =297 nm) are excited, curve 2 (triangles) represents fluorescence when both tryptophan and tyrosine residues are excited (λ_{ex} =280 nm). Each curve is normalized with the fluorescence of unfolded protein. **Insert:** Contribution of tyrosine residues to the bulk fluorescence of native and unfolded GlnBP. Fluorescence spectrum of native (curves 1 and 2) and unfolded (curves 3 and 4) GlnBP excited at 280 nm (curves 1 and 3) and 297 nm (curves 2 and 4). Curves 5 and 6 are the contributions of tyrosine fluorescence in fluorescence spectra of native (5) and unfolded (6) GlnBP.

Table 2.3. Characteristics of the microenvironment of tyrosine residues in *E. coli* DsbC (1EEJ.ent)

Tyr residue	N	Amino acid side chains - potential quenchers of tyrosine fluorescence			R_C (Å)	R_{OH} (Å)	Oxygen atoms of bounded water		
							N_{HOH}	R_C (Å)	R_{OH} (Å)
38	77 (73)	27	MET	SD	5.74		209	4.40	2.90
		51	MET	SD	4.56	4.51			
		61	ASN	OD1		6.04			
		66	MET	SD		6.77			
52	61 (41)	48	GLN	OE1	5.46	4.28	1	6.33	5.71
		48	GLN	NE2		4.18	90	6.74	5.61
		61	ASN	OD1	5.22	5.65	119	4.32	4.77
		61	ASN	ND2		4.09	126	6.10	5.59
81	72 (73)	105	HIS	ND1	4.66	4.54	12	6.49	
		105	HIS	NE2	4.71	3.77	63	6.25	
		112	ASN	OD1	6.44		93	6.16	6.17
		120	TYR	OH		6.31	102	5.80	6.82
		146	ASN	OD1	5.41	5.31			
		146	ASN	ND2	3.49	4.11			
		150	ASP	OD1	4.47	2.99			
		150	ASP	OD2	5.38	4.02			
100	40 (22)	103	LYS	NZ	6.36	6.63	97	5.20	6.94
							138	4.12	2.67
							158	5.18	6.43
							169	5.39	4.67
							196	6.12	6.08
							201		5.12
111	79 (71)	107	GLN	OE1	6.31	5.61			
		107	GLN	NE2	5.70	4.43			
		196	TYR	OH		4.61			
120	79 (72)	81	TYR	OH		6.31	6	3.86	2.77
		105	HIS	ND1	4.35	2.86			
		105	HIS	NE2	6.05	4.93			
		108	MET	SD	5.95	5.01			
		153	MET	SD		6.88			
171	62 (47)						36	4.66	4.22
							107	6.43	5.10
							124	5.96	4.52
							131	4.16	2.64
196	68 (60)	103	LYS	NZ	6.87		99	4.39	3.03
		107	GLN	OE1	6.28	4.09	138	6.25	
		107	GLN	NE2	4.58	2.78	167	5.08	4.89
		111	TYR	OH	4.56		169	6.02	5.97
		201	GLU	OE2		6.75	201	6.48	5.30

N is the number of atoms in the microenvironment of tyrosine residue
N_{HOH} is the number of the molecule of bounded water
R_C is distance from the geometrical center of tyrosine residue
R_{OH} is distance from the oxygen atom of hydroxyl group of tyrosine residue

2.7. TYROSINATE FLUORESCENCE IN PROTEINS

In the vicinity of OH groups of several tyrosine residues of GlnBP there are effective proton acceptor groups (Table 2.5). In particular, worthy of mention is the proximity of the carboxyl group of Asp 28 and the hydroxyl group of Tyr 85 (the distances between atoms OD1 and OD2 of Asp 28 and oxygen of the hydroxyl group of Tyr 85 are 2.9 Å and 3.6 Å, respectively). It is well known that the existence of proton-acceptor groups in the vicinity of hydroxyl groups of tyrosine residues gives the opportunity for the proton to tear off from the tyrosine residue with the tyrosinate formation. Tyrosinate fluorescence with λ_{max} = 345 nm was first recorded by Cornog and Adams (1963) for solutions of tyrosine- and tryptophan-free proteins in 0.12 M NaOH (i.e., at pH > pKa (S_0), when tyrosinate is formed in the ground state (Martin et al., 1985). Under these conditions tyrosinate has a low quantum yield (< 0.01). As would be anticipated, the fluorescence excitation spectrum of tyrosinate in this case is red shifted in comparison with tyrosine and almost coincides with the absorption spectrum of tyrosinate (λ_{max} = 292 nm; Mihalyi, 1968). According to the data of Rayner et al. (1978), the pKa value of this group in the excited state is pKa(S_1^*) = 4.2. Nevertheless, the familiar fluorescence of tyrosine with λ_{max} = 304 nm takes place in neutral aqueous solutions. This is a result of the fact that the rate of deprotonation of the phenol ring in tyrosine is significantly slower than the decay rate of the excited state. At the same time, the rate of proton detachment from the hydroxyl group of the phenol ring becomes comparable

Table 2.4. Efficiency of nonradiative energy transfer from tyrosine to tryptophan residues of GlnBP (1GGG.ent) and GlnBP/Gln (1WDN).

Tyr \ Trp	Trp 32		Trp 220	
	GlnBP	GlnBP +Gln	GlnBP	GlnBP +Gln
Tyr 24	0.89	0.91	0.83	0.78
Tyr 43	0.96	1.00	0.88	0.89
Tyr 85	0.95	0.91	0.43	0.31
Tyr 86	0.86	0.86	0.94	0.96
Tyr 123	**0.00**	**0.00**	**0.05**	**0.00**
Tyr 143	0.02	0.02	0.56	0.73
Tyr 163	0.42	0.36	1.00	1.00
Tyr 185	0.89	0.90	0.91	0.92
Tyr 213	0.40	0.34	0.76	0.70
Tyr 217	0.29	0.28	0.99	0.99

to or even exceeds the total rate of other decay paths for the excited state of tyrosine at large concentrations of phosphate ions (Shimizu and Imakubo, 1977; Shimizu et al 1979) or acetate ions (Rayner et al.1978) in the solution. The fluorescence excitation spectra of tyrosinate in this case differ little from the absorption spectrum of tyrosine (Shimizu and Imakubo, 1977) but the quantum yield of fluorescence is significantly higher (0.16) than in an alkaline medium (Rayner et al., 1978).

Several authors have hypothesized that the possible formation of tyrosinate in the excited state as a result of localization of the natural proton-acceptor groups of the protein in the hydroxyl groups of the tyrosine residues (Szabo et al., 1978; Grasiani et al., 1974; Jordano et al., 1983; Prendergast et al., 1984; Libertini and Small, 1985) could be based on the explanation of the mechanism of excitation associated with the anomalous long-wave fluorescence of tryptophan-free proteins. The carboxyl groups of glutamic acid and aspartic acid (Szabo et al., 1978; Grasiani et al., 1974; Jordano et al., 1983; Prendergast et al., 1984; Libertini and Small, 1985; Pundak and Roche, 1984), and the imidazole rings of histidine and the amino groups of lysine (Longworth, 1981) have been suggested as possible proton-acceptor groups (participants in the reaction of tryosinate formation).

Tyrosinate fluoresces in the same spectral region as tryptophan residues. Therefore, there is always the danger that the luminescence of tryptophan-free proteins with λ_{max} in the 330-345 nm region could be the result of the presence of protein admixtures containing tryptophan in the preparations. Thus, the conclusion of Mani et al. (1982) concerning fluorescence of tyrosinate in S-100B protein turned out to be erroneous. It was subsequently shown (Baudier and Gerard, 1983) that the observed fluorescence was due to the presence of admixtures of tryptophan-bearing S-l00a protein which was difficult to separate out within the preparations of this protein. An important argument in favor of the idea that the anomalous long-wave luminescence of tryptophan free proteins is caused by tyrosinate rather than by foreign admixtures is the sensitivity of such luminescence to the changes in the structure of the protein which take place with cytotoxins from the cobra (Szabo et al., 1978), adrenodoxin (Kimura and Ting, 1971; Kimura et al., 1972; Lim and Kimura, 1980), HI histones from calf thymus (Libertini and Small, 1985) and the fruit fly (Jordano et al., 1983), and α-and β-purotionines (Prendergast et al., 1984).

The existence of anomalous fluorescence of tyrosine residues in tryptophan-free proteins gives us reason to suggest that the luminescence of tyrosinate could take place in tryptophan-bearing proteins, although it would be much more difficult to detect against the background of intense fluorescence from the tryptophan residues. A contribution of tyrosinate to the fluorescence of serum albumins from mammals was discovered by the construction of the difference in fluorescence spectra by Longworth (1981). We also obtained a fluorescence difference spectrum ($I(\lambda)_{280}$ - $I(\lambda)_{297}$) with a maximum in the 330-340 nm region for RNase C2. This provides a basis for assuming that, together with the serum albumin of mammals, RNase C2 is still another tryptophan-bearing protein with a contribution of tyrosinate involved in its fluorescence. As a consequence of the fact that the tryptophan residue of RNase C2 has a fairly short-wave fluorescence spectrum (λ_{max} = 325 nm), the participation of tyrosinate in the emission of this protein leads to the creation of anomalous spectral-polarization characteristics of this protein (Agekyan et al., 1988).

Table 2.5. Characteristics of the microenvironment of tyrosine residues in *GlnBP* (1GGG.ent).

tyrosine residue	N	Amino acid side chains - potential quenchers of tyrosine fluorescence			R_{OH} (Å)	Oxygen atoms of bounded water	
						N_{HOH}	R_{OH} (Å)
24	(52)	11	Thr	O	3.1		
		11	Thr	OG1	5.6		
43	(46)	32	Trp	NE1	4.2	317	5.3
85	(83)	28	Asp	OD1	2.9	225	3.1
		28	Asp	OD2	3.6		
86	(68)	217	Tyr	OH	3.8		
123	(35)	127	Asn	OD2	3.6		
143	(61)	156	His	ND1	3.9	281	2.7
163	(47)	167	Thr	OG1	5.6		
185	(56)	88	Ser	OG	3.7	332	3.1
213	(64)	85	Tyr	O	2.7	231	4.3
		85	Tyr	N	3.5		
217	(66)	86	Tyr	OH OX	3.8		
		224	Glu	T	4.5		

The assumption of the appearance of tyrosinate in the excited state of tyrosine residues cannot explain the experiments on GlnBP fluorescence. At the same time there are evidences that in some proteins, the pK_a of tyrosine residues can be as low as pH 8.1 – 8.3 (Ibarra et al., 2001). We have no strong evidence for the tyrosinate (in the ground state) existence in GlnBP. Nonetheless it is not unlikely that tyrosine residue which has the proton-acceptor groups in the immediate vicinity also has red-shifted absorption spectrum like that of tyrosinate. Among tyrosine residues of GlnBP the most probable candidate to have such spectrum is Tyr 85 which has the carboxyl group of Asp 28 in the proximity of its hydroxyl group. Although there are ten tyrosine residues in GlnBP, it is likely that one residue would significantly change bulk absorption spectrum of protein. At the same time it is worth mentioning an unusual character of the absorption spectrum of GlnBP, namely the shoulder in the range of long-waves and the lengthy optical density decreases in the range over 300 nm (Figure 2.14). So the assumption that Tyr 85 has red-shifted absorption spectrum provides an explanation for the fact that the fluorescence intensity of tryptophan residues of native protein (normalized to that of the completely unfolded state) excited at 280 nm is not greater but even smaller than the fluorescence intensity excited at 297 nm (Figure 2.12). Furthermore, this assumption is in good agreement with the results of Axelsen et al. (1991) on the fluorescence of mutant recombinant forms of GlnBP with one tryptophan residue. Trp 32 has no quenching groups in its microenvironment, however its non-significant contribution to the bulk fluorescence of GlnBP could be explained we assume that excitation energy from this residue is effectively transferred to Tyr 85.

2.8. INTRAMOLECULAR MOBILITY OF TRYPTOPHAN RESIDUES.

The modern concept of the structure of biological macromolecules includes the idea that all elements of the structure (separate atoms, groups of atoms, side chains of amino acids, segments of polypeptide chain etc.) take part in the intramolecular mobility. The dynamics of protein structure attracts attention because it is considered of special importance for protein function (enzyme activity, protein-protein interaction, etc.). The dynamics of proteins is studied by a variety of methods (NMR, X-ray analysis, oxygen exchange, fluorescent methods, etc.).

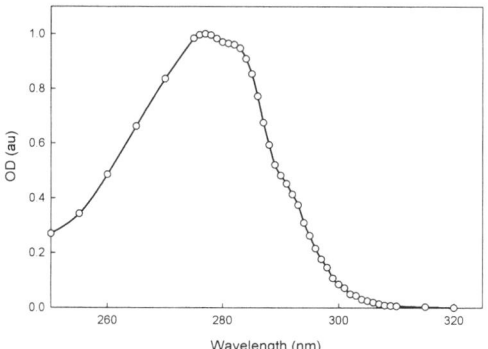

Figure 2.14. Absorption spectrum of GlnBP.

The development of computer facilities allows fulfill numerical simulations of molecular dynamics. Such molecular dynamics simulations need adequate experimental data. Such data can be obtained from B factors of X-ray crystallography, even if they contain information only on the amplitude and not on the frequencies of the processes. Furthermore, B factors refer to the protein in crystal state, and the dynamics of crystal state of proteins can differs from that of proteins in solution. One of the methods which gives us the direct information about the amplitude and frequency of protein intramolecular mobility is intrinsic fluorescence. The intrinsic fluorescence is one of the methods which give the most direct information about dynamic structure of proteins. Protein concentration needed for intrinsic fluorescence is 1 – 2 order smaller than that needed for NMR which can also give direct information of protein dynamics. At the same time the number of centers which give information about intramolecular mobility of protein is limited by the number of tryptophan residues, which are responsible for intrinsic fluorescence of proteins and their location in the structure of macromolecule.
A large experimental material about intramolecular mobility in the side chains of aromatic residues in proteins obtained by steady-state and time-resolved fluorescence experiments is accumulate (Bucci and Steiner, 1988). The existence of some freedom of intramolecular mobility of tryptophan residues is undoubtedly shown both experimentally and by molecular dynamic simulation (Ichiye and Karplus, 1983; Chen et al., 1988;

MacKerell et al., 1988). The problem is how the characteristics of intramolecular mobility depend upon the location of tryptophan residue in the structure of macromolecule.

Table 2.6. Efficiency of energy transfer between tyrosine residues in GlnBP (file 1GGG.ent)

	Tyr 24	Tyr 43	Tyr 85	Tyr 86	Tyr 123	Tyr 143	Tyr 163	Tyr 185	Tyr 213	Tyr 217
Tyr 24		12.8	16.39	18.13	33.04	23.94	14.85	16.83	21.00	20.58
Tyr 43	0.59		14.63	20.03	37.34	32.95	22.40	19.34	18.23	21.53
Tyr 85	0	0		7.05	23.25	23.61	16.11	6.36	7.95	10.57
Tyr 86	0.01	0	0.49		18.16	17.51	11.58	5.15	7.95	5.63
Tyr 123	0	0	0.02	0		17.96	23.57	18.08	23.97	20.47
Tyr 143	0	0	0.02	0.29	0.04		11.86	18.50	24.34	18.61
Tyr 163	0.08	0.02	0	0.78	0	0.72		13.78	16.32	11.63
Tyr 185	0.06	0	0.97	0.99	0.09	0.26	0.36		11.57	10.76
Tyr 213	0.01	0.03	0.89	0.95	0	0.05	0.18	0.58		6.24
Tyr 217	0.06	0.06	0.28	0.99	0	0.15	0.58	0.36	0.01	

We have shown that tryptophan residues in the inner hydrophobic environment can be even more mobile than that nearer to the surface of macromolecule but in the environment of polar side chains of amino acids (Kuznetsova and Turoverov, 1983). In this connection very interesting are the experimental data of Munro et al. (1979), which show the intramolecular mobility of the single tryptophan residue of azurin located in the hydrophobic core of protein.

Munro et al. (1979) drew a conclusion about the subnanosecond (0.5 ns) oscillations of the tryptophan residue within the cone of semiangle 34^0. These results were confirmed by Limkeman and Gratton (1984) and Creri and Gratton (1986). They succeeded to registering the initial part of the Perrin plot, which characterizes the fast rotational motion of the aminoacid residue with respect to the protein structure. And the estimations of the rotational rate of the Trp 48 in the protein interior given by Limkeman and Gratton (1984) and Creri and Gratton (1986) are in close agreement with the results determined by Munro et al. (1979).

At the same time Petrich et al. (1987) came to opposite conclusion anout Trp 48 motion in the interior of azurin macromolecule. In this work azurins from three bacteria types, which differ by the location of tryptophan residues, were studied. The internal motion was detected only for the surface Trp 118 and was not for buried Trp 48 in all these proteins.

Dynamic simulation studies of one of these azurins, which contain both surface Trp 118 and buried Trp 48 were carried out by Chen et al. (1988). Though the authors of this work consider that their results are concordant with the experimental findings of Petrich et al. (1987), they were more cautious in their conclusions. It was only stated that Trp 48 is less mobile than Trp 118.

The very short time scale of the present molecular dynamics simulations generally prevents the direct comparison between simulation results and experimental data. To some extent it is also impeded by various simulation strategies intended to reduce the problem. Stochastic boundary molecular dynamic method with vacuum approximation was used by Chen et al. (1988). The results obtained with and without solvent for Trp 118 differ greatly from each other, even more than the results obtained for Trp 118 and Trp 48 under the same conditions – vacuum simulations. Solvent simulations were not performed for Trp 48. It was considered that the molecule is closely packed around Trp 48, so it is unlikely to allow penetration of solvent molecules. Hence, no water molecules were included, and the reaction and buffer region were taken inside the protein. At the same time Careri and Gratton (1986) and Linkeman and Gratton (1984) found that the value of the anisotropy obtained for azurin in dry polymer film corresponds to the limiting anisotropy and significantly decreases after film hydration.

It is necessary to mention that though Trp 48 is not accessible to the solvent molecules, there is nothing extraordinary in molecule packing density in its vicinity. The comparison of a number of tryptophan residues in different proteins revealed that there can be much more atoms in their microenvironment than in the close vicinity of Trp 48 of azurin (Turoverov et al., 1985).

For verification the possibility of Trp 48 oscillations, which can cause the registered fluorescence depolarization (Munro et al., 1971), static simulation of tryptophan oscillations was proposed.

The simulations consists of the energy refinement of the azurin macromolecule in which the indole ring is turned to a certain angle about CB – CG bond and fixed in this position. It is evident that, due to the appearance of close contacts between the indole ring and some atoms of its microenvironment, the energy of the system would increase. If after the energy refinement 1) the energy of the system would be higher than that of the control one; or 2) essentially bad bonds, bad angles and close contacts would appear in the new structure; or 3) the new structure would differ greatly from the initial one, then it must be concluded that the motion is impossible. Only if all of these conditions are not met, it must be admitted that such motion is possible.

Prior to the statistical simulations the initial structure of azurin from the Protein Data Bank was refined. Otherwise in the process of energy refinement the elimination of the strains that exists in the initial structure of the macromolecule could prevail over the elimination of the strains and appears after the turn of the indole ring.

The value of the indole ring turn was chosen as follows. The subnanosecond motion of Trp 48 in azurin, the correlation time of which has been determined by Munro et al. as 0.5 ns (Munro et la., 1979), is caused by the rotational oscillations of the indole ring about CB-CG bond (Turoverov and Kuznetsova, 1983). If such mechanism of the mobility is suggested, the amplitude of the indole ring oscillations $\sqrt{\langle \varphi^2 \rangle}$, necessary for the appearance of the experimentally recorded depolarization, can be determined as follows (Anifrieva et al., 1976; Gotlib and Rystov 1983):

$$\frac{r_0}{r_0'} = \left[1 - \left(3 Sin^2 \alpha\right) \cdot \langle \varphi^2 \rangle\right]^{-1} \tag{1}$$

where r_0 is the value of the fundamental anisotropy of fluorescence, r_0' is the value of the fluorescence anisotropy obtained by extrapolation of the curve $r = f(t)$ to $t \to 0$;

α is the angle between the oscillator ^1L_a and axis of the indole ring rotation (Fig. 2.15).

The calculations show that for Trp 48 of azurin the value of $\sqrt{\langle \varphi^2 \rangle}$ equals nearly to 30^0.

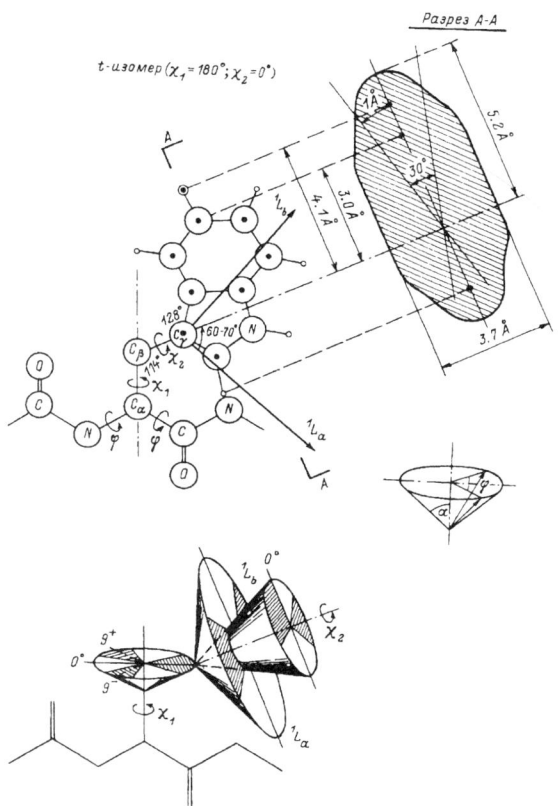

Figure 2.15. The illustration of the mechanism of indole ring oscillations.

The turn of the indole ring about CB-Cg bond causes an increase in the system energy due to the appearance of close contacts between the atoms of the indole ring and those of its microenvironment. This increase is significant when the angle of the turn is $+30^0$ and comparatively small at $+15^0$ and -30^0. Comparison of the close contacts that exist in the structure of azurin before and after the indole ring turn shows that the turn to the positive angles is hindered by CG1, CG2 and CB atoms of Val 31 and, to a smaller extent by CG and CD2 atoms of Phe 110. The turn to the negative angles is hindered by CD1 atom of Ile 7, CB and CG atoms of Leu 50 as well as by CB atom of Phe 110 and C and O atoms of the backbone of Trp 48 itself.

After 64 iterations the energy of azurin structure with tryptophan residue turned to -30^0 and $+15^0$ is close to that of the azurin in which tryptophan residue was remained in its position, while the energy of the structure with tryptophan residue turned to $+30^0$ is much higher (Fig. 2.16). It is evident that the indole ring turn to $+30^0$ is scarcely probable. A drop of the energy of the system with the turned tryptophan residue is achieved at the expense of comparatively small deviations of certain atoms. Mean square deviation is about 0.23 A in all cases. In the system with the indole ring turned to $+15^0$ and $+30^0$ the greater shift was registered for atoms that directly contact the indole ring. These are CG1, CG2 and CB atoms of Val 31. In the structure with the indole ring turned to $+30^0$ the displacement of these atoms are 1.67, 1.19 and 1.02 Å, respectively, while in the system with the indole ring turned to $+15^0$ these values are diminished to 0.70, 0.72 and 0.58 Å.

The displacement of the atoms located far from Trp 48, such as OD1 and CB of Asn 18 and CG2 of Thr 113 are practically the same as for the structure with the indole ring turned to -30^0 (0.73, 0.89 and 0.72 Å) and $+30^0$ (0.94, 0.70 and 0.82 Å). So it is quite clear that the deviation of these atoms results from the elimination of the strains remaining in the control system after the preliminary 64 iterations. The displacement of CB atom of Asn 18 is the largest in the system with the indole ring turned to -30^0.

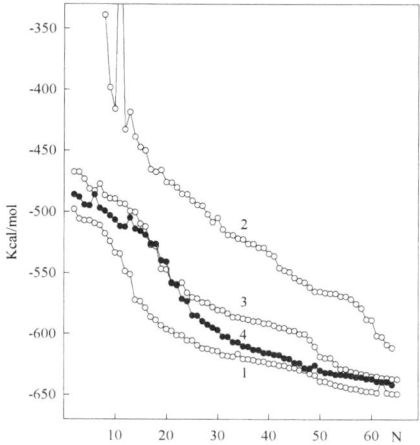

Figure 2.16. The decrease of the energy of azurin structure in the process of 64 iterations of energy refinement. Curve 1, the refinement of the intact azurin structure; Curves 2, 3, and 4 the refinement of azurin structure with the indole ring turned to $+30^0$, -30^0 and $+15^0$, respectively. The structure of azurin (1AZU.ent Adman et al., 1980) was preliminary refined by 64 iterations.

As the result of 64 iterations, the close contacts that appeared after the indole ring turn are practically eliminated. The close contact between CE3 and C atom of Trp 48 remain in the system with the indole ring turned to -30^0. In the system with the indole ring turned to $+30^0$ close contacts between CG2 of Val 31 and CH2 and CZ3 of Trp 48 also remain. At the same time the values of the force between these atoms are close to those in the control system. Furthermore a comparatively small deformation of bonds and bond angles occurred in the residues located on the pathway of the indole ring motion and in the adjacent residues.

The elimination of close contacts provoked by the indole ring displacement was accompanied by some deformation of the bonds and bond angles in the residues located on the pathway of the indole ring motion in the adjacent residues and in Trp 48 itself. This is especially manifested when the ring is turned to $+30^0$ (Table 2.3, 2.4): CA-CB bond and N-CA-CB and CA-CB-CG bond angles of Asn 32, CA-CB, N-CA and CB-CG1 bonds of Val 31, and NE1-CE2 bond of Trp 48 , CA-CB-CG2 bond angle of Thr 30 were distorted. In the system with the indole ring turned to -30^0 the major strains appeared in the Trp 48 itself, namely in the NE1-CE2, CE3-CD2 bonds and N-CA-C bond angles.

Interestingly, the strains in the mostly distorted bonds increase both when the indole ring is turned to positive and negative angles (NE1-CE2 and CG-CD2 bonds of Trp 48). At the same time in some bonds (CA-CB of Thr 17, C-O of Asn 16, and CB-CG of Asn 32) the strains are the highest when the indole ring is turned to $+15^0$ and significantly lower in the system with the indole ring turned to $+30^0$ and -30^0.

The energy of maximum bonds and bond angles distortion in the system with the tryptophan ring turned to $+30^0$ is a little higher, and in the system with the turn angle $+15^0$ and -30^0 is practically the same as in control system. The list of the mostly strained torsion angles in all systems with the turned ring of the Trp 48 and in the control one is the same and the values of strains are also very close. The adaptation of Trp vicinity to its turn for a rather large angle does not need a significant displacement of atoms and strains in bonds. In any case the system disturbance caused by the tryptophan turn to the angles $-30^0 - +15^0$ (and even $+30^0$) does not lead to its destruction as it may be expected.

The analyses of the microenvironment if Trp 48 does not claim to prove the existence of subnanosecond oscillations of the indole ring in azurin, they merely indicate a general possibility of the intramolecular motion of this type according to theoretical group results obtained by Munro (Munro et al., 1979).

2.9. CONCLUSIONS

In conclusion, the location of aromatic ring clusters and the conformation of the side chain of tryptophan residues in protein matrices are of great importance for the occurrence of blue fluorescence spectrum. At the same time high density of microenvironment is not indispensable condition for the occurrence of blue fluorescence spectrum. It is important to point out that when analyzing microenvironment of tryptophan it is necessary to consider that macroscopic dielectric properties of tryptophan residue microenvironments are determined not only by its accessibility to the solvent, but also from their location in different regions of protein matrices.

2.10. ACKNOWLEDGEMENT

This project was realized in the frame of CRdC-ATIBB POR UE-Campania Mis 3.16 activities (S.D., M.R). This work was supported by grants from F.I.R.B. (S.D., M.R.), the Italian National Research Council (S.D, M.R), INTAS 2001-2347 (K.T.), and from Presidium of Russian Academy of Sciences for the program "Molecular and Cell Biology" (K.T.).

2.11. REFERENCES

Agekyan T. V., Bezborodova S. I., Kuznetsova I. M., Polyakov K. M., and Turoverov K. K. (1988) Spectral and polarizational characteristics of ribonuclease C_2. Unusual fluorescence of tyrosine residues. Mol. Biol. 22, 612-623.

Atkins W. M., Stayton P. S., and Villafranca J. J. (1991) Time-resolved fluorescence studies of genetically engineered Escherichia coli glutamine synthetase. Effects of ATP on the tryptophan-57 loop. Biochemistry 30, 3406-3416.

Axelsen P., Prendergast F.G. (1989) Molecular dynamics of tryptophan in ribonuclease-T1. II Correlations with fluorescence. Biophys.J, 56, 43-46.

Axelsen, P.H., Bajzer, Z., Prendergast,F.G., Cottam, P.F., and Ho, C. (1991). Resolution of fluorescence intensity decays of the two tryptophan residues in glutamine-binding protein from *Escherichia coli* using single tryptophan mutants. Biophys.J. 60, 650-659.

Baudier, J., and Gerard, D. (1983). The S-100 protein: tyrosine residues do not exhibit an abnormal fluorescence spectrum. J. Neurochem. , 40, 1765-1767.

Beechem, J.M., E. Gratton, M., Ameloot, J.R. Knutson, and L. Brand 1991. The global analysis of fluorescence intensity and anisotropy decay data: second-generation theory and programs. In: Topics in fluorescence spectriscopy. Ed. Lakowicz J.R. Vol.2: 241-305.

Bernstein F.C, Koetzle T.F., Williams G.J.B., Meyer Jr.E.F., Brice M.D., Rodgers J.R., Kennard O.,Shimanouchi T., Tasumi M (1977) The Protein Data Bank: a computer-based archival file for macromolecular structures. J.Mol.Biol.,112, 535-542.

Burstein E. A. (1976) Intrinsic Protein Fluorescence: Origin and Applications. Ser. Biophysics, vol.7, VINIITI, Moscow.

Burstein E. A., S. M. Abornev, and Y. K. Reshetnyak (2001) Decomposition of protein tryptophan fluorescence spectra into log-normal components. I. Decomposition algorithms. *Biophys J.* 81(3), 1699-1709.

Chen L.X.-Q., Engh R.A. and Brunger A.T. (1988) Dynamics simulation studies of apoazurin of Alcaligenes denitrificans. Biochemistry.. Vol.27. P. 6908-6921.

Careri G., Gratton E. (1986) The statistical time correlation approach to enzyme action: the role of hydration // The fluctauting enzyme.New York Plenum P.227-262.

Cornog Jr., J. L., and Adams, W. R. (1963). The fluorescence of tyrosine in alkaline solution. Biochim. Biophys. Acta, 66, 356-365.

Dale R.E., and Eisinger J. (1974) Intramolecular distances determined by energy transfer. Dependence on orientational freedom of donor and acceptor. Biopolymers., 13, 1573-1605.

Demchenko A. P. (1986) Fluorescence analysis of protein dynamics. Essay Biochem. 22, 120-157.

Diamond R. (1974) Real-space refinement of the structure of hen egg-white lysozyme. J.Mol.Biol., 82, 371-386.

Doyle T. C., Hansen J. E., and Reisler E. (2001) Tryptophan fluorescence of yeast actin resolved via conserved mutations. Biophys. J., 80, 427-434.

Eftink M., Ghiron C.A. (1977) Exposure of tryptophanyl residues and protein dynamics. Biochemistry, 16, 5546-5551.

Eftink,M., and C.A.Ghiron. (1981) Fluorescence quenching studies with proteins. Anal.Biochem., 114, 199-227.

Eftink M. R. (1991) In, Methods of Biochemical Analysis: Protein Structure Determination. C. H. Suelter, Ed., Wiley: New York, 35, 127-205.

Eftink M. R. (1998) The use of fluorescence methods to monitor unfolding transitions in proteins. Biokhimiya (Moscow) 63, 327-337.

Eisinger J., Feuer B., Lamola A.A. (1969) Intramolecular singlet excitation transfer. Applications to polypeptides. Biochemistry., 8, 3908-3915.

Finazzi-Agro A., Rotilio G., Avigliano L., Guerrieri P., Boffi V., Mondovi B. (1970) Environment of copper in Pseudomonas fluorescens azurin: fluorometric approach. Biochemistry, 9, 2009-2014.

Finazzi-Agro A., Giovagnoli C., Avigliano L., Rotilio G., Mondovi B. (1973) Luminescence quenching in azurin. Eur.J.Biochem., 34, 20-24.

Formoso C., and Forster L. (1975) Tryptophan fluorescence lifetimes in lysozyme. J.Biol.Chem., 250, 3738-3745.

Förster Th. (1960) Transfer mechanisms of electronic excitation energy. Rad. Res., Suppl.2, 326-339.

Graziani, M. T., Finazzi-Agro, A., Rotillio, G., Barra, D., and Mondovi, B. (1974).Parsley plastocyanin. The possible presence of sulfhydryl and tyrosine in the copper environment. Biochemistry, 13, 804-809.

Gopalan V., Golbik R., Schreiber G., Fersht A. R., and Altman S. (1997) Fluorescence properties of a tryptophan residue in an aromatic core of the protein subunit of ribonuclease P from Escherichia coli. J.Mol.Biol. 267, 765-769.

Grinvald A., SchlessingerJ., Pecht I., Steinberg I.Z. (1975) Homogeneity and variability in the structure of azurin molecules studied by fluorescence decay and circular polarization. Biochemistry, 14, 1921-1929.

Ibarra, C., Nieslanik, B.S., and Atkins, W.M. (2001). Contribution of aromatic-aromatic interactions to the anomalous pK(a) of tyrosine-9 and the C-terminal dynamics of glutathione S-transferase A1-1. Biochemistry 40, 10614-10624.

Ichiye T., Karplus M. (1983) Fluorescence depolarization of tryptophan residues in proteins: a molecular dynamics study. Biochemistry, 22, 2884-2893.

Imoto T., Forster L.S., Rupley J.A., Tanaka F. (1971) Fluorescence of lysozyme: emission from tryptophan residues 62 and 108 and energy migration. Proc.Nat.Acad.Sci.USA., 69, 1151-1155.

Itani N., Kuramitsu S., Ikeda K., Hamaguchi K.(1975) Effects on tryptophyl absorption of the ionization of the catalytic carboxyls in hen and turkey lysozymes. J.Biochem.,78, 705-711.

Jordano, J, Barbero, J. L., Montero, F., and Franco, L. (1983). Fluorescence of histones H1. A tyrosinate-like fluorescence emission in Ceratitis capitata H1 at neutral pH values. J. Biol. Chem., 258, 315-320.

Kabsch,W., Mannherz H.G., Suck D., Pai E.F., and Holmes H.C.(1990) Atomic structure of the actin: Dnase I complex. Nature, 347, 37-44.

Kirmura, T., Ting, J.J., and Huang, J. J. (1972). Studies on adrenal steroid hydroxylases. J. Biol. Chem., 2A7, 4476-4479.

Knutson, J.R., D.G. Walbridge, and L. Brand. (1982) Decay-associated fluorescence spectra and the heterogeneous emission of alcohol dehydrogenase. *Biochemistry.* 21: 4671-4679.

Kuramitsu S., Ikeda K., Hamaguchi K., Fujio H., Amano T., Miwa S., Nishina T. (1974) Ionization constants of Glu 35 and Asp 52 in hen, turkey and human lysozymes. J.Biochem.,76, 671-683.

Kuramitsu S., Kurihara S., Ikeda K., Hamaguchi K. (1978) Fluorescence spectra of hen, turky, and human lysozymes excited at 305 nm. J.Biochem., 83, 159-170.

Kuznetsova I.M., Khaitlina S.Yu., Konditerov S.N., Surin A.M., Turoverov K.K.(1988) Changes of structure and intramolecular mobility in the course of actin denaturation. Biophys.Chem., 32, 73-78.

Kuznetsova I.M., Khaitlina S.Yu., Turoverov K.K. (1995) A comparison of actin structurak states by time-resolved fluorescence studies. J.Muscle Res. Cell Motility, 16, 150.

Kuznetsova,I., Antropova,O., Turoverov, K., Khaitlina,S. (1996) Conformational changes in subdomain I of actin induced by proteolytic cleavage within the DNase I-binding loop: energy transfer from tryptophan to AEDANS. FEBS Lett. Vol.383. P.105-108.

Kuznetsova I. M., and Turoverov K. K. (1998) What determines the characteristics of protein intrinsic fluorescence? Analysis of tryptophan residue localization in proteins. Tsitologia, 40, 747-762.

Kuznetsova I. M., Yakusheva T. A., and Turoverov K. K. (1999) Contribution of separate tryptophan residues to intrinsic fluorescence of actin. Analysis of 3D structure. FEBS Lett. 452, 205-210.

Kuznetsova I.M., Biktashev A.G., Khaitlina S.Yu., Vassilenko K.S., Turoverov K.K.and Uversky V.N. (1999a). Effect of self-association on the structural organization of partially folded proteins: inactivated actin. *Biophys.J.* 77, 5:2788-2800.

Kuznetsova I. M., Biktashev A. G., Malova L. N., Bushmarina N. A., Uversky V. N., and Turoverov K. K. (2000) Understanding the contribution of individual tryptophan residues to intrinsic lysozyme fluorescence. Prot. Pept. Lett. 7, 411-420.

Lakowicz J. R. (1999) Principles of Fluorescence Spectroscopy, 2nd Ed. Kluwer Academic/Plenum Publishers, New York.

Lehrer S.S., Fasman G.D. (1967) Fluorescence of lysozyme and lysozyme substrate complexes. Separation of tryptophan contributions by fluorescence difference method. J.Biol.Chem., 242, 4644-4651.

Libertini, L. J., and Small, E. W. (1985). The intrinsic tyrosine fluorescence of histone H1. Steady state and fluorescence decay studies reveal heterogeneous emission. Biophys. J., 47, 765-772.

Lim, B. T., and Kimura, T. (1980). Conformation-associated anomalous tyrosine fluorescence of adrenodoxin. J. Biol. Chem., 255, 2440-2444.

Longworth, J. W. (1981). A new component in protein fluorescence. Ann. N. Y. Acad. Sci. 366. 237-245.

MacKerell A.D.Jr., Nilsson L., Rigler R., Saenger W. (1988) Molecular dynamics simulations of ribonuclease T1: analysis of the effect of solvent on the structure, fluctuations and active site of free enzyme. Biochemistry, 27, 4547-4556.

Mallinson R., Carter R., Ghiron C.A. (1981) Acrylamide quenching studies with azurin B. Biochim. Biophys.Acta., 671, 117-122.

Mani, R. S., Boyers, B. E., and Kay, C. M. (1982). Physicochemical and optical studies on calcium- and potassium-induced conformational changes in bovine brain S-100b protein. Biochemistry, 21, 2607-

2612.

Munro I., Pecht I., Stryer L. (1979) Subnanosecond motions of tryptophan residues in proteins. Proc. Natl.Acad. Sci.USA, 76, 56-60.

Martensson L. G., Jonasson P., Freskgard P. O., Svensson M., Carlsson U., and Jonsson B. H., (1995) Contributionof individual tryptophan residues to the fluorescence spectrum of native and denatured forms of human carbonic anhydrase II. Biochemistry 34, 1011-1021.

Martin, R. B., Edsall, J. T., Wetlaufer, D. B., and Hollingworth, J. (1958). A complete ionization scheme for tyrosine, and the ionization constants of some tyrosine derivatives. J. Biol. Chem., 233, 1429-1435.

Mihalyi, E. (1968). Numerical values of the absorbances of the aromatic amino acids in acid, neutral, and alkaline solutions. J. Chem. Eng. Data, 13, 179-182.

Permyakov E.A., Yarmolenko V.V., Emelyanenko V.I., Burstein E.A., Closset J., Gerday C. (1980) Fluorescence studies of the calcium binding to whiting (Gadus merlangus) parvalbumin. Eur.J.Biochem., 109, 307-315.

Petrich J.W., Longworth J.W., Fleming G.R. (1987) Internal motion and elecron transfer in proteins: a picosecond fluorescence study of three homologous azurins. Biochemistry, 26, 2711-2722.

Philips, A.V., M.S. Coleman, K. Maskos, and M.D. Barkley. (1989) Time-resolved fluorescence spectroscopy of human adenosine deaminasse: effects of enzyme inhibitors on protein conformation. *Biochemistry.* 28: 2040-2050.

Prendergast, F. G. , Hampton, P. D., and Jones, B. (1984). Characteristics of tyrosinate fluorescence emission in alpha- and beta-purothionins. Biochemistry, 23, 6690-6697.

Pundak, S., and Roche, R. S. (1984). Tyrosine and tyrosinate fluorescence of bovine testes calmodulin: calcium and pH dependence. Biochemistry, 23, 1549-1555.

Rao MV, Atreyi M, Rajeswari MR. (1981) Fluorescence spectra of lysozyme excited at 305 NM in presence of urea. Int J Pept Protein Res., 17, 2, 205-10.

Rayner, D. M., Krajcarski, D. T., and Szabo, A. G. (1978). Excited state acid-base equilibrium of tyrosine. Can. J. Chem., 56, 1238-1245

Reshetnyak Y. K., and E. A. Burstein (2001) Decomposition of protein tryptophan fluorescence spectra into log-normal components. II. The statistical proof of discreteness of tryptophan classes in proteins. *Biophys. J.* 81(3), 1710-1734.

Reshetnyak Y. K., Y. Koshevnik, and E. A. Burstein (2001) Decomposition of protein tryptophan fluorescence spectra into log-normal components. III. Correlation between fluorescence and microenvironment parameters of individual tryptophan residues. *Biophys. J.* 81(3), 1735-1758.

Shimisu, O., and Imakubo, K. (1977). New emission band of tyrosine induced by interaction with phosphate ion. Photochem. PhotobioL., 26, 541-543.

Shumizu, O., Watanabe, J., and Imakubo, K. (1979). Effect of phosphate ion on fluorescencent characteristics of tyrosine and ots conjugate base. Photochem. Photobiol, 29, 915-919.

Shinitzky M., and Goldman R. (1967) Fluorometric detection of histidine-tryptophan complexes in peptides and proteins. European.J.Biochem., 3, 139-144.

Shopova M. , Genov N. (1983) Protonated form of histidine 238 quenches the fluorescence of tryptophan 241 in subtilisin Novo. Int.J.Peptide Protein., 21, 475-478.

Steinberg I. (1971) Long-range nonradiative transfer of electronic excitation energy in proteins and polypeptides. Ann.Rev.Biochem., 40, 83-114.

Szabo, A.G, Lynn, K., Krajcarski, D., and Rayner, D.M. (1978). Tyrosinate fluorescence maxima at 345 nm in proteins lacking tryptophan at pH 7. FEBS Lett., 94, 249-252.

Szabo A.G., Stepanik T.M., Wayner D.M., Young N.M. (1983) Conformational heterogeneity of the copper binding site in azurin. Biophys.J., 41, 233-244.

Szabo A.G., Rayner D.M. (1980) Fluorescence of tryptophan conformers in aqueous solution // J.Am.Chem.Soc., 102, 554-563.

Szabo A.G. and Faerman C. (1992) Dilemma of correlating fluorescence quantum yields and intensity decay times in single tryptophan mutant proteins. Time-res. Laser Spectrosc. Biochem., 1640, 1-11.

Staiano M., Scognamiglio V., Rossi M., D'Auria S., Stepanenko Olga V., Kuznetsova I.M., Turoverov K.K.(2004) Intrinsic fluorescence of GlnBP and its complex with Gln. Tryptophan and tyrosine residues location, characteristics of their microenvironment and contribution to the bulk fluorescence of GlnBP Biochem.J., (In press)

Stepanenko Olga V., Kuznetsova I.M., Turoverov K.K., Chunjuan Huang and Wang C.-C. (2004) Conformational change of dimeric DsbC molecule induced by GdnHCl - A study by intrinsic fluorescence. Biochemistry, 43, 5296-5303.

Teichberg V.I., Sharon N.N. (1970) A spectrofluorometric study of tryptophan 108 in hen egg-white lysozyme. FEBS Letters, 7, 171-174.

Teichberg V.I., Shinitzky M. (1973) Fluorescence polarization studies of lysozyme and lysozyme-saccharide complexes. J.Mol.Biol., 74, 519-531.

Turoverov K.K., Kuznetsova I.M. and Zaitsev V.N. (1985) The environment of the tryptophan residue in pseudomonas aeruginosa azurin and its fluorecsence properties. Biophys.Chem., 23, 79-89.

Turoverov K.K., and Kuznetsova I.M. (1986) What causes the depolarization of trypsin and trypsinogen fluorescence. Intramolecular mobility or non-radiative energy transfer? Biophys.Chem., 25, 315-323.

Turoverov K.K., and Kuznetsova I.M. (2003) Intrinsic fluorescence of actin. J.Fluorescence 13, 1, 41-57.

Umetskai V.N. and Turoverov K.K. (1978) Examination of the excited state of indole. CD spectra of the polyethylene film activated by indole. Optika and spectoskopia, 44, 12, 1090-1096.

Weber G. (1960) Fluorescence polarization spectrum and electronic-energy transfer in tyrosine, tryptophan and related compounds. Biochem.J., 75, 335-345.

Yamane T., Andou T. and Ashida T. (1977) N-acetyl-L-tryptophan. Acta cryst., B33, 1650-1653.

Yamamoto Yu., Tanaka J. (1970) Spectroscopic studies on the configurational structures of ribonuclease T1. Biochim.Biophys.Acta., 207, 522-531.

Yamamoto Y. and Tanaka J. (1972) Polarized absorbtion spectra of crystall of indole and its related compounds. Bull.Chem.Soc.Japan., 5, 1362-1366.

TAPERED FIBERS FOR CELL STUDIES

P. M. Shankar[1] and Raj M. Mutharasan[2]

3.1. INTRODUCTION

Optical fiber sensors are extensively used in monitoring physical, chemical and biochemical conditions (Mignanai and Baldini, 1995; Pilevar et al., 1998; Cullum et al., 2000). Specifically, these sensors have found a niche in the field of biosensors for the detection and classification of various tissue properties such as cell growth and decay, and presence or absence of pathogens, and others. Their potential use is important in obtaining information at a single cell or molecule level because such information may hold the key to the early detection and identification of cancer (Alfano et al., 1987; Tang et al. 1989; Pradhan et al. 1995) and early warning on bioterrorism brought on by pathogens (Vo-Dinh et al., 2000). For practical applications, these sensors must be small, must have the capability of being incorporated into an existing system such as an endoscope, and possess the ability to operate in the in vivo and in situ mode. The optical fibers fit these specific requirements and they can be engineered to be in the micro to nano size ranges for the localized sensing of biomaterials of interest (Lambelet et al., 1998).

These sensors are fabricated by making the fiber to be active so that the light being carried by the fiber will respond decisively to the presence or absence of the biomaterials. One of the efficient ways to engineer highly sensitivity is to take advantage of the evanescent field of the fiber (Marcuse, 1988; Okamoto, 2000; Shankar et al., 2001). Evanescent field is the exponentially decaying component of the propagating wave in a fiber outside its core. The strength of the evanescent field depends on several factors, the index of the core, index of the surrounding medium, the radius of the fiber and the operating wavelength. One can manipulate the evanescent field strength for sensing, characterizing, and quantifying optical properties of samples in the evanescent region. Tapering the fiber provides an easy access to the evanescent field in fibers, enabling stronger interaction with the sample, thus making the sensors versatile (Henry, 1994; Khijwania and Gupta, 1999; Nath et al., 1997; Nath and Anand, 1998; Prince et al., 2001). Tapered fiber tips have been used as nanodelivery devices because of the ability of

[1] Department of Electrical and Computer Engineering, [2]Department of Chemical Engineering, Drexel University Philadelphia, PA 19104

the tips to produce laser beams of extremely small diameters (< 100 nm). They have also been used in fluorescence sensing of biomolecules (Mononobe et al., 1997; Vo-Dinh et al., 2000). However, in these applications, the fiber tips were used as light delivery devices, rather than as sensing elements. Thus, these tapered tips may not constitute *in situ* sensors, where the three features, namely, light delivery, interaction with the sample, and sensing are accomplished by the same fiber. The tapered fibers also possess important advantages over the traditional method of studying the properties of the biomaterials, typically using a cuvette. A comparison of the tapered fiber, tapered tips and cuvette-based techniques is provided in Table 3.1.

Table 3.1. Comparison of characteristics of three modes of measurement using tapered fibers, tapered tips and the traditional cuvette geometry.

	In situ taper	Tapered tips	Cuvette
Capability for localization	Completely localized. Delivery of the radiation is localized. Reception is of the fluorescence radiation is also localized.	Delivery of the radiation is localized. But, reception of the fluorescence radiation may not be localized	No
Robustness	Use of the same fiber (tapered region) for delivery and reception makes the system very robust.	Use of separate fibers (tip) for delivery and reception makes it less robust.	No
Sensitivity	High	High	Low
Active or passive	Active. The full optical properties of the fiber and taper are used. The sensor can therefore be engineered to enhance the sensing capability.	Passive. Fiber tip does not use any optical property. Merely used as a conduit.	Passive
Scalability and array sensing	Very scalable. Can be used form an array of sensors to provide information at different locations in the volume.	Array sensing may not be possible.	Not scalable

The work reported here describes the concept, fabrication and use of the tapered fibers in biosensing for the measurement of cell concentration and fluorescence of biomaterials. The approaches used here take advantage of the evanescent absorption by the biomaterials in the tapered region (Littlejohn et al., 1999; Prince et al., 2001; Woerderman and Parnas, 2001). The absorption can induce further scattering (evanescent induced scattering) and fluorescence (evanescent induced fluorescence). These three unique features of the tapered fibers were exploited in our research.

3.2. CONCEPT OF TAPERED FIBERS

Optical fibers which have become an integral part of the telecommunications and sensing applications consist of a material of higher refractive index (core) surrounded by a material of lower refractive index (Saleh and Teich, 1991). In most of the fibers, the main component is pure silica (SiO_2) which is doped with slight amounts of germanium (Ge) to produce a small increase in index in the core while pure silica is used as the cladding. If the indices of the core and cladding respectively are given by n_1 and n_2, the propagation of light in the fiber can be expressed in term of the dimensionless quantity known as the V number given by

$$V_{core} = \frac{2\pi a}{\lambda} \sqrt{n_1^2 - n_2^2}$$ (1)

where a is the radius of the core and λ is the operating wavelength. When the radius of the core is small enough such that

$$V_{core} < 2.405$$ (2)

the fiber is referred to as a single mode fiber. It supports the fundamental mode, LP_{01} which is radially symmetric (LP stands for Linearly Polarized). As the value of V_{core} increases, the number of modes supported by the fiber also increases. These modes may be radially symmetric (LP_{0m}, m=1,2,3...) or radially asymmetric (LP_{km}, k=1,2,3..; m=1,2,3,..).

When V is close to 2.405 about 85% of the power in the single mode fiber stays in the core (guided power) and remaining power flows through the cladding (evanescent power). Since the thickness of the cladding is typically 10-20 times that of the core, this evanescent power does not reach outside the cladding. In fact, the cladding power penetrates a few microns into the cladding and thus, a fully clad fiber is virtually useless in sensing applications.

The best way to access the evanescent field outside the fiber is to taper the fiber (Amitay and Presby, 1989; Bobb et al., 1990; Bobb et al., 1991; Brophy et al., 1993). This can be accomplished by heating the fiber to its softening point and pulling it under controlled conditions. We can also use a commercially available fiber puller or a fusion splicer to taper the fiber. When the fiber is tapered as shown in Figure 3.1, the waist region would have a diameter on the order of a few microns and the core would have almost vanished. It has been shown that the light guidance is controlled by the original cladding

index acting as the core and the medium surrounding the taper acting as the cladding. If this tapered region is surrounded by cells, tissue or biomelcules, the evanescent light interacts with the medium outside making it possible to alter the properties of the light inside the fiber resulting in the fiber acting as an *in situ* sensor. The modes will now be determined by the V number in the tapered region beyond $V_{core} < 1$ of Equation (1). In this region, the V value can be evaluated as

$$V(z) = \frac{2\pi b(z)}{\lambda} \sqrt{n_2^2 - n_{out}^2} .$$

(3)

In Equation (3), b is the radius of the fiber in the tapered region which is a function of the local radius and consequently V will also be a function of the local radius as determined by the length (z). The index of the medium surrounding the fiber is n_{out}. Since the V

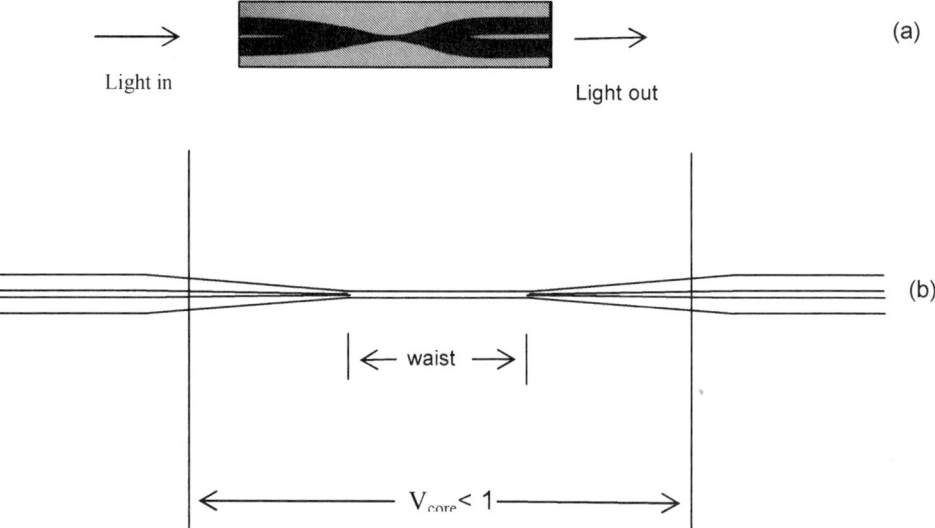

Figure 3.1. The photograph of a tapered fiber fabricated in our laboratories is shown in (a). The tapered region is expanded in (b) to show the contracting, expanding and waist regions of the taper. It also shows the region where the physics of the taper is primarily taking place ($V_{core} < 1$).

value in Equation (3) will be larger than 2.405, this region will support many modes. In a straight fiber, modes will travel without interfering with others. However, in a tapered fiber, the contracting and expanding regions of the tapering introduce perturbations, leading to coupling among the modes supported by the fiber based on the value of V in Equation (3). Since the tapering action is a radially symmetric perturbation, the coupling of modes will take place among the LP_{0m} modes. This process can be explained as follows. When the value of V_{core} as per Equation (1) decreases below unity, the light guidance is controlled by Equation (3) and light in the LP_{01} mode begins coupling to the LP_{0m} modes in the region governed by Equation (3). The waist region contributes a steady change in phase (Shankar et al., 1991). The coupling starts once again on the other side of the waist in the expanding region of the taper. When V_{core} given by Equation (1) once again becomes equal to 1, light goes back to the LP_{01} mode associated with the original fiber which is detected at that end.

The power distribution of a few LP_{0m} modes is plotted in Figure 3.2 as a function of the V value given in Equation (3). It shows the fractional power in the cladding (evanescent power) as V number increases (Gambling et al., 1973; Okamoto, 2001). If the tapering can be engineered so that in the waist region most of the light is in the highest order mode that can be supported (for the case shown in Figure 3.2, this will be the LP_{05} mode), it will be possible to have a significant power in the evanescent component. For example, consider a V value of 7.2. If the tapering is such that coupling of light takes place and the power is in the LP_{03} modes, almost 60-70% of the light will be outside the fiber. This will be the ideal condition for the case of any form of evanescent sensing. If the analyte in the tapered region absorbs this light, this will impact the light remaining in the fiber as measured at the output end. This is the principle behind evanescent absorption. If the analyte contains particulate matter, scattering can take place along with

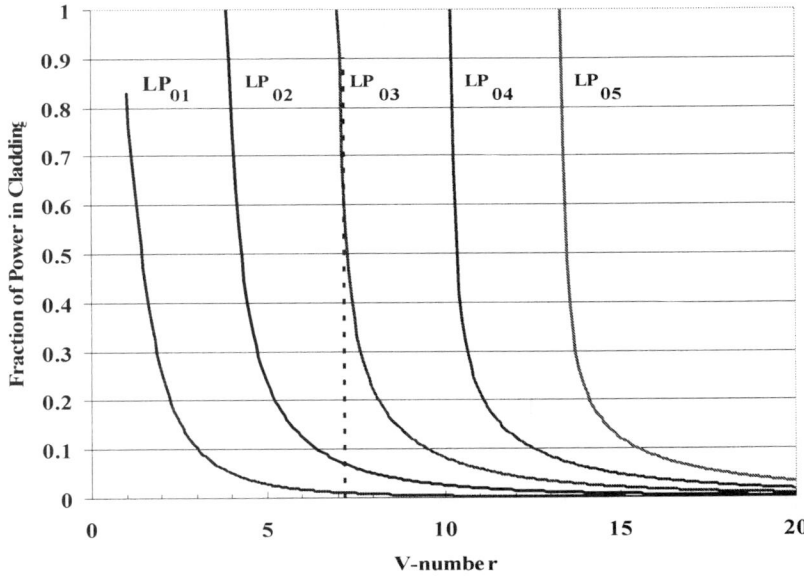

Figure 3.2. The fractional power in the cladding (evanescent power) is shown as a function of V number. At a specific value of V (7.2), the power in the LP_{03} mode is about 70% evanescent and will be available outside the fiber.

the evanescent absorption. If the refractive index of the analyte changes, this will further alter the evanescent characteristics of the taper producing changes in the light remaining in the core.

One of the most interesting applications of this concept of the engineered taper in the evanescent regime is the detection of fluorescence by the tapered fiber in situ. When the medium surrounding the tapered region is fluorescent and the wavelength of the light propagating through the fiber is in the absorption region of the fluorophore, the evanescent light outside the tapered region will be absorbed by the fluorophore. The fluorescence results in emission outside the tapered fiber. This light serves as the evanescent component for the light inside the fiber and thus, will couple back into the fiber and transmitted to the output end of the fiber. This concept is demonstrated in Figure 3.3. As the excitation light (λ_{ex}) enter the region Vcore<1, coupling among modes start. By appropriately tapering the fiber, it is possible to have more and more of light of this light move outside the fiber. As the strength of λ_{ex} decreases inside the fiber, fluorescence emission (λ_{em}) is created outside the fiber and slowly the strength of this light inside the fiber increases.

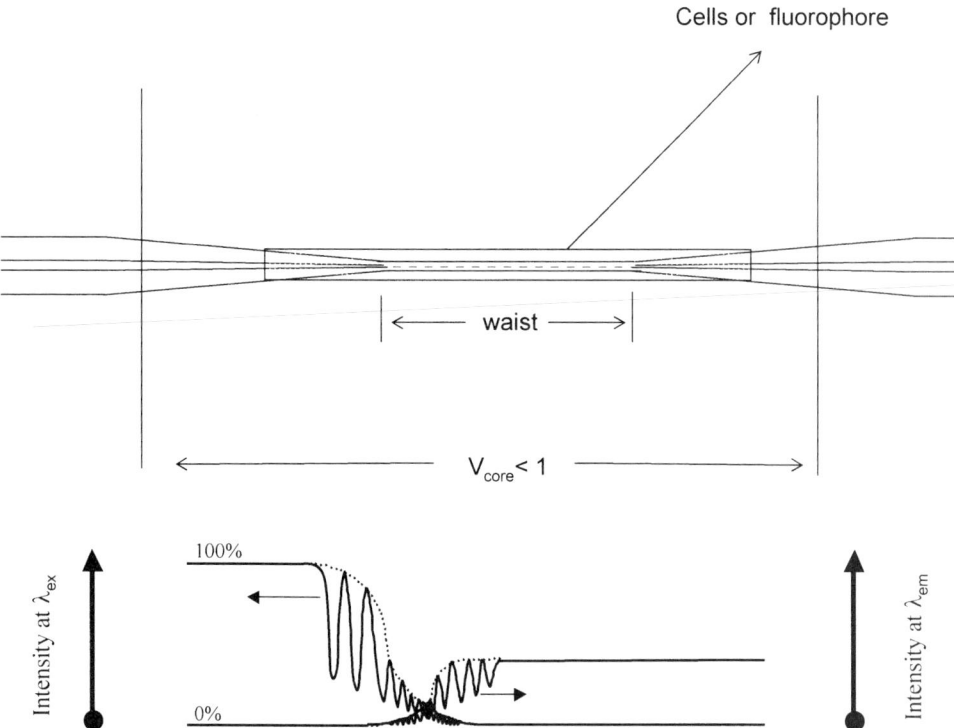

Figure 3.3. The concept of evanescent fluorescence is shown. The analyte will be kept around the tapered region as shown. The bottom portion of the figure shows where the excitation field exits the fiber and the emission field enters the fiber.

3.3. DETAILS OF EXPERIMENTAL ARRANGEMENTS AND RESULTS

All the results reported here used Corguide fiber (Corning Glass Works, NY, attenuation at 1300nm and 1500nm of 0.36 and 0.26 dB/km, respectively) with a core diameter of 8 μm and total diameter of 125 μm. The tapering was done by heating and pulling using a small jig made in our laboratories. Waist diameter of 3.1 mm was the smallest one we obtained in our laboratories (Haddock et al., 1993; Wiejata et al. 2003).

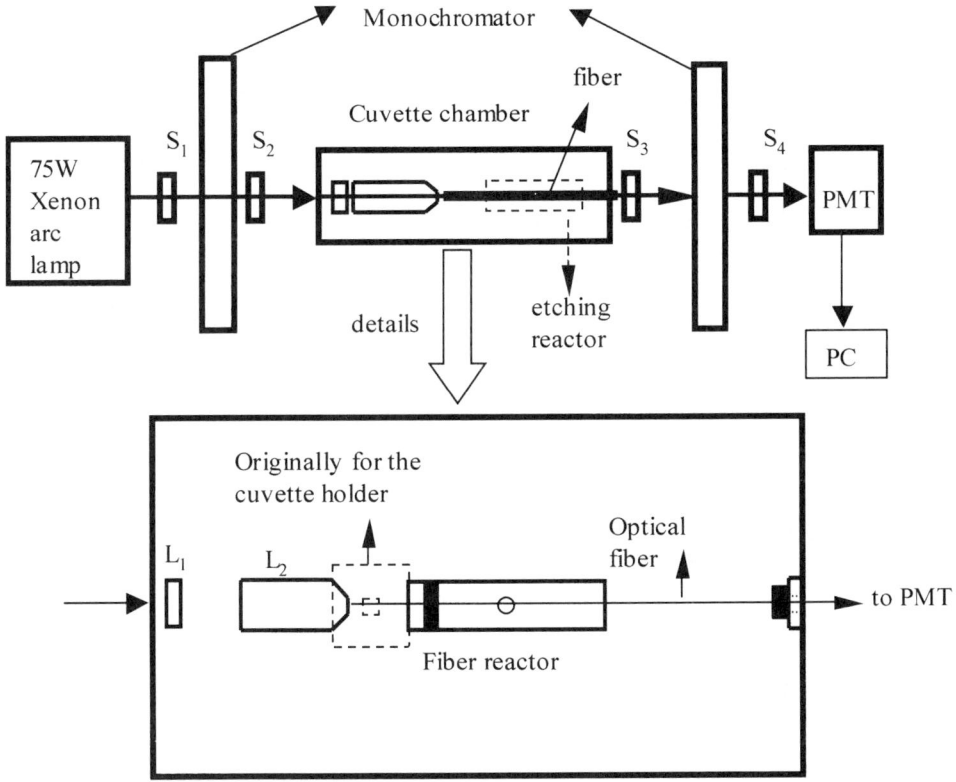

Figure 3.4. Apparatus used for monitoring fiber etching in an analytical fluorometer. Slits S_2 and S_4 were set at 2 nm, while slits S_1 and S_3 were opened fully. Convex lens (L1) and a microscope objective (L2) were used to launch light from monochromator into the fiber.

A computer-coupled spectrofluorometer (Model QM-1, Photon Technology International Inc., PTI) was used in the fiber-based measurements. This instrument is equipped with a 75 W Xenon arc lamp (Ushio Inc., Japan) with a monochromator and a photomultiplier tube (PMT, housing model 710, tube model R1527P, PTI Inc.) coupled through a monochromator. Setting of the monochromator wavelengths and recording of PMT outputs (in counts/s) were done through instrument software. A thick black cloth was also used to cover the entire fluorometer to reduce leakage of ambient light. A reactor made of acrylic (Plexiglas) was designed to mount the tapered fiber, and to provide a chamber (150 µl) adjacent to the taper for loading analyte. The fiber holder is made by gluing two Plexiglas (100 x 30 x 3.2 mm thick) plates together. A 0.2 mm diameter line was scored lengthwise in the center to accommodate the sheathed fiber. A 6.4 mm hole was pre-drilled in the top plate to serve as the chamber for contacting analyte with the tapered region of the fiber. A third piece of Plexiglas (30 mm^2 with a 6.4 mm hole in center) was attached to the base with four screws, after a thin layer of silicone vacuum gel was applied to its underside. The gel eliminated capillary outflow of the analyte from the central chamber. All the vertical surfaces of the fiber holder were painted black to reduce measurement errors.

The spectrofluorometer was modified as shown in Figure 3.4 to accommodate the fiber holder *in lieu* of the cuvette holder. The focusing arrangement and cuvette holder were removed. The fiber holder was mounted on a manual X-Y-Z positioning platform in the cuvette chamber along the axis of the light source and the PMT. A microscope objective (20X UV capable) was mounted to focus the input light into the fiber. In order to make sure that the signal picked up from the PMT came only from the fiber, a black Delran cylindrical piece was machined to fit into the opening leading to the PMT. A 125 µm diameter hole was drilled at the center of the Delran block to accommodate the fiber tip with Play-doh (available from any toy store) to seal the center hole so as to block leakage of stray light from the cuvette chamber.

The full range of the PMT is 3.5 x 10^6 counts/s. The noise floor in measurement is 30 counts/s.

3.4. EVANESCENT ABSORPTION

The fiber taper used in this experiment had a waist diameter of 6.8 µm. The analyte chosen to demonstrate the potential application of evanescent absorption in the characterization of biomaterials was NA(D)PH(tetrasodium salt, approx. 95% purity, Sigma). The powdered NADPH was dissolved in 0.02 M tris(hydroxymethyl)-aminomethane (TRIZMA) buffer with a pH of 7.70 at room temperature. The TRIZMA buffer was made from mixing TRIZMA hydrochloride (reagent grade, Sigma) and TRIZMA base (reagent grade, Sigma) powders, and dissolving the mixture in deionized(DI) water to a concentration of 0.05 M, pH of 7.80 at 25 °C, and then diluting to 0.02 M with a resulting pH of 7.70. Pure buffer was used as zero concentration of the analytes. The transmission and absorption were measured at 350 nm (Siano and Mutharasan, 1989; Siano and Mutharasan, 1991; Asali et al.; 1992; Haddock et al. 2003).

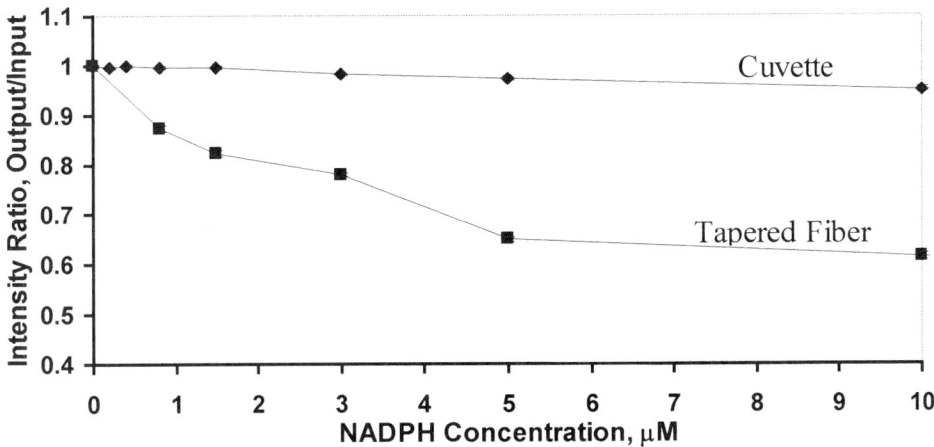

Figure 3.5. The normalized transmission is plotted against the low concentration of NADPH. The tapered fiber shows high sensitivities at low concentrations compared to a cuvvette (Haddock et al., 2003).

The concentration of NADPH was varied from 0 to 500 μM. For each concentration, eighteen experiments on NADH absorption on tapered fibers were done. Typical variation in measurement was less than 2.5%, based on three independent measurements. Only five measurements in cuvette were done because of the stability in these measurements. Figure 3.5 shows the normalized transmission (and hence absorption behavior) seen NADPH studies. The normalization is done by comparing the transmission at different concentrations to that at zero concentration. Only low values of the concentration are shown. Very high sensitivities at low concentrations are seen with the tapered fibers compared to what is seen in the cuevette.

3.5. EVANESCENT ABSORPTION AND SCATTERING

Actively growing Chinese Hamster Ovary (CHO, DG44) cells were harvested from T flasks, and then suspended in phosphate buffered saline solutions at various concentrations up to 4×10^6 cells/ml. The transmission (and consequently absorption) was measured at 450 nm. The evanescent light is absorbed by the cells leading to loss in transmitted optical power. Seven sets of measurements were conducted and the results are shown in Figure 3.6. The output is normalized to the value at zero concentration. The standard deviation was less than 5% over the seven sets of measurements. At a low concentration, cuvette displayed no sensitivity, while the tapered fiber showed good sensitivity, in the range of 0 to 0.25 million cells/ml. At low concentration, the sensitivity was 1% change in transmission per 10,000 cells/ml. On the other hand, the cuvette did not show any change in transmission below a cell concentration of 1 million/ml. That is, the concentration measurement capability with a tapered fiber is well over ten times that obtained in cuvette geometry.

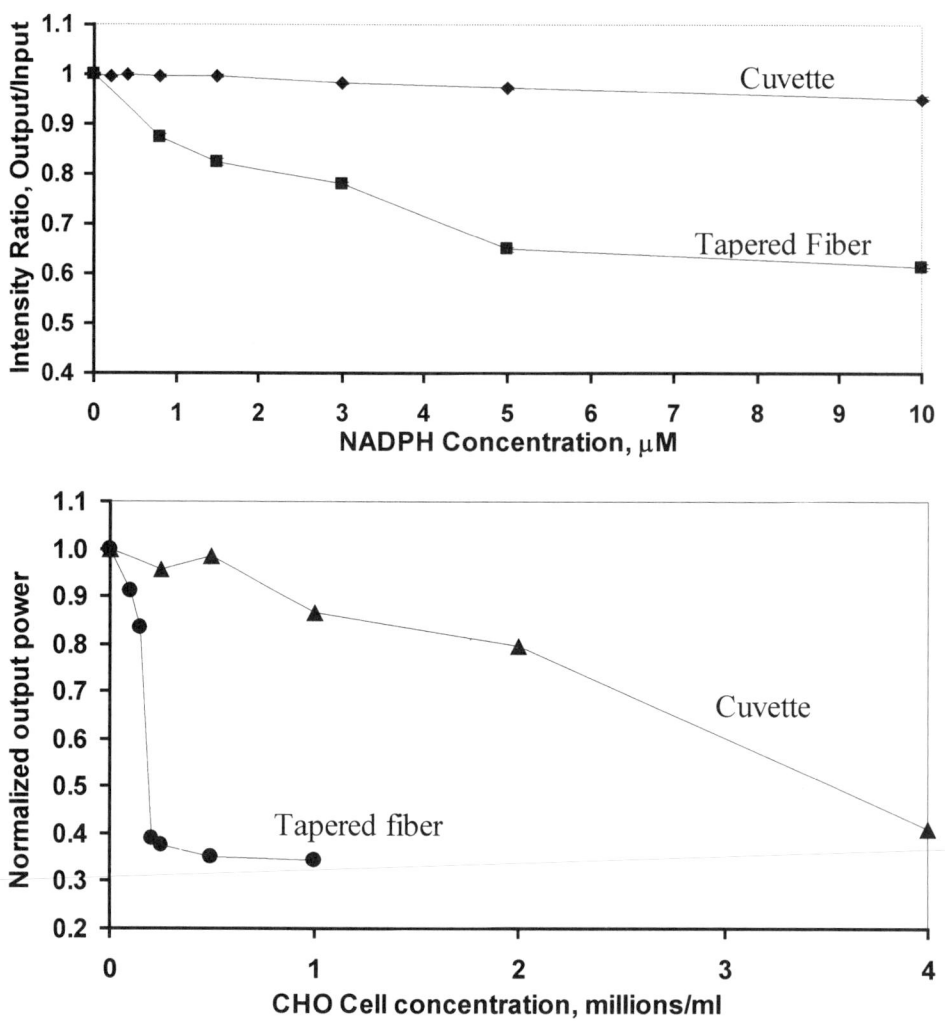

Figure 3.6. The normalized output power is plotted against CHO cell concentration in millions/ml. Tapered fiber shows excellent discrimination at very low concentrations of cells (Haddock et al., 2003).

3.6. EVANESCENT FLUORESCENCE

Once it was established that the evanescent absorption takes place and sufficient evanescent power is available, experiments were done to excite fluorescence outside the tapered region and pick up the fluorescence.

The experiments were undertaken with the taper with the smallest possible waist diameter fabricated in our laboratories (3.11 μm). The cuvette assembly was removed and replaced with the fiber holder. All alignments were done as described in earlier

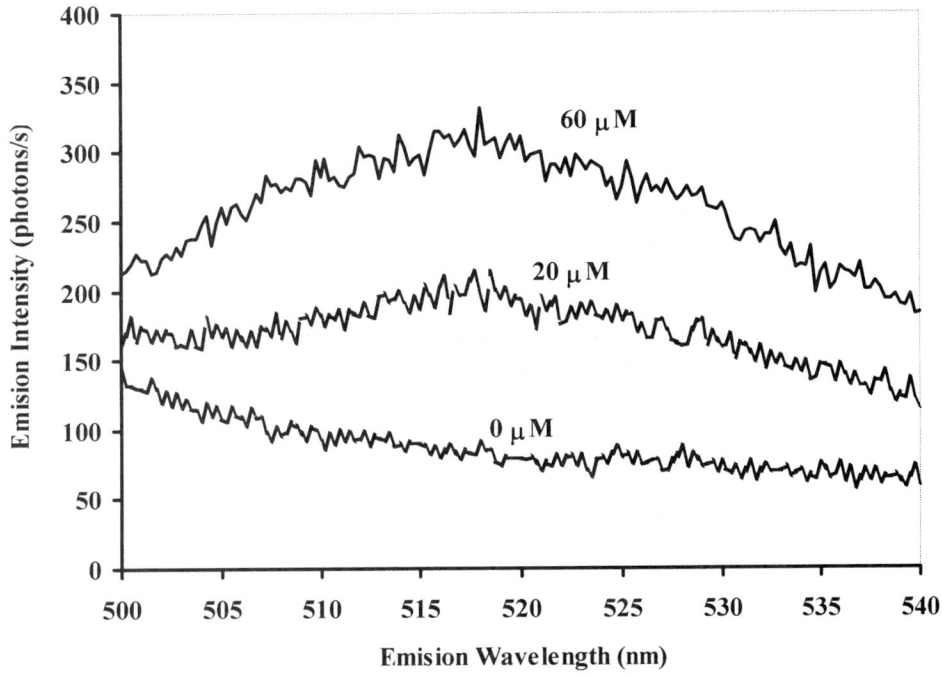

Figure 3.7. The fluorescence emission measured in a tapered fiber is shown for three different concentrations of the fluorophore, fluorescein (Wiejata et al., 2003).

paragraphs on absorption measurements. The fluorophore used was fluorescin with absorption around 460 nm and emission around 510 nm. Results are shown in Figure 7. Increased fluorescence is seen with increased Fluorescin concentration (Wiejata et al., 2003). The observed fluorescence could have been excited only from the light transmitted through the fiber interacting with the solution surrounding the taper via the evanescent field. Since the tapered fiber was continuous and all the possible avenues of light leakage into the PMT were eliminated, the fluorescence observed came only from the fluorescence emission light being guided into the fiber through the evanescent coupling mechanism. The results establish that the tapered fiber can act both as a transmitter and a delivery mechanism of the excitation energy to the optically active solution surrounding the fiber and simultaneously pick up the longer wavelength fluorescence light. Even though the emission counts are much lower than those seen in the cuvette, the fluorescence emission is of sufficient magnitude to show the optical functioning of the sensor.

It should be noted that fluorescence measured using the tapered geometry is along the axis of the fiber, while in cuvette geometry, it is orthogonal. Because the source of fluorescence signal is from the tapered region, one can be certain as to the location of the fluorescent sample, while in open optical path geometry, one can be certain of the direction, but not the location. Thus sensing fluorescence using tapered geometry offers the added advantage of localization. It must be stated that the emission signal in tapered fibers was smaller than what could be obtained in cuvette geometries. This is due to the fact that in tapered fiber sensing, the source of excitation light is the evanescent filed, and thus is of low intensity compared to cuvette.

3.7. CONCLUDING REMARKS

The evanescent field characteristics of a tapered fiber show remarkable capability for sensing applications in bioengineering. Specifically, the tapers can be engineered to bring the light to a designated point outside the fiber without cutting or breaking the fiber. This light can interact with the analyte outside which can be a pathogen or tissue whose chemistry responds to the presence of the specific wavelength of light. Our results showed that evanescent absorption is a starting point of a several important secondary effects that can be utilized in studying the properties of biomaterials. The most interesting of these properties is the phenomenon of evanescent induced fluorescence which allows the interrogation of biomaterials in situ. It may also be possible to use the tapered fiber to characterize biomaterials in the dynamic regime by measuring lifetime fluorescence properties (Pradhan et al., 1995; Lakowicz, 1999).

3.8. ACKNOWLEDGEMENT

This work was partially supported by Ben Franklin Partnership's Nanotechnology Technology Institute and the National Science Foundation (Award No: BES 0329793).

3.9. REFERENCES

Alfano, R. R., Das, B.B., Cleary, J., Prudente, R., and Celmer, E. J., 1987, Fluorescence spectra from cancerous and normal human breast and lung tissues, *IEEE Journal of Q. Electr.* **QE-23**: 1806.

Amitay, N and Presby, H. M., 1989, Optical fiber Up-tapers modeling and performance analysis, *J. Lightwave Technol.* 7:131.

Asali, E. C., Mutharasan, R. and Humphrey, A. E., 1992, Use of NAD(P)H-fluorescence for monitoring the response of starved cells of *Catharanthus roseus* in suspension to metabolic perturbations, J. Biotechnol. 23:83.

Bobb, L.C., Krumboltz, H., and P.M. Shankar, 1991, A novel pressure sensor using bent biconically tapered single mode fibers, Optics Lett. **16**:112.

Bobb, L.C., Shankar, P. M. and and Krumboltz, H.,1990, Bending effects in biconically tapered single mode fibers, *J. of Lightwave Tech.* **LT-8**: 1084.

Brophy T.J, Shankar, P.M., and Bobb, L.,1993, Formation and measurement of tapers in optical fibers, *Rev. Sci. Instrum.* **64**:2650.

Cullum B.M., Griffin, G.D., Miller, G.H., and Vo-Dinh, T., 2000, Intracellular measurements in mammary carcinoma cells using fiber-optic nanosensors, *Anal. Biochem.* **277**:25.

Gambling, W. A., Payne, D. N. and Matsumara, H., 1973, Mode excitation in a multimode optical fiber waveguide, *Elect. Lett.* 9:412.

Haddock, H, Shankar, P. M., Mutharasan, R., 2003. Evanescent Sensing of Biomolecules and Cells, *Biosensors and Actuators – Chem.* **B88**:67.

Haddock, H., Shankar, M., Mutharasan, R., 2003. Fabrication of Biconical Tapered Optical Fibers Using Hydrofluoric Acid, *Materials Sci. & Eng.* **B97**:87.

Henry, W., 1994, Evanescent field devices: a comparison between tapered optical fibers and polished or D fibers, *Opt. Quan. Electron.* **26**:S261.

Khijwania, S.K. and Gupta, B.D., 1999, Fiber optic evanesent field absorption sensor: Effect of fiber parameters and geometry of the probe, *Opt. and Qua. Electr.* **31**: 625.

Lambelet, P., Sayah, A., Pfeffer, P., Philopona, C., and Marquis-Weible, F., 1998, Chemically etched fiber tips for near-field optical microscopy: a process for smoother tips, *Appl. Opt.* **37**:7289.

Littlejohn, D., Lucas, D., and L. Han, 1999, Bent silica fiber evanescent absorption sensors for near-infrared spectroscopy, *Appl. Spectr.* **53**(7): 845.

Lakowicz, J. R., Principles of Fluorescence Spectroscopy, Kluwer Academic/Plenum, New York, 1999

Marcuse, D., 1988, Launching of light into fiber cores from sources located in the cladding, *J. Lightwave Technol.* **6**:1273.

McCulloch, S. and Uttamchandani, D., 1997, Development of fibre optic micro-optrode for intracellular pH measurements, IEE Proc.-Optoelectron. **144**:162.

Mignanai, A. and Baldini, F., 1995, In vivo biomedical monitoring by fiber-optic systems, *J. Lightwave Technol.* **7**:1396.

Mononobe, S., Naya, M., Saiki, T., Ohtsu, M., 1997, Reproducible fabrication of a fiber probe with a nanometric protrusion for near-field optics. *Appl. Opt.* **36**:1496.

Nath N., Jain S.R. and Anand, S., 1997, Evanescent wave fibre optic sensor for detection of L. donovani specific antibodies in sera of kala azar patients, *Biosen. and Bioelect.***12**: 491.

Nath, N. and Anand, S., 1998, Evanescent wave fiber optic fluorosensor: effect of tapering configuration on the signal acquisition, *Opti. Eng.* **37**:220.

Okamoto, K., 2000, Fundamentals of Optical Waveguides, Academic, San Diago, pp.47-124.

Pilevar, S., Davis, C.C. and Portugal, F., 1998, Tapered optical fiber sensor using near-infrared fluorophores to assay hybridization, *Anal. Chem.* **70**:2031.

Pradhan, A., Pal, P., Durocher, G., Villeneuve, L., Balassy, A., Babai, F., Gaboury, L., Blanchard, L., 1995, Steady state and time-resolved fluorescence properties of metastatic and non-metastatic malignant cells from different species, *J. of Photochem. and Photbiol.* **B31**: 101.

Prince, B. J., Kaltcheva, N. T., Schwabacher, A. W. and Geissinger, P., 2001, Fluorescent Fiber-Optic Sensor Arrays Probed Utilizing Evanescent Fiber-Fiber Coupling, *Appl. Spectr.* **55**:1018.

Saleh, B E A and Teich, M. C., 1991, Fundamentals of Photonics, John Wiley, New York, pp. 272-307.

Shankar, P. M., L. Bobb, L., and Krumboltz, H., 1991, Coupling of modes in bent biconically tapered single mode fibers, *J. Lightwave Technol.* **9**:832.

Siano, S. A. and Mutharasan, R., 1989, NADH and flavin fluorescence responses of starved yeast cultures to substrate additions, *Biotechnol. Bioeng.* **34**; 660.

Siano, S. A., Mutharasan, R., 1991, NADH fluorescence and oxygen uptake responses of hybridoma cultures to substrate pulse and step changes, *Biotechnol. Bioeng.* **37**:141.

Tang, C. C., Pradhan, A., Sha, W., Chen, J., Liu, H., Wahl, S. J., and Alfano, R. R., 1989, Pulsed and cw laser fluorescence spectra from cancerous, normal, and chemically treated normal human breast and lung tissues, *Appl. Opt.*, **28**:2337.

Vo-Dinh, T., Alarie, J-P, Cullam, B. M., and Griffin, G., 2000, Antobody based nanoprobe for measurement of a fluorescent analyte in a single cell, *Nature Technol.* **18**:764.

Wiejata, P. J., Shankar, P. M., and Mutharasan, R., 2003, Fluorescent Sensing Using Biconical Tapers," *Sensors and Actuators – Chem.* **B 96**:315.

Woerdeman D. L. and Parnas, R. S., 2001, Model of a fiberoptic evanescent wave fluorescent sensor, *Appl. Spectr.* **55**:331.

MULTI-DIMENSIONAL TIME-CORRELATED SINGLE PHOTON COUNTING

Wolfgang Becker[1] and Axel Bergmann

4.1. INTRODUCTION

Optical spectroscopy techniques have found a wide range of applications in biomedical imaging and sensing because they are non-destructive and deliver biochemically relevant information about the systems investigated (Mycek and Pogue, 2003; Lakowicz, 1999). Typical applications are one- and two-photon fluorescence laser scanning microscopy, fluorescence endoscopy, control of drug delivery in photodynamic therapy, dynamics of protein-dye complexes on the single molecule level, chlorophyll fluorescence dynamics, and diffuse optical tomography of thick tissue. Most of these techniques use the fluorescence of exogenous or endogenous fluorophores to obtain information about the systems investigated. In the majority of the applications the fluorescence intensity, fluctuations of the fluorescence intensity, or the fluorescence spectra are recorded. However, the fluorescence of organic fluorophores is not only characterised by its intensity or spectrum, it has also a characteristic fluorescence lifetime. The fluorescence lifetime is useful as a separation parameter to distinguish the fluorescence components of endogenous fluorophores in cells and tissues. These components often have poorly defined fluorescence spectra but can be distinguished by their fluorescence lifetime (König and Riemann, 2003; Schweitzer, 2001; Urayama and Mycek, 2003). Moreover, the fluorescence lifetime of a fluorophore depends on the local environment of the molecules. Because the lifetime is widely independent of the concentration its measurement is a direct approach to quenching and energy transfer effects (Lakowicz, 1999). Typical examples are the mapping of cell parameters such as pH, ion concentrations, oxygen saturation or the binding state to proteins, lipids, or DNA (Gerritsen et al.,1997; Knemeyer et al., 2002; Lakowicz, 1999; Rück et al., 2003; Sanders et al., 1995, Van Zandvoort et al., 2002).

The fluorescence lifetime can be used to improve the existing techniques of fluorescence resonance energy transfer (FRET). FRET occurs if the fluorescence

---

[1] Becker & Hickl GmbH, Berlin, Germany

emission band of one fluorophore, the donor, overlaps the absorption band of a second one, the acceptor. In this case the energy from the donor is transferred directly into the acceptor, resulting in an extremely efficient quenching of the donor fluorescence (Lakowicz, 1999, Periasamy et al., 2001). The energy transfer rate from the donor to the acceptor depends on the sixth power of the distance. FRET occurs at distances of the order of a few nm and therefore happens only if the donor and acceptor are physically linked. With fluorescence lifetime imaging (FLIM) techniques, FRET results are obtained from a single lifetime image of the donor (Bacskai et al., 2003, Becker et. al. 2004; Biskup et al., 2004, Chen and Periasamy, 2004; Tramier et al., 2002).

Fluorescence lifetime detection is a powerful tool in single molecule spectroscopy. Single molecules in a femtoliter sample volume can be investigated by direct multi-parameter spectroscopy (Eggerling et al., 2001, Kühnemuth and Seidel, 2001; Prummer et al., 2004; Ying and Xie, 2002) or by fluorescence correlation techniques (Berland et al., 1995, Rigler et al., 1993, Rigler and Elson, 2001, Thomson, 1991). Time-resolved detection delivers a wealth of information about the molecules and can be used to identify the molecules or to characterise their local environment.

Diffuse optical tomography (DOT) sends near-infrared light through thick tissue and investigates the diffusely transmitted or reflected light. Illumination by pulsed or modulated light and detection of the waveform of the transmitted or reflected light delivers the reduced scattering and absorption coefficients (Choi et al., 2004, Cubeddu et al., 1996, McBride et al., 1999, Patterson and Chance, 1989; Schmidt et al. 2000; Grosenick et al. 1999, Gròsenick et al. 2004). These, in turn, can be used to derive biochemically relevant parameters, such as oxygen saturation, or haemoglobin contents.

Biological systems often show dynamic changes in their fluorescence, absorption or scattering behaviour. Dynamic changes must be taken into account in plants (Maxwell and Johnson 2000), live cells, experiments of photodynamic therapy (Rück et al., 2003), and brain imaging by diffuse optical tomography (Liebert et al., 2004).

At the same time, the excitation power must be kept low to avoid light-induced changes, photobleaching or photodamage in the sample. A fluorescence technique for biomedical application should therefore have a high recording efficiency, i.e. should not discard any photons, neither by gating off photons on the time axis, nor by blocking a part of the emission spectrum by filters.

The time resolution must be high enough to resolve the fluorescence lifetimes of the typical endogenous and exogenous fluorophores, and the diffusion times of photons through tissue. Typical lifetimes and photon diffusion times are of the order of a few nanoseconds. However, the lifetimes in presence of strong fluorescence quenching, the lifetime of the quenched donor fraction in FRET experiments, and the lifetimes of short autofluorescence components can be as short as 100 ps (Becker et al., 2001; König and Riemann, 2003; Marcu et al., 2003; Urayama and Mycek, 2003, Schweitzer et al., 2001). Lifetimes down to 50 ps are found in dye aggregates (Kelbauskas and Dietel, 2002) and complexes of fluorophores and metallic nano-particles (Geddes et al., 2003; Malicka, 2003; Lakowicz et al., 2001). Due to the mixture of different chromophores or inhomogeneous quenching the fluorescence decay functions found in cells and tissue are normally multi-exponential.

Therefore, a time-resolved spectroscopy technique for biomedical application should be able to record fast changes in the sample, resolve the components of multi-exponential fluorescence decay functions down to less than 100 ps, have imaging capability, multi-wavelength capability, and a near-ideal detection efficiency.

Time-resolved optical recording techniques

Time-resolved optical recording techniques are normally classified into frequency domain techniques, and time-domain techniques. Frequency domain techniques measure the phase shift between the high-frequency modulated or pulsed excitation and the fluorescence signal at the fundamental modulation frequency or its harmonics (Carlsson and Liljeborg, 1997; Pepperkok, 1999; So et al., 1994; So et al., 1995; Squire et al., 2000). Time-domain techniques record the fluorescence decay functions directly (Buurman et al., 1992; Cole et al., 2001; O'Connor and Phillips, 1984). Although the frequency-domain and the time domain are generally equivalent, the corresponding signal recording techniques and instruments may differ considerably in efficiency, i.e. in the lifetime accuracy obtained for a given number of detected photons.

For fluorescence lifetime imaging, both the frequency-domain and time-domain FLIM techniques can be further classified into camera techniques and point-detector scanning techniques. The benefit of the camera techniques is that they are relatively easy to use. Point-detector scanning techniques have the benefit of depth resolution, optical sectioning capability, and deep tissue imaging capability by two-photon excitation (Denk et al., 1990; White et al., 1987).

Frequency domain camera techniques are based on modulated image intensifiers, frequency domain point-detector techniques on modulated photomultiplier tubes (PMTs) or on photomultiplier tubes or avalanche photodiodes with subsequent electronic mixers. Time domain camera techniques use gated image intensifiers, time-domain point-detector scanning techniques time-correlated single photon counting (TCSPC), or gated photon counting in several parallel time-gates.   Among all these techniques, TCSPC yields the highest recording efficiency and the highest time resolution (Ballew and Demas, 1989; Carlsson and Philip, 2002; Philip and Carlsson, 2003; Köllner and Wolfrum, 1992, O'Connor and Phillips, 1984). Classic time-correlated single photon counting (Cova et al., 1973; Hartig and Sauer, 1976; Leskovar and Lo, 1976; Lewis and Ware, 1973; Meiling and Stary, 1963; O'Connor and Phillips, 1984; Schuyler and Isenberg, 1971) is based on the detection of single photons of a periodical light signal, the measurement of the detection times of the individual photons within the signal period, and the reconstruction of the waveform from the individual time measurements. The TCSPC technique makes use of the fact that for low level, high repetition rate signals the light intensity is normally low enough that the detection of several photons in one signal period can be neglected. It is therefore sufficient to detect the photons, measure their time in the signal period, and build up a histogram of the photon times.

The principle shown in Figure 4.1. There are many signal periods without photons, other signal periods contain one photon. Periods with more than one photons are very rare. When a photon is detected, the arrival time of the corresponding detector pulse is measured. The events are collected in a memory by adding a '1' at an address proportional to the detection time. After many photons, the distribution of the detection times, i.e. the waveform of the optical pulse builds up.

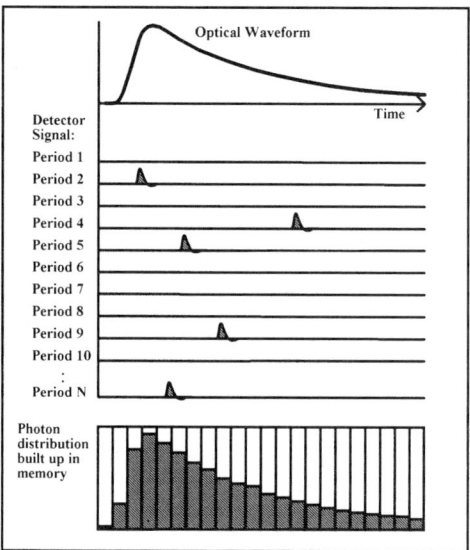

Figure 4.1: Principle of classic time-correlated single-photon counting

The TCSPC technique does not gate off any photons and therefore, as long as the light intensity is not too high, reaches a counting efficiency close to one. The time resolution is limited by the transit time spread in the detector. With fast detectors, a width of the instrument response function (IRF) of 25 ps can be achieved. Moreover, the width of the time channels of the recorded photon distribution can be made as small as 1 ps. The small time bin width in conjunction with the high number of time channels available makes it possible to sample the signal shape adequately according to the Nyquist theory. Therefore standard deconvolution techniques can be used to determine fluorescence lifetimes much shorter than the IRF width and to resolve the components of multi-exponential decay functions.

The drawback of classic TCSPC instruments was that their counting capability was very limited. The instruments were therefore restricted to very low light intensities, which resulted in extremely long acquisition times. Moreover, the conventional TCSPC technique was intrinsically one-dimensional, i.e. only the waveform of the light signal in one spot of a sample and at one wavelength was recorded at a time.

4.2. MULTI-DIMENSIONAL TCSPC

New conversion principles in the time-measurement block of TCSPC devices have increased the counting capability by more than two orders of magnitude (Becker et al., 2001). Therefore, results can be obtained in much shorter acquisition times than with the classic instruments, or a much higher number of photons can be recorded within a given acquisition time. Moreover, advanced TCSPC devices use programmable logic arrays (FPGAs) to implement all the digital control and signal processing functions. Compared to the very limited capacity of the more or less discrete control electronics of classic TCSPC systems the use of FPGAs initiated a breakthrough in functionality. Advanced TCSPC devices use a multi-dimensional histogramming process, i.e. they record the photon density not only as a function of the time in the signal period, but also over other parameters, such as wavelength, spatial coordinates, location within a scanning area, the time from the start of the experiment, or other externally measured variables. The general architecture of a TCSPC device for multi-dimensional data acquisition is shown in Figure 4.2.

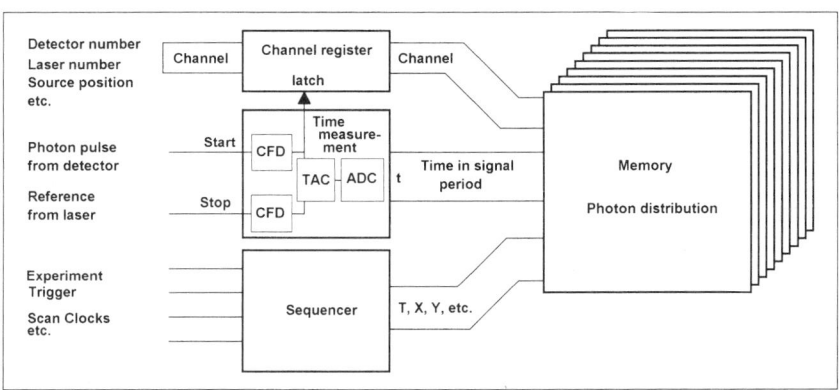

Figure 4.2: Architecture of multi-dimensional TCSPC

The times of the photons are measured in the time-measurement block. Two constant fraction discriminators, CFD, are receiving the photon pulses and the reference pulses from the light source. A time-to-amplitude converter, TAC, is started by the photon pulse and stopped by the next reference pulse. A subsequent analog-to-digital converter delivers the digital equivalent of the photon time.

In addition to the time measurement block multi-dimensional TCSPC contains a channel register and a sequencer logic. The memory is much larger than for classic TCSPC. When a photon is recorded its destination in the device memory is controlled by the time-measurement block, by the channel register, and the sequencer logic. Consequently, the device builds up a multi-dimensional photon distribution over the time of the photons in the signal period, over the data word at the 'channel' input, and over one or several additional coordinates generated in the sequencer. The additional address bits provided by the channel register and the sequencer are often called 'routing bits', and the technique is termed 'routing'. Depending on the information fed into the 'channel' input and generated by the sequencer, a number of novel signal acquisition principles are available.

4.2.1. Multi-Detector Operation

The general condition of TCSPC is that the detection of several photons per signal period must be negligible. This implies automatically that simultaneous detection of photons in different detector channels is negligible as well. Multi-detector operation makes use of this fact. Several detectors are detecting simultaneously. The detector pulses of all detectors are sent through a single time-measurement channel but stored in different memory blocks. Multi-detector operation was first introduced by Birch et al. (1988).

They used several detectors with individual CFDs. Due to the increasing number of CFDs, the number of detectors was very limited.

The modern implementation uses a single CFD for all detector channels. A 'router' combines the single photon pulses into one common timing pulse line, and generates a 'channel' signal that indicates at which of the detectors the current photon arrived (Becker et al., 2001). The principle is shown in Figure 4.3.

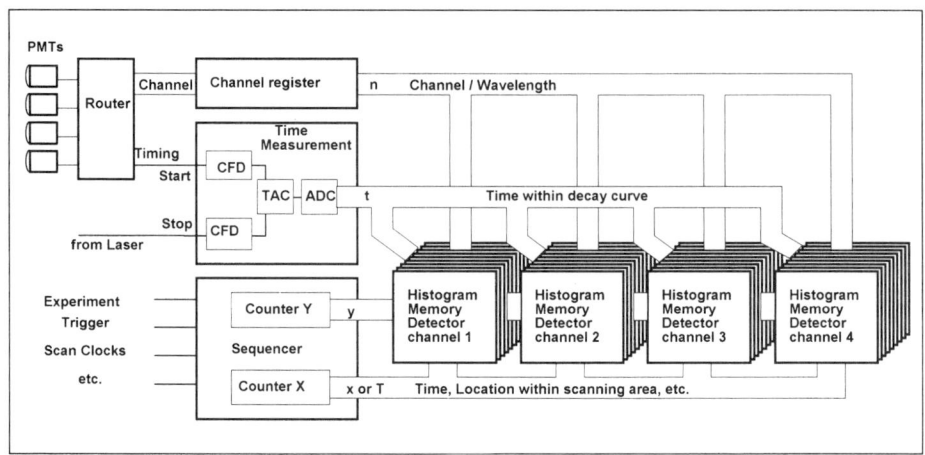

Figure 4.3: Multi-detector operation. All detectors are detecting simultaneously. The photons are routed into different TCSPC memory blocks by the 'channel' signal.

At the input of the detection system are a number of photomultipliers (PMTs), or a multi-anode PMT with 16 or more detection channels. The router combines the photon pulses and delivers the 'channel' signal. The timing pulse is sent through the normal time-measurement block of the TCSPC device.

The detector channel number from the router is used as a second dimension of the photon distribution. Together with additional parameters generated in the sequencer, a number of data blocks for the individual detectors is built up. These data blocks can contain a single waveform, a sequence of waveforms, or an array of waveforms scanned by an external scanner.

The multi-detector technique was first implemented in 1993 in the SPC-300 module of Becker & Hickl. Although an amazingly efficient technique, it was rarely used until DOT demanded for a large number of time-resolved detection channels (Ntziachristos et al., 1998). Until now, multi-detector operation is often equated to multiplexing. It should, therefore, be pointed out that multi-detector operation does not discard any photons, neither by gating in time, nor by detector multiplexing or wavelength scanning. Any photon detected by any of the detectors is routed into a memory location depending on its time in the laser pulse sequence, and its detector channel number.

Currently multi-detector operation is used for spectrally resolved fluorescence decay measurements, fluorescence anisotropy measurements, multi-spectral lifetime imaging in laser scanning microscopes (Becker et al., 2002, Becker et al. 2004), single molecule spectroscopy, and for diffuse optical tomography (Becker et al., 2001).

4.2.2. Multiplexed Detection

The 'channel' input bits of multi-dimensional TCSPC can be used for multiplexed recording of signals originating from several lasers of different wavelength, signals from different samples, or signals from different sample positions. The principle is shown in Figure 4.4.

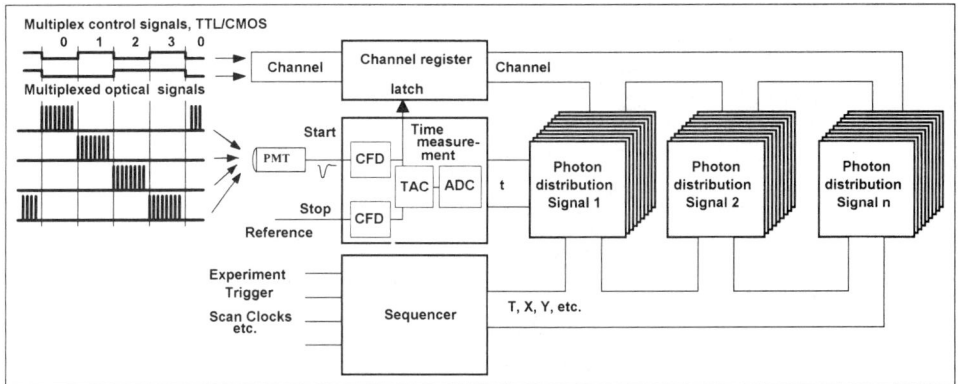

Figure 4.4: Multiplexed TCSPC operation. Several signals are actively multiplexed into the detector. The destination in the TCSPC memory is controlled by a multiplexing signal at the 'channel' input. For each multiplexing channel a separate photon distribution over the time in the signal period and the sequencer coordinates is built up.

Optical multiplexing is accomplished by electronically multiplexing several diode lasers, by fibre switches, or by rotating elements in an optical system (Becker et al., 1991). The signal at the 'channel' input of the TCSPC device indicates which laser was active or which optical signal was sent to the detector when a photon was detected. Consequently, the photons of the multiplexed optical signals are routed into separate photon distributions.

In most applications multiplexing has advantages compared to consecutive recording of the same signals. One advantage is that slow changes in the sample are less likely to cause artefacts in the results. Moreover, fast multiplexing can be combined with sequencing on a slower time scale, and with multi-detector operation. In this case some of the 'channel' bits are used for multiplexing, other for the detector channel information delivered by the router.

A potential application of multiplexed multi-detector TCSPC is DOT (Cubeddu et al., 1999; Grosenick et al. 2003; Schmidt et al., 2000). Lasers of different wavelength and a large number of source positions are multiplexed, and the diffusely transmitted or reflected light is recorded in a large number of detector channels.

4.2.3. Sequential Recording

The sequencer of the TCSPC module can add one or more additional dimensions to the recording process. One of these can be the time from the start of the experiment or from a trigger pulse. The principle is shown in Figure 4.5.

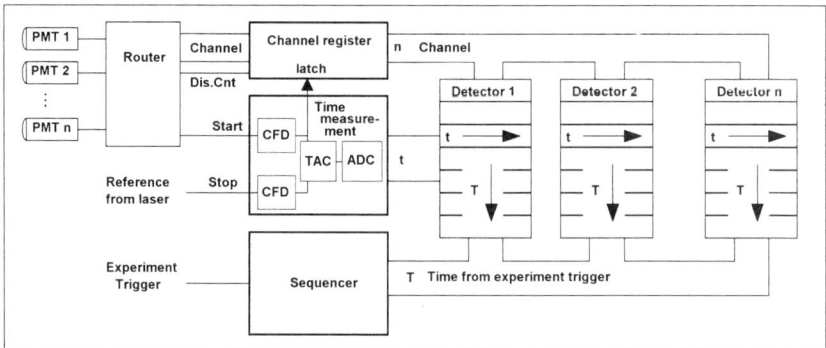

Figure 4.5: Sequential recording. Triggered by the experiment trigger, the sequencer switches through a large number of memory blocks. Each block contains the photon distribution over the time in the signal period, t, and the detector channels.

The sequencer counts through a range of subsequent address words. The result is a sequence of measurements. Triggered by an 'experiment trigger, a large number of extremely fast sequences can be accumulated. The individual measurements can be multi-dimensional themselves, due to the control capabilities of the 'channel' input. The sequencing technique is also termed 'double kinetic mode' or 'time lapse recording'. It can be combined with a memory-swapping architecture to record virtually unlimited sequences of waveforms (Becker et al., 1999). Sequential recording in conjunction with laser multiplexing and multi-detector operation is used for functional brain imaging by diffuse optical tomography techniques (Liebert et al, 2003).

4.2.4. Scanning

The sequencer can be used to synchronise the recording process with the action of an external scanner. The sequencer then delivers two additional dimensions, X and Y. Synchronisation with the scanner is obtained by receiving clock pulses from the scanner (see Figure 4.3). The result is a number data blocks for the individual detectors. Each data block can be interpreted as a sequence of images for different times in the laser period, or as an image with a full fluorescence decay curve in each pixel (Becker et al., 2004).
TCSPC scanning has become a standard lifetime imaging technique in confocal and two-photon laser scanning microscopes. These microscopes use optical beam scanning with pixel dwell times down to a few 100 ns (König, 2000, Pawley, 1995). Several individual detectors or the channels of a multi-anode PMT detect the fluorescence in different wavelength intervals. In typical applications the pixel rate is higher than the photon count rate. This makes the recording process more or less random. When a photon is detected,

the TCSPC device measures its time in the laser pulse period, t, determines the detector channel number, n, i.e. the wavelength of the photon, and the current beam position, X and Y, in the scanning area. These data are used to build up the photon distribution over t, n, X, and Y.

The data acquisition into the pixels of the image is controlled essentially by the scanner. Therefore, the scan rate, zoom, or region of interest selected in the microscope automatically acts on the TCSPC recording. Scanning can simply be started and repeated until a sufficient number of photons has been collected. Of course, the TCSPC scanning is not restricted to confocal and two-photon laser scanning microscopy or high-speed scanning. Due to the simple interface between the TCSPC device and the scanner it can be used for other scanning applications as well (Schweitzer et al. 2001).

4.2.5. Time-Tag Recording

A variant of the multi-dimensional TCSPC technique does not build up photon distributions but stores information about each individual photon. The mode is called 'time tag', 'time stamp', 'list', or 'FIFO' mode. For each photon, the time in the signal period, the 'channel' word, and the time from the start of the experiment, or 'macro time' is stored in a first-in-first-out (FIFO) buffer. During the measurement, the FIFO is continuously read, and the photon data are transferred in the main memory or on the hard disc of a computer. The structure in the time-tag mode is shown in Figure 4.6.

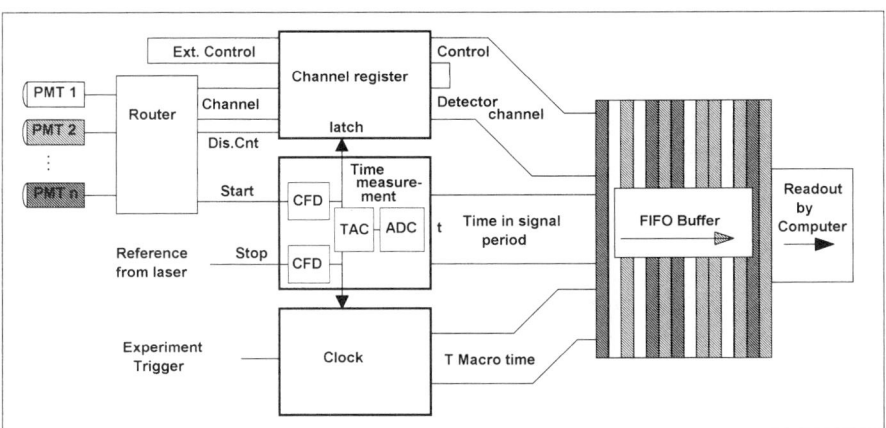

Figure 4.6: Architecture of a TCSPC module in the time-tag mode. For each photon, the time in the signal period, the time from the start of the experiment, and the detector channel number are recorded.

The TCSPC device contains a channel register, a time-measurement block, a 'macro time' clock, and a FIFO buffer for a large number of photons. The structure has some similarity with the multi-dimensional TCSPC architecture described in the paragraphs above. In fact, many advanced TCSPC modules have both the photon distribution and the time-tag mode implemented, and the configuration can be changed by a software command. The sequencer then turns into the macro-time clock, and the memory into the FIFO (first-in-first-out) buffer.

In principle, many multi-dimensional recording problems can be solved also in the time-tag mode. Synchronisation with the experiment can be achieved via the experiment trigger, the macro time clock, and additional experiment control bits fed into the 'channel' register. The drawback of the time-tag mode is the large amount of data that has to be transferred into the computer and processed or stored. A single FLIM or DOT recording may deliver 10^8 to 10^9 photons, resulting in several Gigabytes of data. At high count rates the bus transfer rate of the computer may be still sufficient to transfer the data, but the computer may be unable to process the data on-line or write them to the hard disc. The transfer rate problem is even worse if several TCSPC modules are operated in parallel.

Nevertheless, the time-tag mode is sometimes used for imaging (Gratton et al., 2003) and for standard fluorescence lifetime experiments. This is not objectionable as long as possible count rate limitations by the bus transfer rate, the enormous file sizes, and possible synchronisation problems are taken into regard. The true realm of the time-tag mode is single molecule spectroscopy and FCS. FCS records the autocorrelation and cross-correlation functions of the fluorescence intensity (Berland et al., 1995; Rigler et al., 1993; Schwille et al., 1999). By using the TCSPC technique, FCS can be combined with fluorescence lifetime detection. Fluorescence correlation spectra are calculated by correlating the macro times of the photons of one detector. Cross correlation spectra are calculated by correlating the macro times of the photons of different detectors. Fluorescence lifetime data are obtained from histograms of the micro times.

The time-tag mode in conjunction with multi-detector capability and MHz counting capability was introduced in 1996 (SPC-431, Becker & Hickl). Despite of its large potential in single molecule spectroscopy the mode did not attract much attention until sufficiently fast computers with large memories and hard discs became available.

4.3. APPLICATIONS OF MULTI-DIMENSIONAL TCSPC

4.3.1. Multi-Spectral Fluorescence Lifetime Detection

In biomedical applications of time-resolved fluorescence the excitation power has to be kept low to avoid photobleaching or photodamage, and to meet the requirements of laser safety standards for human patients. Recording efficiency is therefore an important issue. Moreover, biological systems often show dynamic effects. This requires the signals to be acquired in a short time. Scanning a time-resolved fluorescence spectrum by a monochromator is certainly not a good solution to these problems. It is by far better to use multi-detector TCSPC and record the fluorescence decay functions simultaneously over the entire emission spectrum.

Multi-wavelength operation of a TCSPC device can be obtained by splitting the fluorescence light into several spectral intervals by a system of dichroic mirrors and recording these signals by individual detectors via a router. A setup with dichroic beamsplitters yields an exceptionally high optical efficiency. However, the number of wavelength channels is very limited, and the wavelength intervals are fixed.

A more detailed fluorescence spectrum is obtained by using a polychromator and recording the spectrum by a multi-anode PMT. The principle of a multi-wavelength fluorescence lifetime experiment is shown in Figure 4.7.

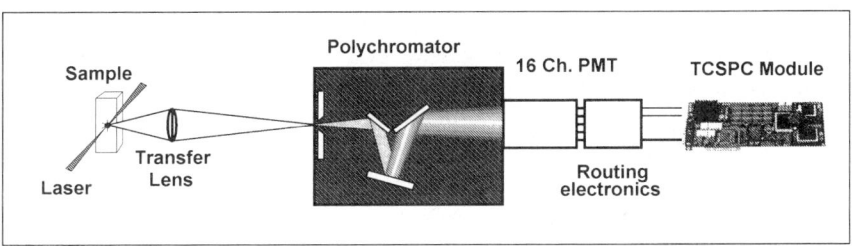

Figure 4.7: Multi-wavelength fluorescence lifetime experiment

The fluorescence is excited by a 405 nm ps diode laser (Becker & Hickl) at a repetition rate of 20 MHz. The detector is a 16 channel multi-anode PMT (R5900-L16, Hamamatsu) integrated in a PML-16 detector head (Becker & Hickl). The detector head contains the routing electronics, i.e. delivers the detector channel number and the timing pulse to the TCSPC module. The fluorescence signal is spread spectrally by a polychromator (MS 125-8M, LOT Oriel) over the cathode area of the R5900-L16.

As an example, Figure 4.8 shows decay curves of a mixture of rhodamine 6G and fluorescein, both at $5 \cdot 10^{-4}$ mol/l. The recorded spectral range was 520 to 600 nm. The different lifetimes are clearly visible. The high concentration causes re-absorption of the rhodamine 6G fluorescence, with a corresponding increase of lifetime and deviation from a single-exponential decay.

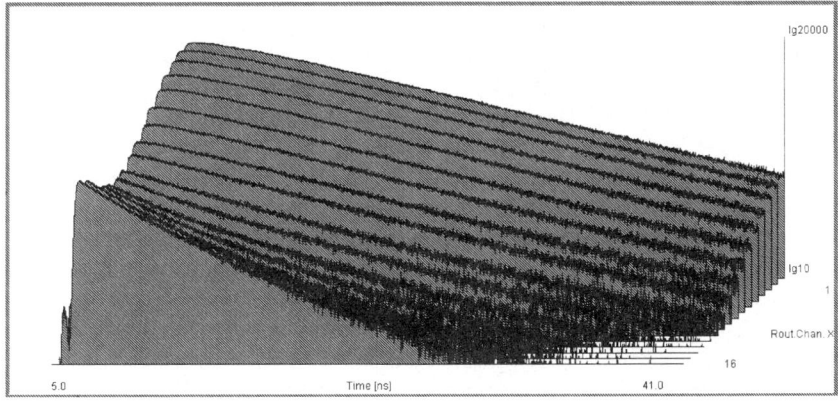

Figure 4.8: Fluorescence of a mixture of rhodamine 6G and fluorescein, simultaneously recorded over time and wavelength

The benefit of multi-spectral detection is obvious. Because the complete photon distribution is recorded simultaneously, the efficiency of the measurement is 16 times higher than for a wavelength scan of comparable resolution. The acquisition time and the sample exposure for a given excitation power are therefore correspondingly smaller.

A potential application is the detection of tissue autofluorescence. Endogenous fluorophores have broad, poorly defined and often variable spectra (Mycek and Pogue, 2003). Moreover, the fluorescence lifetime of some endogenous fluorophores depend on their binding state (Lakowicz 1999, Lakowicz et al., 1992, Paul and Schneckenburger, 1996). Multi-spectral lifetime detection does not only help to disentangle the fluorescence components, it yields also information about the state of the tissue. An application of multi-wavelength detection to time-resolved laser scanning microscopy is described in (Becker et al., 2002).

4.3.2. Recording dynamic changes of the fluorescence lifetime

A typical example of dynamic fluorescence lifetime changes are the chlorophyll transients found 1931 by Kautsky and Hirsch. When a dark-adapted leaf is exposed to light the intensity of the chlorophyll fluorescence starts to increase. After a steep rise the intensity falls again and finally reaches a steady-state level (Maxwell and Johnson, 2000). The rise time is of the order of a few milliseconds to a second, the fall time can be from several seconds to minutes. The initial rise of the fluorescence intensity is attributed to the progressive closing of reaction centres in the photosynthesis pathway. Therefore the quenching of the fluorescence by the photosynthesis decreases with the time of illumination, with a corresponding increase of the fluorescence intensity. The fluorescence quenching by the photosynthesis pathway is termed 'photochemical quenching'. The slow decrease of the fluorescence intensity at later times is termed 'non-photochemical quenching'. Non-photochemical quenching seems to be essential in protecting the plant from photodamage, or may be even a result of moderate photodamage.

Figure 4.9 shows a photochemical quenching sequence of fluorescence decay curves obtained from a dandelion leaf by TCSPC. A picosecond diode laser of 650nm wavelength, 50 MHz repetition rate, and 60 ps pulse width was used for excitation. The average power was 250 µW. The beam was focused to obtain a power density of about 1 mW/mm^2. The laser was periodically switched on for 13 ms at a frequency of 2 Hz. A Hamamatsu H5773P-1 photosensor module and a Becker & Hickl SPC-630 TCSPC module were used for fluorescence detection. Starting with each off-on transition of the laser, a sequence of 128 fluorescence decay curves was recorded with an acquisition time of 100 microseconds per curve. 10,000 off-on periods were accumulated to obtain enough photons in each of the curves.

Figure 4.9 clearly shows the increase of the fluorescence lifetime with the rate of exposure. The peak intensity is the same for all curves, which indicates that the intensity of photochemical quenching changes, not the concentration of the fluorescing chlorophyll molecules.

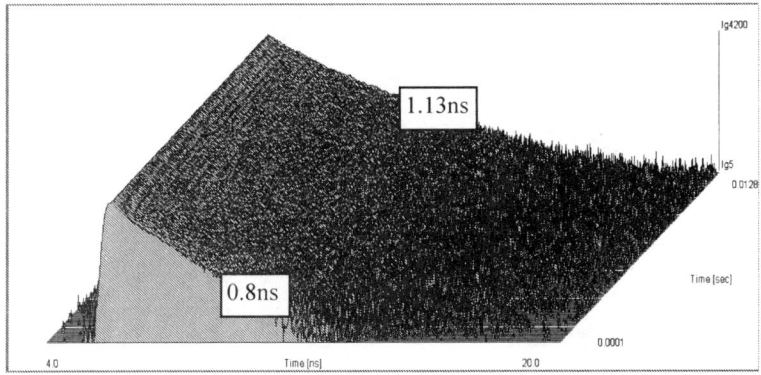

Figure 4.9: Photochemical quenching of the fluorescence of a dandelion leaf. Triggered sequence of fluorescence decay curves obtained by switching the excitation laser. The time per curve is 100 µs, 10,000 off-on periods were accumulated.

The non-photochemical quenching is shown in Figure 4.10. A single sequence was started when the laser was switched on. 50 decay curves with a collection time of 2 s each were recorded. For better display, the sequence starts at the back. A considerable decrease in lifetime is observed in a fresh leaf (left) but not in a dry leaf (right). Also here, the change is mainly in the quenching intensity, not in the concentration of the chlorophyll.

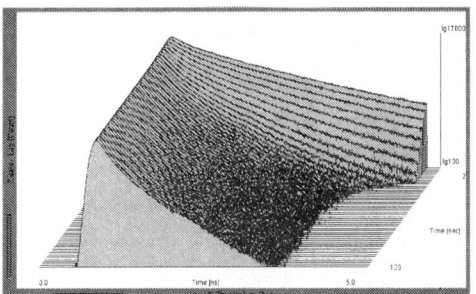 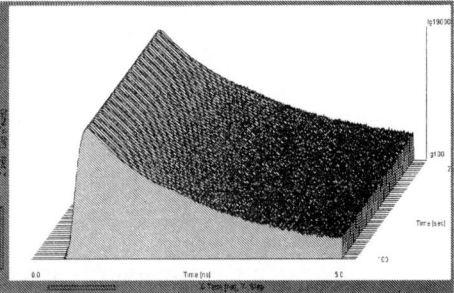

Figure 4.10: Non-photochemical quenching in a dandelion leaf. Sequence of fluorescence decay curves obtained from a fresh leaf (left) and a dry leaf (right) after switching on the excitation laser. A single sequence of 50 curves was recorded with 2 s per curve. Sequence starts at the back.

4.3.3. Time-resolved laser scanning microscopy

Confocal and two-photon laser scanning microscopes have initiated a breakthrough in fluorescence microscopy of biological samples (Denk et. al., 1990, Minski et. al, 1988, White et al., 1987). The applicability of multi-photon excitation, the optical sectioning capability, out-of-focus light suppression, and the multi-wavelength capability (Dickinson et. al., 2002) of these instruments make them an ideal choice for steady-state fluorescence imaging of cells and tissue (Herman, 1998; Pawley, 1995; Periasami, 2001).

Because of its x-y-z and multi-wavelength capability laser scanning microscopy is often termed 'multi-dimensional' microscopy. The optical principle of these instruments is, strongly simplified, shown in Figure 4.11.

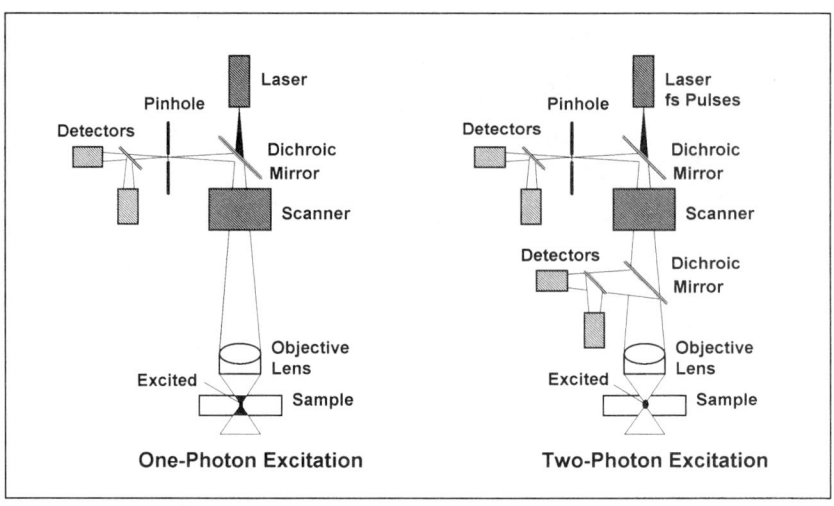

Figure 4.11: Optical principle of a one-photon (left) and two-photon (right) laser scanning microscope

The laser is fed into the optical path via a dichroic mirror and focused into the sample by the microscope objective lens. In the traditional confocal setup used for one-photon excitation (Figure 4.11, left) the light from the sample is fed back through the objective lens, through the scanner, is diverted by a dichroic mirror and finally goes through a pinhole in the upper image plane of the objective lens. Light from outside the focal plane is not focused into the pinhole and therefore substantially suppressed. Usually the fluorescence light is split into several spectral components and detected by several PMTs. X-Y imaging is achieved by optically scanning the laser spot over the sample, Z imaging (optical sectioning) by moving the sample up and down. Typical x-y scan rates are of the order of a few 100 ns to 10 us per pixel.

With a fs Ti:Sa laser the sample can be excited by two-photon absorption. Two-photon excitation occurs only in the focus, so that no pinhole is required to reject light from outside the focal plane. Therefore, the fluorescence light need not be fed back through the scanner and through the pinhole. Instead, it can be diverted by a dichroic mirror directly behind the microscope objective. This setup is termed 'non-descanned detection' in contrast to the 'descanned detection' used for one-photon excitation.

Two-photon excitation in conjunction with non-descanned detection can be used to image tissue layers several 100 μm deep (König, 2000; Masters et al., 1997; Sherman et al., 2001; Wallace et al, 2001). Since the scattering and the absorption at the wavelength of the two-photon excitation are small the laser beam penetrates through relatively thick tissue. The loss on the way to the focus it can easily be compensated by increasing the laser power. The increased power does not cause much photodamage because the power density outside the focus is small. Of course, the fluorescence photons are heavily scattered on their way out of the tissue and therefore emerge from a relatively large area of the sample surface which is out of the focus of the objective. However, for non-descanned detection there is no need to focus the fluorescence light anywhere. Therefore the fluorescence photons can be efficiently transferred to the detectors.

FLIM techniques used in scanning microscopes include modulated image intensifiers (Lakowicz and Berndt, 1991; So et al, 1994; Squire et al., 2000; Vervier at al., 2001), modulated single-point detectors (Carlsson and Liljeborg, 1997; Carlsson and Liljeborg, 1998; So et al., 1995), gated image intensifiers (Cole et al., 2001; Dowling et al., 1997; Paul and Schneckenburger, 1996, Straub and Hell, 1998), photon counting in several subsequent time gates (Buurman et al., 1992, Gerritsen, 2002; Syrtsma et al., 1998), and multi-dimensional TCSPC.

The introduction of multi-dimensional TCSPC into laser scanning microscopy opened the way to high accuracy and high resolution recording of multi-exponential decay functions. TCSPC works with near-ideal efficiency at low and moderate intensities. Such intensities are typical for autofluorescence, FRET experiments, and other situations where the fluorophore concentration is low or the fluorophore is localised in highly specific subunits of the cells. Multi-dimensional TCSPC is compatible with the fast scan rate of the microscopes, and does not restrict the depth resolution capability of these instruments. It records in several detection channels simultaneously and delivers a near-ideal counting efficiency. Photobleaching or photodamage (Hopf and Neher, 2002; König et al., 1996; Patterson and Piston, 2000) is therefore reduced to a minimum. The time resolution is sufficient to resolve multi-exponential fluorescence decay curves down to 30 ps.

TCSPC FLIM options are available for the Zeiss LSM-510, the Biorad Radiance 2000 and the Leica SP 2. The technique has also been implemented in a large number of other microscopes, including the Biorad MRC 600 and 1024, the Olympus FV 300 and the Nikon PCM 2000.

Applications include tissue autofluorescence (König and Riemann, 2003), histological imaging (Eliceri et al., 2003), FRET experiments (Ammeer-Beg et al., 2003; Bacskai et al., 2003; Becker et al., 2004; Bereszovska et al, 2003; Chen et al 2004), and the internalisation (Kelbauskas and Dietel, 2002) and conversion (Rück et al., 2003) of photodynamic therapy agents. TCSPC FLIM has also been used to confirm laser induced cell transfection with GFP (Tirlapur and König, 2002). Another potential application is the use of the environment-dependence of the lifetime of many fluorophores as a probe function.

Lifetime images obtained by a Leica TCS SP2 D-FLIM confocal microscope are shown in Figure 4.12. This microscope uses a 405 nm picosecond diode laser for excitation, descanned confocal detection, and multi-dimensional TCSPC.

Figure 4.12, left, shows a fluorescence lifetime image of a plant tissue sample. The fluorescence decay function in a selected pixel of the recorded data array, and a double exponential Levenberg-Marquardt fit are shown in Figure 4.13. The decay is clearly double-exponential and cannot be reasonably approximated by a single exponential decay. Therefore, the lifetime image shows an average of both lifetime components weighted with the intensity coefficients. This 'mean lifetime' is displayed as colour in the lifetime image.

Merging the components of the double exponential decay into a mean lifetime does, of course, discard useful information. An example of using multi-exponential decay data is shown in Figure 4.12, right. It shows an image that displays the colour-coded ratio of the intensity coefficients of both lifetime components. The coefficient-ratio image has some similarity with the image of the mean lifetime. However, the ratio of the intensity coefficients directly represents the concentration ratio of the molecules emitting the fast and slow decay component and can therefore be used to separate different fluorescent species of more or less similar spectra.

The probably most sophisticated application of TCSPC-FLIM is fluorescence resonance energy transfer (FRET). FRET occurs if the fluorescence band of one molecule - the donor - overlaps the absorption band on another - the acceptor. If the distance is of the order of a few nanometers the energy is transferred directly from the donor and the acceptor. The result is a considerable decrease of the donor fluorescence intensity and lifetime (Lakowicz 1999). The relative decrease is proportional to the 6th power of the distance and is used to determine distances at the molecular scale. Steady state FRET measurements have to calibrate for 'donor bleadthrough' and 'acceptor bleadthrough' (Wallrabe et al., 2003). This requires up to six measurements, including results from cells containing only the donor or the acceptor. Another way to obtain FRET results is to record a donor image, then photobleach the acceptor, and record a second donor image. The FRET intensity is obtained from the intensity ratio of the images.

By FLIM techniques the FRET efficiency can, in principle, be obtained from a single donor lifetime image. This is a considerable advantage compared to steady-state techniques. A general problem of FRET experiments in cells is that by far not all donor molecules are actually linked to an acceptor molecule. Moreover, the dipoles of the donor and acceptor molecules are more or less randomly oriented so that not all donor and acceptor pairs are interacting. Therefore, obtaining a FRET efficiency and calculating distances from the decrease of the donor intensity or from a single-exponential approximation of the donor lifetime is questionable.

Figure 4.14, left, shows a TCSPC donor lifetime image of a human embryonic kidney (HEK) cell expressing two interacting proteins labelled with CFP and YFP. The image was recorded in a Zeiss LSM 510 NLO two-photon microscope upgraded with a Becker & Hickl SPC-830 TCSPC module. FRET is expected in the regions where the proteins are linked. **Figure 4.14**, right, shows the fluorescecne decay in a selected pixel. The fluorescence decay is indeed double exponential, with a fast lifetime component t_1 of about 660 ps and a slow component t_2 of about 2.3 ns. The lifetimes do not change appreciably throughout the image. There is, however, a considerable change in the intensity coefficients. The ratio a_1 / a_2 varies from about 0.2 in the bulk of the cell to more than 1 in the region of strongest FRET. A similar behaviour is found in almost all TCSPC FLIM-FRET results (Bacskai et al., 2003, Becker et al., 2001, Becker et al., 2004) and is confirmed by streak-camera measurements (Biskup et al., 2004). It appears reasonable to relate the fast and the slow lifetime component to the interacting and non-interacting donor molecules, respectively.

Because TCSPC FLIM separates the donor lifetime components it also separates the effect of the distance and the variable ratio of interacting and non-interacting donor molecules. This is a considerable advantage over steady state or single exponential FLIM-FRET techniques.

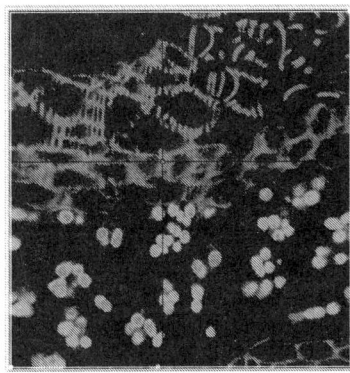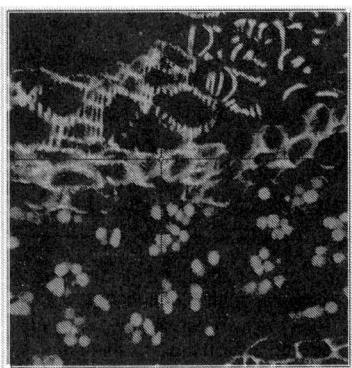

Figure 4.12 (Please see color inserts section): TCSPC lifetime image recorded in a confocal laser scanning microscope, 512 x 512 pixel scan. Left: Colour represents the mean lifetime of the double exponential decay, blue to red = 200 ps to 2 ns. Right: Colour represents the ratio of the intensity coefficients of the fast and slow decay component, a_{fast} / a_{slow}. Blue to red = 1 to 10.

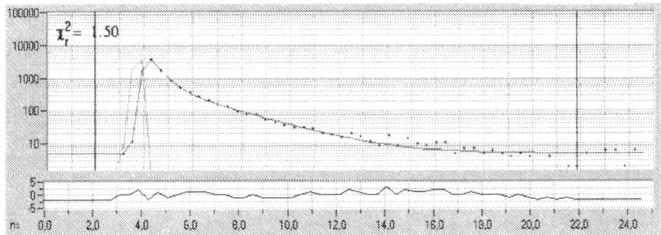

Figure 4.13 (Please see color inserts section): Fluorescence decay function in a selected pixel of the image, double exponential Levenberg-Marquardt fit and residuals of the fit. Data points are shown blue, the fitted curve red, and the instrument response function green. The pixel is marked in Figure 4.12.

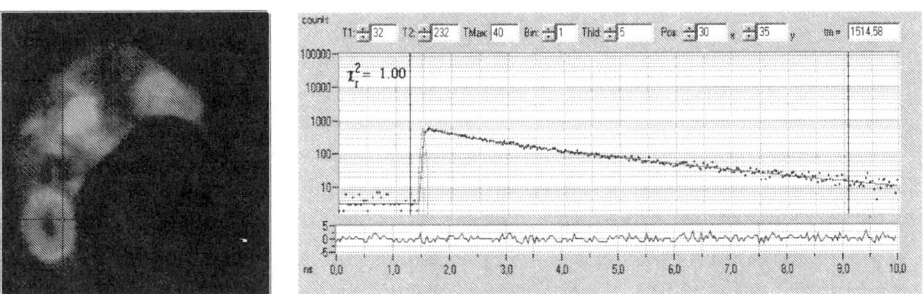

Figure 4.14(Please see color inserts section), left: Lifetime image of an HEK cell expressing two interacting proteins labelled with CFP and YFP (left). Colour represents weighted mean lifetime, red to blue = 1500 to 2300 ps. Right: Fluorescence decay function in a selected spot. The lifetime components are 660 ps and 2.3 ns.

4.3.4. Single-Molecule Spectroscopy

Single-molecule spectroscopy is based on confocal detection or two-photon excitation in a femtoliter sample volume. Single-molecule-experiments can be based on correlation techniques or multi-parameter spectroscopy. Correlation can be measured between subsequent photons on the picosecond and nanosecond scale. In this case the investigated effects are driven by the absorption of a single photon of the excitation light. On a longer time scale, FCS measures the correlation of fluctuations in the fluorescence light intensity. The effects investigated by FCS are driven by Brownian motion, rotation, diffusion effects, intersystem crossing, or conformational changes. Because of these random, and essentially sample-internal stimulation mechanisms correlation techniques do not depend on a pulsed laser. The second way to obtain information on single molecules is direct time-resolved multi-parameter spectroscopy with pulsed excitation at high repetition rate. The molecules are embedded in a polymer matrix, and the fluorescence is detected by several detectors in different wavelength intervals and under different angles of polarisation.

Anti-Bunching

The classic photon correlation - or 'anti-bunching' - experiment records a histogram of the time intervals between the photons of the investigated signal. The basis of all photon correlation experiments is the Hanbury-Brown-Twiss setup (Hanbury-Brown and Twiss, 1956). The principle is shown in Figure 4.15.

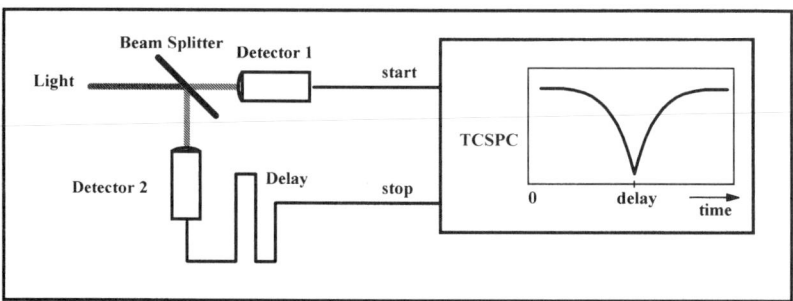

Figure 4.15: Dual detector (Hanbury-Brown Twiss) photon correlation setup

The investigated light signal is split optically by a 1:1 beam splitter, and the two light signals are fed into separate detectors. One detector delivers the start, the other the stop pulses of a TCSPC device. The stop pulses are delayed by a few ns to place the coincidence point in the centre of the recorded time interval. The setup delivers a histogram of the time differences between the photons at both detectors.

The Hanbury-Brown-Twiss setup can be used with continuous excitation or with excitation by picosecond pulses. Continuous excitation of a small number of emitters delivers the typical antibunching curve (Basche et al., 1992). An example is shown in

Figure 4.16. It was recorded with two Hamamatsu H7422-40 PMT modules connected to one channel of an SPC-134 TCSPC package (Becker & Hickl, Berlin).

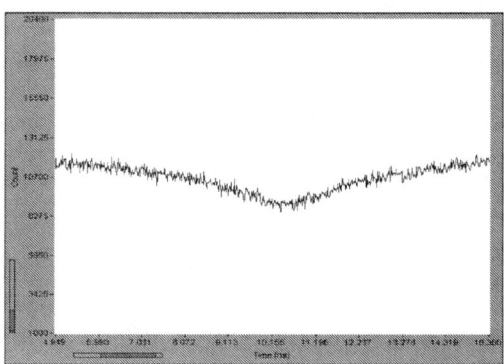

Figure 4.16: Antibunching curve for Rhodamine 110, CW excitation

The antibunching curve delivers the average number of emitters in the excited volume, and the excited state lifetime. Antibunching can also be used to confirm controlled emission of single photons from quantum dots and other semiconductor nanostructures (Michler et al., 2000; Thompson et al., 2001; Yuan et al., 2002; Zwiller et al., 2001).

Antibunching experiments excited by high repetition rate picosecond lasers deliver a number of correlation peaks which are spaced by the laser pulse period, see Figure 4.17. If the laser pulse width is much shorter than the fluorescence lifetime there is almost no chance that a single molecule is excited several times within one laser pulse. Consequently, the emission of several photons from a single molecule within one laser period becomes extremely unlikely. Therefore, the ratio of the height of the central coincidence peak and the adjacent peaks is a indicator of the number of the molecules in the excited volume (Weston et al., 2002).

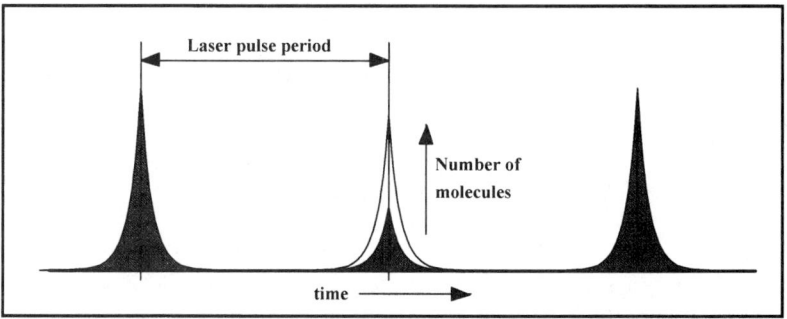

Figure 4.17: Anti-bunching with pulsed excitation. The result is a train of correlation peaks spaced by the laser pulse interval. The size of the central correlation peak depends on the number of molecules in the focus.

Fluorescence Correlation Spectroscopy

Fluorescence correlation spectroscopy, or FCS, is based on the excitation of a small number of molecules in a femtoliter volume and the recording of the intensity fluctuations caused by diffusion, Brownian motion, intersystem crossing, conformational changes, or other random effects. The small sample volume is obtained by confocal detection or by two-photon excitation. Conventional FCS records the intensity fluctuations of the fluorescence signals and correlates the intensity fluctuations. If several detectors are used in different wavelength intervals the autocorrelation of the signals of the individual detectors, or the cross-correlation between the signals of different detectors can be obtained.

FCS experiments can be performed by TCSPC in the time-tag mode. For each detected photon, the TCSPC module records the time from the start of the experiment, the time in the laser pulse sequence, and the number of the detector at which the photon arrived. From these data fluorescence decay functions, fluorescence correlation spectra, and - if several detectors are used - fluorescence-cross-correlation spectra can be calculated. An example for a 10^{-9} molar solution of GFP is shown in Figure 4.18. The solution was excited by two-photon excitation. The signal was recorded by a Perkin Elmer SPCM-AQR module connected to a Becker & Hickl SPC-830 TCSPC module.

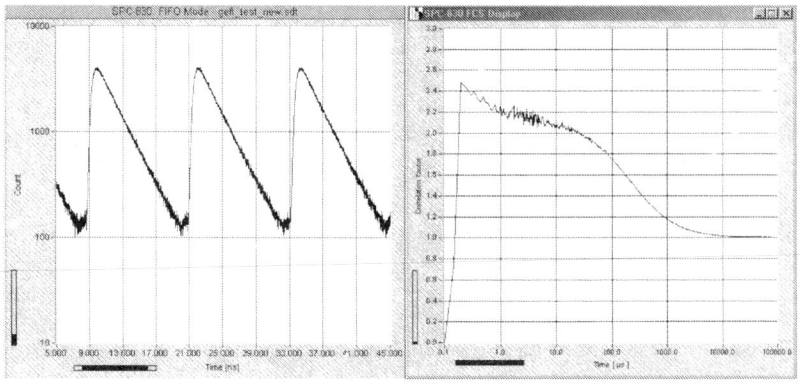

Figure 4.18: Fluorescence decay curve (left) and FCS curve (right) of a GFP solution. Count rate 5 to 7 kHz, acquisition time 980 s. The increase below 1 μs correlation time is due to afterpulsing in the detector. Courtesy of Petra Schwille, Dresden, Germany.

Single-Molecule Multi-Parameter Spectroscopy

Multi-detector TCSPC can be used to obtain several spectroscopic parameters simultaneously from a single molecule (Eggerling et al., 2001, Kühnemuth and Seidel, 2001; Prummer et al., 2004; Ying and Xie, 2002). Prummer et al. (2004) give a detailed description of the optical setup. A high high-repetition-rate frequency-doubled Nd-YAG laser is coupled into the beam path of a microscope and focused into the sample. The sample is mounted on a piezo-driven sample stage. The fluorescence light from the sample is collected by the microscope lens, and separated from the excitation light by a dichroic mirror and a notch filter. The light is split into a 0° and a 90° component. The 0°

component is further split into a short-wavelength and a long-wavelength part. The signals are detected by single-photon APD modules and recorded by a single TCSPC module via a router. The TCSPC module records the photons in the FIFO mode.

The molecules to be investigated are embedded in polymethylmathacrylate (PMMA). An image of the sample is obtained by scanning and assigning the photons to the individual pixels by their macro-time. Based on this image, appropriate molecules are selected for further investigation.

From the macro times of the recorded photons traces of the emission intensity of a single molecule can be build up. The traces show bright periods when the molecule cycles between the ground state and the excited singlet state, and dark periods, when the molecule is in the triplet state.

A large number of spectroscopic parameters can be derived from the photon data of the individual detectors. The intersystem crossing yield can be obtained from a histogram of the number of photons in the bright periods. The histogram of the dark periods reflects the triplet decay and delivers the triplet lifetime. The fluorescence lifetime is obtained by building up the fluorescence decay functions from the micro times during the bright periods.

The polarisation of the fluorescence is obtained by comparing the counts of the APDs detecting under $0°$ and $90°$ polarisation. The anisotropy decay can be calculated from the micro times of the photons in these APD channels. Even spectral relaxation of Alexa 546 molecule conjugated to a protein could be observed by APDs detecting in different spectral ranges (Prummer et al., 2004).

Spectroscopic parameters of single molecules can be used to identify single fluorophore molecules or characterise their binding state to proteins, DNA, or lipids. Identification can be based on the fluorescence lifetime, the fluorescence anisotropy, the burst size and duration, on spectral information, or several of these parameters (Becker et al., 1999; Kästner et al., 2003; Müller et al., 1996, Prummer et al., 2000; Sauer et al., 1997, Schaffer et al., 1999; Zander et al., 1996; Zander et al., 1998).

4.3.5. Diffuse Optical Tomography

Diffuse optical tomography (DOT) is based on the illumination of thick tissue by NIR light and the detection of the diffusely transmitted or reflected light, or the fluorescence of endogenous or exogenous fluorophores (Patterson and Chance, 1989). Typical applications of DOT techniques are optical mammography, brain imaging, and non invasive investigations of drug effects in small animals.

DOT uses the absorption window of biological tissue from approximately 650 to 1300 nm. In this wavelength interval light can be transmitted through tissue layers as thick as 10 cm. However, for tissue thicker than a few millimeters there are practically no ballistic (unscattered) photons (Dunsby and French, 2003), and the photons must be considered to diffuse through the tissue. Consequently, the resolution of DOT images is extremely poor and cannot compete with positron emission, X ray and MRI techniques. Nevertheless, DOT in the NIR has the benefit that the measured absorption coefficients are related to the biochemical constitution of the tissue (Mc Bridge et al., 1999, such as haemoglobin concentration and blood oxygenation. If exogenous markers are used the absorption or fluorescence delivers additional information about blood flow, or blood leakage, ion concentrations, or the binding state to proteins (Desmettre et al., 2000, Mordon et al., 1998, Sevick-Muraca et al., 2003).

The effects of scattering and absorption are hard to distinguish in simple steady state images. However, if pulsed or modulated light is used to transilluminate the tissue, the diffusion of the light through the tissue can be directly observed. Both increased scattering and increased absorption decrease the output intensity. However, stronger scattering increases the pulse width while stronger absorption tends to decrease it. Therefore, the shape of the 'time-of-flight distribution' of the photons can be used to obtain scattering and absorption coefficients of the tissue.

Time-resolved detection also improves the depth resolution in diffuse reflection measurements. Photons arriving at the detector later are more likely to have travelled deeply through the tissue. On average, late photons contain information from deeper tissue layers than early photons.

Time-resolved detection is almost mandatory if DOT is combined with fluorescence detection (Farrell and Patterson, 2003; O'Leary et al., 1996). DOT fluorescence techniques are commonly used in small-animal imaging (Gallant et al, 2004). Due to the moderate tissue thickness in a mouse or rat the fluorescence lifetimes can be separated from the photon migration effects with reasonable data analysis effort.

The complete reconstruction of tissue structures and optical properties from time resolved data is extremely demanding, and there is no general solution yet (Arridge, 1999; Chernomordik et al., 2002, Farrel and Patterson, 2003; Gao et al, 2000, Laques, 2003; Ntziachristos et al., 2001; Torricelli et al, 2003, Wabnitz and Rinneberg, 1997). It turns out that a large number of time-resolved detection channels, different source-detector distances or different transillumination angles are helpful to obtain useful optical tomography images. Moreover, several lasers of different wavelength and a large number of source positions are multiplexed.

Optical tomography techniques for human medicine are currently at the stage of clinical tests. Frequency domain instruments using modulation techniques and time domain instruments using TCSPC and are competing. TCSPC is superior in terms of efficiency and sensitivity. The effective detection bandwidth is much higher than for modulation systems. Moreover, the IRF is extremely stable, and the waveform is correctly sampled according to the Nyquist theory. The count rate of TCSPC is limited to a few MHz per TCSPC channel, which is often considered a drawback. It should, however, be taken into account that TCSPC obtains a better signal-to-noise ratio (SNR) from a given number of detected photons than any other technique. Therefore, the limited count rate is less important than commonly believed. The excellent results obtained with TCSPC-based DOT instruments under clinical conditions (Grosenick et al, 2003; Grosenick et al, 2004, Taroni et al, 2004) demonstrate the applicability of TCSPC.

Many DOT instruments are based on multiplexing several lasers and recording in a single channel TCSPC channel of high count rate (Cubeddu et al., 1998, Grosenick et al., 1999; Grosenick et al., 2003; Torricelli et al., 2001). Multi-detector operation of up to eight detectors connected to a single TCSPC channel was used by Cubeddu et al. (1999 and 2001), Ntziachristos et al. (1998, 1999 and 2000), and Torricelli et al. (2004). A system with 32 fully parallel TCSPC channels based on NIM modules was described by Schmidt et al. (2000) and used for breast and brain imaging (Hebden et al., 2001; Hebden et al., 2004). Recent TCSPC based instruments often use packages of four parallel multi-detector TCSPC devices operated in a single PC (Becker et al., 2001; Liebert, 2003; Pifferi et al., 2003, Taroni et al., 2004; Torricelli et al, 2003; Wabnitz et al., 2002).

Some typical DOT time-of-flight distributions are shown in Figure 4.19.

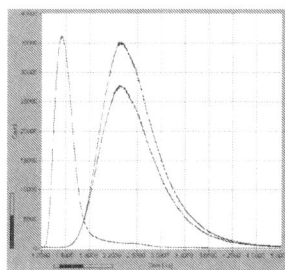 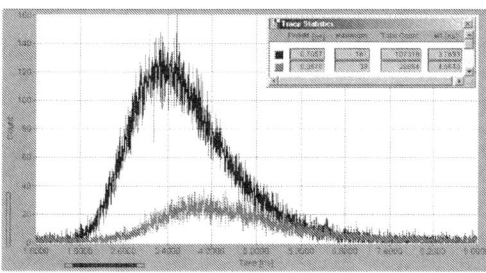

Figure 4.19: Time-of flight distributions measured at the adult human forehead. Laser 2.5 mW, 785 nm, 50 MHz, detector H5773-20. Left: Instrument response function and time-of flight distributions for 6 cm source-detector distance. Count rate about 1 MHz, acquisition time 20s. Right: Two time-of flight curves detected at 12 cm source-detector distance from temple to temple through an adult head. Count rates 896 s^{-1} and 2688 s^{-1}.

The curves were measured at different source and detector positions at the adult human forehead. The curves were obtained with a laser of 2.5 mW at a wavelength of 785 nm. The repetition rate was 50 MHz. The detector was a Hamamatsu H5773-20 photosensor module cooled down to 10° C. One channel of a Becker & Hickl SPC-134 TCSPC system was used to record the data. Figure 4.19, left, shows the two curves obtained at a source-detector distance of 6 cm at different positions of the forehead together with the instrument response function. A count rate of about 1 MHz was obtained so that about $2 \cdot 10^7$ photons were recorded in the acquisition time of 20 s.

At larger source-detector distance the count rate drops dramatically. The curves in Figure 4.19, right, were measured at two different positions from temple to temple, at a source-detector distance of 12 cm. The count rates were 896 s^{-1} and 2688 s^{-1}. A total number of 29,000 and 107,000 photons was acquired within an acquisition time of 60 s.

Figure 4.19 shows that the available count rates and consequently the data quality and acquisition time may vary considerably for different locations and source-detector distances. For short distances the available rates may easily exceed the counting capability of TCSPC. On the other hand, for large distance the rates may be so low that only TCSPC is able to record reasonable data.

For distances shorter than 8 cm reasonable time-of-flight curves are obtained within an acquisition time of 100 ms or less. Figure 4.20 shows 20 time-of-flight curves selected from a longer sequence recorded by the sequential mode of an SPC-134 channel. The acquisition time is 100 ms per curve, the ADC resolution 1024 channels. The light source was a diode laser of 2.5 mW average power, 785 nm wavelength, and 50 MHz repetition rate. The left curve was detected at a source-detector distance of 8 cm, the right curve at a distance of 5 cm. The count rates where $1.8 \cdot 10^5$ s^{-1} and $4.5 \cdot 10^6$ s^{-1}, respectively. The resolution of the sequences is sufficient to resolve dynamic effects in the brain. Typical time-of-flight changes induced by brain stimulation are of the order of 1 ps. Changes of this order can indeed be derived from sequential TCSPC data. Details are described by Liebert at al. (2003 and 2004).

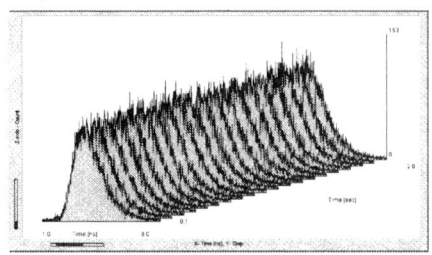 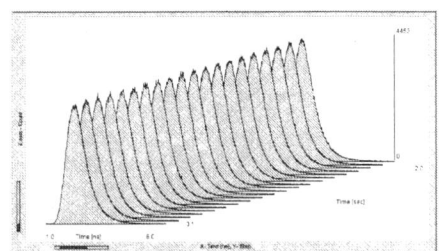

Figure 4.20: 20 steps of a TOF sequence recorded by TCSPC memory swapping. Acquisition time 100 ms per curve, ADC resolution 1024 channels. Diode laser 785 nm, 2.5 mW, detector H5773-20. Left: source-detector distance 8 cm, count rate $1.8\ 10^5\ s^{-1}$. Right 5 cm, count rate $4.5\ 10^6\ s^{-1}$.

4.4. LIMITATIONS OF THE TCSPC TECHNIQUE

Classic TCSPC devices had the reputation of being extremely limited in the applicable count rate and therefore unable to record any dynamic fluorescence lifetime effects. Indeed, the count rate of TCSPC is limited by pile-up (O'Connor and Phillips, 1984), counting loss (Becker et al., 2004), and detector effects.

The term 'pile-up' is used for the distortion of the recorded waveform by detecting a second photon within one signal period. Pile-up is almost undetectable for count rates smaller than 1% of the laser repetition rate. It remains tolerable of to about 5% and correctable up to 10%. Classic TCSPC often used ns flashlamps as excitation sources. Maximum pulse repetition rates were about 100 kHz, which limited the count rate to a few kHz. Modern excitation sources run at 50 or 80 MHz, which pushes the pile-up limit to a count rate of 2.5 to 8 MHz.

Counting loss is caused by the signal processing time, or dead time, of a TCSPC device. At higher count rates an increasing fraction of the detected photons is lost in the dead time. For reversed start-stop operation at a signal period much shorter the efficiency, i.e. the ratio of the recorded and the detected rate is:

$$E = r / r_d = 1 / (1 + r_d \cdot t_d)$$

E = efficiency, r = recorded rate, r_d = detected rate, t_d = dead time

Currently the fastest TCSPC modules have a dead time of 100 ns. That means, a detector count rate of 10 MHz is recorded with 50% efficiency. This efficiency is still better than for most other time-resolved signal recording techniques.

The third limitation of the count rate is set by detector effects. The specified maximum output current of the detector, timing shifts, and reversible and non reversible gain changes set a limit to the count rate. For MCP PMTs a reasonable limit is about 1 MHz. Conventional PMTs can be operated at count rates of several 10 MHz. However, the distortion of the dynode voltage distribution by the dynode currents causes a noticeable shift of the TCSPC instrument response function. Excellent results at high count rates are obtained with the Hamamatsu photosensor modules, such as the H5773, H5783, or H7422. These modules use a Cockroft-Walton voltage divider that keeps the dynode voltages stable. Figure 4.21 shows the TCSPC IRF for a H5773-20 operated with a 20dB preamplifier at a Becker & Hickl SPC-144 TCSPC device. A number of IRFs were measured at 30 kHz, 300 kHz, and 4 MHz recorded count rate.

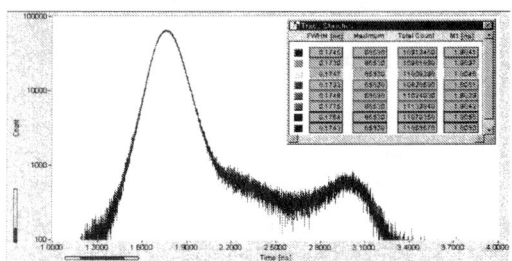

Figure 4.21: IRF of an H5773-20 photosensor module for count rates of 30 kHz, 300 kHz, and 4 MHz in logarithmic scale (left). The shift between 30 kHz and 2 MHz is <2 ps and not discernible in the IRF curves.

The shift in the first moment, M1, of the IRF remains smaller than 2 ps up to 4 MHz and is not discernible in the IRF recordings. The FWHM (full width at half maximum) is about 174 ps.

Figure 4.22 compares the typical count rates of TCSPC applications with the counting loss curves of a single TCSPC channel of 100 ns dead time. The count rate ranges are given under reasonable assumptions. For microscopy and single-molecule spectroscopy it is assumed that photobleaching and saturation are avoided. For all fluorescence applications it is important that the fluorophore concentration is at a level where secondary effects like concentration dependent quenching or re-absorption can be neglected. For DOT a source-detector distance is assumed that gives a reasonable penetration depth into the human head or a reasonable breast compression in scanning mammography.

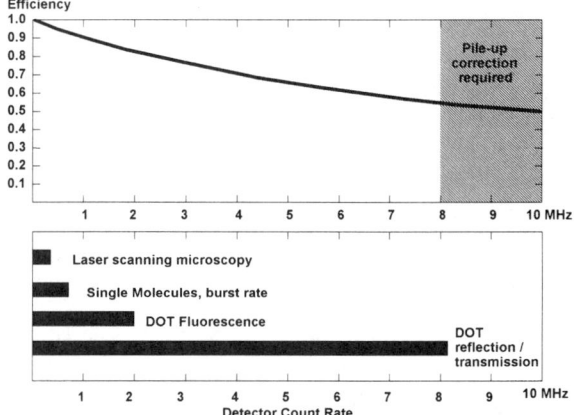

Figure 4.22: Efficiency of a single TCSPC channel of 100 ns dead time versus detector count rate, and typical count rates of the most common applications

Figure 4.22 shows that the rates available in DOT can be up to the limit of the counting capability of a single TCSPC channel. However, normally only a few detectors close to the source position are operating at the highest count rate. The acquisition time has to be selected to obtain sufficient photons for all channels, including the ones at large source-detector distance. It is therefore not a severe drawback if the photons in the high-intensity channels cannot be fully exploited. A correction for the counting loss, should, however, be made in any case.

Further reduction of the dead time may improve the counting efficiency at high count rates, but will not reduce the pile-up error. A breakthrough in the counting capability can only be achieved by operating several TCSPC channels with separate detectors in parallel (Becker et al., 2001). Such devices are now commonly used in DOT and found first applications in laser scanning microscopy (Becker et al, 2004). Typically four TCSPC channels are operated in parallel, resulting in a push in the total count rate of a factor of four compared to Figure 4.22.

4.5. CONCLUSIONS

Time-correlated single photon counting has developed from a slow and intrinsically one-dimensional fluorescence lifetime technique into a fast multi-dimensional optical recording technique. The photon distribution can be recorded simultaneously over the time within the excitation pulse sequence, the wavelength, the time from the start of an experiment, and over the coordinates of a scanning area. The multi-dimensional recording process does not use any time-gating or wavelength scanning and therefore works with near-ideal efficiency. Moreover, the TCSPC technique is able to record spectral and temporal information about each individual photon. These features make advanced TCSPC technique suitable for a large number of biomedical spectroscopy applications, such as autofluorescence detection, diffuse optical tomography, fluorescence lifetime microscopy, and single-molecule spectroscopy.

4.6. REFERENCES

1.   Ameer-Beg, S.M., Edme, N., Peter, M., Barber, P.R., Ng, T. and Vojnovic, B., 2003, Imaging protein-protein interactions by multiphoton FLIM, *Proc. SPIE* **5139**: 180.
2.   Arridge, S.R., 1999, Optical tomography in medical imaging, *Inverse Problems* **15**: R41.
3.   Bacskai, B.J., Skoch, J., Hickey, G.A., Allen, R. and Hyman, B.T., 2003, Fluorescence resonance energy transfer determinations using multiphoton fluorescence lifetime imaging microscopy to characterize amyloid-beta plaques, *J. Biomed. Opt* **8**: 368.
4.   Ballew, R.M. and Demas, J.N., 1989, An error analysis of the rapid lifetime determination method for the evaluation of single exponential decays, *Anal. Chem.* **61**: 30.
5.   Basche, T., Moerner, W.E., Orrit, M. and Talon, H., 1992, Photon antibunching in the fluorescence of a single dye molecule trapped in a solid, *Phys. Rev. Lett.* **69**: 516.
6.   Becker, W., Stiel, H. and Klose, E., 1991, Flexible instrument for time-correlated single photon counting, *Rev. Sci. Instrum.* **62**: 2991.
7.   Becker, W., Hickl, H., Zander, C., Drexhage, K.H., Sauer, M., Siebert, S. and Wolfrum, J., 1999, Time-resolved detection and identification of single analyte molecules in microcapillaries by time-correlated single photon counting, *Rev. Sci. Instrum.* **70**: 1835.
8.   Becker, W., Bergmann, A., Wabnitz, H., Grosenick, D. and Liebert, A., 2001, High count rate multichannel TCSPC for optical tomography, *Proc. SPIE* **4431**: 249.
9.   Becker, W., Benndorf, K., Bergmann, A., Biskup, C., König, K., Tirlapur, U. and Zimmer, T., 2001, FRET measurements by TCSPC laser scanning microscopy, *Proc. SPIE* **4431**: 94.

10. Becker, W., Bergmann, A., Biskup, C., Zimmer, T., Klöcker, N. and Benndorf, K., 2002, Multi-wavelength TCSPC lifetime imaging, *Proc. SPIE* **4620**: 79.
11. Becker, W. and Bergmann, A., 2004, High-speed FLIM data acquisition by time-correlated single photon counting, *Proc. SPIE* **5323**: 27.
12. Becker, W., Bergmann, A., Hink, M.A., König, K., Benndorf, K. and Biskup, C., 2004, Fluorescence lifetime imaging by time-correlated single photon counting, *Micr. Res. Techn.* **6**: 58.
13. Bereszovska, O., Ramdya, P., Skoch, J., Wolfe, M.S., Bacskai, B.J. and Hyman, B.T., 2003, Amyloid precursor protein associates with a nicastrin-dependent docking site on the presenilin 1-γ-secretase complex in cells demonstrated by fluorescence lifetime imaging, *J. Neurosc.* **23**: 4560.
14. Berland, K.M., So, P.T.C. and Gratton, E., 1995, Two-photon fluorescence correlation spectroscopy: Method and application to the intracellular environment, *Biophys. J.* **68**: 694.
15. Birch, D.J.S., Holm, A.S., Imhof, R.E., Nadolski, B.Z. and Suhling, K., 1988, Rapid communication: Multiplexed array fluorometry, *J. of Phys. E., Sci Instrum.* **21**: 415.
16. Biskup, C., Zimmer, T. and Benndorf, K., 2004, FRET between cardiac Na$^+$ channel subunits measured with a confocal microscope and a streak camera, 2004, *Nature Biotechnology* Vol. **22**, 2: 220.
17. Buurman, E.P., Sanders, R., Draaijer, A., Gerritsen, H.C., van Veen, J.J.F., Houpt, P.M. and Levine, Y.K., 1992, Fluorescence lifetime imaging using a confocal laser scanning microscope, *Scanning* **14**: 155.
18. Carlsson, K. and Liljeborg, A., 1997, Confocal fluorescence microscopy using spectral and lifetime information to simultaneously record four fluorophores with high channel separation, *J. Microsc.* **37**: 185.
19. Carlsson, K. and Liljeborg, A., 1998, Simultaneous confocal lifetime imaging of multiple fluorophores using the intensity-modulated multiple-wavelength scanning (IMS) technique, *J. Microsc.* **191**: 119.
20. Carlsson, K. and Philip, J.P., 2002, Theoretical investigation of the signal-to-noise ratio for different fluorescence lifetime imaging techniques, *Proc. SPIE* **4622**: 70.
21. Chen, Y. and Periasamy, A., 2004, Characterization of two-photon excitation fluorescence lifetime imaging microscopy for protein localization, *Microsc. Res. Tech.* **63**, 1: 72.
22. Chernomordik, V., Hattery, D.W., Grosenick, D., Wabnitz, H., Rinneberg, H., Moesta, K.T., Schlag, P.M. and Gandjbakhche, A., 2002, Quantification of optical properties of breast tumor using random walk theory, *J. Biomed. Opt.* **7**: 80.
23. Choi, JH., Wolf, M., Toronov, V., Wolf, U., Polzonetti, C., Hueber, D., Safonova, L.P., Gupta, R., Michalos, A., Mantulin, W. and Gratton, E., 2004, Noninvasive determination of the optical properties of adult brain: near-infrared spectroscopy approach, *J. Biomed. Opt.* Vol. **9**, 1: 221.
24. Cole, M.J., Siegel, J., Dowling, R., Dayel, M.J., Parsons-Karavassilis, D., French, P.M., Lever, M.J., L.O. Sucharov, Neil, M.A, Juskaitas, R. and Wilson, T., 2001, Time-domain whole-field lifetime imaging with optical sectioning, *J. Microsc.* **203**: 246.
25. Cova, S., Bertolaccini, M. and Bussolati, C., 1973, The measurement of luminescence waveforms by single-photon techniques, *Phys. Stat. Sol.* **18**: 11.
26. Cubeddu, R., Pifferi, A., Taroni, P., Torricelli, A., Valentini, G., 1996, Time-resolved imaging on a realistic tissue phantom: μ_s and μ_a images versus time-integrated images, *Appl. Opt.* **35**, 22: 4533.
27. Cubeddu, R., Pifferi, A., Taroni, P., Torricelli, A. and Valentini, G., 1998, Imaging with diffusing light: an experimental study of the effect of background optical properties, *Appl. Opt.* **37**, No. 16: 3564.
28. Cubeddu, R., Pifferi, A., Taroni, P., Torricelli, A. and Valentini, G., 1999, Compact tissue oximeter based on dual-wavelength multichannel time-resolved reflectance, *Appl. Opt.* **38**, No. 16: 3670.
29. Cubeddu, R., Giambattistelli, E., Pifferi, A., Taroni, P. and Torricelli, A., 2001, Portable 8-channel time-resolved optical imager for functional studies of biological tissues, *Proc. SPIE* **4431**: 260.
30. Denk, W., Strickler, J.H. and Webb, W.W.W., 1990, Two-photon laser scanning fluorescence microscopy, *Science* **24**: 73
31. Desmettre, T., Devoisselle, J.M. and Mordon, S., 2000, Fluorescence Properties and Metabolic Features of Indocyanine Green (ICG) as Related to Angiography, *Elsevier Science: Survey of Ophthalmology* Vol. **45**, 1: 15.
32. Dickinson, M.E., Waters, C.W., Bearman, G., Wolleschensky, R., Tille, S. and Fraser, S.E., 2002, Sensitive imaging of spectrally overlapping fluorochromes using the LSM 510 META BIOS, *Proc. SPIE* **4620**: 123.
33. Dowling, K., Hyde, S.C.W., Dainty, J.C., French, P.M.W. and Hares, J.D., 1997, 2-D fluorescence liefetime imaging using a time-gated image intensifier, *Elsevier: Opt. Communications* **135**: 27.
34. Dunsby, C. and French, P. M. W., 2003, Techniques for depth-resolved imaging through turbid media including coherence-gated imaging, *J. Phy. D: Appl. Phys.* **63**: R207.
35. Eggerling, C., Berger, S., Brand, L., Fries, J.R., Schaffer, J., Volkmer, A. and Seidel, C.A., 2001, Data registration and selective single-molecule analysis using multi-parameter fluorescence detection, *J. Biotechnol.*, **86**: 163.

36. Eliceiri, K.W., Fan, C.H., Lyons, G.E. and White, J.G., 2003, Analysis of histology specimens using lifetime multiphoton microscopy, *J. Biom. Opt.* **8**: 376.
37. Farrel, T. J. and Patterson, M. S., 2003, Diffusion modelling of fluorescence in tissue, in: *Handbook of Biomedical Fluorescence*; ed. by Mycek and Pogue, Marcel Dekker, New York Basel, pp. 29-60.
38. Gallant, P., Belenkov, A., Ma, G., Lesage, F., Wang, Y., Hall, D, and McIntosh, L., 2004, A quantitative time-domain optical imager for small animals in vivo fluorescence studies, *OSA Biomedical Optics* Topical meeting, Technical digest.
39. Gao, F., Poulet, P. and Yamada, Y., 2000, Simultaneous mapping of absorption and scattering coefficients from a three-dimensional model of time-resolved optical tomography, *Appl. Opt.* **39**: 5898.
40. Geddes, C.D., Parfenov, A. and Lakowicz, J.R., 2003, Photodeposition of silver can result in metal-enhanced fluorescence, *Appl. Spectr.* **57**, 5: 526.
41. Geddes, C.D., Cao, H., Gryczynski, I., Fang, J. and Lakowicz, J.R., 2003, Metal-enhanced fluorescence (MEF) due to silver colloids on a planar surface: Potential applications of indocyanine green to in vivo imaging, *J. Phys. Chem.* A **107**: 3443.
42. Gerritsen, H.C., Sanders, R., Draaijer, A. and Levine, Y.K., 1997, Fluorescence lifetime imaging of oxygen in cells, *J. Fluoresc.* **7**: 11.
43. Gerritsen, H.C., Asselbergs, M.A.H., Agronskaia, A.V. and van Sark, W.G.J.H.M., 2002, Fluorescence lifetime imaging in scanning microscopes: acquisition speed, photon economy and lifetime resolution, *J. Microsc.* **206**: 218.
44. Gratton, E., Breusegem, S., Sutin, J., Ruan, Q. and Barry, N., 2003, Fluorescence lifetime imaging for the two-photon microscope: Time-domain and frequency domain methods, *J. Biomed. Opt.* **8**, 3: 381.
45. Grosenick, D., Wabnitz, H., Rinneberg, H., Moesta, K.T. and Schlag, P.M., 1999, Development of a time-domain optical mammograph and first in-vivo applications, *Appl. Opt.* **38**: 2927.
46. Grosenick, D., Moesta, K.T., Wabnitz, H., Mucke, J., Stroszcynski, C., MacDonald, R., Schlag, P.M. and Rinneberg, H., 2003, Time-domain optical mammography: initial clinical results on detection and characterization of breast tumors, *Appl. Opt.* **42**, No 16: 3170.
47. Grosenick, D., Wabnitz, H., Moesta, K.T., Mucke, J., Möller, M., Stroszcczynski, C., Stößel, J., Wassermann, B., Schlag, P.M. and Rinneberg, H., 2004, Concentration and oxygen saturation of haemoglobin of 50 breast tumours determined by time-domain optical mammography, *Physics in Medicine & Biology* **49**, 7: 1165.
48. Hanbury-Brown, R.and Twiss, R.Q., 1956, *Nature*, **177**: 27.
49. Hartig, P.R. and Sauer, K., 1976, Measurement of very short fluorescence lifetimes by single photon counting, *Rev. Sci. Instrum.* **47**: 122.
50. Hebden, J.C., Veenstra, H., Dehghani, H., Hillman, E.M.C., Schweiger, M., Arridge, S.R. and Delpy, D.T., 2001, Three-dimensional time-resloved optical tomography of a conical breast phantom, *Appl. Optics* **40**: 3278.
51. Hebden, J.C., Gibson, A., Austin, T., Yusof, R.M., Everdell, N., Delpy, D.T., Arridge, S.R., Meek, J.H. and Wyatt, J.S., 2004, Imaging changes in blood volume and oxygenation in the newborn infant brain using three-dimensional optical tomography, *Physics in Medicine & Biology* **49**, 7: 1117.
52. Herman, B., 1998, *Fluorescence Microscopy*, 2nd. edn., Springer, New York.
53. Hopf, A. and Neher, E., 2002, Highly nonlinear photodamage in two-photon fluorescence microscopy, *Biophys. J.* **80**: 2029.
54. Kästner, C.N., Prummer, M., Sick, B., Renn, A., Wild, U.P. and Dimroth, P., 2003, The citrate carrier CitS probed by single molecule fluorescence spectroscopy, *Biophys. J.* **84**: 1651.
55. Kelbauskas, L. and Dietel, W., 2002, Internalization of aggregated photosensitizers by tumor cells: Subcellular time-resolved fluorescence spectroscopy on derivates of pyropheophorbide-a ethers and chlorin e6 under femtosecond one- and two-photon excitation, *Photochem. Photobiol.* **76**: 686.
56. Knemeyer, J.-P., Marmé, N. and Sauer, M., 2002, Probes for detection of specific DNA sequences at the single-molecule level, *Anal. Chem.* **72**: 3717.
57. Köllner, M. and Wolfrum, J., 1992, How many photons are necessary for fluorescence-lifetime measurements? *Phys. Chem. Lett.* **200**: 199.
58. König, K., So, P.T.C., Mantulin, W.W., Tromberg, B.J. and Gratton, E., 1996, Two-photon excited lifetime imaging of autofluorescence in cells during UVA and NIR photostress, *J. Microsc.* **183**: 197.
59. König, K., 2000, Multiphoton microscopy in life sciences, *J. Microsc.* **200**: 83.
60. König, K. and Riemann, I., 2003, High-resolution multiphoton tomography of human skin with subcellular spatial resolution and picosecond time resolution, *J. Biom. Opt.* **8**: 432.
61. Kühnemuth, R. and Seidel, C.A.M., 2001, Principles of single molecule multiparameter fluorescence spectroscopy, *Single Molecules* **2**: 251.
62. Lakowicz, J.R. and Berndt, K., 1991, Lifetime-selective fluorescence lifetime imaging using an rf phase-sensitive camera, *Rev. Sci. Instrum.* **62**: 1727.

63. Lakowicz, J.R., Szmacinski, H., Nowaczyk, K. and Johnson, M.L., 1992, Fluorescence lifetime imaging of free and protein-bound NADH, *Proc. Natl. Acad. Sci. USA Biochem.* Vol. **89**: 1271.

64. Lakowicz, J.R., 1999, *Principles of Fluorescence Spectroscopy*, 2nd ed., Plenum Press, New York.

65. Lakowicz, J.R., Shen, B., Gryczynski, Z., D'Auria, S. and Gryczynski, I., 2001, Intrinsic fluorescence from DNA can be enhanced by metallic particles, *Biochem. and Biophys. Research Communications* **286**: 875.

66. Laques, S. L., 2003, Monte Carlo simulations of fluorescence in turbid media, in: *Handbook of Biomedical Fluorescence*; Mycek and Pogue, ed., Marcel Dekker, New York Basel, pp. 61-107.

67. Leskovar, B. and Lo, C.C., 1976, Photon counting system for subnanosecond fluorescence lifetime measurements, *Rev. Sci. Instrum.* Vol. **47**: 9.

68. Lewis, C. and Ware, W.R., 1973, The measurement of short-lived fluorescence decay using the single photon counting method, *Rev. Sci. Instrum.* Vol. **44**: 2.

69. Liebert, A., Wabnitz, H., Steinbrink, J., Obrig, H., Möller, M., Macdonald, R. and Rinneberg, H., 2003, Intra- and extracerebral changes of hemoglobin concentrations by analysis of moments of distributions of times of flight of photons, *SPIE* **5138**: 126.

70. Liebert, A., Wabnitz, H., Grosenick, D., Möller, M., Macdonald, R. and Rinneberg, H., 2003, Evaluation of optical properties of highly scattering media by moments of distributions of times of flight of photons, *Appl. Opt. Vol.* **42**: 28.

71. Liebert, A., Wabnitz, H., Steinbrink, J., Obrig, H., Möller, M., Macdonald, R., Villringer, A. and Rinneberg, H., 2004, Time-resolved multidistance near-infrared spectroscopy of the adult head: intracerebral and extracerebral absorption changes from moments of distribution of times of flight, *Appl. Opt.* **43**: 2037.

72. Malicka, J., Gryczynski, I., Geddes, C.D. and Lakowicz, J.R., 2003, Metal-enhanced emission from indocyanine green: a new approach to in vivo imaging, *J. Biomed. Opt.* **8**: 472.

73. Marcu, L., Grundfest, W. S. and Fishbein, C., 2003, Time-resolved laser-induced fluorescence spectroscopy for staging atherosclerotic lesions, in: *Handbook of Biomedical Fluorescence*; Mycek and Pogue, ed., Marcel Dekker, New York Basel, pp. 397-430.

74. Masters; B.R., So, P.T.C. and Gratton, E , 1997, Multiphoton excitation fluorescence microscopy and spectroscopy of in vivo human skin, *Biophys. J.* Vol. **72**: 2405.

75. Maxwell, K. and Johnson, G.N., 2000, Chlorophyll fluorescence - a practical guide, *J. Experimental Botany* Vol. **51**, 345: 659.

76. McBride, T.O., Pogue, B.W, Gerety, E.D., Poplack, S.B., Österberg, U.L. and Paulsen, K.D., 1999, Spectroscopic diffuse optical tomography for the quantitative assessment of hemoglobin concentration and oxygen saturation in breast tissue, *Appl. Opt* **38**: 5480.

77. McBride, T.O., Pogue, B.W, Jiang, S., Österberg, U.L. and Paulsen, K.D., 2001, A parallel-detection frequency-domain near-infrared tomography system for hemoglobin imaging of the breast in vivo, *Rev. of Sci. Instrum.* Vol. **72**, No. 3: 1817.

78. Meiling, W. and Stary, F., 1963, *Nanosecond pulse techniques*, Akademie-Verlag, Berlin.

79. Michler, P., Imamoglu, A., Mason, M.D., Carson, P.J., Strouse, G.F. and Buratto, S.K., 2000, Quantum correlation among photons from a single quantum dot at room temperature, *Nature* **406**: 968.

80. Minsky, M., 1988, Memoir on inventing the confocal microscope, *Scanning* **10**: 128.

81. Mordon, S., Devoiselle, J.M. and Soulie-Begu, S., 1998, Indocyanine green: physicochemical factors affecting its fluorescence in vivo, *Microvascular Res.* **55**: 146.

82. Müller, R., Zander, C., Sauer, M., Deimel, M., Ko, D.-S., Siebert, S., Arden-Jacob, J., Deltau, G., Marx, N.J., Drexhage, K.H. and Wolfrum, J., 1996, Time-resolved identification of single molecules in solution with a pulsed semiconductor diode laser, *Chem. Phys. Lett.* **262**: 716.

83. Mycek, M.-A., Pogue, B.W., 2003, *Handbook of Biomedical Fluorescence*, Marcel Dekker, New York - Basel.

84. Ntziachristos, V., Ma, X.H. and Chance, B., 1998, Time-correlated single-photon counting imager for simultaneous magnetic resonance and near-infrared mammography, *Rev. Sci. Instrum.* **69**, No. 12: 4221.

85. Ntziachristos, V., Ma, X.H., Yodh, A.G. and Chance, B., 1999, Multichannel photon counting instrument for spatially resolved near infrared spectroscopy, *Rev. Sci. Instrum.* **70**, No. 1: 193.

86. Ntziachristos, V., Yodh, A.G., Schnall, M. and Chance, B., 2000, Concurrent MRI and diffuse optical tomography of breast after indocyanine green enhancement, *PNAS* **97**: 2767.

87. Ntziachristos, V. and Chance, B., 2001, Accuracy limits in the determination of absolute optical properties using time-resolved NIR spectroscopy, *Med. Phys.* **28**, No. 6: 410.

88. O'Connor, D.V. and Phillips, D., 1984, *Time Correlated Single Photon Counting*, Academic Press, London.

89. O'Leary, M.D., Boas, D.A., Li, X.D., Chance., B. and Yodh, A.G., 1996, Fluorescence lifetime imaging in turbid media, *Opt. Letters* **21**, 2: 158.

90. Patterson, M.S., Chance, B. and Wilson, B.C., 1989, Time-resolved reflectance and transmittance for the noninvasive measurement of tissue optical properties, *Appl. Opt.* **28**: 2331.
91. Patterson, G.H. and Piston, D.W., 2000, Photobleaching in two-photon excitation microscopy, *Biophys. J.* **78**: 2159.
92. Paul, R.J. and Schneckenburger, H., 1996, Oxygen concentration and the oxidation-reduction state of yeast: Determination of free/bound NADH and flavins by time-resolved spectroscopy, *Springer Naturwissenschaften* **83**: 32.
93. Pawley, J., 1995, *Handbook of Biological Confocal Microscopy*, Plenum, New York.
94. Pepperkok, R., Squire, A., Geley, S. and Bastiaens, P.I.H., 1999, Simultaneous detection of multiple green fluorescent proteins in live cells by fluorescence lifetime imaging microscopy, *Curr. Biol.* **9**: 269.
95. Periasamy, A., Elangovan, M., Wallrabe, H., Barroso, M., Demas, J.N., Brautigan, D.L. and Day, R.N., 2001, Wide-field, confocal, two-photon, and lifetime resonance energy transfer imaging microscopy, in: *Methods in Cellular Imaging*, A. Periasamy, ed., Oxford University Press, pp. 295-308.
96. Philip, J.P. and Carlsson, K., 2003, Theoretical investigation of the signal-to-noise ratio in fluorescence lifetime imaging, *J. Opt. Soc. Am. A* **20**: 368.
97. Pifferi, A., Taroni, P., Torricelli, A., Messina, F., Cubeddu, R. and Danesini, G., 2003, Four-wavelength time-resolved optical mammography in the 680-980-nm range, *Opt. Lett.* **28**: 1138.
98. Prummer, M., Hübner, C., Sick, B., Hecht, B., Renn, A. and Wild, U.P., 2000, Single-molecule identification by spectrally and time-resolved fluorescence detection, *Anal. Chem.* **72**: 433.
99. Prummer, M., Sick, B., Renn, A. and Wild, U.P., 2004, Multiparameter microscopy and spectroscopy for single molecule Analysis, *Anal. Chem.* **76**: 1633.
100. Rigler, R., Mets, Ü, Widengren, J. and Kask, P., 1993, Fluorescence correlation spectroscopy with high count rate and low background: analysis of translational diffusion, *European Biophys. J.* **22**: 169.
101. Rigler, R. and Elson, E.S., 2001, *Fluorescence Correlation Spectroscopy*, Springer, Berlin, Heidelberg.
102. Rück, A., Dolp, F., Scalfi-Happ, C., Steiner, R. and Beil, M., 2003, Time-resolved microspectrofluorometry and fluorescence lifetime imaging using ps pulsed diode lasers in laser scanning microscopes, *Proc. SPIE* **5139**: 166.
103. Sanders, R., Draaijer, A., Gerritsen, H.C., Houpt, P.M. and Levine, Y.K., 1995, Quantitative pH imaging in cells using confocal fluorescence lifetime imaging microscopy, *Anal. Biochem.* **227**, 2: 302.
104. Sauer, M., Zander, C., Müller, R., Ullrich, B., Kaul, S., Drexhage, K.H. and Wolfrum, J., 1997, Detection and identification of individual antigen molecules in human serum with pulsed semiconductor lasers, *Appl. Phys. B* **65**: 427.
105. Schaffer, J., Volkmer, A., Eggerling, C., Subramaniam, V., Striker, G. and Seidel, C.A.M., 1999, Identification of single molecules in aqueous solution by time-resolved fluorescence anisotropy, *J. Phys. Chem. A*, **103**: 331.
106. Schmidt, F.E.W., Fry, M.E., Hillman, E.M.C., Hebden, J.C. and Delpy, D.T., 2000, A 32-channel time-resolved instrument for medical optical tomography, *Rev. Sci. Instrum.* **71**: 256.
107. Schuyler, R. and Isenberg, I., 1971, A Monophoton Fluorometer with Energy Discrimination. *Rev. Sci. Instrum.*, Vol. **42**: 6.
108. Schweitzer, D., Kolb, A., Hammer, M. and Thamm, E., 2001, Basic investigations for 2-dimensional time-resolved fluorescence measurements at the fundus, *Int. Ophthalmol.* **23**: 399.
109. Schwille, P., Kummer, S., Heikal, A.H., Moerner, W.E. and Webb, W.W.W., 2000, Fluorescence correlation spectroscopy reveals fast optical excitation-driven intramolecular dynamics of yellow fluorescent proteins, *PNAS* **97**: 151.
110. Schwille, P., Haupts, U., Maiti, S. and Webb, W.W.W., 1999, Molecular dynamics in living cells observed by fluorescence correlation spectroscopy with one- and two-photon excitation, *Biophys. J.* Vol. **77**: 2251.
111. Sevick-Muraca, E.M., Godavarty, A., Houston, J.P., Thompson, A. B. and Roy, R., 2003, Near-infrared imaging with fluorescecent contrast agents, in: *Handbook of Biomedical Fluorescence*; Mycek and Pogue, ed., Marcel Dekker, New York Basel, pp. 445-527.
112. Sherman, L., Ye, J.Y., Albert, O. and Norris, T.B., 2002, Adaptive correction of depth-induced aberrations in multiphoton scanning microscopy using a deformable mirror, *J. Microsc.* **206**: 65.
113. So, P.T.C., French, T. and Gratton, E., 1994, A frequency domain microscope using a fast-scan CCD camera, *Proc. SPIE* **2137**: 83.
114. So, P.T.C., French, T., Yu, W.M., Berland, K.M., Dong, C.Y. and Gratton, E., 1995, Time-resolved fluorescence microscopy using two-photon excitation, *Bioimaging* **3**: 49.
115. Squire, A., Verveer, P.J. and Bastiaens, P.I.H., 2000, Multiple frequency fluorescence lifetime imaging microscopy, *J. Microsc.* **197**: 136.
116. Straub, M., Hell, S.W., 1998, Fluorescence lifetime three-dimensional microscopy with picosecond precision using a multifocal multiphoton microscope, *Appl. Phys. Lett.* **73**: 1769.

117. Syrtsma, J., Vroom, J.M., de Grauw, C.J. and Gerritsen, H.C., 1998, Time-gated fluorescence lifetime imaging and microvolume spectroscopy using two-photon excitation, *J. Microsc.* **191**: 39.
118. Taroni, P., Pifferi, A., Torricelli, A., Spinelli, L., Danesini, G.M. and Cubeddu, R., 2004, Do shorter wavelengths improve contrast in optical mammography? *Physics in Medicine & Biology* **49**, 7: 1203.
119. Taroni, P., Danesini, G., Torricelli, A., Pifferi, A., Spinelli, L. and Cubeddu, R., 2004, Clinical trial of time-resolved scanning optical mammography at 4 wavelengths between 683 and 975 nm, *J. Biomed. Opt.* **9**: 464.
120. Tirlapur, U.K. and König, K., 2002, Targeted transfection by femtosecond laser, *Nature* **418**: 290.
121. Thomson, N.L., 1991, Fluorescence correlation spectroscopy, Topics in: *Fluorescence Spectroscopy*, J.R. Lakowicz, ed., Plenum Press, Vol. 1, New York, pp. 337.
122. Thompson, R.M., Stevenson, R.M., Shields, A.J., Farrer, I., Lobo, C.J., Ritchie, D.A., Leadbeater, M.L. and Pepper, M., 2001, Single-photon emission from exciton complexes in individual quantom dots, *Phys. Rev. B* **64**: 201302.
123. Torricelli, A., Pifferi, A., Taroni, P., Gambattiste, E. and Cubeddu, R., 2001, In vivo optical characterization of human tissue from 610 to 1010 nm by time-resolved reflectance spectroscopy, *Phys. Med. Biol.* **46**: 2227.
124. Torricelli, A., Spinelli, L., Pifferi, A., Taroni, P. and Cubeddu, R., 2003, Use of a nonlinear perturbation approach for in vivo breast lesion characterization by multi-wavelength time-resolved optical mammography, *Optics Express* **11**: 853.
125. Torricelli, A., Quaresima, V., Pifferi, A., Biscotti, G., Spinelli, L., Taroni, P., Ferrari, M. and Cubeddu, R., 2004, Mapping of calf muscle oxygenation and haemoglobin content during plantar flexion exercise by multi-channel time-resolved near-infrared spectroscopy, *Phys. Med. Biol.* **49**: 685.
126. Tramier, M., Gautier, I., Piolot, T., Ravalet, S., Kemnitz, K., Coppey, J., Durieux, C., Mignotte, V. and Coppey-Moisan, M., 2002, Picosecond-hetero-FRET microscopy to probe protein-protein interactions in live cells, *Biophys. J.* **83**: 3570.
127. Urayama, P., Mycek, M-A., 2003, Fluorescence lifetime imaging microscopy of endogenous biological fluorescence, in: *Handbook of Biomedical Fluorescence*; Mycek and Pogue, ed., Marcel Dekker, New York Basel, pp. 211-236.
128. Van Zandvoort, M.A.M.J., de Grauw, C.J., Gerritsen, H.C., Broers, J.L.V., Egbrink, M.G.A., Ramaekers, F.C.S. and Slaaf, D.W., 2002, Discrimination of DNA and RNA in cells by a vital fluorescent probe: Lifetime imaging of SYTO13 in healthy and apoptotic cells, *Cytometry* **47**: 226.
129. Verveer, P.J., Squire, A., Bastiaens, P.I.H., 2001, Frequency-domain fluorescence lifetime imaging microscopy: A window on the biochemical landscape of the cell, in: *Methods in Cellular Imaging*, A. Periasamy, ed., Oxford University Press, pp. 273-294.
130. Wabnitz, H. und Rinneberg, H., 1997, Imaging in turbid media by photon density waves: spatial resolution and scaling relations, *Appl. Opt.* **36**: 64.
131. Wabnitz, H., Liebert, A., Möller, M., Grosenick, D., Model, R. und Rinneberg, H., 2002, Scanning laser-pulse mammography: matching fluid and off-axis measurements, *Biomedical Topical Meeting*, Technical Digest, OSA 686.
132. Wallace, V.P., Dunn, A.K., Coleno, M.L. and Tromberg, B.J., 2001, Two-photon microscopy in highly scattering tissue, in: *Methods in Cellular Imaging*, A. Periasamy, ed., Oxford University Press, pp. 180-199.
133. Wallrabe, H., Stanley, M., Periasamy, A. and Barroso, M., 2003, One- and two-photon fluorescence resonance energy transfer microscopy to establish a clustered distribution of receptor-ligand complexes in endocytic membranes, *J. of Biomed. Opt.* **8**: 1.
134. Weston, K.D., Dyck, M, Tinnefeld, P., Müller, C., Herten, D.P. and Sauer, M., 2002, Measuring the number of independent emitters in single molecule fluorescence images and trajectories using coincident photons, *Anal. Chem.* **74**: 5342.
135. White, J.G., Amos, W.B. and Fordham, M., 1987, An evaluation of confocal versus conventional imaging of biological structures by fluorescence light microscopy, *J. Cell. Biol.* **105**: 41.
136. Ying, H. and Xie, X.S., 2002, Probing single-molecule dynamics photon by photon. *J. Chem. Phys.* **117**:10965.
137. Yuan, Z., Kardynal, B., Stevenson, R.M., Shields, A.J., Lobo, C.J., Cooper, K., Beattie, N.S., Ritchie, D.A. and Pepper, M., 2002, Electrically driven single-photon source, *Science* **295**: 102.
138. Zander, C., Sauer, M., Drexhage, K.H., Ko, D.-S., Schulz, A., Wolfrum, J., Brand, L., Eggerling, C. and Seidel, C.A.M., 1996, Detection and characterisation of single molecules in aqueous solution, *Appl. Phys. B* **63**: 517.
139. Zander, C., Drexhage, K.H., Han, K.-T., Wolfrum, J. and Sauer, M., 1998, Single-molecule counting and identification in a microcapillary, *Chem. Phys. Lett.* **286**: 457.

140. Zwiller, V., Blom, H., Jonsson, P., Panev, N., Jeppesen, S., Tsegaye, T., Goobar, E., Pisto, M.E., Samuelson, L. and Björk, G., 2001, Single quantum dots emit single photons at at time: Antibunching experiments, *Appl. Phys. Lett.* **78**: 2476.

SOME ASPECTS OF DNA CONDENSATION OBSERVED BY FLUORESCENCE CORRELATION SPECTROSCOPY

Teresa Kral[1,2], Aleš Benda[1], Martin Hof[1] and Marek Langner[3,4*]

5.1. INTRODUCTION

Conformational changes of macromolecules are a crucial issue that needs to be comprehended before the processes occurring in living systems are understood. They decide about the adaptability of protein tertiary and quaternary structures, which in turn determines their functions and how they will interact with other elements of the system. Protein folding is one of the leading research areas in the modern biology (Arnold et al., 2001; Ellis and Hartl, 1999; Grantcharova et al., 2001). Similar problems are encountered when the conformation of other macromolecules need to be considered, i.e. supramolecular aggregates. Phenomena related to the structure and dynamics of such aggregates form the basis for the function of larger structures, starting from the biological membrane and ending on the processing of genetic information. Despite the fact that the properties of a number a single molecular identities are being successfully tackled, their behavior in large functional ensembles is barely been touched (Lenart and Ellenberg, 2003; Matouschek, 2003; Nakai, 2001; Nelson and Yeaman, 2001; Parak, 2003).

Nucleic acids are macromolecules whose primary function is to store genetic information. Access to this information and the control of transcription and translation

---

* 1- J. Heyrovský Institute, Academy of Sciences of Czech Republic and Center for Complex Molecular Systems and Biomolecules, Dolejškova 3, Cz-18223 Prague 8, Czech Republic; 2- Department of Physics and Biophysics, Agricultural University, Norwida 25, 50-375 Wroclaw, Poland. tek@ozi.ar.wroc.pl; phone: +48-71-3205274; fax: +48-71-3205172; 3- Wrocław University of Technology, Institute of Physics, Wyb. Wyspiańskiego 27, 50-370, Wrocław, Poland; 4- Academic Centre for Biotechnology of Lipids Aggregates, Przybyszewskiego 63/67; 51-148 Wrocław, Poland.

processes depend on how nucleic acids associate with assisting molecules. They also depend on how this association is spatially and temporally controlled. Precisely controlled continuous conformational alterations in molecules are intrinsically correlated with access to the particular genetic information. Depending on the actual needs of the cell, conformational changes may occur on a global as well as a local level. A lot of effort has been invested in elucidating the relations between specific nucleic acid conformation, the state of an associated molecule and the actual overall cell state (Ahringer, 2003; Chevet et al., 2001; Conti and Izaurralde, 2001; Janicki and Spector, 2003; Stein et al. 2003). This knowledge has been acquired using sophisticated experimental methodologies, but unfortunately predominantly in *in vitro* systems. The next step is to test the relevance of these data for *in vitro* situation. This, however, is not a trivial issue, since it requires single molecule studies *in situ*. The other major limitation is to obtain temporal and spatial statistical distributions of such molecules and/or their conformations in living cells. Only then can a correlation between the state of a particular type of macromolecule and cell state be determined (Iino at al., 2001; Janicki and Spector, 2003; Verkman, 2002). It seems that such problems will not be experimentally addressed without techniques that enable massive data acquisition and analysis without significant interference in the living system.

Similar types of problems are encountered in the development of new generation pharmaceuticals based on supramolecular ensembles (Torchilin and Lukyanov, 2003; Zarif et al., 2000; Barenholz, 2001; Marcucci and Lefoulon, 2004) This include targeted drug delivery systems (Barenholz, 2001; Derycke and de-Witte, 2004; Garcia et al., 2000; Langner, 2000a; Langner and Kral, 1999; Lian and Ho, 2001) and of the formulations required for gene therapies (Brown et al., 2001; Davis, 2002; Mhashilkar et al., 2001; Molema, 2002). Those new products require new preparation protocols and quality control procedures, which are needed to determine the structure and conformation of the involved macromolecules and their ensembles in a statistical manner. This yet unresolved issue is a prerequisite for the successful development of efficient genetic information carrier devices. The lack of such methodologies hampers the progress of the technology (Byk and Scherman, 2000; Langner, 2000b). In this paper we present a series of experiments that focus specifically on studies of DNA and its interactions with small molecules. However, the results obtained and approaches proposed are of general importance for complex multicomponent systems. In order for such structures to become commercially acceptable products, their production needs to be strictly controlled; their particle size and the conformation of their constituent macromolecules need to be precisely controlled. This issue is vividly manifested in attempts to develop the optimal supramolecular ensemble for gene therapy (Langner, 2000b). In this application, the nucleic acid molecule needs to be associated with assisting compounds in a predictable and reproducible meaner.

Different sets of data are required when the interaction of such aggregates with living matter is to be determined. Here, the multicomponent ensemble may dissociate and the fate of each molecule should be followed individually in the cell or tissue (Hui et al., 1996). Again, due to the statistical character of the process, simultaneous measurements of the localization of a large number of species as a function of time and spatial distribution are needed. Currently available methodologies address these issues to some extent, but the available data is fragmentary and frequently uncorrelated. This is largely due to the fact that multi-experiment procedures need to be performed, and these do not ensure coherent conditions and are rarely performed in *in situ* situations.

Techniques based on the analysis of signals produces by labeled relevant molecules may provide solutions to these problems. Such signals carry information regarding the distribution of molecules and/or their ensembles in the sample. Here, we present a study on the conformation of the DNA molecule upon changing conditions using Fluorescence Correlation Spectroscopy (FCS). This single molecule technique contains mainly information about translational diffusion of the fluorescently labeled molecules and their concentration. The principles of this method are outlined in the next paragraph of this article.

As mentioned before, the construction of desired supramolecular aggregates containing DNA is a prerequisite for the success of gene therapy (de-Gennes, 1999; Langner, 2000b). The formation of such aggregates requires changes in nucleic acid conformation upon their interaction with other compounds to be determined and the completion of the process evaluated. So far, one of the major obstacles of such a procedure is due to sample heterogeneity, which makes the correlation of structure and function impossible. The DNA molecule is a relatively simple polymer, which can be considered to be a charged poly-ion with a large persistence length (about 50 nm (Bloomfield, 1996). In the double helix, the outer surface is negatively charged, whereas the hybridized complementary bases form a hydrophobic interior. Such a structure causes double helix conformation to be very sensitive to variety of compounds. Screening by ions in solution lowers the intensity of electrostatic interactions, which in turn alter a DNA persistence length (Baumann et al., 1997; Mel'nikov et al., 1999). Hydrophobic and amphiphilic compounds on the other hand may alter DNA conformation in a different way, namely by affecting the organization of the bases. This is exemplified by intercalating compounds, which exhibit a flat molecular structure and are predisposed to insert between bases (Arndt-Jovin and Jovin, 1989; Hortobagyi, 1997). Many fluorescent probes and anticancer drugs are of this type. This mode of dye interaction alters overall DNA conformation therefore the selection of an appropriate fluorescent probe is crucial for a successful experimental outcome.

For the construction of the supramolecular ensemble, the usually large DNA molecule needs to be condensed with cationic compound (Bloomfield, 1996; 1997). The condensation process needs to be performed in a manner that ensures a uniformity of the resulting aggregates.

The methodology developed, which is based on the single molecule technique FCS for monitoring this process is the subject of this paper. First, the effect of selected fluorescent dyes on DNA conformation characterized by FCS will be presented. Next, FCS measurements on the effect of cationic compound (spermine and hexadecyltrimethyl ammonium bromide, HTAB) and amphiphilic molecules (cationic lipids) on DNA conformation will be discussed.

5.2. PRINCIPLES OF FLUORESCENCE CORRELATION SPECTROSCOPY

The first experimental realization of Fluorescence Correlation Spectroscopy (FCS) that studied the reversible binding to DNA of ethidium bromide and the first theoretical background was presented already thirty years ago (Magde et al., 1972, 1974, 1978; Elson and Magde, 1978). FCS is based on a statistical analysis of temporal behavior of detected spontaneous fluorescence intensity fluctuations. In other words, by means of normalized autocorrelation function $G(\tau)$ it studies the relation between the detected intensity in certain time t and time τ later.

In order to be able to subtract any information from the fluctuations, the magnitude of the fluctuations compared to the averaged intensity must be as high as possible. In reality it means to detect signal from single molecules, as only than any change of state of the measured molecule highly influences the overall fluorescence intensity. The first reasonable realization of favorable signal-to-noise setup came only in 1993 (Rigler at al., 1993). It makes use of stable lasers, aberrations free epifluorescence confocal microscopes with high numerical aperture, interference filters, sensitive single-photon detectors and fast hardware correlators. This setup is still one of the most widely used. Further technical development later brought in a pulsed two photon excitation, pulsed diode lasers, a combination of several lasers, polarizators and detectors in one setup, and fast computers enabling online software correlation with saving all information about every detected photon, thus enabling later more elaborate analysis.

The highly focused laser beam creates diffraction limited Gaussian-Lorentzian intensity profile with minimum lateral radii of order 200 – 300 nm. The detection in axial dimension is limited by pinhole in image plane of microscope to value between 2 and 4 µm. The size of the detection volume is than 0.3 to 1 fl. If such a volume is combined with nM concentration of fluorescent molecules, one comes to a single-molecule level.

There are different sources of fluctuations in the studied system. The decrease of autocorrelation towards shorter times on nanosecond time scale (antibunching) is caused by finite fluorescence lifetime, but with usual 50 - 200 ns time resolution is not observable. On a microsecond time scale, the shape of $G(\tau)$ is governed by triplet state, possible photo-isomerization and blinking dynamics, or chemical equilibrium among different fluorescent species. And finally the longest observable autocorrelation decay is caused by translational diffusion, flow or photobleaching.

To obtain real physical parameters from the autocorrelation function, one needs to apply a proper physical model. At this step first approximations take place. The detection volume is for bulk measurements approximated as a 3D Gaussian. Assuming small point-like noninteracting molecules freely diffusing in a space much larger than the detection volume, showing up only triplet state dynamics, the $G(\tau)$ takes form,

$$G(\tau) = 1 + \left(1 - T + Te^{-\tau/\tau_{tr}}\right) \left(\frac{1}{PN[1-T]}\right) \cdot \frac{1}{1 + (\tau/\tau_D)} \left(\frac{1}{1 + (\tau/\tau_D)(\omega_0/\omega_Z)^2}\right)^{1/2}$$

where T is a triplet fraction, τ_{tr} is a triplet decay time, PN is the apparent particle number, τ_D is a diffusion time and ω_0 and ω_Z are lateral and axial radii of detection volume. The derivation of equations for the applied models makes use of the natural laws applied in classical methods of perturbation kinetics as the only difference is in the source of fluctuations.

The parameters PN and τ_D are related with macroscopic values of concentration c and diffusion coefficient D via:

$$\tau_D = \frac{w_0^2}{4D} \qquad \text{and} \qquad PN = \pi c w_0^2$$

The diffusion coefficient for spherically symmetric molecules is related to hydrodynamic radius r_h via Einstein-Stokes equation:

$$D = \frac{k_B T}{6 \pi \eta r_h}$$

,where k_B is Boltzmann constant, T is thermodynamic temperature, η is dynamic viscosity and r_h is hydrodynamic radius. The hydrodynamic radius can be calculated from molecular mass M using:

$$r_h = \sqrt[3]{\frac{3M}{4\pi\rho N_A}}$$

where ρ is mean density of the molecule and N_A is Avogadro's number.

The translational diffusion coefficient depends largely on the shape of the molecule. For rod molecules, such as a DNA, it can be estimated as:

$$D = \frac{A k_B T}{3 \pi \eta L}$$

where L corresponds to the length of the rod (for a DNA it's the rise per base pair (0.34nm) multiplied by the number of base pairs), d is a diameter of the rod (2.38 nm for DNA) and A represents a correction factor:

$$A = \ln(L/d) + 0.312 + 0.565/(L/d) - 0.1/(L/d)^2$$

It means that a diffusion coefficient of a 1000 BP DNA is approximately 5 times smaller and diffusion times 5 times larger for a rod-like shaped molecule than for a spherical one.

Figure 5.1 presents a selected example of the autocorrelation functions for a DNA molecule in two different conformational states. The ACF with the lower amplitude represents a 10 000 BP large DNA molecule labeled by propidium iodide. The ACF with the large amplitude (equal lower apparent PN) results from a FCS measurements of the same DNA molecule in the condensed state induced by spermine. Even visual inspection of those ACF indicates a significant decrease in τ_D due to the condensation process. In the shown example fitting by the above mentioned 3 dimensional diffusion model reveals a difference of a factor 8 in the τ_D values. It should be mentioned that this demonstrated large sensitivity for conformational changes in DNA is one of the major advantages of the FCS method in this research field.

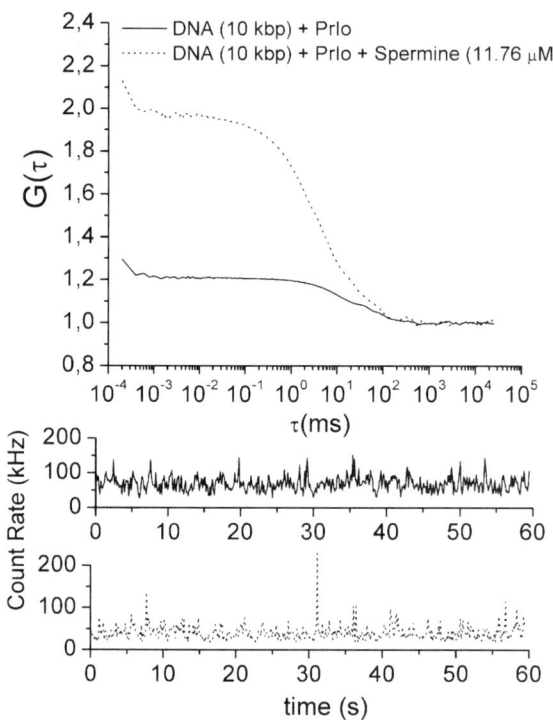

Figure. 5.1 Normalized fluorescence autocorrelation functions for plasmid (βApr-1-Neo, 10 kbp contour length 3,4 µm) labeled with propidium iodide ($C_{DYE}/C_{DNA\ phosphate} = 0.2$). Solid lines show ACF and time course of corresponding fluorescence count rates for DNA in its native form. Dashed lines show ACF and time course of corresponding fluorescence count rates for condensed DNA induced by spermine.

As mentioned above, the used mathematical model used to fit the measured ACF assumes point-like molecules. In the herein reported FCS measurements, the DNA molecules are multiple labeled and their size is comparable with the detection volume. Thus the above mentioned assumption of point-like molecules is an oversimplification of the observed real situation and most likely the reason for the obtained later mentioned high values of the apparent particle number. As to derive an exact analytical model properly describing the effect of multiple labeled large molecules diffusing through the detection volume is extremely difficult, we believe that Monte-Carlo simulations of FCS experiment (Bunfield et al., 1998; Davies et al., 2002; Wohland et al., 2001) combined with a suitable description of DNA movement can bring a new look into this problem.

5.3. MATERIALS AND METHODS

5.3.1 Chemicals

Propidium iodide and PicoGreen dyes were purchased from Molecular Probes (Leiden, The Nederlands). All experiments were performed in a TE buffer (pH 7.95, 10mM Tris, 1mM EDTA). Spermine and hexadecyltrimethyl ammonium bromide (HTAB) were obtained from Sigma. Cationic lipid N-(1-(2,3-dioleoyloxy) propyl)-N, N, N-trimethyl ammonium methylsulfate (DOTAP) and egg yolk phosphatydyl choline (egg PC) were purchased from Avanti lipids (Oregon).

5.3.2 DNA

A 10 000 BP plasmid DNA (pHβAPr-1-neo, 3,4 µm contour length) was used in the experiments below. Plasmid was a generous gift from Dr. Maciej Ugorski's Laboratory (Department of Immunochemistry, Ludwik Hirszfeld Institute of Immunology and Experimental Therapy, Polish Academy of Sciences, Wrocław, Poland). It was prepared as described elsewhere (Sombrook et al., 1989), with slight modifications during the final purification stage (Kral et al., 2002a, 2002b).

5.3.3 Liposome formulation

Multilamellar liposomes (MLVs) were prepared by mixing appropriate amounts of the egg PC, DOTAP or mixture stock solutions in dry chloroform to obtain the desired compositions. Thereafter the solvent was removed by evaporation under a stream of nitrogen. For removal of residual amounts of solvent the samples were further maintained under high vacuum for at least 2 h. The resulting dry lipid films were then hydrated with TE buffer to obtain 1mM concentration of the lipid mixture To obtain small unilamellar vesicles (SUV) the hydrated lipid dispersions were vortexed vigorously and then extruded with a LiposoFast small volume homogenizer (Avestin, Ottawa, Canada) by subjecting to 19 passes through polycarbonate filter (100-nm pore size, Nucleopore, Pleasanton, CA).

5.3.4. Design of experiment

A central composite experimental design was implemented to determine the optimal conditions to identify a phase space for condensation of the plasmid DNA. The central composite design incorporated four internal factors as: dye/DNA phosphate ratio, spermine/DNA or HTAB/DNA phosphate ratio, cationic lipid percentage in the liposomes and lipid/DNA phosphate ratio.

5.3.5. Experimental setup of FCS

Fluorescence measurements were performed on a ConfoCor® 1 (Carl Zeiss Jena, Germany), as described elsewhere (Kral et al., 2002a, 2002b; Benes et al., 2001). In short, ConfoCor 1 is a PC-controlled fluorescence correlation-adapted AXIOVERT 135 TV microscope, equipped with an x-y-z adjustable pinhole, avalanche Photodiode SPCM-200-PQ, ALV-hardware correlator, and CCD camera. The Ar^+-laser beam (excitation wavelength - 514 nm) was focused by a water-immersion microscope objective at an open focal light cell. The size of the confocal volume element was determined by calibration measurements using rhodamine-6G.

5.4. FCS EXPERIMENTS ON DNA

5.4.1. Effect of dye on DNA conformation

There are a number of fluorescent dyes which are used for nucleic acid staining (Arndt-Jovin and Jovin, 1989). Depending on the mode of interaction with DNA they may affect its conformation to a different extend. The dyes like propidium iodide or ethidium bromide intercalate between bases therefore altering the structure of double helix (Eckel et al., 2003). This effect should alter the overall conformation of DNA molecule which will likely result in a change in a molecule hydrodynamic radius. If this change is big enough it should be possible to evaluate it using FCS. The effect of nucleic acid interactions with different concentrations of a fluorescent dye on parameters determined with FCS are shown on the two examples of propidium iodide and PicoGreen (Figure 5.2). When certain amount of a 10 000 BP plasmid (pHβAPr-1-neo, 3,4 μm contour length) was titrated with propidium iodide the emitted fluorescence intensity per single detected event increase with rising dye concentration. Such dependence reflects the rising amount of bound fluorescent molecules. The fluorescence intensity saturates at concentration about 0.1 – 0.2 dye molecules (propidium iodide) per phosphate group giving the size of binding sit of about 10 - 5 bases, in agreement with available literature data (Waring, 1965). The fluorescence intensity increase is accompanied by the decrease of the diffusion constant showing that the rising amount of bound fluorescent dyes changes the plasmid mobility by factor of three.

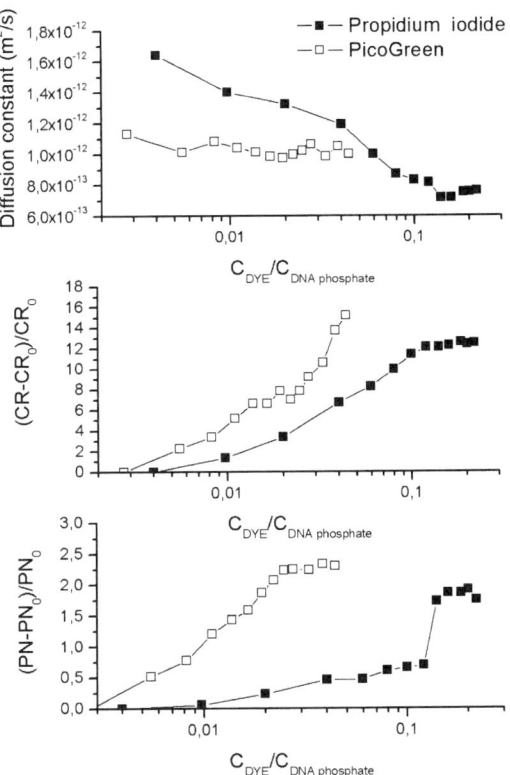

Figure 5.2. The dependence of the βApr-1-Neo plasmid (10 kBP, contour length 3,4μm) diffusion constant, relative count rate $(CR–CR_0)/CR_0$ and relative apparent particle number $(PN-PN_0)/PN_0$ on the $C_{DYE}/C_{DNA PHOSPHATE}$ ratio for propidium iodides as well as for PicoGreen.

This change of plasmid mobility can result only from the altered conformation of the nucleic acid molecule and/or increasing mass of plasmid molecule upon dye association. The second possibility is unlikely since even when assuming that one dye molecule binds to four base pairs will result in the mass increase not bigger that by factor of two, far too small to cause observed decrease of the diffusion constant. The direction of the diffusion constant change suggests that hydrodynamic radius of the plasmid is increasing what indicates that the molecule adopts the relax conformation. This conclusion is in agreement with theoretically predicted effect of intercalator on the nucleic acid molecule (Reese, 1994).

An unambiguous interpretation of the dependence of the apparent particle number on the dye concentration is hampered by the fact that the obtained values for PN are significant larger than PN calculated from the used DNA concentrations and the lateral radius of detection volume (ω_0). As mentioned earlier this discrepancy might be due the treatment of multiple labeled DNA molecules as point-like molecules. Nevertheless, the abrupt increase of the apparent particle number by factor of four above the concentration of 0.2 propidium iodide molecule per phosphate group of DNA is remarkable. There might be two possible explanations for such a jump. First possibility requires the assumption that plasmids are in two distinctly different conformational states which have different affinity towards the propidium iodide. The difference should be so large that one of the populations should be totally inaccessible for the dye, in contrary to the generally observed labeling patterns, numerous data show that both supercoil and relax forms of plasmid are equally visible on gels stained with fluorescent dyes. The other explanation is that there are fragments of DNA molecule which conformation differs locally causing the preferential dye binding. Initially the effect of the dye on local nucleic acid conformation is such that the molecule overall properties are not different from that where dye associate. As the concentration of the fluorophore increases the local disturbances become significant to the point that nucleic acid cannot be consider as a uniform identity any longer. The single plasmid molecule passing the confocal volume is not a single fluorescent event. It is perceived now as a source of multiple fluorescence signals. The last hypothesis is supported by recent AFM and microscopic data showing such a situation induced upon spermine association (Dunlap et al.,1997; Lin et al., 1998; Mel'nikov et al., 1999; Mel'nikov et al.,1995a; Mel'nikov et al., 1995b). This issue is also discuss in our recent paper (Kral et al., 2004 in press).

When plasmids are labeled with PicoGreen the FCS parameters dependences on dye concentrations looks different. The diffusion constant is constant throughout the whole dye concentrations range showing that the plasmid conformation is not altered by the dye association. The fluorescence intensity per plasmid particle is rising monotonically. The apparent particle number saturates at 0.02 dyes molecules per phosphate group. The two last plots shows that the number of detected plasmids increases monotonically with the number of bound dye molecules. Such dependences indicate that there are no preferences for certain of plasmid subpopulation. Furthermore, at concentrations where apparent particle number stabilizes the fluorescence intensity is still rising indicating that all particles are labeled only after certain amount of is added and that further increase of the fluorophore quantity rises only the fluorescence intensity. These results show that the labeling of the DNA molecule with fluorescent dye depends on both the type of the dye and its concentration.

5.4.2. Effect of cationic compounds on the DNA condensation

In subsequent studies the condensation processes itself and the effect of fluorescent dye on the process was examined. We have shown that the condensation process induced by cationic hydrophilic compounds can be effectively monitor with FCS technique (Kral et al., 2002a; Kral et al., 2002b; Kral et al., 2002c). The high sensitivity of the FCS technique for the DNA condensation process is also illustrated in Figure 5.1. Data obtained in the course of those experiments provide information on the amount of the condensing compound needed to complete the process when analyzed particle by particle (see Figure 5.3). Such information is difficult to obtain by other means. Since the monitoring of the condensation process requires the DNA labeling the effect of the fluorescent dye should be accounted for. As mentioned above the fluorescent labeling itself affect the nucleic acid conformation therefore it is expected that the labeling will influence the condensation itself. Indeed we have shown that the plasmid condensation induced by cationic compounds (for example spermine or HTAB) depend on the fluorescent dye used for the monitoring purposes. This effect is strong when propidium iodide or ethidium bromide is used. On the other hand, this affect appears to be neglect able in the case of PicoGreen. Thus, again, PicoGreen appears to be superior to the priopidium iodide (Cosa et al., 2001, Kral et al., 2004 in press). Another result of that investigation is the clear difference in DNA-dye-condenser interactions when comparing spermine with HTAB (Figure 5.3). In the case of spermine the increase in the diffusion coefficient (since around 1 x 10^{-12} till 8 x 10^{-12} m^2/s), indicating condensation, occurs at the same critical spermine concentration (9 x 10^{-6} M) as the drop in the relative count rate (indicating dye release). The diffusion coefficient dependence on HTAB concentration appears to be qualitatively similar to the one observed for spermine, showing a smaller increase in the diffusion constant (since around 1 x 10^{-12} till 5 x 10^{-12} m^2/s). The latter finding confirms the known hypothesis that HTAB is a weaker condensing agent than spermine. The relative count rate, on the other hand, is decreasing constantly within the HTAB titration range and no "drop" as in the case of spermine is observed. These results indicate that condensation and dye release are connected processes in the case of spermine, while for HTAB a different mechanism of the DNA-dye-condenser interactions has to be proposed.

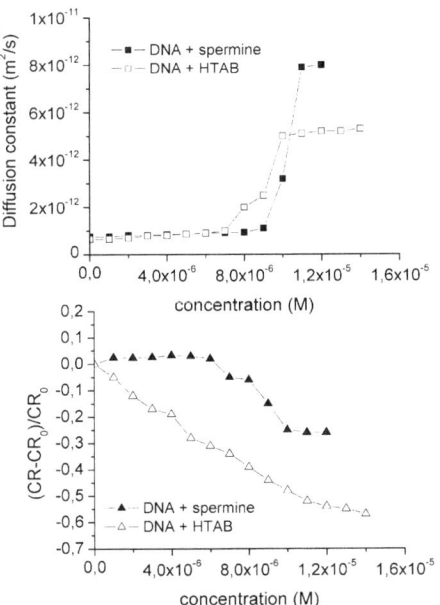

Figure 5.3. Changes in the diffusion constant and relative count rate (CR-CR$_0$)/CR$_0$ of the plasmid βApr-1-Neo plasmid (10 kbp, contour length 3,4μm) as a dependence of a concentration of the spermine and HTAB compounds. Plasmid was labeled by the propidium iodide. The C$_{DYE}$/C$_{DNA\ PHOSPHATE}$ was equal 0.2.

At this point it should be mentioned that the group of Yves Mely has been using the FCS technique with two-photon excitation to characterize the complexes formed by rhodamine-labeled 25 kDa polyethylenimine (PEI) and DNA plasmid molecules (Clamme et al., 2003a; 2003b). As an application, FCS was used to monitor the purification of PEI/DNA complexes by ultrafiltration as well as the heparin-induced dissociation of the complexes.

5.4.3. Effect of cationic lipids on the DNA condensation

Having all issues listed above in mind we have attempted to evaluate the effect of cationic lipids on the DNA condensation. Cationic lipids are frequently used for the transfection formulations in vitro (Byk et al., 1998) as well as potential adjuvant for gene therapies (Miller, 1998). Here the correlation between the lipid properties and DNA containing topology, sample uniformity and transfection efficiency needs to be precisely established. We employed the FCS technique to evaluate the interaction between lipids and nucleic acids. First we attempt to establish the minimum amount positive charge on the lipid surface needed to obtain complete nucleic acid binding. In this experiment oligonucleotides labeled with fluorescein was used. When such oligonucleotides were exposed to liposomes with various amount of cationic lipids the binding efficiency can be easily evaluated (Jurkiewicz et al., 2003). When large nucleic acid molecules (plasmids) interactions with lipid aggregates are studied the obtained results are no longer straightforward. There are a number of processes taking place simultaneously and resulting aggregates distributions are complex. It has been shown previously DNA molecule induces liposomes aggregation along with condensation process itself (Eastman and et al., 1997; Kennedy et al., 2000). In this case the experimental design is complicated and the clear protocol has not been worked out yet. The example of such an attempt is presented on Figure 5.4 where plasmids were titrated with increasing amount of liposomes (small unilamellar vesicles) with various cationic lipids (DOTAP) content. DNA was labeled with PicoGreen.

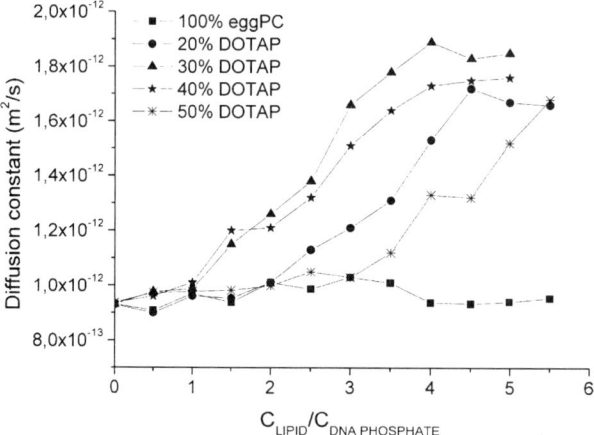

Figure 5.4. The dependence of the diffusion constant of PicoGreen labeled βApr-1-Neo plasmid (10 kbp, contour length 3,4μm) on the ratio between lipid concentration and DNA phosphate concentration ($C_{LIPID}/C_{DNA\ PHOSPHATE}$). The dependence found for small unilamellar egg phosphatidylcholine vesicles containing 0%-15% and 55%-100% DOTAP (not shown) was almost identical as the shown dependence for 100 % egg PC SUV.

It is certainly difficult to draw definite conclusions from such experiments. However, it becomes clear that the diffusion coefficient of the labeled DNA molecule does not change in presence of egg PC vesicles with a DOTAP concentration range of 0% – 15% as well as 55% -100%. Thus in this DOTAP concentration range there is apparently no interaction between DNA and liposomes. On the other hand, there is a clear dependency of the DNA diffusion coefficients on vesicle concentration for the DOTAP range between 20% and 50%. Here we observe the formation of DNA lipid aggregates. The observed profiles, however, differ within this concentration range, indicating the formation of different sized aggregates. We believe that the shown FCS data indicates DNA-lipid aggregate formation, but is by itself not sufficient for a precise characterization of topology and uniformity of these systems.

5. 5. CONCLUSIONS

Changes in diffusion times determined by FCS give certainly straight forward information on conformational changes in DNA down to a single molecular level. The validity of this approach is demonstrated for conformational changes, induced either by dyes or condensing agents. When using this method for controlling the formation of supramolecular aggregates containing DNA, like for example DNA-lipid-aggregates, one has to be aware that the used dye might change the DNA conformation and/or influence the aggregate formation process. There are of course many challenges for the future in this field. We would like to name two as examples: a) Understanding of the ACF resulting from multiple labeled large (DNA) molecules diffusing through the detection volume, in order to extract all information contained in those ACF's. b) Combining FCS of supramolecular aggregates containing DNA with complementary techniques, like dynamic light scattering, AFM, or other microscopic techniques on absolutely identical aggregate systems.

5. 6. ACKNOWLEDGEMENTS

This study was possible thanks to the financial support of the Institute of Physics, Wroclaw University of Technology (M.L.) and the Biocenter for Biotechnology and Protection of Lower Silesia Ecosystem (M.L.), grant No. PBZ-04/PO4/98 from the State Committee for Scientific Research, Warszawa, Poland (M.L.), and to the financial support of the Ministry of Education, Youth and Sports, Czech Republic (via LN 00A032 (T.K. and M.H.).

5.7. REFERENCES

Ahringer, J., 2003, Contral of cell polarity and mitotic spindle positioning in animal cells. *Curr. Opin. Cell Biol.* **15**:73.
Arndt-Jovin, D. J., and Jovin, T. M., 1989, Fluorescence labeling and microscopy of DNA. *Meth. Cell Biol.* **30**:417.
Arnold, F. H., Wintrode, P. L., Miyazaki, K., and Gershenson, A., 2001, How enzymes adapt: lessons from directed evolution. *Trends Biochem. Sci.* **26**:100.
Barenholz, Y., 2001, Liposome application: problems and prospects. *Curr. Opin. Coll. Interface Sci.* **6**:66.

Baumann, C. G., Smith, S. B., Bloomfield, V. A., and Bustamante, C., 1997, Ionic effects on the elasticity of single DNA molecules. *Proc Natl Acad Sci USA* **94**:6185.

Beneš, M., Hudeček, J., Anzenbacher, P., and Hof, M., 2001, Coumarin 6, hypericin, resorufins, and flavins: suitable chromophores for fluorescence correlation spectroscopy of biological molecules. *Coll. Czech. Chem. Commun.* **66**: 855..

Bloomfield, V. A., 1996, DNA condensation. *Curr. Opinion Struct. Biol.* 6:334.

Bloomfield, V. A., 1997, DNA condensation by multivalent cations. *Biopolymers.* **44**:269.

Brown, M. D., Schatzlein, A. G., and Uchegbu, I. F., 2001, Gene delivery with synthetic (non-viral) carriers. *Inernatl. J. Pjarmaceut.* **229**:1.

Bunfield, D. H., and Davis, L. M., 1998, Monte Carlo simulation of a single-molecule detection experiment. *Applied Optics.***37(12)**:2315.

Byk, G., Dubertret, C., Escriou, V., Frederic, M., Jaslin, G., Rangara, R, Pitard, B., Crouzet, J., Wils, P.,and Schwartz, B., 1998, Synthesis, activity, and structure-activity relationship studies of novel cationic lipids for DNA transfer. *J Med Chem.* **41**:224.

Byk, G., and Scherman, D., 2000, Genetic chemistry: tool for gene therapy coming from unexpected directions. *Drug Develop. Res.* **50**:566.

Chevet, E., Cameron, P. H., Pelletier, M. F., Thomas, D. Y., and Bergeron, J. J. M., 2001, The endoplasmic reticulum: integration of protein folding, quality control, signaling and degradation. *Curr. Opin. Struc. Biol.* **11**:120.

Clamme, J. P., Azoulay, J., and Mely, Y., 2003a, Monitoring of the formation and dissociation of polyethylenimine/DNA complexes by two photon fluorescence correlation spectroscopy. *Biophys. J.* **84**:1960.

Clamme, J. P., Krishnamoorthy, G., and Mely, Y., 2003b, Intracellular dynamics of the gene delivery vehicle polyethylenimine during transfection: investigation by two- photon fluorescence correlation spectroscopy. *Biochim. Biophys. Acta-Biomembr.* **1617**:52.

Conti, E, and Izaurralde, E., 2001, Nucleocytoplasmic transport enters the atomic age. *Curr. Opin. Cell Biol.* **13**:310.

Cosa, G., Focsaneanu, K. -S., McLean, J. R. N., McNamee, J. P., and Scaiano, J. C., 2001, Photophysical properties of fluorescent dna-dyes bound to single- and double-stranded dna in aqueous buffered solution *Photochemistry and Photobiology.* **73(6)**:585.

Davis, L. M., Williams, P. E., Cain, H. M., Ball, D. A., Parigger, C. G., Matayoshi, E. D., and Swift, K. M., 2002, Comparison of fluorescence correlation spectroscopy (FCS) and other single-molecule data analysis methods for assay of protein-ligand interactions. Biophysical Journal. **82(1)**:43A.

Davis, M. E., 2002, Non-viral gene delivery systems. *Curr. Opin. Biotech.* **13**:128.

de-Gennes, P. G., 1999, Problems of DNA entry into a cell. *Physica A.* **274**:1.

Derycke, A. S. L., and de-Witte, P. A. M., 2004, Liposomes for photodynamic therapy. *Adv. Drug Delivery Rev.* **56**:17.

Dunlap, D. D., Maggi, A., Soria, M. R., and Monaco, L., 1997, Nanoscopic structure of DNA condensation for gene delivery. *Nucleic Acids Res.* **25**:3095.

Eastman, S. J., Siegel C., Tousignant, J., Smith, A. E., Cheng, S. H., and Scheule, R. K., 1997, Biophysical characterization of cationic lipid:DNA complexes. *Biochim Biophys Acta* **1325**:41.

Eckel, R., Ros, R., Ros, A., Wilking, S. D., Sewald, N., and Anselmetti, D., 2003, Identification of binding mechanisms in single molecule-DNA complexes. *Biophys. J.* **85**:1968.

Ellis, R. J., and Hartl, F. U., 1999, Principles of protein folding in the cellular environment. *Curr. Opin. Struct. Biol.* **9**:102.

Elson, E., Magde, D., 1974, Fluorescence Correlation Spectroscopy. I. Conceptual Basis and Theory. *Biopolymers.***13**:1.

Garcia, M., Alsina, M. A., Reig, F., and Haro, I., 2000, Liposomes as vehicles for the presentation of a synthetic peptide containing an epitope of hepatitis A virus. *Vaccine.* **18**:276.

Grantcharova, V., Alm, E. J., Baker, D., and Horwich, A. L., 2000, Mechanisms of protein folding. *Curr. Opin. Struct. Biol.* **11**:70.

Hortobagyi, G. N., 1997, Antracyclines in the treatment of cancer. *Drugs.* **54**:1.

Hui, S. W., Langner, M., Zhao, Y. L., Ross, P., Hurley, E., and Chan, K., 1996, The role of helper lipids in cationic liposome-mediated gene transfer. *Biophys . J.* **71**:590.

Iino, R., Koyama, I., and Kusumi, A., 2001, Single molecule imaging of green fluorescent proteins in living cells: E-cadherin forms oligomers on the free cell surface. *Biophys. J.* **80**:2667.

Janicki, S. M., and Spector, D. L., 2003, Nuclear choreography: interpretations from living cells. *Curr. Opin. Cell Biol.* **15**:149.

Jurkiewicz, P., Okruszek, A., Hof, M, and Langner, M., 2003, Associating oligonucleotides with positively charged liposomes. *Cell. Molec. Biol. Lett.* **8**:77.

Kennedy, M. T., Pozharski, E. V., Rakhmanova, V. A., and MacDonald, R. C., 2000, Factors governing the assembly of cationic phospholipid-DNA complexess. *Biophys J* **78**:1620.

Kettling, U., Koltermann, A., Schwille, P., and Eigen, M., 1998, Real-time enzyme kinetics monitored by dual-color fluorescence cross-correlation spectroscopy. *Proc. Natl. Acad. Sci. USA* **95**:1416.

Kral, T., Hof, M., Jurkiewicz, P., Langner, M., 2002a, Fluorescence correlation spectroscopy (FCS) as a tool to study DNA condensation with hexadecyltrimethylammonium bromide (HTAB). *Cell. Mol. Biol. Lett.* **7**:203.

Kral, T., Hof, M., and Langner M., 2002b, Effect of spermine on the plasmid condensation and dye release observed by fluorescence correlation spectroscopy *Biol. Chem.* **383**:331.

Kral, T., Langner, M., Benes, M., Baczynska D., Ugorski M., and Hof, M., 2002c, The application of fluorescence correlation spectroscopy in detecting DNA condensation *Biophys. Chem.* **95**:135.

Kral, T., Widerak, K., Langner, M., and Hof, M., 2004, Propidium iodide and PicoGreen as dyes for the DNA Fluorescence Correlation Spectroscopy measurements. *Journal of Fluorescence* (in press).

Langner, M., 2000a, Effect of liposome molecular composition on its ability to carry drugs. *Polish J Pharmacol.* **52**:3.

Langner, M., 2000b, The intracellular fate of non-viral DNA carriers. *Cell. Mol. Biol. Lett.* **5**:295.

Langner, M., and Kral, T., 1999, Liposome-based drug delivery systems. *Pol. J. Pharmacol.* **51**:211.

Lenart, P, and Ellenberg, J., 2003, Nuclear envelope dynamics in oocytes: from germinal vesicle breakdown to mitosis. *Curr. Opin. Cell Biol.* **15**:88.

Lin, Z., Wang, C., Feng, X., Liu, M., Li, J., and Bai, C., 1998, The observation of the local ordering characteristics of spermidine-condensed DNA: atomic force microscopy and polarizing microscopy studies. *Nucleic Acids Res.* **26**:3228.

Lian, T, and Ho, R. J. Y., 2001, Trends and developments in liposome drug delivery systems. *J. Pharmaceut. Sci.* **90**:667.

Magde, D., Elson, E., and Webb, W. W., 1972, Thermodynamic fluctuations in a reacting system - measurement by fluorescence correlation spectroscopy. *Physical Review Letters.* **29**(11):705

Magde, D., and Elson, E., 1974, Fluorescence Correlation Spectroscopy. II. An Experimental Realization. *Biopolymers.* **13**:29.

Magde, D., Webb, W. W., and Elson, E., 1978, Fluorescence Correlation Spectroscopy. III. Uniform Translation and Laminar Flow. *Biopolymers.* **17**:361.

Marcucci, F, and Lefoulon, F., 2004, Active targeting with particulate drug carriers in tumor therapy: fundamentals and recent progress. *DDT* **9**:219.

Matouschek, A., 2003, Protein unfolding - an important process in vivo? *Curr. Opin. Struc. Biol.* **13**:98.

Medina, M. A, and Schwille, P., 2002, Fluorecence correlation spectroscopy for the detection and study of single molecules in biology. *BioEssay* **24**:758.

Mel'nikov, S. M., Sergeyev, and V. G., Yoshikawa, K., 1995a. Discrete coil-globule transition of large DNA induced by cationic surfactant. *J Am Chem Soc* **117**:2401.

Mel'nikov, S. M., Sergeyev, V.G., and Yoshikawa K., 1995b, Transition of double-stranded DNA chains between random coil and compact globule states induced by cooperative binding of cationic surfactant. *J Am Chem Soc* **117**:9951.

Mel'nikov, S. M., Dias, R., Mel'nikova, Y. S., Marques, E. F., Miguel, M. G., and Lindman, B., 1999, DNA conformational dynamics in the presence of cationic mixtures. *FEBS Lett.* **453**:113.

Mhashilkar, A., Chada, S., Roth, J.A., and Ramesh, R., 2001, Gene therapy therapeutic approaches and implications. *Biotech. Adv.* **19**:279.

Miller, A. D., 1998, Cationic liposomes for gene therapy. *Angew. Chem. Int. Ed.* **37**:1768.

Molema, G., 2002, Tumor vasculature directed drug targeting: applying new technology and knowledge to the development of clinically relevant therapies. *Pharmaceut. Res.* **19**:1251.

Nakai, K., 2001, Prediction of in vivo fates of proteins in the era of genomics and proteomics. *J. Struct. Biol.* **134**:103.

Nelson, W. J., and Yeaman, C., 2001, Protein trafficking in the exocytic pathway of polarized epithelial cells. *Trends Cell Biol.* **11**:483.

Parak, F. G., 2003, Proteins in action: the physics of structural fluctuations and conformational changes. *Curr. Opin. Struc. Biol.* **13**:552.

Reese, H.R., 1994, Effects of DNA charge and length on the electrophoretic mobility of intercalated DNA.. *Biopolymers.* **34**(10):1349.

Rigler, R., Mets, U., Widengren, J., Kask, P., 1993, Fluorescence correlation spectroscopy with high count rate and low background: analysis of translational diffusion. *European Biophysics Journal.* **22**:169.

Sombrook, J., Fritsch, E. F., and Maniatis, T., 1989, *Molecular Cloning. A laboratory manual*, Cold Spring Harbor Laboratory Press, New York. p. 1.33.

Stein, G. S., Zaidi, S. K., Braastad, C. D., Montecino, M., Wijnen, A. Jv., Choi, J. Y., Stein, J. L., Lian, J. B., and Javed, A., 2003, Functional architecture of the nucleus: organizing the regulatory machinery for gene expression, replication and repair. *Trends Cell Biol.* **13**:584.

Torchilin, V. P., and Lukyanov, A. N., 2003, Peptide and protein drug delivery to and into tumors: challenges and solutions. *DDT* **8**:259.

Verkman, A. S., 2002, Solute and macromolecule diffusion in cellular aqueous compartments. *Trends Biochem. Sci.* **27**:27.

Waring, M. J., 1965, Complex formation between ethidium bromide and nucleic acids. *J Mol Biol* **13**: 269.

Wohland, T., Rigler, R., and Vogel, H., 2001, The standard deviation in fluorescence correlation spectroscopy. *Biophysical Journal.* **80(6)**:2987.

Zarif, L., Graybill, J. R., Perlin, D., and Mannino, R. J., 2000, Cochleates: new lipid-based drug delivery systems. *J. Liposome Res.* **10**:523.

LUMINESCENCE-BASED OXYGEN SENSORS

B. A. DeGraff and J. N. Demas[+]

6.1. INTRODUCTION

The area of oxygen sensors encompasses a broad range of sensing techniques, devices, and applications. The latter range from monitoring combustion mixtures to use in fish farming.[1] The majority of the techniques and devices involve transduction of the changes in oxygen concentration or pressure to changes in an electrical parameter such as voltage, current, or resistance. This sensing approach is older and has more working devices on the market. A more recent approach involves transduction via changes in luminescence characteristics of both organic and inorganic dyes. This approach has attracted considerable interest for applications such as pressure sensitive paints (PSP) and for situations amenable to use of fiber optic technology.

Good overviews of the various types of sensors are available[2] with one focusing on fiber optics.[3] A brief review of some aspects of luminescent oxygen sensors has also appeared.[4] The purpose of this review is to survey the recent progress in luminescent oxygen sensors and to raise some of the unanswered questions that must be satisfactorily addressed before this technique can achieve its full potential.

6.2. OVERVIEW

Virtually all luminescence-based oxygen sensors use the quenching of the dye's luminescence by oxygen as the transduction mechanism (quenchometric). The electronically excited dye is deactivated by molecular oxygen as shown below.

$$D + h\nu \rightarrow D^* \qquad\qquad I_a \qquad\qquad (1)$$

$$D^* \rightarrow h\nu' \qquad\qquad k_r \qquad\qquad (2)$$

[+] B. A. DeGraff, Chemistry Department, James Madison University, Harrisonburg, VA 22807. J. N. Demas, Chemistry Department, University of Virginia, Charlottesville, VA 22904.

$$D^* \rightarrow heat \qquad\qquad k_{nr} \qquad\qquad\qquad (3)$$

$$D^* + O_2 \rightarrow D + O_2^* \qquad\qquad k_q \qquad\qquad\qquad (4)$$

In this simple scheme, the excited dye, D^*, has only three fates: Emission (Eq. 2), non-radiative loss of energy to the environment (Eq. 3), and loss of energy by collision with ground state O_2 (Eq. 4). The latter is shown here as a collisional process, though static quenching through ground state dye/oxygen pairs could also occur. This static quenching will alter the intensity-based measurements, but will not affect lifetime-based techniques. If Eq. (4) is a simple second order process, and the amount of available ground state oxygen is not significantly depleted by energy transfer from D^*, the scheme can be modeled by standard Stern-Volmer kinetics and results in a set of expressions relating the decreases in either luminescence intensity or lifetime to the available oxygen concentration.

$$I_0/I = \tau_0/\tau = 1 + K_{sv}[O_2] \qquad\qquad K_{sv} = k_q/(k_r + k_{nr}) \qquad\qquad (5)$$

where I_0 and τ_0 are the intensity and lifetime in the absence of oxygen. This expression predicts a linear change in either the intensity or lifetime ratios with oxygen concentration. Further, the intensity changes and lifetime changes should track each other. Such behavior should allow very simple calibration of luminescence-based sensors for oxygen, and this type of behavior is indeed observed for many dyes in homogeneous solutions with dissolved oxygen as the quencher. However, like the ideal gas law, Eq. (5) is the simplest case and rests on a series of assumptions that may not be universally valid. It is the failure of some of these implicit assumptions that generates many of the current problems to be solved for luminescent O_2 sensors.

From Eq. (5) it is clear that the sensitivity of a dye to oxygen quenching is reflected in the magnitude of K_{sv}. The actual k_qs reported for various dyes in solution range from diffusion limited to an order of magnitude or so less. With polymer-supported dyes, the rigidity of the support can significantly reduce the k_qs. However, for similar environments, the wide variation seen in oxygen quenching sensitivity is usually a reflection of variation in the denominator of the K_{sv} expression, $1/(k_r + k_{nr}) = \tau_0$. It is the large range of excited state lifetimes available with various dyes that allows one to tune the oxygen response for a sensor. The impact of τ_0 on the oxygen sensitivity is the reason that many organic dyes have limited application as oxygen sensors. For example, laser dyes are available in a wide range of emission colors and have very high emission efficiencies. However, they typically make terrible oxygen sensors as $\tau_0 \sim 1$ - 5 ns. Even given diffusional quenching, to reduce the luminescence intensity or lifetime by 50% would require $3 \rightarrow 5$ atm of pure oxygen and $15 \rightarrow 25$ atm of air if $\tau_0 \sim 1$ ns. While there are some organic dyes with longer lifetimes such as derivatives of pyrene[5,6,7] or decacyclene[8] ($\tau_0 \sim 100$ ns) that have been used as oxygen sensors, the majority of dyes used for ambient conditions are metal complexes due to their longer τ_0s. Indeed, the range of lifetimes exhibited by metal complexes is from ~100 ns to nearly 1 ms, which is an impressive 4 orders of magnitude.

The idyllic world of Eq. (5) is often shattered by the need to support the dye in some way to make a useful sensor. The need to localize the dye and in some cases control access to it by the analyte is a task often assumed by some type of supporting matrix. The ideal matrix would (1) localize the dye without adversely affecting its photophysical response, (2) protect the dye from leaching and deactivation by materials found accompanying the analyte (i.e. solvents, biological materials, etc.), (3) exhibit a high permeability to oxygen, and (4) have desirable optical and mechanical properties. The hunt for this "holy grail" represents a significant portion of the current research effort on luminescent O_2 sensors. The compromises necessary with current supports are, in large measure, responsible for many of the problems that will be highlighted in this review.

Current luminescence-based oxygen sensors usually employ transition metal complexes, and we will limit our discussion to these systems. We will discuss recent advances in dyes and supports and techniques related to their characterization. We will also cover major ongoing problems in the development and understanding of oxygen sensor systems including non-linear sensor response and photochemistry. We will conclude with a look towards the future.

6.3. DYES

There is no one ideal oxygen sensing dye. Indeed, a number of parameters and attributes must be considered in designing a new dye for a specific application. Of paramount importance is the dye's lifetime so that its O_2 sensitivity falls within the desired range. The dye must be easily optically pumped, preferably using either inorganic or organic LED sources or laser diodes. A significant Stokes shift for the emission insures easy separation of the emission from pump light. Since often only small amounts of the dye are mounted in a support matrix on a fiber optic, photostability is a significant concern. Dyes that have photophysical properties relatively insensitive to temperature over a modest range allow for much simpler calibration. Finally, compatibility with the desired support matrix material is essential. Various compromises are required and, hence, a large number of complexes have been suggested as O_2 sensors.

While numerous complexes have been suggested as potential oxygen sensors, the majority of the most recent work involves either platinum metal porphyrins (e.g. Pt(II), Pd(II), Rh(III)) or Ru(II) α-diimine complexes. Figure 6.1 gives the structure of some of the complexes including the abbreviations used for the α-diimines referenced here. The exceptionally long lifetimes ($0.1 \rightarrow 10$ ms) of the platinum metal porphyrins makes these materials particularly suitable for low oxygen concentration/pressure sensing. Of the α-diimines, the Ph$_2$phen (Figure 6.1) has received the greatest attention. The lifetimes of the Ru(II) α-diimines are typically 10 to 100 times shorter with K_{sv}s on the order of 10^3 M^{-1}. Depending on the exact complex and support matrix, there is some overlap between the sensing ranges for the two classes of materials.

The luminescences of all of these complexes come from heavy atom perturbed triplet states, the lowest excited state in the molecule. Although it is not strictly correct, these states are generally referred to as triplet states, which lead to s the emissions phosphorescences. The emissions of the Ru(II) complexes are metal-to-ligand charge transfer (MLCT) while the porphyrins are ligand localized π-π* in character.

A key parameter for any potential oxygen sensing dye is the dynamic range exhibited. This is simply a measure of the fractional change in intensity or lifetime that occurs between environments that contain differing amounts of oxygen. This parameter has been expressed as the ratio of the emission intensity in air to that in pure oxygen, I_{air}/I_{oxygen}, and also as the ratio of emission intensity for an inert gas (e.g., nitrogen) or vacuum) environment to that in pure oxygen, $I_{nitrogen}/I_{oxygen}$. A similar set of ratios exists

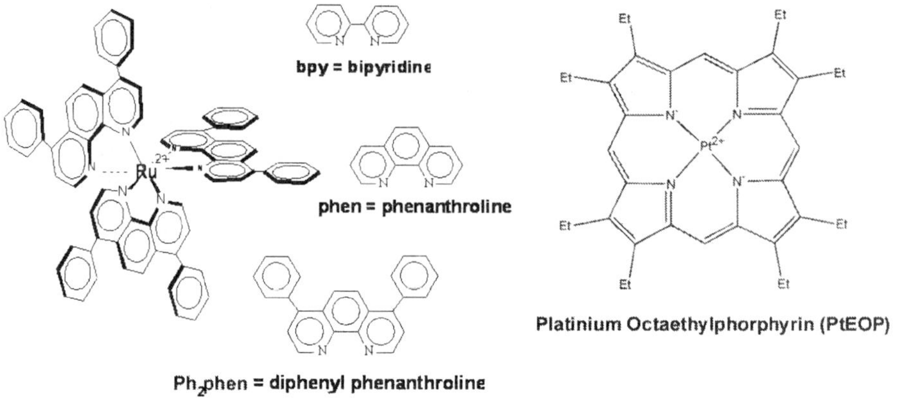

Figure 6.1. Basic structures and abbreviations of complexes and ligands used. The 3-D structure of Ru(Ph₂phen)₃²⁺ complex is shown.

for lifetime measurements. One would like these ratios to be large to insure adequate signal change over the expected range of oxygen concentrations or pressures. If the dye's lifetime is too long for the desired oxygen concentration, either saturation can occur and additional oxygen will not significantly affect the intensity or lifetime, or the signal level will drop to a point where discrimination from the background is impossible. While the dynamic range is very much a function of the environment of the sensor (i.e., solution, polymer matrix, sol-gel, etc.), most viable candidates for oxygen sensing exhibit a $I_{nitrogen}/I_{oxygen}$ ratio of at least 10 in fluid solution. Calculation of k_qs and comparisons among different supported dyes are difficult since the solubility of oxygen in the matrix material is seldom known.

Among the most sensitive materials are the platinum metal porphyrins due to their long lifetimes. Both pump and emission light falls in the visible region and the Stokes shifts are sufficient to allow easy discrimination. They do have a rather significant temperature sensitivity[9,10] and are somewhat photosensitive to decomposition.[11–15] The issue of photochemical robustness has been addressed by oxidation of the porphyrin system to the corresponding ketone, which is more photochemically stable,[14,15] or by substituting fluorine for hydrogen.[16,17] The temperature sensitivity has been addressed by use of a second luminophore, which is insensitive to oxygen but has a significant and known temperature response.[9,10] To provide more reliable intensity measurements for PSP applications, a reference luminophore has been employed.[16] A series of water soluble platinum metal porphyrins has been immobilized in a Nafion membrane to provide an oxygen sensor with good sensitivity.[18] The complexes are based on Pt(II), Pd(II), and Rh(III) complexes of (4-N-methylpyridyl)porphyrin and (4-N,N,N-trimethylaminophenyl)porphyrins. The authors observed a significant increase in the complexes' lifetimes when immobilized on the Nafion film compared to fluid solution.

Values of I_{argon} / I_{oxygen} from 6 to over 80 were reported. The study was based on intensity measurements.

In an effort to address the temperature sensitivity of both porphyrin based and many Ru(II) based complexes,[19] a complex with a bichromophoric luminophore was synthesized and characterized.[20] The complex was based on a Ru(bpy)$_2$ core with a third Ru(II) coordinated bipyridine having a covalently bound pendent pyrene. The dye in degassed solution shows a biexponential luminescent decay with lifetimes of 1.3 μs (MLCT) and 57.4 μs (π-π* phosphorescence). The dye was dispersed in a poly(ethylene-glycol)ethylethermethacrylate and tri(propyleneglycol)diacrylate copolymer. The system's emission properties were virtually temperature independent in the 25 - >55°C range and showed good oxygen sensitivity due to the long effective lifetime. An analysis of the conditions required for this temperature insensitivity is also presented. The system was proposed as an improved PSP. Intensity data only were used to evaluate the oxygen and temperature sensitivity.

An interesting approach to self-calibration for intensity based O_2 sensing involves a Pt(II) complex with a dual emission.[21,22] The complex [(dppe)Pt{S$_2$C$_2$(CH$_2$CH$_2$-N-2-pyridinium)}(BPh$_4$), where dppe is 1,2-bis-(diphenylphosphino)ethane and BPh$_4^-$ is tetraphenyl borate anion, has both fluorescent and phosphorescent emissions. Only the long lived phosphorescence is quenched by oxygen and so the complex is self-referencing. An $I_{nitrogen}$/I_{oxygen} ratio of ~ 10 was observed for this material immobilized in a plasticized cellulose acetate (75% by wt. triethylcitrate plasticizer) film.

A series of Eu(III) complexes has been suggested as oxygen sensors.[23] Since most luminescent rare earth complexes are not subject to oxygen quenching, this is an interesting study. Oxygen sensitivity was conferred by adjusting the metal ion's emissive f→f states so that some coupling with the ligands' π* levels occurs. The heteroleptic Eu(III) complexes involve phen and thenoyltrifluoroacetonato ligands. Homoleptic complexes were also made using the thenoyltrifluoroacetonato, 1,1,1-trifluoro-5,5-dimethyl-2,4-hexanedionato, and 1,1,1,2,2,3,3-heptalfluoro-7,7-dimethyl-4,6-octanedion-ate ligands. The materials were studied in cast polystyrene films by monitoring the intensity changes. While relatively linear Stern-Volmer quenching plots were obtained, the dynamic range exhibited by the series was rather small ($I_{nitrogen}$/I_{oxygen} ratio of ~ 2). These complexes might still be useful as high pressure oxygen monitors in hyperbaric chambers.

6.4. SENSOR SYSTEMS

While the number of reports of totally new dyes suitable for oxygen sensing seems to be tapering off, the interest in finding new ways to support existing dyes in suitable matrices has increased. This is due, in part, to an appreciation of the inherent advantages of lifetime measurements as a basis for O_2 sensing. However, this approach has the sticky problem of complex emission decays that are often encountered with supported dyes. Also, with applications involving *in vivo* or food use, the stringent condition that no dye leaching occurs places strict requirements on dye immobilization in the support.[24] While chemically bonding the dye to the support can eliminate leaching, the competing goal of ease of sensor fabrication has led to exploration of approaches other than covalent attachment of the dye to the matrix. We provide a brief summary of some of the recent reports concerning amelioration of the complex decay and leaching problems encountered with supported dyes.

As discussed elsewhere (*vide infra*), there are several suggested causes for the observed non-ideal emission decay behavior of dyes in host matrices. These, broadly viewed, revolve around variation of host density and polarity that give rise to variations in oxygen permeability and differences in the binding of the dye. Also to be considered is the host/dye compatibility. Poor compatibility may result in dye aggregate formation on either the micro or nano scale. Dye aggregates with micron diameters have been observed for organic dyes dispersed in PMMA using near-field scanning optical spectroscopy.[25] In principle, smaller aggregates can be detected using this technique, but to our knowledge no one has reported detection of nanometer diameter dye aggregates. Depending on whether host density variations or poor host/dye compatibility is deemed operative, a number of different strategies have been tried to minimize the problems.

In the currently used dyes, the problem of multiple decay times and non-linear Stern-Volmer quenching behavior appears most severe for the Ru(II) α-diimine dyes. This is often due to solubility problems with the dye/support combination. One of the best performing dyes is Ru(Ph$_2$phen)$_3$X$_2$ (X= monovalent counter ion) due to its long lifetime ($\sim$ 5 μs). Hosts based on polydimethylsiloxane (PDMS) polymers are often used due to their large oxygen permeabilities. Unfortunately, the dye is virtually insoluble in pure PDMS supports. Selection of a hydrophobic anion, X$^-$, such as lauryl sulfate can offer some improvement, but usually one is forced to use a copolymer based on PDMS and some more polar component.[26,27] Typically the dye is introduced by swelling the polymer host with a dilute solution of the dye followed by slow evaporation of the solvent. The dye is then physically, but not chemically, bound within the host support. Films can also be cast from solution with linear polymers or photopolymerized from monomer solutions of the dye. Even though these different fabrication methods yield films that appear homogeneously luminescent under ordinary magnification, the decays are still multi-exponential and aggregation on the nanometer scale cannot be ruled out.

There appear to be fewer compatibility problems with the porphyrin dyes as Douglas and Eaton[28] reported single exponential decays with porphyrin dyes physically trapped in several very different polymers in the absence of oxygen, but multiple decays in its presence. This suggests that aggregation is not the issue here, but rather differences in local oxygen permeability or solubility in the support.

One approach to improving compatibility is to covalently bond the dye to the polymer backbone as a pendent group. Winnik, et al. have used this approach with a polythionylphosphazene polymer and a covalently tethered Ru(phen)$_3$$^{2+}$ as the dye.[29] These materials were compared to a system in which the same dye was immobilized in the polythionylphosphazene polymer by standard swelling techniques. The emission response (I$_0$/I) of both systems was nearly linear with oxygen pressure, but the Stern-Volmer constant for the covalently linked material was notably greater. Interestingly, the lifetimes of both systems were multi-exponential. However, the Stern-Volmer quenching constant evaluated for the average lifetime matched the intensity value for the covalently linked system, but the two constants did not match for the swelled material.

A cyclometalated Ir(III) complex was studied in two environments: (1) the complex was covalently bound to the polymer support and (2) the complex was swelled into a similar polymer.[30] The complex, Ir(ppy)$_2$(vpy)Cl was the starting point where ppy is 2-phenylpyridine and vpy is 4-vinylpyridine. This complex was either covalently bound to a PDMS polymer or the chloride replaced with a phenyldimethylsilyl group and then physically adsorbed into the PDMS polymer using standard swelling techniques. Evaluation of the oxygen response of cast thin films by intensity measurements showed that the sensitivity of the covalently bound complex was superior. However, all the Stern-Volmer response curves were somewhat non-linear.

Intermediate between physical adsorption and covalent binding is chemisorption. In this approach a dye is held to a polyelectrolyte host by way of ionic charge effects. Amao and Okura used this approach in which several Ru(II) α-diimine dyes were chemisorbed on to a substrate using a polyelectrolyte such as poly(acrylic acid) or poly(sodium-4-styrene sulfonate).[31] In this technique, the Ru(II) complex cation is combined with the polyelectrolyte containing either multiple carboxyl or benzene sulfonate groups. These carboxyl or sulfonate groups become, in effect, the counter ion for the complex that is held to the polyelectrolyte by electrostatic interaction. This material is then coated on an inert support such as alumina. The resulting thin film is removed mechanically and evaluated for oxygen response. The complexes $[Ru(bpy)_3]^{2+}$, $[Ru(phen)_3]^{2+}$, and $[Ru(Ph_2phen)_3]^{2+}$ were used. Only intensity studies were done, but all the systems showed strongly non-linear Stern-Volmer plots and significant compression of the dynamic response. Typical $I_{nitrogen}/I_{oxygen}$ values were $\sim$ 1.5->4. One must conclude that this type of chemisorption exacerbates differences among binding sites and suppresses dynamic response.

Sequestration of the dye into porous, three-dimensional networks has also received considerable attention as a support matrix.[32–36] Sol-gel technology has been the most common approach to this type of dye immobilization. While a number of oxygen sensors have been reported using sol-gel sequestration of the dye, the Stern-Volmer response plots are often non-linear and there are long term stability issues.[37,38] It was found that by using organically modified silicates (ormosils) the cast sol-gel films offered improved response.[39,40,41] Recently Bright, et al.[42] reported linear intensity Stern-Volmer plots using ormosils and cast films with good long term stability. An octyl group was used as the hydrophobic polymer modifier and varying amounts of n-octy-triethoxysilane in the copolymer mixture were tried. The dye used was $[Ru(Ph_2phen)]_3^{2+}$. At sufficiently high n-octyl-triethoxysilane concentration in the copolymer mixture they obtained single exponential intensity decays. This offers the hope that two-point calibration for oxygen sensors based on Ru(II)-α-diimine dyes can be achieved.

6.5. NEW SENSING SCHEMES, TECHNIQUES, AND APPLICATIONS

Our purpose here is not an exhaustive review of recent reports, but rather to provide a flavor of some of the interesting sensing techniques and applications that have appeared. Due both to the smaller diffusion coefficient of dissolved oxygen and the lower oxygen concentration in most solutions, measurement of dissolved oxygen is more challenging than of gaseous oxygen levels due to smaller changes in the signal levels. For many applications, the most common and simplest form of oxygen sensing involves excitation of a supported dye using an intensity modulated LED. The emission phase shift is measured at a single frequency and the oxygen concentration or pressure determined from a suitable calibration. A variety of configurations, dyes, and supports have been reported that use this basic approach. The phase shift approach is also capable of providing modulation information as well, but this is often not used due to problems in making reliable modulation measurements. However, a ratio method has been reported that allows modulation measurements to be made even under adverse conditions.[43]

6.5.1. Self-Referencing Measurements.

One of the problems of conventional intensity-based measurements is that there is no reference. Decomposition, fluctuation in source intensity, optical transmission properties, or aging of the detector can produce spurious analyte measurements. Lifetime measurements to a considerable degree circumvent this problem, although extensive decomposition can still yield errors. Alternative approaches use an internal reference. If a luminophore is added that does not respond to oxygen and has a different emission or excitation spectrum than the analyte-sensitive component, then the analyte signal can be referenced to the oxygen insensitive reference signal. The need for two wavelength measurements adds complexity since two intensities at different excitation or emission wavelengths are required. However, in cases such as nanosensors in cells, it is largely the only approach available.[44]

A variation on a dual wavelength system exploits differences in excited state lifetimes. A single excitation wavelength is used and the intensity contribution of the reference and sensor molecules are separated by their temporal or phase differences. This can be done in a pulsed, AC, or gated AC mode.[45,46,47] Details of both are given in the literature. Long lived transition metal complexes with short lived organic fluorophores are an ideal combination.

6.5.2. Miscellaneous Approaches

With the advent of small, reliable, CCD cameras spatially resolved measurements are now easily made and a number of systems using this technique have been reported.[48] By use of gated detection, lifetime as well as intensity-based measurements can be obtained simultaneously. This technique has seen wide application in biologically related studies ranging from tissue samples to cell culture growth.

The appearance of useful OLED (Organic LED) sources has led to integrated oxygen sensors using this light source.[49,50] In addition to low cost, OLED's can be fabricated in a variety of geometries, are flexible, and generate very little heat. These qualities make integrated oxygen sensors for physiological use easier to fabricate.

One of the major problems facing sensors in both field and physiological use is fouling, which can be just inconvenient or life threatening. In marine applications it was found that including a phosphorylcholine substituted methacrylate component in the support copolymer significantly reduced adhesion of marine bacteria without a significant loss of sensor performance.[51]

Given the importance of oxygen in physiological processes, there has been considerable interest in monitoring oxygen at the cellular level. The challenge is to do so without disrupting the normal cell functions. This requires sensors on the micron scale. The first approach to this task involved fabricating very small optodes often mounted on pulled fibers that were inserted into the cell.[52] More recently sensors have been fabricated by binding the sensing material or dye to very small particles of cross-linked polymer or sol-gel.[44] These PEBBLEs (probes encapsulated by biologically localized embedding) show a narrow size range and can be fabricated to be physiologically inert and non-fouling. Oxygen sensors using both Ru(II)-α-diimine[53] and metal porphyrin complex[54] dyes with ormosil supports have been evaluated. The latter were deemed superior due to greater oxygen sensitivity and lack of interference from light scattering and autofluorescence of the cellular material in the detection wavelength region, although α-diimines can be prepared with longer wavelength emissions.[55]

6.5.3. Diffusion Measurements

A critical parameter in the development of new sensors is identifying supports with controllable and typically high oxygen diffusion coefficients. High diffusion coefficients (Ds) allow more rapid response of film of a given thickness. Also, high Ds generally correlate with large k_qs and bigger K_{SV}s for a given excited state lifetime. Thus, both sensor response time and sensitivity can be affected by the support's D. Since many of the materials used for supports are new, *a priori* knowledge of the D is largely a matter of luck, although glass transition temperatures (Tgs) below room temperature and large void volumes (e.g., polydimethysiloxanes) generally enhance D.

Luminescence-based methods using the actual sensor molecule support now readily permit measuring Ds from the temporal response of the emission following a step function change in the oxygen concentration. These techniques are simple to implement for laboratories designing sensors and allow direct measurement of the composite system avoiding any ambiguities from having to take D from other preparations. Further, it was well known experimentally that the rise and fall times and the shapes of the luminescence curves for step changes were different on going from low to high oxygen concentrations and for the reciprocal high to low oxygen concentration. The diffusion equations completely clarify this difference. Figure 6.2 shows the asymmetric fall and rise times and the computed signal.[56]

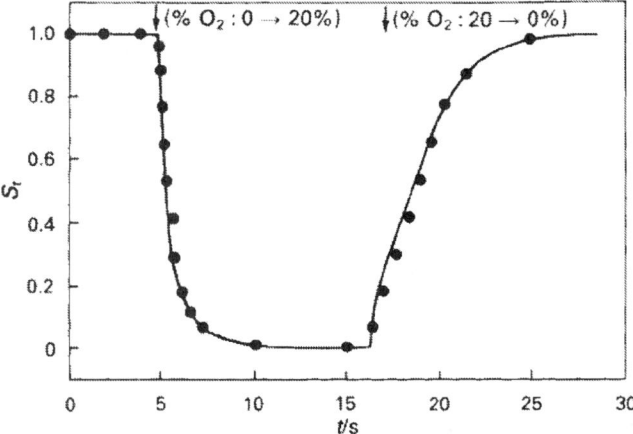

Figure 6.2. Variation in the relative optical signal as a function of time for an optical film sensor on changing the ambient O_2 level from 0 to 20% oxygen and then back to nitrogen. The solid line is the best fit. Adapted with permission from Reference 56. Reproduced by permission of The Royal Society of Chemistry.

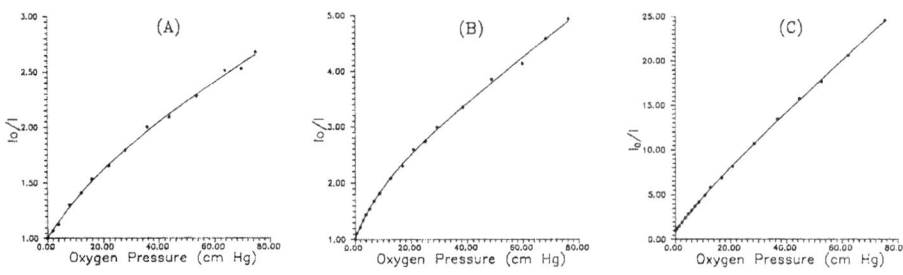

Figure 6.3. Stern-Volmer quenching data for three Ru complexes in RTV-118 silicon rubber. The solid lines are the best experimental fits to the points. Data adapted from Reference 57.

The oxygen diffusion measuring method has traditionally been intensity based,[58] but more recently a phase shift lifetime method has been developed.[59] The results agree well with the intensity based method. As with other lifetime versus intensity measurements, the former is more immune to deviations from changes in source intensity and photodecomposition. The mathematics is somewhat more complex since lifetimes do not sum in the way intensities do. However, where long measurement times from thick samples or low Ds make stability an issue, the lifetime method provides a viable alternative to intensity measurements.

One unexpected result concerned the effect of plasticizer on a Ru(II) oxygen sensor.[58] As expected, a plasticizer greatly increased D, but the degree of quenching dropped substantially. This is contrary to expectations as an increase in D would be expected to increase quenching. It is possible that the plasticizer greatly reduced the oxygen solubility with a concomitant reduction in the $k_q[O_2]$ term. However, given the magnitude of the effect, this seems unlikely. A more plausible explanation is that the microenvironment around the Ru(II) complex was altered to affect shielding or diffusion coefficient. Thus, at least in some systems, one has to worry about the effect of changes in both the bulk D and in the nanoenvironment for additives and fillers around the probe.

6.6. NON-IDEAL SENSOR RESPONSE

The vast majority of oxygen sensors show some deviation from linear Stern-Volmer behavior. The deviation is a function of luminophore, polymer, additives, and the method of sample preparation. Even a minor change in a casting solvent can radically alter quenching behavior. Figure 6.3 shows varying degrees of non-linear intensity quenching behavior for several ruthenium complexes embedded in RTV 118 silicone rubber. Similar variations can be found for a single complex dispersed in different polymers. Since more work has been done with polymer supported dyes as sensors, we will focus on the non-ideal responses of these systems.

Important questions that arise from such results are:
- What are the sources of the deviations?
- How can they be controlled to give reproducible behavior?
- What mathematical models can be used to describe these deviations?

- What mathematical models can be used to characterize the sensors for actual field calibration?
- Can new dye-support systems be found to solve these problems?

The general consensus of opinion is that the non-linear behavior arises from heterogeneity in the sensor environments. The complexes can exist in sites of different accessibility to the oxygen, regions of different oxygen solubility, and regions of different diffusion coefficients for the oxygen. Particularly in polymers, one can envision more than one of these factors being operative at the same time. Many of these features have their counterparts in luminescence probes of biomolecules. For example, a binding pocket for a probe may block access of the quencher from some directions, the local solubility and mobility of a quencher in a membrane can greatly affect quenching, or the probe may exist in multiple conformations or environments with different lifetime or quenching properties.[60]

The use of additives or solid filler to modify the mechanical, optical, or luminescence properties can further increase the complexity. Additives can increase the complexity of the domains present with the extreme case being a polymer blend. For solid particle fillers such as fumed silica, the luminophores can partition between the surface of the filler and the bulk polymer. The partitioning can depend on the concentrations of the components, the binding coefficients, or in some case the kinetics during preparation and post-equilibration time. In each domain, there is a distribution of binding sites. An excellent review of the complexities of composite systems is available.[61]

Real sensors turn out in general to be very complex, and the meaningful fitting of these systems can be most difficult. What you seem to measure depends very much on the measurement made. Lifetime and intensity quenching data can give completely different results. Intensity measurements made by AC or steady-state measurements do not have to match each other. Pulsed and phase lifetime measurements can each give quite different sensing properties depending on how the data are collected and worked up.

As is well known in data fitting, complex systems are frequently poorly poised and a wide range of parameters and even completely different models can give indistinguishable fits. Thus, workers wishing to generate reliable models of the underlying photophysics and structure are likely to be asking different questions and approaching the problem differently than those who want only a robust fitting equation that reliably describes the response of a sensor to an analyte.

In particular, it is critical to recognize that just because a mathematical model fits the data does not mean that it has any physical reality. To further compound the issues, as is well known from kinetics it is common for two completely different physical models to give exactly the same mathematical expression for a phenomena. Below we will describe some of the widely used models, their successes and pitfalls, and a few of the issues that can arise.

Probably, the most common model for fitting intensity quenching data is the multisite model where the luminophore is assumed to be in two or more sites each being described by a normal Stern-Volmer equation. This yields

$$I_0 / I = \left[\sum_{j=1}^{N} \frac{f_{0j}}{1 + K_{SVj}^{true} k_{Hj} p_{O_2}} \right]^{-1} = \left[\sum_{j=1}^{N} \frac{f_{0j}}{1 + K_{SVj}^{p} p_{O_2}} \right]^{-1} \tag{6a}$$

$$K_{SVj}^{p} = K_{SVj}^{true} k_{Hj}$$

(6b)

$$K_{SVj}^{true} = k_{2j} \tau_{0j}$$

(6c)

where K_{SVj}^{true} is the true Stem-Volmer constant, k_{2j} is the bimolecular quenching constant, k_{Hj} is the Henry's law gas solubility constant, and j corresponds to an index for the different binding sites or regions. We have made the assumption that different binding regions or domains might exhibit different local solubilities (i.e., Henry's law constants). The simplest form that gives a non-linear Stern-Volmer equation is the two-site model.

$$I_0 / I = \cfrac{1}{\cfrac{f_{01}}{1 + K_{SV1}^{p} p_{O_2}} + \cfrac{f_{02}}{1 + K_{SV2}^{p} p_{O_2}}} = \cfrac{1}{\cfrac{f_{01}}{1 + K_{SV1}^{p} p_{O_2}} + \cfrac{1 - f_{01}}{1 + K_{SV2}^{p} p_{O_2}}}$$

(7)

In our platinum metal α-diimine complex work, we have yet to see an intensity Stern-Volmer plot that could not be quantitatively fit with the two-site model. The fits in Figure 6.3 are all based on a two-site model.

Another popular three parameter intensity model is the non-linear gas solubility model based on the known non-linear deviations from Henry's Law solubility as a function of oxygen pressure for some polymers. The non-linear Stern-Volmer behavior is then attributable to non-linearity of the oxygen concentration in the polymer as a function of oxygen pressure coupled to normal Stern-Volmer kinetics within the polymers. The equation for non-linear oxygen solubility is

$$[O_2] = C_H + C_L = k_H p + C'_L bp /(1 + bp)$$

(8)

where C_H and C_L are the Henry's law and Langmuir contributions to the solubility, respectively. k_H, C_L', and b denote the Henry's law parameter, the Langmuir adsorption capacity, and the affinity constant of the gas for the Langmuir sites, respectively. Based on these assumptions, one arrives at a non-linear Stern-Volmer intensity equation[62,63]

$$I_0 / I = 1 + A p_{O_2} + \frac{B p_{O_2}}{1 + b p_{O_2}}$$

(9a)

$$A = K_{SV} k_H$$

(9b)

$$B = K_{SV} C'_L b$$

(9c)

This equation is that of Reference 63, but we have omitted the possibility of dissociation of the encounter complex without energy transfer. However, omission of this complication has no effect on the functional form of their equation. Depending on the relative Henry's law and Langmuir contributions to quenching, the Stern-Volmer plots will range from linear to concave downward and tend towards a plateau at high pressures. The equation also fits a variety of intensity quenching data as well as the two-site model.

It should. Eqs. (7) and (9) are identical mathematically.[64] In spite of the completely different physical significance of the model, both equations have been shown to be the same along with the relationships between the fitting parameters. In short, either one can be used for generating calibration curves, but one cannot base a belief in either model solely on the basis of the quality of the fits.

Using other physical evidence, we have shown that the non-linear solubility model is less appropriate for some of our metal complex based systems.[64] However, the non-linear gas solubility model is simpler and mathematically faster to evaluate, and we use it for just fitting experimental data in our diffusion measurements without any regard for the physical significance of the parameters.

In reality, it is easy to show that for our systems, and most others, both models are wrong. One need only look at the lifetime quenching data. The decays are non-exponential at all oxygen pressures, which rules out the non-linear solubility model. In many cases the decays can be fit to two exponentials, but three is common. However, if the data were actually described by a two-exponential decay, the ratio of the pre-exponential factors for the two decays would be independent of oxygen concentration, and each lifetime would be given by a lifetime Stern-Volmer equation. Neither condition is satisfied, so the systems are more complex.

As far as fitting intensity quenching data, a variety of discrete and continuous functions of lifetime have been explored.[65,66,67] The two-site system with one component being weakly quenched or unquenched has been applied successfully to some systems. Other functions include a Gaussian distribution, log normal distribution, Freundlich adsorption isotherm, and a variety of other functions. The Freundlich distribution has worked for adsorption on silica surfaces.[68] All models generate downward curved Stern-Volmer plots. Truncated single Gaussian distributions (no probability of negative lifetimes) have difficulty giving high degrees of non-linearity, and a second Gaussian generally had to be added.[69] Log normal curves were introduced by Mills and do a credible job with a variety of intensity quenching systems.

Before turning to lifetime fitting, we explore one of the complexities that arises in lifetime measurements. For simple diffusional quenching in solution, the lifetime and intensity Stern-Volmer plots will be coincident (Eq. 5). If the two plots differ, static quenching is present. This analysis depends on the assumption that the luminescence decay curves are single exponentials, which they will be in well-behaved solutions systems. However, how does one make a comparison in microheterogeneous systems where the decays are not single exponential and not characterized by a single decay time? Complex decay curves are usually fit by a sum of exponentials.

$$D(t) = \sum_{j=1}^{N} \alpha_j \exp(-t/\tau_j) \tag{10}$$

One needs to reduce these parameters to a single effective lifetime in order to use Eq. (5). The commonly reported mean lifetime[60] is not an acceptable substitute for a single lifetime. However, using the preexponential weighted or median lifetime as it is sometimes called, allows a direct comparison of intensity and lifetime data. The preexponential weighted lifetime is given by

$$\tau_M = \sum_{j=1}^{N} \alpha_i \tau_j \Bigg/ \sum_{j=1}^{N} \alpha_i \tag{11}$$

The necessary αs and τs are extracted from the emission decay of Eq. (10). The number of exponentials is varied until a successful fit is obtained as judged by the magnitude of the chi square and the appearance of the weighted residuals plot. In our experience, no more than three exponentials have been required for experimental data. The computed best fit αs and τs are used to calculate τ_M from Eq. (11).

Even if no physical significance can be ascribed to the τs and αs, the following is true if there is no static quenching

$$I_0/I = \tau_{M0}/\tau_M \tag{12}$$

which is equivalent to Eq. (5). Thus, to detect the presence of static quenching, the preexponential weighted lifetime Stern-Volmer data can be compared with the intensity data. Agreement between the two establishes the absence of static quenching. In all of the common oxygen sensors examined to date, there has been no evidence for static quenching. This is what one would expect on the basis of the chemical association of ground state oxygen with these complexes. So, in general, unless there is unusual upward curved intensity Stern-Volmer behavior, one can safely assume that all quenching is dynamic.

It should be pointed out that the quality of the intensity-lifetime comparison in Eq. (12) is critically dependent on the quality of the decay time data. It should only be attempted with the best single photon counting decay data or multi-frequency phase shift data.

We turn now to the much more complex lifetime fitting. In contrast to a simple intensity quenching plot, which is essentially an intensity versus oxygen pressure curve, there really is not much information to fit, and it should come as no surprise that a variety of models work well. However, lifetime measurements basically add another dimension to the measurement. One has the shape of the decay curve at each oxygen concentration that must now be fit. Indeed, if one can accurately reconstruct the decay curves as a function of oxygen concentration, one should be able to reconstruct the steady state emission intensities by integrating the area under these decay curves. Ultimately, accurate modeling of the information rich lifetime data would be the best approach.

Again, a number of models have been tried. The most extensive analysis was done on a Pt octaethylporphine-ketone in polystyrene with a variety of discrete and continuous distribution functions used to model both phase shift lifetime data and AC intensity measured on their phase shift instrument.[65,66] This is an excellent system to model because the lifetimes are long enough to measure accurately and the decay curves are single exponential in the absence of oxygen, which allows the analysis to begin with a homogeneous initial distribution of lifetimes. The log normal and other asymmetric distributions gave the best fits. However, the best fit to both intensity and lifetime data was not with the same model or parameters. This leaves little doubt that the selected models were in fact not rigorously correct. Since the authors were only looking at the phase shift and amplitude at one frequency, they were clearly not exploring the experimental data space in its entirety, and inspection of different ways of collecting the data would no doubt yield different results.

Another elegant example used pulsed decay data on Pt and Pd octaethylporphyrins (OEP) in ethyl cellulose (EC).[67] Again, even though the Stern-Volmer intensity plots were relatively linear, the decay data as a function of oxygen concentration revealed underlying complexity. The decay curves were distinctly non-exponential at non-zero oxygen concentrations. They examined impulse excited decay curves and fit the full decays versus oxygen concentration. They tried three different functions for the quenching rate constant: a truncated Gaussian, a log normal stretched distribution, and a constant probability over a fixed range. The three distributions using the parameters giving the best fit are shown in Figure 6.4. The Gaussian gave the worst fit. Surprisingly, the physically unrealistic constant probability distribution gave the best fit. The decay curves and their residuals for the linear best fit are given in Figure 6.5. While the fits are quite reasonable, it is clear that they are not perfect and the model is physically unreasonable as well. Intensity quenching data were not computed from their fitting function and compared with intensity quenching data, so that we do not have the whole picture.

From the two detailed modeling studies, it is clear that even for the "simple" case of platinum porphyrins, the modeling is making inroads, but there is still a ways to go before it can be claimed that physically realistic models fit the data quantitatively. Once this is achieved, the even more complex issue of trying to explain the model can begin.

While it has not yet been used, it seems that an approach that may yield useful insights is the Maximum Entropy Method (MEM). MEM makes no detailed assumptions about the physical model describing the data. It generates smooth, continuous distributions of lifetimes that fit the data.[70,71,72] These sets of distribution can then be subsequently modeled.

In summary, there are good mathematical models that can be used to fit intensity or lifetime calibration curves. The more information rich lifetime measurements are by far the hardest to fit, but by limiting the information (e.g., phase shift at a single modulation frequency), relatively simple equations provide good data descriptors. Good models for fitting complete sets of intensity and lifetime data are still lacking even for the relatively simple porphyrin systems. The majority of α-diimine systems exhibit more complex kinetics starting with non-exponential decays at zero oxygen concentration, and as such will prove much more demanding. Until plausible mathematical models are available, any attempt to try to describe the underlying structure of the systems is premature. We turn now to various approaches that will help to understand various micromechanistic aspects that generate non-ideal behavior.

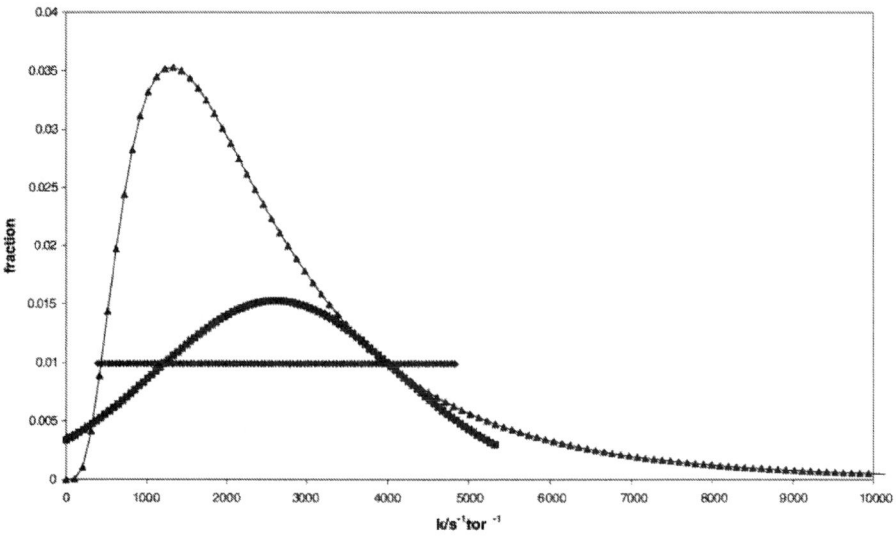

Figure 6.4. The three distributions giving the best fit to the experimental PdOEP/EC data. Adapted with permission from Reference 67.

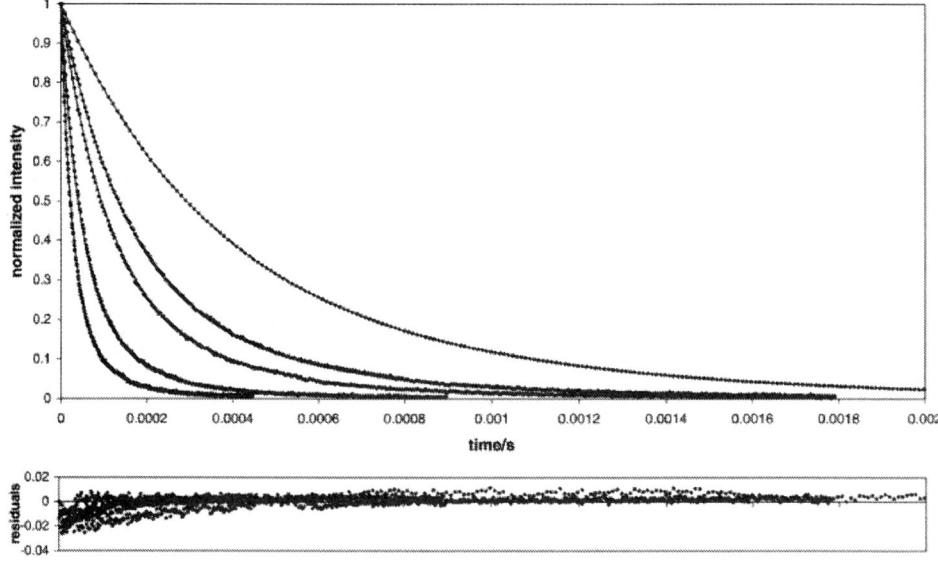

Figure 6.5. Experimental decay curves for a PdOEP/EC film (symbols) overlaid with modeled decay curves from a uniform distribution (full lines) at the following partial pressures of oxygen (left to right): 12.9, 7.1, 2.8, 1.9 and 0.68 Torr. Adapted with permission from Reference 67.

6.7. SOURCES OF HETEROGENEITY AND METHODS OF STUDYING

We can envision heterogeneity as falling into three categories: 1) macroheterogeneity, 2) microheterogeneity, and 3) nanoheterogeneity. Macroheterogeneity would visibly manifest its presence as inhomogeneous coloration, and turbidity in systems without solid additives, or visible crystals. Such problems are a clear demonstration of unsatisfactory fabrication that would have to be remedied before further work. Microheterogeneity would arise at a microscopic level and could arise from crystallization of the polymer or dye, cracking, or phase separation of the polymer. These issues would be directly amenable to study by conventional and, more directly, fluorescence microscopy (FM). Nanoheterogeneity would arise at below the optical resolution of a conventional microscope. For example, if nearby molecules were in different orientations in the polymer, or felt different environmental perturbations of their electronic properties, or the nanostructure of the polymer differed enough to cause local variation in the oxygen solubility or diffusion coefficient, two nearby molecules could easily differ in their quenching properties. In addition, nanoaggregates could also affect behavior. The only way to quantify such differences would be with direct measurements of the properties of single molecules or nanoaggregates.

There have been several studies clearly demonstrating the suitability of FM for detecting and quantitating microheterogeneity. In FM one can readily see differences in fluorescence intensity between different regions of a polymer. Variations in the intensity alone do not establish the different Stern-Volmer kinetics for the different regions that is necessary for non-linear macroscopic Stern-Volmer quenching. However, by examining luminescence under differing conditions of oxygen pressure, one can readily measure the Stern-Volmer quenching constant at every point in the image. Thus, it becomes a straightforward matter to determine variations in the Stern-Volmer quenching in different domains. Two groups have used this technique. In one case, a Pt porphyrin in a silicone was shown to be homogeneous with respect to region while a Ru(II) complex showed heterogeneity in Stern-Volmer quenching constants over micrometer distances.[73] Our group has also demonstrated heterogeneous quenching with Ru complexes in silicones.[74,75,76] At least some of this heterogeneity was ascribed to incompatibility of the charged metal complex with the very hydrophobic PDMS silicone support.

Studies that rely on a two point Stern-Volmer plot do not provide any indication of heterogeneity at the optical resolution of the microscope. By taking intensity measurements at three oxygen concentrations, it was possible to assess the linearity of the Stern-Volmer plots of each image pixel, resolution that is smaller than the optical resolution. In some cases, there was clear non-linear Stern-Volmer behavior even at the pixel level. Thus, in systems exhibiting microheterogeneity, there was evidence of an underlying nanoheterogeneity.[74,75]

In the data of Figure 6.3, there is decreasing non-linear Stern-Volmer behavior on going from $[Ru(bpy)_3]^{2+}$ to $[Ru(phen)_3]^{2+}$ to $[Ru(Ph_2phen)_3]^{2+}$. This correlates with the compatibility of the complex with the largely hydrophobic PDMS support. Greater hydrophobicity in the dye gives better linearity. A possible interpretation of this result is that there is microcrystallization. Bowman and Kneas made a comprehensive study of different types of FM for detecting microcrystallization.[76] They found that conventional FM works, but finding the crystals is difficult. Conventional 1-photon confocal microscopy works, but because of the high light exposures throughout the sample, photobleaching was a pervasive problem, particularly for the longer experiments required to do a multilayer imaging of a sample. Infrared 2-photon confocal FM proved far and away the best approach to study microcrystals.

Figure 6.6 shows microcrystals with the three different complexes in the same polymer. The bright stars on the black background are the microcrystals. As predicted from solvent-solute compatibility models, the $[Ru(bpy)_3]^{2+}$ showed the greatest degree of microcrystallization. The crystals were inhomogeneous with respect to quantity and size at different depths in the polymer. $[Ru(phen)_3]^{2+}$ also exhibited a significant but lesser degree of microcrystallization, but the crystals were much more uniformly distributed with respect to size and density. $[Ru(Ph_2phen)_3]^{2+}$, which has a nearly linear Stern-Volmer curve showed no crystallization beyond the surface 10 μm layer.

The one problem with two photon confocal microscopy was that it was difficult to get quenching data. The stage and alignment were too easily disturbed during the gas changes to always get reliable registration of images.

While single molecule detection by luminescence in solution and on surfaces has been known for some time, the technique has not been applied to the phosphorescent molecules typically used in oxygen sensors. A major problem has been their long lifetimes. To get good signal-to-noise ratio on single molecules the molecule must undergo many excitations and emissions to detect enough photons. For a fluorescent molecule with a low nanosecond lifetime, the excited molecules emit and are ready for re-excitation in a few nanoseconds. For a typical Ru complex the lifetime is 1-5 μs and Pt porphyrins are longer still. Thus, the recharge time for each molecule is much longer, and it is not possible to get such high rates of photon production from a phosphorescent molecule.

In spite of these problems, single molecules of a Pt octabutoxycarbonyl porphyrin in PMMA and surface bound $[Ru(Ph_2phen)_3]^{2+}$ have been measured.[77] Further, it has been possible to measure the quenching of individual molecules. In the case of the Pt complex, the quenching is similar to the bulk measurement and there was little or no detected heterogeneity. In contrast for $[Ru(Ph_2phen)_3]^{2+}$, while the Stern-Volmer quenching curve for each molecule is linear, the Stern-Volmer quenching constants vary widely. There was a broad distribution of Stern-Volmer quenching constants for single molecules bound to the surface. Figures 6.7 and 6.8 show the luminescence and oxygen quenching of single $[Ru(Ph_2phen)_3]^{2+}$ molecules on a silica surface along with the distribution of Stern-Volmer quenching constants measured for 377 single molecules. It was not clear which of the several factors such as geometric shielding, excited state lifetime, or variations in the local oxygen concentration and quenching constant were operative. Since the complex was surface bound rather than dissolved in a polymer, there is no correlation with the polymer data. It is well known that Ru complexes bound to hydrophilic silicas show grossly non-linear Stern-Volmer plots and very non-exponential decays.[68] Since such surfaces probably represent the best model for binding to a silica surface, the broad distribution of quenching constants comes as no surprise.

The single molecule work is still in its infancy. However, what is clear is that such single molecule studies will provide an entry into making nanoscale measurements and will allow the correlation of the results for ensembles of single molecules with the bulk properties. For the first time it may be directly possible to correlate the nano with macro behavior. Further, from nano measurements, the modelers will have an unambiguous distribution of properties to use. We consider this one the most exciting areas in the search for a fundamental understanding of sensors.

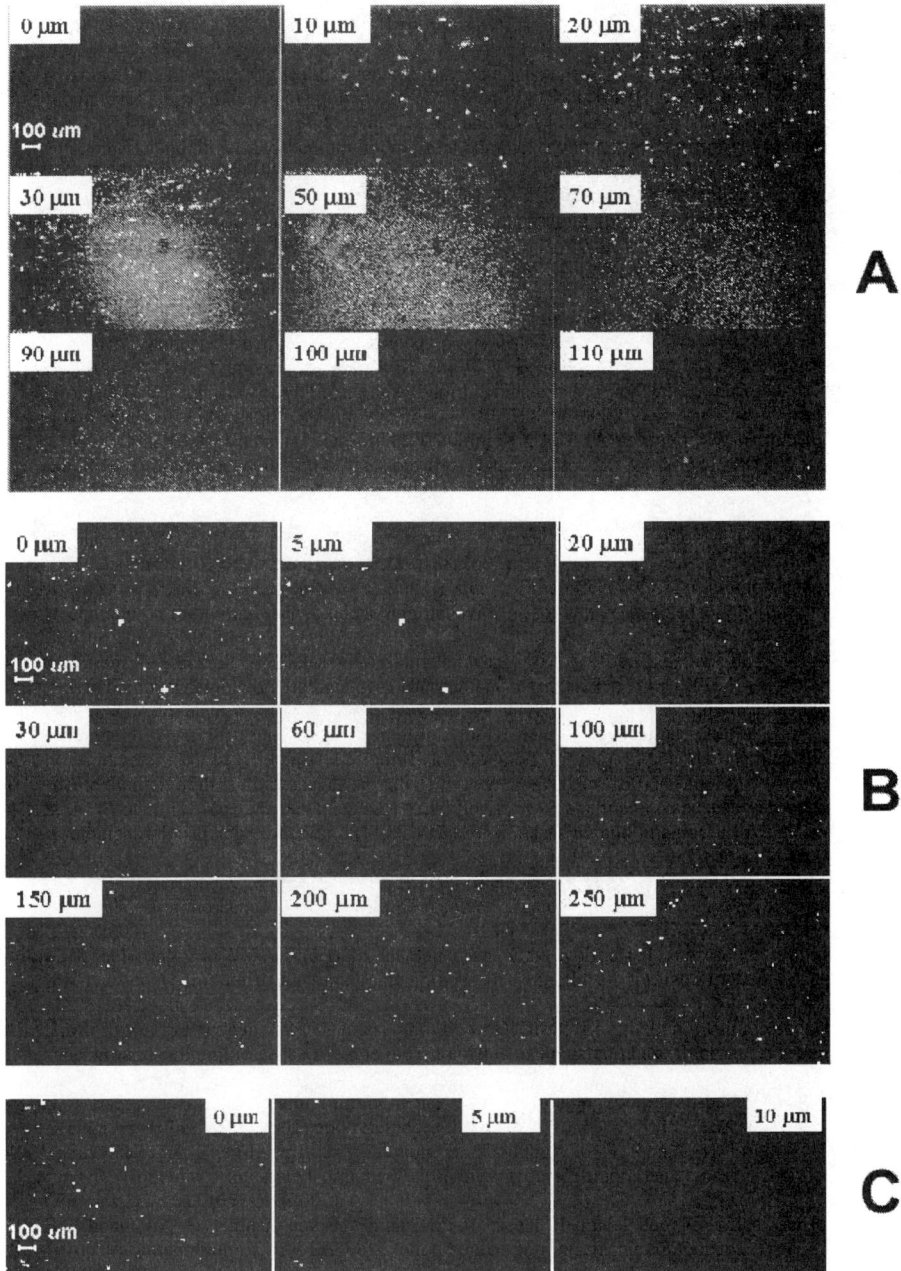

Figure 6.6. Two photon cross-sectional images of (A) [Ru(bpy)₃]Cl₂, (B) [Ru(phen)₃](ClO₄)₂ and (C) [Ru(Ph₂phen)₃]Cl₂ microcrystals in RTV-118. The depth of each measurement from the surface is indicated in the box. Adapted with permission from Reference 76.

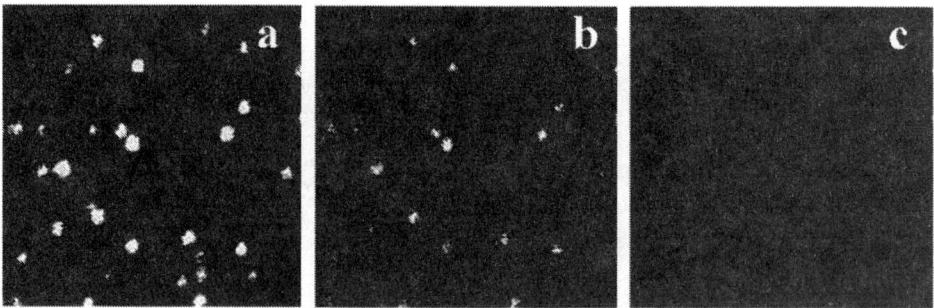

Figure 6.7. Single molecule phosphorescence images of [Ru(Ph$_2$phen)$_3$]Cl$_2$ immobilized on a quartz surface in pure Ar (a), in Ar with 0.5% O$_2$ (b), and in Ar with 3% O$_2$ (c). Adapted with permission from Reference 77.

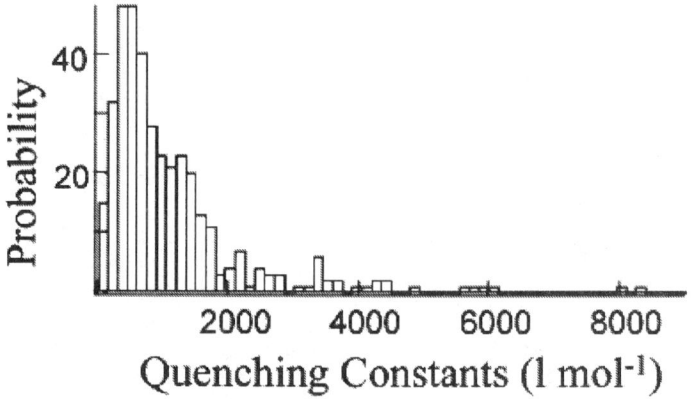

Figure 6.8. Distribution of quenching constants of 377 single molecules of [Ru(Ph$_2$phen)$_3$]Cl$_2$ on a quartz surface. Adapted with permission from Reference 77.

6.8. PHOTOCHEMISTRY

A major issue is the robustness of the sensors. Problems such as leaching into the environment or the sample package, photochemical stability with time, and dye/support aging are all of importance. Since the sensing properties are critically dependent on the stability of the polymer and the environment around the sensor molecule, changes in properties such as crosslinking, crystallization, cracking, plasticizer migration, sensor migration and crystallization, and phase transformations can all contribute to long term instability. While important, even critical, these are not issues that we will deal with here. We limit our discussion to photochemical instability of the sensor dye.

In many instances the photostability of the sensor is of little importance. Before use sensors can be kept in the dark. Single or short term use sensors will have minimal light exposure before being discarded. For example, single use sensors for blood glucose monitoring need only last for a few seconds. Long term industrial or environmental monitoring and some shorter term monitors at the end of fiber optic probes require significantly greater photostability. Repeated light exposure, even by relatively weak sources, over long times could expose sensors to relatively high photon doses. Fiber optic systems with small probe volumes can also suffer from damage since they are frequently illuminated by intense laser sources for good signal-to-noise. This could lead to degradation over relatively short periods due to the high photon flux per unit volume of the sensor.

There are huge numbers of photochemical reactions possible for dyes. The triplet states of organic molecules are notoriously reactive in part because of their long lifetimes.[78] This reactivity is probably one of the reasons that relatively small numbers of molecules possess the long lifetimes and long term stabilities necessary for good sensors. Porphyrin photochemistry generally goes through some type of reactive radical. Use of certain ketones and elimination of reactive sites by fluorination has significantly reduced the photostability problems of early porphyrins (vide supra).[14,16] We will focus most of our discussion on the photochemical stability of platinum metal α-diimine complexes, which is still very much a work in progress.

For platinum metal α-diimine complex dyes, there are two primary decomposition modes. 1) Direct reactivity of the excited state either in an intramolecular reaction or a reaction with solvent or other nearby molecules. 2) Generation of a reactive secondary species by energy transfer such as singlet oxygen formation in oxygen quenching by ground state triplet oxygen. Both mechanisms are present in α-diimine systems, but with some interesting complications.

The prototype Ru(II) complex $[Ru(bpy)_3]^{2+}$ is photochemically and thermally robust. The luminescence was first reported by Paris and Brandt who used it analytically for ruthenium, but they noticed no instability.[79] In deoxygenated water the complex decomposes with a quantum efficiency on the order of 10^{-5}, which is comparable to the organic laser dyes. Acid enhances the decomposition.[80,81] The photochemistry is somewhat reversible and has been attributed to excited state dissociation of one nitrogen of the bpy with solvent substitution to form a monodentate bpy (Figure 6.9).[82] Over a period of time the singly coordinated bpy can displace the solvent to form the original luminescent complex, or a second solvent molecule can displace the bpy to form a di-solvato complex. Neither the mono- nor di-solvato form are luminescent and both have absorption bands shifted substantially to the red of the $[Ru(bpy)_3]^{2+}$ MLCT absorption band.[82] Solvents that enhance ion pairing with good entering group counter ions such as bromide or thiocyanate can substantially increase the photodecomposition yield.[83] Elevated temperatures enhance the thermal activation of the dissociative d-d excited states that can lead to photodecomposition.[82,84] We have discussed in detail the excited state properties of these systems and the criteria that can be used to rationally design photochemically stable luminescent complexes.[85]

Less obvious are the inherent seeds of destruction planted by the oxygen sensing reaction itself. Oxygen quenching of the α-diimine excited states was well known from the earliest days of the use of these complexes as photosensitizers and in analytical methods. Our group was the first to demonstrate that oxygen quenching of the excited state produced excited singlet oxygen in near unity yield.[86,87] Near unity efficient singlet

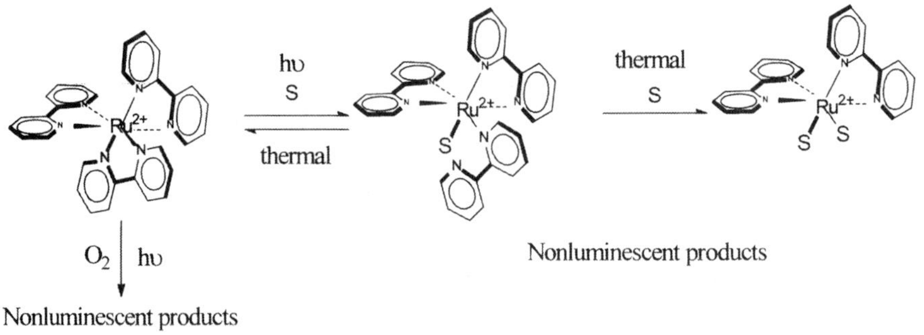

Figure 6.9. Photochemical paths for a typical Ru(II) α-diimine complex showing both potentially reversible photosubstitution and irreversible photooxygenation. S can be either a solvent molecule or a good entering counter ion.

oxygen production was observed in both solution and in heterogeneous cation exchange resins where the authors were able to probe interesting substructures of the system by quenching and solvent deuteration studies.[88]

Singlet oxygen is a very reactive molecule and is widely used in synthesis, waste treatment, and photodynamic cancer therapy.[89,90,91] It reacts with unsaturated molecules, proteins, and nucleic acids. It was reasonable to expect that singlet oxygen could also attack the ligands of α-diimine complexes. The first semi-quantitative study of oxygen effects on oxygen sensors involved several metal complexes in an RTV silicone rubber.[57] Oxygen greatly enhanced the decomposition rate. The photochemistry showed a high degree of heterogeneity with there being a luminescence component that was essentially insensitive to photoreaction. There are well known reactive patterns for singlet oxygen and organics.[92] Based on ligand structure we expected increasing reactivity on going from bpy to phen to 5,6-dimethyl-1,10-phenanthroline complexes. However, the photochemical reactivity did not correlate with this expected singlet oxygen reactivity, and we suggested that another reactive oxygen intermediate might be involved, which was a surprising result given the efficient production of reactive singlet oxygen. Since it is now known that at least some of these systems are microheterogeneous with varying levels of microcrystallization,[75,76] the heterogeneous photochemistry and the absence of a correlation with expected singlet oxygen reactivity may be accounted for by other factors.

More recently the direct interaction of singlet oxygen in the photochemistry of $[Ru(Ph_2phen)_3]^{2+}$ has been shown by stabilization of the complex on addition of a singlet oxygen quencher DABCO (1,4-Diazabicyclo[2.2.2]octane).[93] DABCO reduced the efficiency of the photoreaction by roughly a factor of two to three in lifetime and absorption measurements. Mass spectral analysis of the reaction products indicated the presence of oxygenated ligands. Thus, there now seems little doubt that singlet oxygen is a player in the photodecomposition of α-diimine complexes, and that judicious design of ligands to reduce singlet oxygen attack could be used to improve photostability.

The other significant mode of decomposition is photodissociation. We have carried out a comprehensive study of the effects of oxygen on photodecomposition of $[Ru(bpy)_3]^{2+}$ in solution and polymers in the presence and absence of oxygen.[94] Our expectation was that singlet oxygen attack would enhance decomposition in solution, since oxygen greatly enhances the rate of decomposition in polymers (up to a factor of 10 compared to nitrogen). In fact, oxygen stabilized the complex in solution, which was the opposite of the polymer data. We observed 1) oxygen insensitive photochemistry that we attribute to the known photodissociation of the complex to form solvated species, and 2) a photooxidation chemistry introduced in the presence of oxygen. Since the photodissociation goes through the emitting excited state, oxygen quenching of this state will reduce the photodissociation by quenching the state before it dissociates. If the singlet oxygen chemistry is less efficient than the photodissociation, then adding oxygen results in a net stabilization of the complex in solution as is observed. The higher efficiency of oxygen enhanced decomposition in polymers was attributed to a cage effect; the singlet oxygen, once formed, had difficulty escaping the polymer cage around the metal complex that formed it. This trapping led to multiple encounters with the complex and greater chances of reaction.

6.9. FUTURE WORK

We expect continued work on the development of new materials, both dyes and supports. Major areas will include the development of dye-polymer systems with better compatibility and reduced heterogeneity. While covalent attachment holds some promise for reduction in heterogeneity, fabrication is more difficult. Longer wavelength excitation with higher quantum yields will be important for *in vivo* systems. Improving photostability will be an important area of study. A better understanding of the polymer stability is also critical for long term applications. For many biological and environmental monitoring applications, fouling is an issue coming to the forefront, and much more work is needed.

While our present knowledge allows some rational design of improved sensor systems hopefully many of the above issues will be guided by a more fundamental understanding of the underlying processes. Microscopy will be useful for understanding microheterogeneity. New single molecule methods promise a deep understanding of the fundamental processes involved.

New low cost and miniaturized instrumentation, especially in the lifetime area, will be critical to new applications. These includes a variety of uses such as embedded glucose analyzers and external rapid glucose blood analyzers.

6.10. ACKNOWLEDGEMENTS

We gratefully acknowledge the University of Virginia, the National Science Foundation (grant CHE-0097659 and CHE-0410061), and the W.M. Keck Foundation.

6.11. REFERENCES

1. J. N. Demas, B. A. DeGraff, and P. Coleman, Oxygen sensors based on luminescence quenching, *Anal. Chem.*, **71**, 793A-800A (1999).
2. R. Ramamoorthy, P. K. Dutta, and S. A. Akbar, Oxygen sensors: materials, methods, designs and applications, *J. Materials Sci*, **38**, 4271-82 (2003).

3. O. S. Wolfbeis, Fiber optic chemical sensors and biosensors, *Anal.Chem.*, **76**, 3269-8, (2004).
4. Y. Amao, Probes and polymers for optical sensing of oxygen, *Microchim. Acta*, **143**, 1-12 (2003).
5. Wenying Xu, R. Schmidt, M. Whaley, J. N. Demas, B. A. DeGraff, E. K. Karikari, and B. A. Farmer, Oxygen sensors based on luminescence quenching: Interactions of pyrene with the polymer supports, *Anal. Chem.* **67**, 3172-80(1995.)
6. B. J. Basu, and K. S. Rajam, Comparison of the oxygen sensor performance of some pyrene derivatives in silicone polymer matrix, *Sens. Actuators B*, **99**, 459-467 (2004).
7. Y. Amao, and Y.Fujiwara, Optimizing oxygen-sensitivity of optical sensor using pyrene carboxylic acid by myristic acid co-chemisorption onto anodic oxidized aluminum plate, *Talanta*, **62**, 655-660 (2004).
8. O. S. Wolfbeis, *Fiber Optic Chemical Sensors and Biosensors*, (Vol. 2, CRC Press, Boca Raton, FL., 1991).
9. L. M. Coyle, D. Chapman, G. Khalil, E. Schibili, and M. Gouterman, Non-monotonic temperature dependence in molecular referenced pressure sensitive paint, *J. Lumin.*, **82**, 33-39 (1999).
10. L. M. Coyle, and M. Gouterman, Correcting lifetime measurements for temperature, *Sens. Actuators B*, **61**, 92-99 (1999).
11. F. A. Nwachukwu, and M G. Baron, Polymeric matrices for immobilising zinc tetraphenylporphyrin in absorbance based gas sensors, *Sens. Actuators B*, **90**, 276-285 (2003).
12. P. Hrdlovic, Luminescence quenching by oxygen in polymer matrices: The role of polymer structure, *Polymer News*, **28**, 79-90(2003).
13. G. DiMarco, and M. Lanza, Optical solid-state oxygen sensors using metalloporphyrin complexes immobilized in suitable polymer matrices, *Sens. Actuators B*, **63**, 42-48 (2000).
14. D. B. Papkovsky, G. V. Ponomarev, W. Trettnak, and P. O'Leary, Phosphorescent complexes of porphyrin ketones: Optical properties and application to oxygen sensing, *Anal. Chem.*, **67**, 4112-4117 (1995).
15. P. Hartmann, and W. Trettnak, Effects of polymer matrixes on calibration functions of luminescent oxygen sensors based on porphyrin ketone complexes. *Anal. Chem.*, **68**, 2615-2620 (1996).
16. G. E. Khalil, C. Costin, J. Crafton, G. Jones, S. Grenoble, M. Gouterman, J. B. Callis, and L. R. Dalton, Dual-luminophor pressure-sensitive paint I. Ratio of reference to sensor giving a small temperature dependency, *Sens. Actuators B*, **97**, 13-21(2004).
17. B. Zelelow, G. E. Khalil, G. Phelan, B. Carlson, M. Gouterman, J. B. Callis, and L. R. Dalton, Dual-luminophor pressure sensitive paint II. Lifetime based measurement of pressure and temperature, *Sens. Actuators B*, **96**, 304-14(2003).
18. V. V. Vasil'ev, and S. M. Borisov, Optical oxygen sensors based on phosphorescent water-soluble platinum metals porphyrins immobilized in perfluorinated ion-exchange membrane, *Sens. Actuators B*, **82**, 272-76(2002).
19. B. F. Carroll, J. P. Hubner, K. S. Schanze, and J. M. Bedlek-Anslow, Principal component analysis of dual-luminophore pressure/temperature sensitive paint, *J. Visualization*, **4**, 121-129 (2001).
20. H-F. Ji, Y. Shen, J. P. Hubner, B. F. Carroll, R. Schmehl, J. A. Simon, and K. A. Schanze, Temperature-independent pressure-sensitive paint based on a bichromophoric luminophore, *Appl. Spectroscopy*, **54**, 856-63(2000).
21. Y. Kostov, and G. Rao, Ratio measurements in oxygen determinations: wavelength ratiometry, lifetime discrimination, and polarization detection, *Sens. Actuators B*, **90**, 139-142(2003).
22. Y. Kostov, K. A. Van Houten, P. Harms, R. A. Pilato, and G. Rao, Unique oxygen analyzer combining a dual emission probe and a low-cost solid-state ratiometric fluorometer, *Appl. Spectroscopy*, **54**, 864-68 (2000).
23. Y. Amao, I. Okura, and T. Miyashita, Optical oxygen sensing based on the luminescence quenching of europium(III) complex immobilized in fluoropolymer film, *Bull. Chem. Soc. Japan*, **73**, 2663-68(2000).
24. M. Smiddy, N. Papkovskaia, D. B. Papkovsky, and J. P. Kerry, Use of oxygen sensors for the non-destructive measurement of the oxygen content in modified atmosphere and vacuum packs of cooked chicken patties; impact of oxygen content on lipid oxidation. *Food Research International*, **35**(6), 577-584(2002).
25. D.Birnbaum, S. Kook, and R. Kopelman, Near-field scanning optical spectroscopy: spatially resolved spectra of microcrystals and nanoaggregates in doped polymers, *J. Phys. Chem.*, **97**, 3091-94 (1993).
26. Wenying Xu, R. C. McDonough, B. Langsdorf, J. N. Demas, and B. A. DeGraff, Oxygen sensors based on luminescence quenching: Interactions of metal complexes with the polymer support, *Anal. Chem.*, **66**, 4133-4141 (1994).
27. I. Klimant, and O. S. Wolfbeis, Oxygen-sensitive luminescent materials based on silicone-soluble ruthenium diimine complexes, *Anal. Chem.*, **67**, 3160-6 (1995).
28. P. Douglas, and K. Eaton, Response characteristics of thin film oxygen sensors, Pt and Pd octaethylporphyrins in polymer films, *Sens. Actuators B*, **82**, 200-208 (2002).

29. Z. Wang, A. R. McWilliams, C. E. B. Evans, X. Lu, S. Chung, and M. A. Winnik, I. Manners, Covalent attachment of RuII phenanthroline complexes to polythionylphosphazenes: The development and evaluation of single-component polymeric oxygen sensors, *Adv. Funct. Mater.*, **12**, 415-419 (2002).

30. M. C. DeRosa, P. J. Mosher, G. P. A. Yap, K-S. Focsaneanu, R. J. Crutchley, and C. E. B. Evans, Synthesis, characterization, and evaluation of [Ir(ppy)2(vpy)Cl] as a polymer-bound oxygen sensor, *Inorganic Chem.*, **42**, 4864-72 (2003).

31. Y. Amao, and I. Okura, Optical oxygen sensing materials: chemisorption film of ruthenium(II) polypyridyl complexes attached to anionic polymer, *Sens. Actuators B*, **88**, 162-167 (2003).

32. K. F. Mongey, J. G. Vos, B. D. MacCraith, and C. M. McDonagh, The photophysical properties of monomeric and dimeric ruthenium polypyridyl complexes immobilized in sol-gel matrices, *Coord. Chem. Rev.*, **185-186**, 417-29 (1999).

33. O. S. Wolfbeis, I. Oehme, N. Papkovskaya, and I. Klimant, Sol-gel based glucose biosensors employing optical oxygen transducers, and a method for compensating for variable oxygen background, *Biosens. Bioelectron.*, **15**, 69-76 (2000).

34. M. R. Shahriari,. In *Optical Fiber Sensor Technology*, K. T. V. Grattan, B. S. Meggitt, Eds.; Kluwer Academic: London, 1998; Vol. 4.

35. J. T. Bradshaw, S. B. Mendes, and S. S. Saavedra, A simplified broadband coupling approach applied to chemically robust sol-gel, planar integrated optical waveguides, *Anal.Chem.*, **74**, 1751-59 (2002).

36. G. A. Baker, B. R. Wenner, A. N. Watkins, and F. V. Bright, On the origin of the heterogeneous emission from pyrene sequestered within tetramethylorthosilicate-based xerogels: a decay-associated spectra and O2 quenching study, *J. Sol-Gel Sci. Technology*, **17**, 71-82 (2000).

37. N. A. Watkins, B. R. Wenner, J. D. Jordan, Wenying Xu, J. N. Demas, and F. V. Bright, Portable, low cost, solid-state luminescence-based O2 sensor, *App. Spectroscopy*, **52**, 750-754 (1998).

38. C. McDonagh, C. Kolle, A. K. McEvoy, D. L. Dowling, A. A. Cofolla, S. J. Cullen, B. D. MacCraith, Phase fluorometric dissolved oxygen sensor, *Sens. Actuators B*, **74**, 124-130 (2001).

39. X. Chen, Z. Zhong, Z. Li, Y. Jiang, X. Wang, K. Wong, Characterization of ormosil film for dissolved oxygen-sensing, *Sens. Actuators B*, **87**, 233-38(2002).

40. M. T. Murtagh, M. R. Shahriari, M. Krihak, A study of the effects of organic modification and processing technique on the luminescence quenching behavior of sol-gel oxygen sensors based on a ru(ii) complex, *Chem. Mater.*, **10**, 3862-69 (1998).

41. I. Klimant, F. Ruckruh, G. Liebsch, A. Stangelmayer, O. S. Wolfbeis, Fast response oxygen micro-optodes based on novel soluble ormosil glasses, *Mikrochimica Acta*, **131**, 35-46(1999).

42. Y.Tang, E. C. Tehan, Z. Tao, F. V. Bright, Sol-gel derived sensor materials that yield linear calibration plots, high sensitivity, and long term stability, *Anal. Chem.*, **75**, 2407-2413(2003).

43. D. Andrzejewski, I. Klimant, H. Podbielska, Method for lifetime-based chemical sensing using the demodulation of the luminescence signal, *Sens. Actuators B*, **84**, 160-66 (2002).

44. S. M. Buck, H. Xu, M. Brasule, M. A. Philbert, R. Kopelman, Nanoscale probes encapsulated by biologically localizing embedding (PEBBLEs) for ion sensing and imaging in live cells, *Talanta*, **63**, 41-59(2004).

45. W. Lei, A. Duerkop, Z. Lin, M.Wu, and O.S. Wolfbeis, Detection of hydrogen peroxide in river water via a microplate luminescence assay with time-resolved ("gated") detection, *Microchimica Acta* **143**(4), 269-274 (2003).

46. M. Schaeferling, M. Wu, J. Enderlein, H. Bauer, and O.S. Wolfbeis, Time-resolved luminescence imaging of hydrogen peroxide using sensor membranes in a microwell format, *Applied Spectroscopy*, **57**(11), 1386-1392 (2003).

47. H.M. Rowe, S.P. Chan, J.N. Demas, and B.A. DeGraff, Elimination of fluorescence and scattering backgrounds in luminescence lifetime measurements using gated-phase fluorometry, *Anal. Chem.*, **74**(18), 4821-4827 (2002).

48. G. Holst, and B. Grundwald, Luminescence lifetime imaging with transparent oxygen optodes, *Sens. Actuators B*, **74**, 78-90 (2001) and references therein.

49. V. Savvate'ev, Z. Chen-Esterlit, J. W. Aylott, B. Choudhury, C-H. Kim, J. H. Friedl, R. Shinar, J. Shinar, and R. Kopelman, Integrated organic light-emitting device/fluorescence-based chemical sensors, *Appl. Physics Lett.*, **81**, 4652-54 (2002).

50. B. Choudhury, R. Shinar, and J. Shinar, Luminescent chemical and biological sensors based on the structural integration of an OLED excitation source with a sensing component, *Proc. SPIE*, **5214**, 64-72 (2004).

51. F. Navarro-Villoslada, G. Orellana, M. C. Moreno-Bondi, T. Vick, M. Driver, G. Hildebrand, and K. Liefeith, Fiber-optic luminescent sensors with composite oxygen-sensitive layers and anti-biofouling coatings, *Anal. Chem.*, **73**(21), 5150-56 (2001).

52. B. M. Cullum, T. Vo-Dinh, The development of optical nanosensors for biological measurements, *Trend Biotechnol.*, **18**, 288-93(2000).

53. H. Xu, J. W. Aylott, R. Kopelman, T. J. Miller, M. A. Philbert, A real-time ratiometric method for the determination of molecular oxygen inside living cells using sol-gel-based spherical optical nanosensors with applications to rat C6 glioma *Anal. Chem.*,**73**, 4124-4133(2001).

54. Y-E. L. Koo, Y. Cao, R. Kopelman, S. M. Koo, M. Brasuel, M. A. Philbert, Real-time measurements of dissolved oxygen inside live cells by organically modified silicate fluorescent nanosensors, *Anal. Chem.*, **76**, 2498-2505 (2004).

55. W. Xu, K.A. Kneas, J.N. Demas, and B.A. DeGraff, Oxygen sensors based on luminescence quenching of metal complexes: Osmium complexes suitable for laser diode excitation, *Anal. Chem,* **68**, 2605-2609 (1996).

56. A. Mills, and Q. Chang, Modeled diffusion-controlled response and recovery behavior of a naked optical film sensor with a hyperbolic-type response to analyte concentration. *Analyst* (Cambridge, United Kingdom), **117**, 1461-6 (1992).

57. E.R. Carraway, J.N. Demas, B.A. DeGraff, and J.R. Bacon,. Photophysics and photochemistry of oxygen sensors based on luminescent transition-metal complexes, *Anal. Chem.*, **63**, 337-42 (1991).

58. K. A. Kneas, J. N. Demas, B. Nguyen, A. Lockhart, W. Xu, and B. A. DeGraff, Simple method for measuring oxygen diffusion coefficients of polymer films by luminescence quenching, *Anal. Chem.*, **74**, 1111-1118 (2002).

59. W.J. Bowyer, W. Xu, and J.N. Demas, Determining oxygen diffusion coefficients in polymer films by lifetimes of luminescent complexes measured in the frequency domain, *Anal. Chem.*, in press.

60. J.R. Lakowicz,. *Principles of Fluorescence Spectroscopy.* 2nd ed.; Plenum Press: New York, 1999.

61. X. Lu, and M.A. Winnik, Luminescence quenching in polymer/filler nanocomposite films used in oxygen sensors, *Chemistry of Materials* **13**, 3449-3463 (2001).

62. X.M. Li, and K.Y. Wong, Luminescent platinum complex in solid films for optical sensing of oxygen. *Anal. Chim. Acta,* **262**, 27-32 (1992).

63. X.M. Li, F.C. Ruan, and K.Y. Wong, Optical characteristics of a ruthenium(II) complex immobilized in a silicone rubber film for oxygen measurement, *Analyst*, **118**, 289-92 (1993).

64. J.N. Demas, B.A. DeGraff, and W. Xu, Modeling of luminescence quenching-based sensors: Comparison of multisite and nonlinear gas solubility models, *Anal. Chem.* **67**, 1377-80 (1995).

65. V.I. Ogurtsov, and D. Papkovsky, Modeling of luminescence-based oxygen sensors with non-uniform distribution of excitation and quenching characteristics inside active medium, *Sens. Actuators B*, **B88**, 89-100 (2003).

66. V.I. Ogurtsov, D. B. Papkovsky, N.Y. Papkovskaia, Approximation of calibration of phase-fluorometric oxygen sensors on the basis of physical models, *Sens. Actuators B,* **B81**, 17-24 (2001).

67. K. Eaton, B. Douglas, and P. Douglas, Luminescent oxygen sensors: time-resolved studies and modelling of heterogeneous oxygen quenching of luminescence emission from Pt and Pd octaethylporphyrins in thin polymer films, *Sens. Actuators B*, **B97**, 2-12 (2004).

68. E.R. Carraway, J.N. Demas, and B.A. DeGraff, Photophysics and oxygen quenching of transition-metal complexes on fumed silica, *Langmuir*, **7**, 2991-8 (1991).

69. J. N. Demas, and B. A. DeGraff, Luminescence sensors: Modelling of microheterogeneous systems and model differentiation, *SPIE, Optically Based Methods for Process Analysis*, Vol. **1681**, 2-11, (1992).

70. J.W. Hofstraat, H.J. Verhey, J.W. Verhoeven, M.U. Kumke, G. Li, S.L. Hemmingsen, and L.B. McGown, Fluorescence lifetime studies of labeled polystyrene latexes, *Polymer*, **38**, 2899-2906 (1997).

71. A.A. Istratov, and O.F. Vyvenko, Exponential analysis in physical phenomena, *Rev. Sci. Instrum.*, **70**, 1233-1257 (1999).

72. J.K. Kamal, B. Amisha; and D.V. Behere, Spectroscopic studies on human serum albumin and methemalbumin: optical, steady-state, and picosecond time-resolved fluorescence studies, and kinetics of substrate oxidation by methemalbumin, *J. Biol.ogical Inorg. Chem.*, **7**, 273-283 (2002).

73. J.M. Bedlek-Anslow, J.P. Hubner, B.F. Carroll, and K.S. Schanze, Micro-heterogeneous oxygen response in luminescence sensor films, *Langmuir*, **16**, 9137-9141 (2000).

74. K.A. Kneas, J.N. Demas, B.A. DeGraff, and A Periasamy, Fluorescence microscopy study of heterogeneity in polymer-supported luminescence-based oxygen sensors, *Microscopy and Microanalysis*, **6**, 551-561 (2000).

75. K.A. Kneas, J.N. Demas, B.A. DeGraff, Jr., and A. Periasamy, Comparison of conventional, confocal, and two-photon microscopy for detection of microcrystals within luminescence-based oxygen sensor films. *Proceedings of SPIE-The International Society for Optical Engineering*, **4262** (Multiphoton Microscopy in the Biomedical Sciences), 89-97 (2001).

76. R.D. Bowman, K.A. Kneas, J.N. Demas, and A. Periasamy, Conventional, confocal and two-photon fluorescence microscopy investigations of polymer-supported oxygen sensors, *Journal of Microscopy* (Oxford, United Kingdom), **211**, 112-120 (2003).

77. Mei, Erwen; Vinogradov, Sergei; Hochstrasser, Robin M.. Direct Observation of Triplet State Emission of Single Molecules: Single Molecule Phosphorescence Quenching of Metalloporphyrin and Organometallic Complexes by Molecular Oxygen and Their Quenching Rate Distributions. J. Am. Chem. Soc. (2003), 125(43), 13198-13204.

78. N.J.Turro, *Modern Molecular Photochemistry*. (W. A. Benjamin, Inc. 1978).

79. J.P. Paris, and W.W. Brandt, Charge transfer luminescence of a ruthenium(II) chelate, *J. Am. Chem. Soc,.* **81**, 5001-2 (1959).

80. A.W. Adamson, and J.N. Demas, New photosensitizer. Tris(2,2'-bipyridine)ruthenium(II) chloride, *J. Am. Chem. Soc.*, **93**, 1800-1 (1971).

81. J.N. Demas, and A.W. Adamson, Tris (2,2'-bipyridine)ruthenium(II) sensitized reactions of some oxalato complexes, *J. Am. Chem. Soc.* **95**, 5159-68 (1973).

82. J. Van Houten, and R.J. Watts, Temperature dependence of the photophysical and photochemical properties of the tris(2,2'-bipyridyl)ruthenium(II) ion in aqueous solution, *J. Am. Chem. Soc.*, **98**, 4853-8 (1976).

83. B. Durham, J.V. Caspar, J.K. Nagle, and T.J. Meyer, Photochemistry of tris(2,2'-bipyridine)ruthenium(2+) ion, *J. Am. Chem. Soc.* **104**, 4803-10 (1982).

84. A. Vaidyalingam, and P.K. Dutta, Analysis of the photodecomposition products of Ru(bpy)$_3^{2+}$ in various buffers and upon zeolite encapsulation, *Anal. Chem.*, **72**, 5219-5224 (2000).

85. J.N. Demas, and B.A. DeGraff, Design and applications of highly luminescent transition metal complexes, *Anal. Chem.*, **63**, 829A-837A (1991).

86. J.N. Demas, D. Diemente, and E.W. Harris, Oxygen quenching of charge-transfer excited states of ruthenium(II) complexes. Evidence for singlet oxygen production, *J. Am. Chem. Soc,* **95**, 6864-5 (1973).

87. J.N. Demas, E.W. Harris, and R.P McBride, Energy transfer from luminescent transition metal complexes to oxygen, *J. Am. Chem. Soc.*, **99**, 3547-51 (1977).

88. S.L. Buell, and J.N. Demas, Heterogeneous preparation of singlet oxygen using an ion-exchange-resin-bound tris(2,2'-bipyridine)ruthenium(II) photosensitizer, *J. Phys. Chem.*, **87**, 4675-81 (1983).

89. M.C. DeRosa, and R.J. Crutchley, Photosensitized singlet oxygen and its applications, *Coord. Chem. Rev.*, **233-234**, 351-371 (2002).

90. K. Lang, J. Mosinger, and D.M. Wagnerova, Photophysical properties of porphyrinoid sensitizers non-covalently bound to host molecules; models for photodynamic therapy. *Coord. Chem. Rev.*, **248**, 321-350 (2004).

91. S. Wang, R. Gao, F. Zhou, and M. Selke, Nanomaterials and singlet oxygen photosensitizers: potential applications in photodynamic therapy. *J. Mat. Chem.*, **14**, 487-493 (2004).

92. P.A. Schapp, Editor, *Benchmark Papers in Organic Chemistry, Vol. 5: Singlet Molecular Oxygen,* (Dowden, Hutchinson&Ross, Inc., Stroudsburg, PA., 1976).

93. P. Hartmann, M. Leiner, J.P. Marc, and P. Kohlbacher, Photobleaching of a ruthenium complex in polymers used for oxygen optodes and its inhibition by singlet oxygen quenchers, *Sens. Actuators B*, **B51**, 196-202 (1998).

94. Z.J. Fuller, W.D. Bare, K.A. Kneas, W. Xu, J.N. Demas, and B.A. DeGraff, Photostability of luminescent ruthenium(II) complexes in polymers and in solution, *Anal. Chem.*, **75**, 2670-2677 (2003)

TIME-RESOLVED FLUORESCENCE IN BIOMEDICAL DIAGNOSTICS

Herbert Schneckenburger[1] and Michael Wagner

7.1. INTRODUCTION

The application of time-resolving fluorescence techniques in biomedical research has been a challenge for many years. Since radiative and non-radiative transitions from the excited state of a molecule to its ground state are competing, the fluorescence lifetime (given as the reciprocal of the sum of all transition rates) of a molecule is sensitive to numerous parameters, e.g. its conformation or interaction with adjacent molecules. Therefore, fluorescence lifetime measurements can give numerous informations about the microenvironment of specific molecules.

As depicted in Figure 7.1, electronic excitation of organic molecules usually occurs by absorption of light from the singlet ground state S_0 to an excited state S_n and its vibrational levels. Commonly, fast non-radiative transitions ("internal conversion") occur from S_n within the femtosecond time range to the lowest excited state S_1, from where various subsequent transitions can be distinguished: fluorescence to the ground state S_0 (including its vibrational states) with a rate k_F, internal conversion to the ground state S_0 (rate k_{IC}), intersystem crossing from the singlet to the triplet state T_1 (rate k_{ISC}) and non-radiative energy transfer to adjacent molecules (rate k_{ET}). All these rates sum up according to

$$k = k_F + k_{IC} + k_{ISC} + k_{ET} = 1/\tau \ (1),$$

where τ is the lifetime of the excited state S_1, whereas the ratio $\eta = k_F/k$ corresponds to the fluorescence quantum yield. Although by optical spectroscopy only radiative transitions can be monitored, changes of k_{IC} or k_{ET} may be deduced from fluorescence lifetime measurements according to Equation 1.

Fluorescence lifetime measurements date back to the 1960's,[1] when either the phase shift between excitation light and fluorescence or the time course of a fluorescence signal following a short light pulse were measured. Since that time both, time-resolved[2] and

---

[1] Hochschule Aalen, Institut für Angewandte Forschung, Beethovenstraße 1, 73430 Aalen, Germany, E-mail: Herbert.Schneckenburger@fh-aalen.de

phase-modulated[3] techniques were improved considerably, and fluorescence lifetimes down to about 100 ps could be resolved.

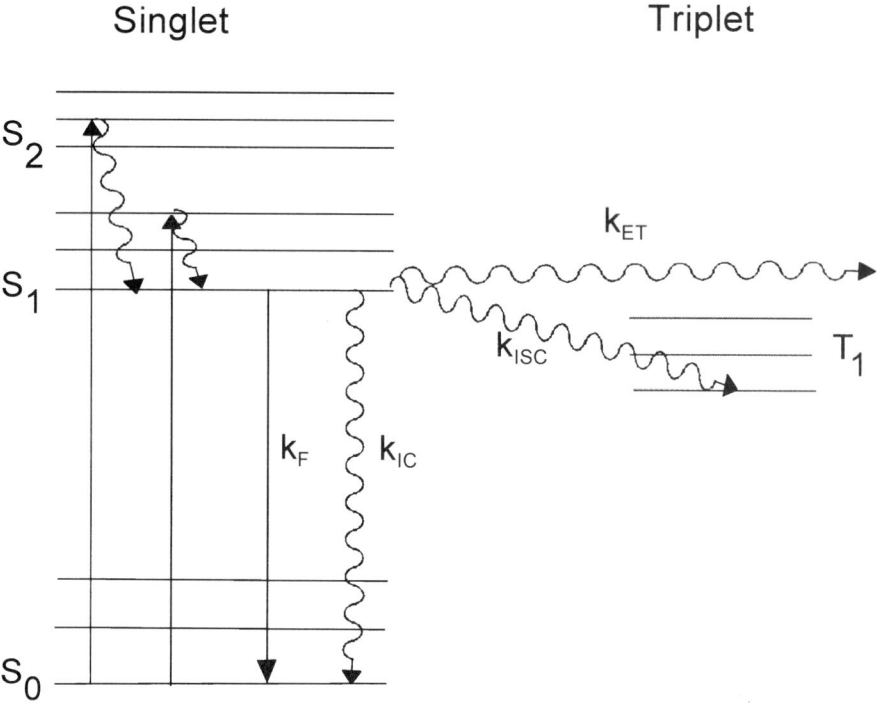

Figure 7.1. Molecular energy levels and transition rates. Straight lines: radiative transitions; waving lines: non-radiative transitions.

Spatial resolution was introduced in the field of life sciences around 1985, when cells or tissue samples were scanned first in one[4, 5] and later in two dimensions.[6] Since that time the implementation of time-resolution in laser scanning microscopy (LSM) has been improved continuously,[7] and one of the most recent developments consists of a detection device, where fluorescence decay curves are measured for each pixel of a LSM image.[8] Parallel developments include spatially resolved phase fluorometry[9–11] as well as time-gated image intensifying camera systems. Time-resolution of image intensifiers has recently been increased from about 5 ns[12, 13] to the picosecond range,[14] and detection sensitivity of image intensifying camera systems has been improved concomitantly.

Fluorescence intensity (number of photon counts per time unit) excited by short light pulses usually shows an exponential behaviour according to

$$I(t) = A\,e^{-kt} = A\,e^{-t/\tau} \quad (2)$$

with k corresponding to the total rate of deactivation of the excited electronic state S_1 and τ to the fluorescence lifetime. If several molecular species contribute to the fluorescence decay, their intensities sum up according to

$$I(t) = \Sigma_i \; A_i \; e^{-k_i t} = \Sigma \; A_i \; e^{-t/\tau_i} \quad (3)$$

with A_i being the amplitude and τ_i the fluorescence lifetime of an individual component. In fluorescence lifetime imaging (FLIM) the effective fluorescence lifetime τ_{eff} is often used instead of the individual lifetimes τ_i. As depicted in Figure 7.2, τ_{eff} can be calculated for each pixel of the image from the intensities I_A and I_B measured within two time gates (A,B) that are shifted by an interval Δt according to the equation[6]

$$\tau_{eff} = \Delta t \; / \; \ln \; (I_A/I_B) \quad (4).$$

Only when the fluorescence decay curve is monoexponential, τ_{eff} corresponds to the real fluorescence lifetime (as measured by decay kinetics); if this curve is multiexponential, τ_{eff} depends on all lifetimes.

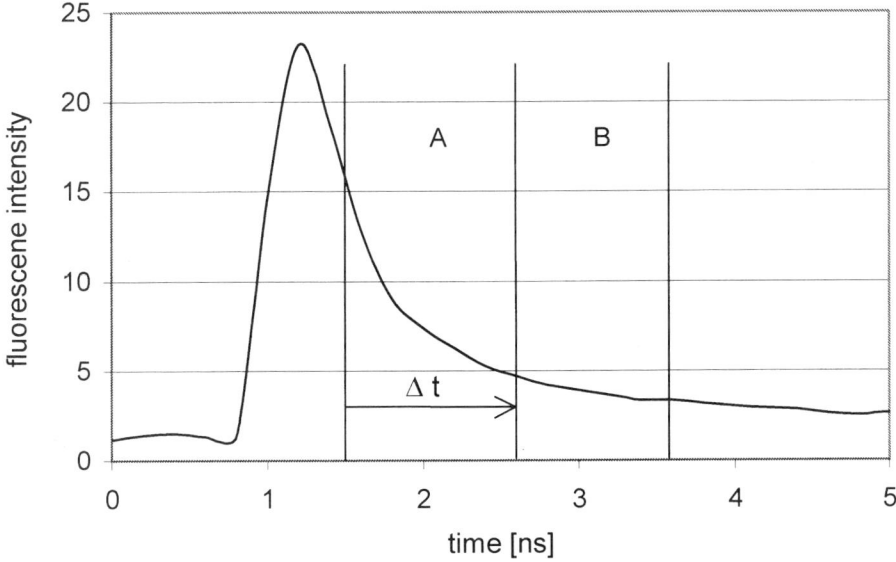

Figure 7.2. Decay profile of intrinsic fluorescence of U373-MG glioblastoma cells. Time gates A, B as well as the time interval Δt used for fluorescence lifetime imaging are indicated. Reproduced from Ref. 30 with modifications.

Since optical transition dipole moments have defined orientations within a molecule, polarization spectroscopy can be used to determine molecular rotation during the lifetime of an excited electronic state. Upon irradiation of polarized light, one preferentially excites those molecules, whose transition dipoles are parallel to the electric field vector of incident light. This selective excitation of an oriented population of molecules results in partially polarized fluorescence, which is described by the degree of polarization

$$P = (I_{||} - I_{\perp}) / (I_{||} + I_{\perp}) \quad (5)$$

or by the fluorescence anisotropy

$$r = (I_{||} - I_{\perp}) / (I_{||} + 2 I_{\perp}) \quad (6).$$

with $I_{||}$ and $I_{\perp}$ being the fluorescence intensities of light polarized parallel or perpendicular to the exciting electric field vector, respectively. Usually P and r depend on the time interval between excitation and fluorescence detection, since during the lifetime of their excited states, many molecules change their orientation by rotation ("rotational diffusion"). Time-resolved measurements of fluorescence anisotropy permit to calculate the time constant τ_r of rotational diffusion according to

$$r(t) = I_0 \, e^{-t/\tau r} \quad (7)$$

which for spherical molecules is correlated with their volume V and the viscosity η of the environment following Einstein's equation[2]

$$\tau_r = \eta \, V / k_B \, T \quad (8)$$

with k_B corresponding to Boltzmann's constant and T to the absolute temperature. According to Equation 7 τ_r appears to be an appropriate parameter to determine viscosities or fluidities of cell membranes. In the case of various subpopulations of fluorescent molecules r (t) may show multiexponential behaviour (similar to Equation 3).

The present article gives an overview of the authors' recent work on time-resolving fluorescence methods (including kinetics, FLIM, polarization and energy transfer measurements) and applications.

7.2. MATERIALS AND METHODS

For the experiments described below various cell cultures were used, in particular

- BKEz-7 endothelial cells from calf aorta[15] cultivated in Eagle's minimum essential medium (MEM) supplemented with 10% fetal calf serum, glutamine and antibiotics;
- CHO-K1 hamster ovary cells cultured in F-10 HAM nutrient mixture supplemented with 10% fetal calf serum, sodium bicarbonate and antibiotics;
- U373MG human glioblastoma cells cultivated in medium RPMI 1640 again supplemented with 10% fetal calf serum and antibiotics.

In all cases 150 cells/mm² were seeded on microscope object slides and grown for 24h or 48h, thus resulting in smaller cell clusters or sub-confluent monolayers, prior to addition of fluorescent dyes (s. below) and microscopic experiments. In some measurements of CHO-K1 cells subcultures (SC's) of "young cells" (SC 12-23) and "aged cells" (SC 35-38) were distinguished. In addition, samples of dextran embedded in agarose at low concentration (10^{-7}–10^{-8} M) and linked to fluorescein isothiocyanate (FITC) were used for test experiments.

A fluorescence microscope (Axioplan 1, Carl Zeiss Jena, Germany) was equipped with a novel illumination device[16], which allowed all samples to be illuminated under dark field conditions with a variable angle of incidence Θ. For cultivated cells total internal reflection (TIR) occurred at $\Theta \geq 64.6°$. Therefore, when using the angles $\Theta = 62°$ and $\Theta = 66°$ in the experiments, whole cells were illuminated in the first case, whereas the plasma membrane and adjacent cellular sites were illuminated selectively in the second case (with a penetration depth $d \approx 160$ nm of the evanescent wave). In all cases the electric field vector was polarized perpendicular to the plane of incidence. As picosecond light sources we used either a modelocked argon ion laser (Innova 90, Coherent, Palo Alto, USA) at a wavelength of 476 nm, a pulse duration of 200 ps and a repetition rate of 99 MHz (reduced to 2.5 MHz by a Pockels cell), or a picosecond laser diode (LDH 400, Picoquant, Berlin, Germany) at a wavelength of 391 nm, a pulse duration of 55 ps and a repetition rate of 40 MHz. A monomode quartz fiber with collimation optics at its entrance and exit (KineFLEX-P-3; Point Source, Southampton, U.K.) was used to couple excitation light to the illumination device, thus permitting a small divergence angle of $\pm 0.25°$ in the plane of the sample. When using light pulses with an energy of 40 pJ (Ar^+ laser) or 12 pJ (laser diode) no phototoxicity and only little photobleaching of the samples occurred. The electrical output of the modelocker was used to trigger the time gate of the image intensifier (Picostar HR 12, LaVision, Göttingen) which was coupled to a cooled ICCD camera with 640×480 pixels. Time gates between 200 ps and 1000 ps as well as variable delay times with respect to the trigger pulse could be adjusted. The whole setup is depicted in Figure 7.3.

Fluorescence decay kinetics and images were recorded from object fields of 220 µm x 160 µm (using a 63x/0.90 water immersion objective lens) or 320 µm × 250 µm (using a 40x/1.30 oil immersion lens). These fields typically contained 5–15 individual cells. Fluorescence decay kinetics were measured by shifting the time gate in intervals of 200 ps over a scale of 10–20 ns (dwell time: 1s for each interval), and lifetimes were obtained from mono- or biexponential curve fitting according to the Equations 2 or 3 using a least-square fitting algorithm. Fluorescence lifetime images were calculated according to Equation 4 using the fluorescence intensities measured in the time gates A and B. From individual lifetime images, the frequency of all fluorescence lifetimes was calculated and displayed as a histogram.

7.3. RESULTS

7.3.1. Mitochondrial Energy Metabolism

Part of the authors' recent work was concentrated on studies of energy metabolism of the cell, in particular measurements of mitochondrial malfunction using either the intrinsic fluorescence of some coenzymes, (e.g. nicotinamid adenine dinucleotide, NADH, and flavin mononucleotide, FMN) or the fluorescence of the well established mitochondrial marker rhodamine 123 (R123).

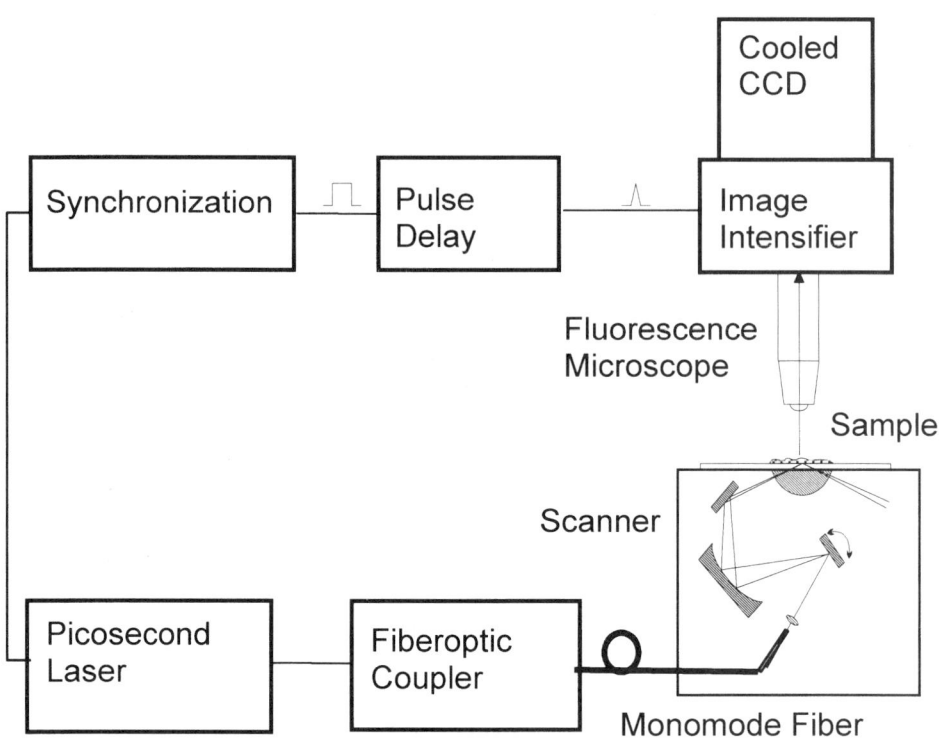

Figure 7.3. Experimental setup for fluorescence lifetime imaging microscopy (FLIM) with variable-angle excitation.

Since the oxidation of NADH may be impeded upon inhibition of the mitochondrial respiratory chain, its fluorescence is expected to increase. Therefore, NADH fluorescence has been used to measure mitochondrial malfunction[17] or hypoxia[18, 19] and has also been suggested as a parameter for tumour detection.[20–23] So far, free NADH and protein bound NADH have been distinguished, the first species (together with nicotinamid adenine dinucleotide phosphate, NADPH) being mainly localized in the cytoplasm and the second species within mitochondria.[24] In their previous work the authors could show that upon inhibition of the mitochondrial respiratory chain[17, 25] or deprivation of oxygen[26, 27] not only protein bound NADH, but also free NADH increased considerably in various cell lines and organisms.

Fluorescence spectra of free NADH, protein bound NADH and flavins with emission maxima around 440 nm, 470 nm and 520 nm are overlapping.[28, 29] However, it has been shown previously that these species can be distinguished on the basis of their fluorescence lifetimes of about 0.5 ns, 2.5 ns and 5-6 ns, respectively.[25] Therefore, when using time-gated fluorescence spectroscopy, the emission bands of free NADH, protein bound NADH and flavins could be easily separated and quantified.[25–27, 29] Recent FLIM measurements showed a rather homogenous distribution of effective fluorescence lifetimes over the cells with a maximum of the lifetime distribution around 800 ps and a slight reduction of τ_{eff} in some perinuclear granules, possibly mitochondria.[30] As depicted in Figure 7.4, the lifetime distribution shows a slight shift towards lower lifetimes after incubation with rotenone, a well-known inhibitor of the mitochondrial respiratory chain, and a slight shift towards higher lifetimes after incubation with deoxyglucose which inhibits the glycolytic pathway. This indicates that the ratio of free and protein bound NADH depends on the function of both metabolic pathways.

Selectivity in detection of mitochondrial NADH versus cytoplasmic NADH (and NADPH) was further enhanced by incubation of cells with the mitochondrial marker rhodamine 123 (R123).[25] In this case NADH molecules were excited by absorption of near ultraviolet light, and a non-radiative energy transfer to R123 molecules was examined. Since energy transfer according to the Förster mechanism[31] is limited to intermolecular distances of a few nanometers, only those NADH molecules are detected which are in close proximity to the inner mitochondrial membrane. Using Förster resonance energy transfer (FRET) spectroscopy a 4-fold increase of mitochondrial NADH was measured after inhibition of the first enzyme complex of the mitochondrial respiratory chain (by rotenone),[25] whereas total NADH increased only by a factor 1.7.[29]

Intracellular accumulation of R123 was further examined by FLIM measurements.[32] These experiments proved a homogenous distribution of fluorescence lifetimes within the cells at low or moderate R123 concentration. At high concentration (more than 25 µM in the incubation medium) a pronounced shortening of τ_{eff} was observed in the mitochondria. Concomitantly, R123 accumulated in the plasma membrane (as shown by TIR microscopy) and to some extent also in the cytoplasm (as shown by LSM). Due to quenching of R123 fluorescence in the mitochondria and accumulation of R123 in further cellular sites, its concentration in the incubation medium should not be above 25 µM for diagnostic applications, e.g. measurements of mitochondrial malfunction.

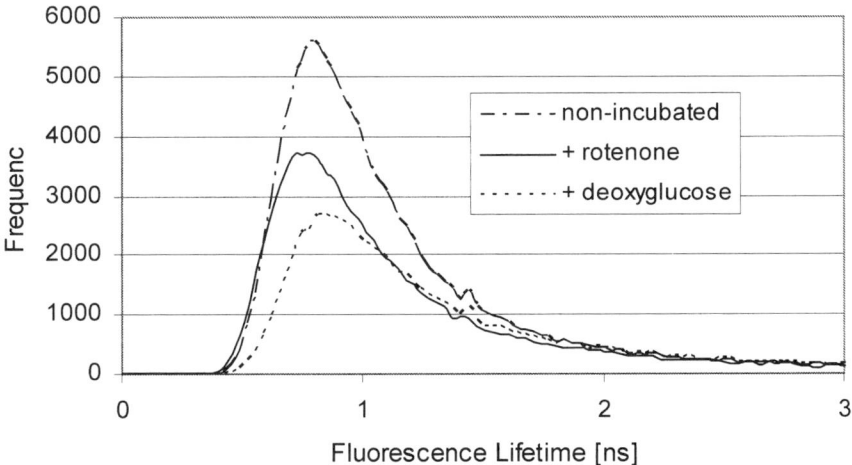

Figure 7.4. Histograms of effective fluorescence lifetimes for BKEz-7 endothlial cells incubated with either rotenone or deoxyglucose in comparison with non-incubated control cells. Histograms were calculated as averages of 5 lifetime images in each case. For excitation of free and protein bound NADH the 391 nm laser diode was replaced by a laser diode emitting at 375 nm. Reproduced from Ref. 30 with modifications.

7.3.2. Membrane Dynamics

Using specific illumination techniques in fluorescence microscopy (dark field and total internal reflection fluorescence microscopy, TIRFM[16]) the authors dedicate a main part of their work to studies of membrane dynamics in living cells. Whereas in the literature numerous applications of TIRFM have been described, e.g. studies of cell-substrate contacts,[33, 34] measurements of dynamics[35] or self-association[36] of proteins, detection of membrane-proximal ion fluxes[37] as well as imaging of endocytosis or exocytosis,[38, 39] the authors' studies are concentrated to the lipid phase of cell membranes using the fluorescent membrane marker 6-dodecanoyl-2-dimethylamino-naphthalene (laurdan). This naphthaline derivative is a polarity-sensitive probe, whose electronic excitation energy is different in polar and non-polar environments.[40, 41] Once incorporated into cell membranes, its fluorescence shows a spectral shift towards longer wavelengths when its molecules get into contact with adjacent water molecules, e.g. when a phase transition from the tightly packed gel phase to the liquid crystalline phase of membrane lipids occurs. Concomitantly, fluorescence kinetics may be affected, if during the lifetime of the excited state of the laurdan molecules the adjacent water dipoles (of the cytoplasm) are re-oriented.[42] This dipole-dipole interaction usually occurs at higher temperature when membrane lipids are in the liquid-crystalline phase, but not within the tightly packed gel phase.

In recent studies fluorescence spectra and lifetime images of cultivated CHO cells incubated with laurdan (8 μM; 60 min.) have been examined as a function of temperature, age and growth phase of the cells.[43, 44] Emission spectra showed an overlap by two broad bands with maxima around 440 nm and 480–490 nm. The long-wave band was most pronounced at higher temperatures (T $\geq$ 35°C), whereas the short wave-band was predominant at T $\leq$ 30°C and was more pronounced upon selective illumination of the plasma membrane (by TIRFM) than upon illumination of whole cells.

Fluorescence lifetimes around 5 ns were observed at temperatures between 24°C and 41°C (24°C: 5.8 $\pm$ 1.0 ns; 35°C: 5.1 $\pm$ 0.6 ns).[44] More pronounced differences were found for the fluorescence rise time, which at T = 24°C was within the experimental resolution of 200 ps, but at T = 35°C increased to 0.5–0.8 ns. This rise time had some impact on the effective fluorescence lifetime τ_{eff} evaluated in the FLIM images. Fluorescence lifetime images of CHO cells incubated with laurdan are compared in Figure 7.5 for temperatures of 24°C (left) and 35°C (right). In both cases the effective lifetime varied between about 2 ns and 8 ns for individual pixels. However, at T = 24°C all effective lifetimes were homogenously distributed over the cells, whereas at temperatures above 30°C domains with shorter and longer lifetimes could be distinguished. Domain patterns were not observed for aged cells (SC 35-38), where fluorescence lifetimes were always homogenously distributed over all cells.[44] For all temperatures (24°C $\leq$ T $\leq$ 41°C) the fluorescence lifetime of aged cells was 3.3 $\pm$ 0.3 ns, i.e. considerably shorter than for younger subcultures.

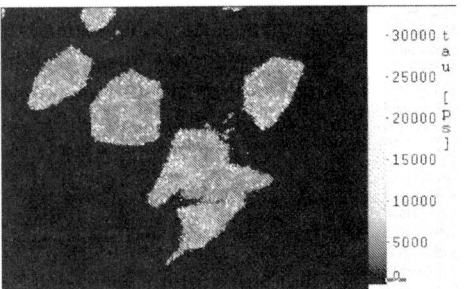 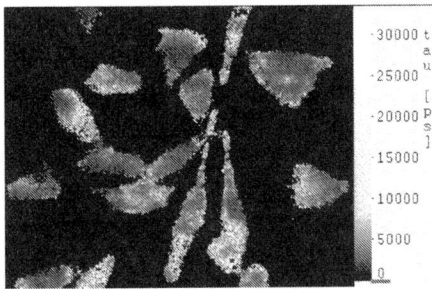

Figure 7.5. Effective fluorescence lifetime τ_{eff} at 24°C (left) and 35° C (right) of CHO cells ("young cells") incubated with laurdan (8 µM; 60 min.) using whole cell illumination ($\Theta = 62°$). Excitation wavelength: 391 nm; detection range: 420–800 nm; image size: 220 µm x 160 µm in each case. Reproduced from Ref. 44 with modifications.

Fluorescence kinetics of CHO cells incubated with laurdan are depicted in Figure 7.6 for polarizations parallel and perpendicular to the exciting electric field vector. The curves were corrected for polarization dependent transmission and detection sensitivity. Fluorescence anisotropy r(t) was evaluated according to Equation 6 and also plotted in the Figure. For T = 24°C a time constant of rotational diffusion $\tau_r \approx 1$ ns was determined. This time constant decreased at higher temperatures, indicating a lower viscosity or higher fluidity of cell membranes.

In view of measuring membrane dynamics in well defined microenvironments (e.g. close to specific proteins) non-radiative energy transfer from laurdan to certain acceptor molecules was examined. For preliminary tests the membrane marker 1,1'-dioctadecyl-3,3,3',3'-tetramethyl-indocarbocyanine perchlorate (DiI) proved to be an appropriate acceptor. Whereas cell incubation with laurdan (8 µM) was kept constant, the concentration of DiI in the incubation medium was varied in the range of 0–8 µM prior to selective illumination of the plasma membrane by TIRFM. In addition to the laurdan emission bands a fluorescence band around 570 nm was detected whose intensity increased almost linearly with DiI concentration.[44] DiI fluorescence was not excited without the presence of laurdan, i.e. it was excited solely by energy transfer. Figure 7.7 shows the fluorescence decay kinetics of laurdan prior and after co-incubation with DiI and proves a shortening of the fluorescence lifetime upon addition of DiI. When using a monoexponential fit a fluorescence lifetime of 3.5 ns is calculated after addition of 8 µM DiI in comparison with 5.4 ns without co-incubation. This results in an energy transfer rate around 1.3×10^8 s^{-1} according to Equation 1. It cannot be excluded that the fluorescence decay curve (upper curve in Figure 7.7) may be multiexponential with different energy transfer rates for specific subpopulations of laurdan molecules; the low fluorescence intensity in TIRFM experiments, however, would presently not justify multiexponential curve fitting.

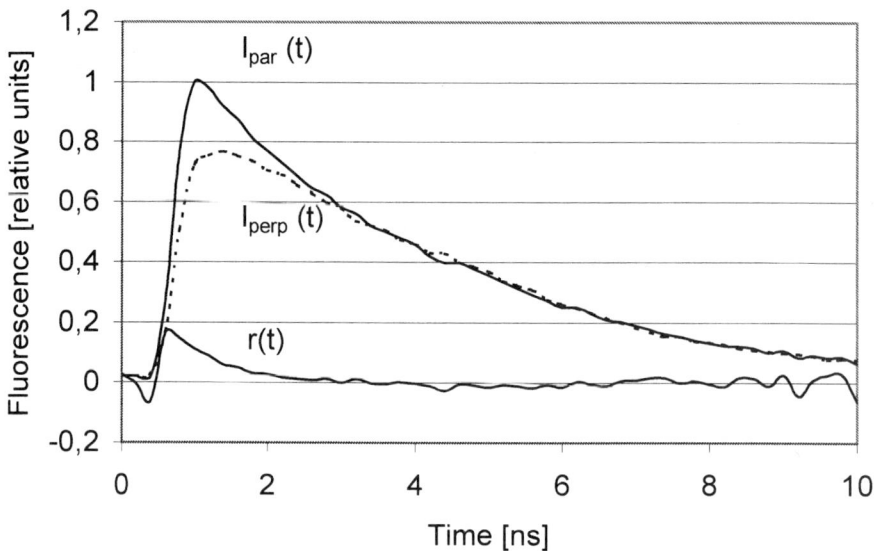

Figure 7.6. Fluorescence decay kinetics of CHO cells incubated with laurdan (8 μM; 60 min.) at T = 24°C for polarizations parallel and perpendicular to the exciting electric field vector; fluorescence anisotropy function r(t) (whole cell illumination by picosecond laser pulses at λ = 391 nm). Profiles are obtained from an image of 320 μm × 250 μm in the spectral range of 420–800 nm.

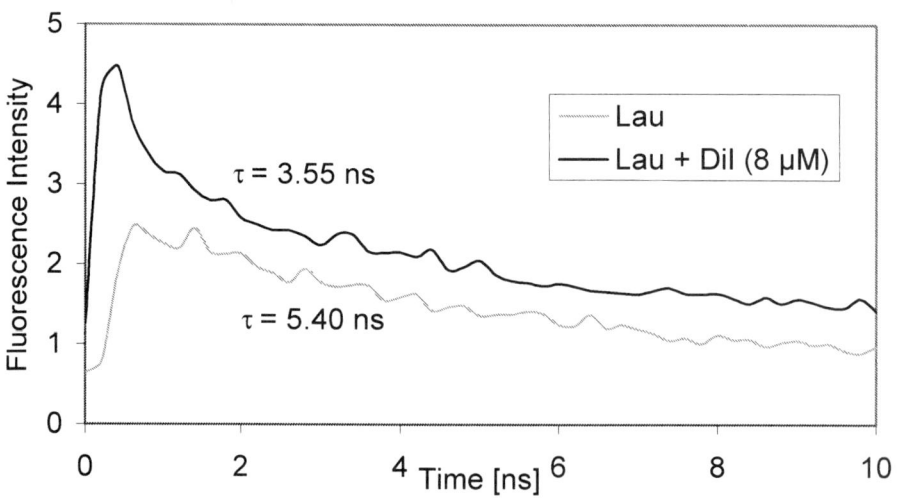

Fi

Figure 7.7. Fluorescence decay kinetics of CHO cells either incubated with laurdan or co-incubated with laurdan and DiI (10 min., 8 μM each) using selective illumination of the plasma membrane (Θ = 66°; λ_{ex} = 391nm).

7.3.3 Localization and Light-Induced Reactions of Photosensitizers

Photodynamic therapy (PDT) is a new and promising method for the treatment of various kinds of tumours. After topical or systemic application of a photosensitizer and its subsequent accumulation in neoplastic tissue, absorption of light may initiate phototoxic processes. Conventional photosensitizers like hematoporphyrin derivative are presently replaced by more selective dyes, and a rather promising approach is the use of protoporphyrin IX (PP IX) synthesized endogenously from the externally administered precurser 5-aminolevulinic acid (5-ALA).[45] 5-ALA-induced PP IX accumulates rather specifically in certain types of tumours and is therefore used in various clinical disciplines. Since a promising new field is PDT treatment of brain cancer with 5-ALA induced PP IX,[46] some studies are now concentrated on the location and light-induced reactions of this photosensitizer in a glioblastoma cell line.

Accumulation of 5-ALA induced PP IX in U373-MG glioblastoma cells was found to be rather moderate in comparison with some cell lines of breast or ovarian cancer.[47] Nevertheless, its photodynamic efficacy (per photosensitizer molecule) was higher than in any other of those cell lines. Time-resolving and microscopic experiments revealed to be helpful in explaining this effect.[47] First, we found different cellular patterns of PP IX fluorescence in U373-MG brain cancer and T47-D breast cancer cells (the latter accumulating PP IX to a very high extent): T47-D cells showed a pattern of brightly fluorescence granules representing subcellular organelles, e.g. lysosomes or mitochondria, whereas U373-MG cells exhibited mainly diffuse fluorescence, probably resulting from intracellular membranes. Second, photobleaching of PP IX was more pronounced in T47-D than in U373-MG cells, since reduction of fluorescence intensity to 50% occurred at a light dose of about 1.5 J/cm² in the first and about 3 J/cm² in the second case (using an excitation wavelength of 391 nm). Third, fluorescence decay times of τ_1=3–4 ns and τ_2=20–25 ns were measured for both cell lines, resulting in an effective fluorescence lifetime around 8 ns. Reduction of τ_{eff} by irradiation, however, was more pronounced for T47-D than for U373-MG cells. Assigning τ_2 to porphyrin monomers and τ_1 to a superposition of porphyrin aggregates and photoproducts,[48] one can conclude that more protoporphyrin monomers are bleached in T47-D than in U373-MG cells. Fluorescence lifetime images proved that this photobleaching effect was most pronounced in cellular sites of highest fluorescence intensity. One can conclude that PP IX is accumulated to a high amount in granules of T47-D breast cancer cells, where it is rapidly photobleached, whereas it is accumulated in membranes of U373-MG brain cancer cells where it is less photobleached and more efficient for photodynamic therapy. This might explain the good perspectives of PDT for brain cancer treatment using 5-ALA induced protoporphyrin IX.

7.3.4 Single Molecule Detection

Single molecule detection is a new analytical tool for measuring conformations or dynamics of well defined molecules, e.g. proteins, which may play a role in tumour genesis or progression as well as in apoptosis. The method is based on multiple excitation (typically 100–1000 times per second) of specific fluorescence markers and their ultra-sensitive detection. Thin layers of samples are selected using either confocal methods,[49–51] total internal reflection fluorescence microscopy (TIRFM)[52] or optical nearfield microscopy.[50] However, small amounts of overlapping scattered light and background luminescence, arising e.g. from glass slides, optical filters or immersion oil, may cause major problems. Time-resolved or time-gated detection of single molecule fluorescence excited by picosecond laser pulses may be helpful for suppressing those background signals, since fluorescence lifetimes of most organic fluorophores are in the range of 1–10 ns, i.e. longer than picosecond light scattering and shorter than glass or quartz luminescence (>100 ns).

Using TIRFM and time-gated detection in a range of 1–2 ns after excitation by picosecond laser pulses allowed us to suppress background signals efficiently and to measure fluorescein isothiocyanate (FITC) bound to dextrane and immobilized in agarose on a single molecule level. Upon excitation of a thin layer of 110 nm thickness the fluorescence of a minimum of 3 FITC molecules corresponding to one dextran molecule could be detected on single pixels of the camera system, if a concentration of 10^{-8} M FITC-conjugated dextrane was used. Figure 7.8 shows a comparison of fluorescence images of agarose (left) and FITC-dextrane in agarose at a concentration of 10^{-8} M (right). The measured signal (right) can be clearly distinguished from the reference (left).

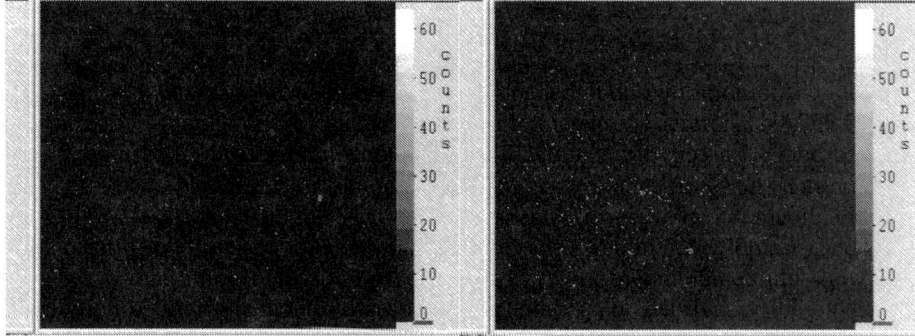

Figure 7.8. Time-gated fluorescence of agarose (left) and 10^{-8} M FITC-conjugated dextran embedded in agarose (right) Excitation wavelength: 476 nm; detection range: ≥ 515 nm. The background (without illumination) was subtracted in both cases.

7.4. DISCUSSION AND PERSPECTIVES

Principles, methods and various applications of time-resolving fluorescence methods for biomedical diagnostics have been described. In all cases specific fluorescent species in well defined microenvironments could be examined and distinguished from background signals. As mentioned above, time-resolved fluorescence proved to be helpful to quantify free and protein bound NADH in the cytoplasm as well as in the mitochondrial respiratory chain and to examine mitochondrial malfunction. The applicability of the mitochondrial marker rhodamine 123 for probing the function of the respiratory chain was limited to concentrations around or below 25 µM, since at higher concentrations fluorescence quenching and accumulation of this marker outside the mitochondria occurred.

Fluorescence lifetime measurements can also give numerous informations about membrane stiffness and fluidity, in particular, if this method is combined with further microscopic (e.g. TIRFM, fluorescence anisotropy) or spectroscopic (e.g. FRET) techniques. These studies can be used to describe changes of membrane properties during cell aging or in the case of various diseases. Preliminary results show that these membrane properties are rather sensitive to the intracellular content of cholesterol and to certain pharmaceutical agents. Therefore, time-resolving fluorescence techniques may also have a large potential in pharmacoloy.

Time-resolving fluorescence techniques can also be used to localize photo-sensitizers in cells or tissues and to deduce their light-induced (phototoxic) reactions. This may be helpful in order to optimize the dose of a photosensitizer as well as the conditions of irradiation during the photodynamic treatment of malignant tumours.

Single molecule detection is a new and promising field with large scientific and diagnostic perspectives. Using mutants of green fluorescent protein (GFP)[53] site-specific tracking has become possible for numerous proteins in cells and tissues, thus permitting to study metabolic pathways or signal transduction on a single molecule level. The appearance of new electron-multiplying CCD cameras[54] on the market may be a milestone in ultrasensitive fluorescence detection.

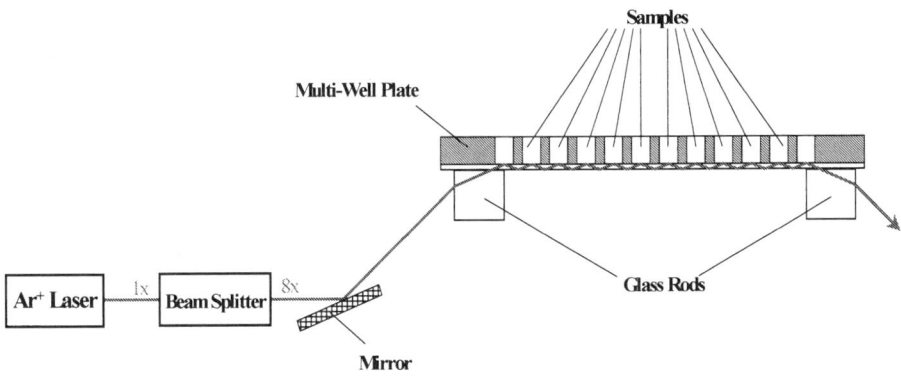

Figure 7.9. TIR screening system for multiwell plates (using an evanescent electromagnetic wave for excitation).

In clinical diagnostics or laboratory medicine simultaneous measurements of samples from a larger number of patients are desired, and glass slides (as used in microscopy) are commonly replaced by multiwell plates for 96, 384 or 1536 individual samples. A setup for total internal reflection (TIR) screening of 96 samples is depicted in Figure 7.9[55]. An Ar$^+$ laser beam is split into 8 individual beams, and each beam is totally reflected on the bottom of 12 cavities containing the samples, e.g. cell monolayers. In this way thin layer of the samples, e.g. cell membranes, are illuminated, and their fluorescence is detected without any superposition from further parts of the sample including walls and supernatant. Even washing of the samples, which is a conventional procedure in laboratory   medicine can be excluded. What is needed is a special glass bottom of the multiwell plate with high optical transmission and an appropriate thickness for total internal reflection. For getting detailed information on individual samples, high throughput screening (HTS) may be combined with high content screening (HCS) in view of specific parameters, e.g. fluorescence spectra, lifetimes or anisotropy. A first screening system with time-gated and phase-resolved fluorescence detection is described in Ref. 56.

7.5. REFERENCES

1. S. Udenfriend, *Fluorescence Assay in Biology and Medicine* (Academic Press, New York – San Francisco – London, 1962).
2. D.V. O'Connor and D. Philipps, *Time-correlated single photon counting* (Academic Press, New York – San Francisco – London, 1984).
3. J.R. Lakowicz, G. Laczko, I. Gryczinski, H. Szmacinski, and W. Wiczk, Gigahertz frequency domain fluorometry: resolution of complex decays, picosecond processes and future developments, *J. Photochem. Photobiol. B:Biol.* **2**, 295–311 (1988).
4. H. Schneckenburger, F. Pauker, E. Unsöld, and D. Jocham, Intracellular distribution and retention of the fluorescent components of Photofrin II, *Photobiochem. Photobiophys* **10**, 61–67 (1985).
5. I. Bugiel, K. König, and H. Wabnitz, Investigation of cells by fluorescence laser scanning microscopy with subnanosecond resolution, *Lasers Life Sci.* **3**, 47–53 (1989).
6. E.P. Buurman, R. Sanders, A. Draijer, H.C. Gerritsen, J.J.F. van Veen, P.M. Houpt, and Y.K. Levine, Fluorescence lifetime imaging using a confocal laser scanning microscope, *Scanning* **14**, 155–159 (1992).
7. H.C. Gerritsen, M.A. Asselbergs, A.V. Agronskaia, and W.G. van Sark, Fluorescence lifetime imaging in scanning microscopes: acquisition, speed, photon economy and lifetime resolution, *J. Microsc.* **206**, 218–224 (2002).
8. W. Becker, A. Bergmann, M.A. Hink, K. König, K. Benndorf, and C. Biskup, Fluorescence lifetime imaging by time-correlated single photon counting, *Micr. Res. Techn.* **63**, 58–66 (2004).
9. J.R. Lakowicz and K. Berndt, Lifetime-selective fluorescence imaging using an rf phase-sensitive camera, *Rev. Sci. Instrum.* **62**, 1727–1734 (1991).
10. T.W.J. Gadella, T.M. Jovin, and R.M. Clegg, Fluorescence lifetime imaging microscopy (FLIM): Spatial resolution of microstructures on the nanosecond time scale, *Biophys. Chem.* **48**, 221–239 (1993).
11. A. Squire, P.J. Verveer, and P.I.H. Bastiaens, Multiple frequency fluorescence lifetime imaging microscopy, *J. Microsc.* **197**, 136–149 (2000).
12. M. Kohl, J. Neukammer, U. Sukowski, H. Rinneberg, D. Wöhrle, H.-J. Sinn, and A. Friedrich, Delayed observation of laser-induced fluorescence for imaging of tumours, *Appl. Phys.* **B56**, 131–138 (1993).
13. H. Schneckenburger, K. König, T. Dienersberger, and R. Hahn, Time-gated microscopic imaging and spectroscopy in medical diagnosis and photobiology, *Opt. Eng.* **33**, 2600–2606 (1994).
14. M.J. Cole, J. Siegel, S.E. Webb, R. Jones, K. Dowling, M.J. Dayel, D. Parsons-Karavassilis, P.M. French, M.J. Lever, L.O. Sucharov, M.A. Neil, R. Juskaitis and T. Wilson, Time-domain whole-field fluorescence lifetime imaging with optical sectioning, *J. Microsc.* **203**, 246–257 (2001).
15. W. Halle, W.-E. Siems, K.D. Jentzsch, E. Teuscher, and E. Göres, Die *in vitro* kultivierte Aorten-Endothelzelle in der Wirkstofforschung - Zellphysiologische Charakterisierung und Einsatzmöglichkeiten der Zellinie BKEz-7, *Die Pharmazie* **39**, 77–81 (1984).

16. K. Stock, R. Sailer, W.S.L. Strauss, M. Lyttek, R. Steiner, and H. Schneckenburger, Variable-angle total internal reflection fluorescence microscopy (VA-TIRFM): realization and application of a compact illumination device, *J. Microsc.*, **211**, 19–29 (2003).

17. M.H. Gschwend, R. Rüdel, W.S.L. Strauss, R. Sailer, H. Brinkmeier, and H. Schneckenburger, Optical detection of mitochondrial NADH content in human myotubes, *Cell. Mol. Biol.* **47**, OL95–OL104 (2001).

18. K.A. Horvath, K.T. Shomacker, C.C. Lee, and L.H. Cohn, Intraoperative myocardial ischemia detection with laser-induced fluoescence, *J. Thorac. Cardiovasc. Surg.* **107**, 220–225 (1994).

19. E.T. Obi-Tabo, L.M. Hanrahan, R. Cachecho, E.R. Beer, S.R. Hopkins, J.C. Chan, J.M. Shapiro, and W.W. LaMorte, Changes in hepatocyte NADH fluorescence during prolonged hypoxia, *J. Surg. Res.* **55**, 575–580 (1993).

20. W. Lohmann and E. Paul, In situ detection of melanomas by fluorescence measurements, *Naturwissenschaften* **75**, 201–202 (1988).

21. G.A. Wagnières, M.S. Willem, and B.C. Wilson (1998). In vivo fluorescence spectroscopy and imaging for oncological applications. *Photochem. Photobiol.*, 68, 603–632.

22. A.C. Croce, A. Spano, D. Locatelli, S. Barni, L. Sciola, and G. Bottiroli, Dependence of fibroblast autofluorescence properties on normal and transformed conditions. Role of the metabolic activity, *Photochem. Photobiol.* **69**, 364–374 (1999).

23. L. Rigacci, R. Alterini, P.A. Bernabei, P.R. Ferrini, G. Agati, F. Fusi, and M. Monici, Multispectral imaging autofluroescence microscopy for the analysis of lymph-node tissues, *Photochem. Photobiol.* **71**, 737–742 (2000).

24. M. Wakita, G. Nishimura, and M. Tamura, Some characteristics of the fluorescence lifetime of reduced pyridine nucleotides in isolated mitochondria, isolated hepatocytes and perfused rat liver in situ, *J. Biochem.* **118**, 1151–1160 (1995).

25. H. Schneckenburger, M.H. Gschwend, W.S.L. Strauss, R. Sailer, M. Kron, U. Steeb, and R. Steiner, Energy transfer spectroscopy for measuring mitochondrial metabolism in living cells, *Photochem. Photobiol.* **66**, 34–41 (1997).

26. R.-J. Paul and H. Schneckenburger, Oxygen concentration and the oxidation-reduction state of yeast: determination of free/bound NADH and flavins by time-resolved spectroscopy, *Naturwissenschaften* **82**, 32–35 (1996).

27. R.J. Paul, J. Gohla, R. Föll, and H. Schneckenburger, Metabolic adaptations to environmental changes in Caenorhabditis elegans, *Comp. Biochem. Physiol.* B **127,** 469–479 (2000).

28. J.-M. Salmon, E. Kohen, P. Viallet, J.G. Hirschberg, A.W. Wouters, C. Kohen, and B. Thorell, Microspectrofluorometric approach to the study of free/bound NAD(P)H ratio as metabolic indicator in various cell types, *Photochem. Photobiol.* **36**, 585–593 (1982).

29. H. Schneckenburger, M.H. Gschwend, R. Sailer, H.-P. Mock, and W.S.L. Strauss, Time-gated fluorescence microscopy in molecular and cellular biology, *Cell. Mol. Biol.* **44**, 795–805 (1998).

30. H. Schneckenburger, M. Wagner, P. Weber, W.S.L. Strauß, and R. Sailer, Autofluorescence lifetime imaging of cultivated cells using a novel uv picosecond laser diode, *J. Fluoresc.* (in press).

31. T. Förster, Zwischenmolekularer Übergang von Elektronenanregungsenergie, *Z. Elektrochem.* **64**, 157–164 (1960).

32. H. Schneckenburger, K. Stock, M. Lyttek, W.S.L. Strauss, and R. Sailer, Fluorescence lifetime imaging (FLIM) of rhodamine 123 in living cells, *Photochem. Photobiol. Sci.* **3**, 127–131 (2004).

33. D. Axelrod, Cell-substrate contacts illuminated by total internal reflection fluorescence, *J. Cell Biol.* **89**, 141–145 (1981).

34. G.A. Truskey, J.S. Burmeister, E. Grapa, and W.M. Reichert, Total internal reflection fluorescence microscopy (TIRFM) (II) Topographical mapping of relative cell/substratum separation distances, *J. Cell Sci.* **103**, 491–499 (1992).

35. S.E. Sund and D. Axelrod, D, Actin dynamics at the living cell submembrane imaged by total internal reflection fluorescence photobleaching, *Biophys. J.* **79**, 1655–1669 (2000).

36. N.L. Thompson, A.W. Drake, L. Chen, and W.V. Broek, Equilibrium, kinetics, diffusion and self-association of proteins at membrane surfaces: Measurement by total internal reflection fluorescence microscopy, *Photochem. Photobiol.* **65**, 39–46 (1997).

37. G.M. Omann and D. Axelrod, Membrane-proximal calcium transients in stimulated neutrophils detected by total internal reflection fluorescence. *Biophys. J.* **71**, 2885–2891 (1996).

38. W.J. Betz, F. Mao, and C.B. Smith, Imaging exocytosis and endocytosis, *Curr. Opin. Neurobiol.* **6**, 365–371 (1996).

39. M. Oheim, D. Loerke, W. Stühmer, and R.H. Chow, Multiple stimulation-dependent processes regulate the size of the releasable pool of vesicles, *Eur. J. Biophys.* **28**, 91–101 (1999).

40. T. Parasassi, G. de Stasio, A. d'Ubaldo, and E. Gratton, Phase fluctuation in phosopholipid membranes revealed by laurdan fluorescence, *Biophys. J.* **57**, 1179-1181 (1990).
41. T. Parasassi, E.K. Krasnowska, L. Bagatolli, and E. Gratton, Laurdan and prodan as polarity-sensitive fluorescent membrane probes, *J. Fluoresc.* **4**, 365-373 (1998).
42. L.A. Bagatolli, T. Parasassi, G.D. Fidelio, and E. Gratton, A model for the interaction of 6-lauroyl-2-(N,N-dimethylamino) naphthalene with lipid environments: implications for spectral properties, *Photochem. Photobiol.*, **70**, 557-564 (1999).
43. H. Schneckenburger, K. Stock, W.S.L. Strauss, J. Eickholz, and R. Sailer, Time-gated total internal reflection fluorescence spectroscopy (TG-TIRFS): application to the membrane marker laurdan, *J. Microsc.* **211**, 30-36 (2003).
44. H. Schneckenburger, M. Wagner, M. Kretzschmar, W.S.L. Strauss, and R. Sailer, Laser-assisted fluorescence microscopy for measuring cell membrane dynamics, *Photochem. Photobiol. Sci.* (in press).
45. Z. Malik and H. Lugaci, Destruction of erythroleukaemic cells by photoinactivaton of endogenous porphyrins, *Br. J. Cancer* **56**, 589-595 (1987).
46. H. Kostron, A. Obwegeser, and M. Seiwald, PDT in neurosurgery; a review, J. Photochem. Photobiol. 36, 157-168 (1996).
47. R. Sailer, W.S.L. Strauss, H. Emmert, M. Wagner, R. Steiner, and H. Schneckenburger, Photodynamic efficacy and spectroscopic porperties of 5-aminolevulinic acid induced portoporphyrin IX in U373-MG glioblastoma cells, in preparation.
48. H. Schneckenburger, K. König, K. Kunzi-Rapp, C. Westphal-Frösch, and A. Rück, Time-resolved in-vivo fluorescence of photosensitizing porphyrins, *J. Photochem. Photobiol. B:Biol.* **21**, 143-147 (1993).
49. S. Brasselet, E.J.G. Peterman, A. Miyawaki, and W.E. Moerner, Single-molecule fluorescence energy transfer in calcium concentration dependent cameleon, *J. Phys. Chem.* **B104**, 3676-3682 (2000).
50. M.F. Garcia-Parajo, G.M.J. Segers-Nolten, J.-A. Veerman, J. Greve, and N.F. Van Hulst, Real-time light-driven dynamics of the fluorescence emission in single green fluorescent protein molecules, *Proc. Natl. Sci. USA* **97**, 7237-7242 (2000).
51. C. Eggelinh, J.R. Fries, L. Brand, R. Günther, and C.A.M. Seidel, Monitoring conformational dynamics of a single molecule by selective fluorescence spectroscopy, *Proc. Natl. Sci. USA* **95**, 1556-1561(1998).
52. Y. Sako, S. Minoguchi, and T. Yabagida, Single-molecule imaging of RGFR signalling on the surface of living cells, *Nature Cell. Biol.* **2**, 168-172 (2000).
53. C.W. Cody, D.C. Prasher, W.M. Westler, F.G. Prendergast, and W.W. Ward, Chemical structure of the hexapeptide chromophore of the Aequorea green-fluorescent protein, *Biochemistry* **32**, 1212-1218 (1993).
54. C.G. Coates, D.J. Denvir, N.G. McHale, K.D. Thornbury, and M.A. Hollywood, Ultra-sensitivity, speed and resolution: optimizing low-light microscopy with the back-illuminated electron-multiplying CCD, in Confocal, Multiphoton and Nonlinear Microscopic Imaging, edited by T. Wilson (Proc. SPIE, Vol. 5139, Belllingham, 2003) pp. 56–66.
55. T. Bruns, Screening of cell surfaces, Master Thesis, Hochschule Aalen (2004).
56. H. Schneckenburger, M.H. Gschwend, R. Sailer, W.S.L. Strauss, M. Lyttek, K. Stock, and P. Zipfl, Time-resolved *in situ* measurement of mitochondrial malfunction by energy transfer spectroscopy, *J. Biomed. Opt.* **5**, 362-366 (2000).

ANALYSIS OF CRUDE PETROLEUM OILS USING FLUORESCENCE SPECTROSCOPY

Alan G. Ryder*

8.1. INTRODUCTION

Crude oil is defined as *"a mixture of hydrocarbons that existed in the liquid phase in natural underground reservoirs and remains liquid at atmospheric pressure after passing through surface separating facilities"* (joint American Petroleum Institute, American Association of Petroleum Geologists, and Society of Petroleum Engineers definition).[1] Crude petroleum oils are complex mixtures of different compounds (mainly organic), which are obtained from an extensive range of different geological sources.[2, 3] Their physical appearance can vary from solid black tars to almost transparent liquids. In their natural state within an oilfield reservoir or entrapped within Hydrocarbon bearing Fluid Inclusions (HCFI), crude oils will also contain varying amounts of gasses (carbon dioxide, methane, etc.).[4] This presents the analyst with considerable challenges when developing methods for the characterisation and analysis of crude oils.[5] The non-contact, non-destructive, quantitative analysis of crude petroleum oils is a highly desirable objective for both research (e.g. study of microscopic HCFI) and industry (e.g. real-time assessment of oil production). Satisfying the needs of both macroscopic and microscopic applications is not straightforward, however, optical methods offer a convenient route to achieving these goals. Fluorescence spectroscopy is the best available optical technique, because it offers high sensitivity, good diagnostic potential, relatively simple instrumentation, and is perfectly suited to both microscopy and portable instrumentation.

This review encompasses a survey of fluorescence techniques used for the analysis of crude petroleum oils. Specific sections focus on the analysis of bulk crude petroleum oils, HCFI, and oil spills. The review focuses mainly on advances and studies reported in the literature from 1990 onwards, and outlines some of the issues that need to be addressed to make fluorescence methods more reproducible and quantitative.

* Alan G. Ryder, Department of Chemistry, and National Centre for Biomedical Engineering Science, National University of Ireland – Galway, Galway, Ireland.
  Tel: 353-(0)91-524411 ext. 2943; *Fax:* 353-(0)91-750596; *Email:* alan.ryder@nuigalway.ie

8.2. PETROLEUM COMPOSITION

The chemical composition of crude petroleum oils varies enormously, and their characterization requires the application of considerable instrumentation and time resources. Both Hunt[2] and North[3] provide good foundations in the basics of petroleum composition, and the geological factors that influence changes in oil composition. Comprehensive, laboratory based chemical analysis of crude oils is done using a variety of gravimetric, solvent extraction, and chromatographic techniques.[6, 7] Details of these methods, their advantages, and disadvantages are outside the scope of this review. However, it should be noted that the different methodologies employed, could provide subtly different values for oil composition parameters. The primary compositional parameters reported in fluorescence studies (if at all, see table 1) tend to be obtained from column chromatography and in chemical terms are: an alkane (or paraffin, or aliphatic) fraction, an aromatic fraction, a polar (or resin) fraction, and an asphaltene fraction.[8] The asphaltene fraction is that portion of the oil, which is insoluble in pentane or heptane and comprises of high molecular weight species.[9] Often the methods used, and the results obtained for chemical composition will differ from laboratory to laboratory (depending on the chromatographic conditions used, and whether the light, volatile fraction has been removed prior to chromatographic analysis). Therefore much care should be taken in comparing fluorescence data and results from different crude oil studies. In general, for accurate characterisation of crude oil fluorescence one should have information on the source rock, its density (API gravity), and the gross chemical composition (alkane, aromatic, polar, and asphaltene concentration).[10]

Table 8.1. Survey of crude oil data provided for fluorescence studies from a selection of literature sources.

Reference	Location	API gravity	Source rock	Chemical data
Abu-Zeid, et al. [11]	Partial	No	No	No
Alpern et al. [12]	No	Yes[a]	No	Partial
Blanchet et al. [13]	Yes	Yes	No	No
Camagni et al. [14]	Partial	No	No	No
Hagemann & Hollerbach, [15]	Yes	Partial	Partial	Partial
Measures et al. [16]	Yes	Yes	No	No
Mullins, Mitra-Kirtley, & Zhu.[17]	No	No	No	No
Mullins and Downare.[18]	Partial	No	No	No
Quinn et al. [19]	No	Yes	No	No
Rayner & Szabo.[20]	Partial	Partial	No	No
Ryder, & Ryder et al. [21, 22]	Partial	Yes	No	Partial
Ryder et al. [23]	Yes	Yes	Yes	Yes
Ryder (2004), [10]	Yes	Yes	Yes	Yes
Stasiuk et al.[24]	Yes	No	No	Yes
Wang & Mullins.[25]	Partial	Yes	No	No
Zhu & Mullins. [26]	Partial	No	No	No

[a] Provided as oil density.

In many cases, the only information available for the crude oil being investigated is its location, source, or API gravity. API gravity is one of the simplest and most wide-ranging parameters used for describing crude oils. It is inversely related to the density by the formula: API gravity = ((141.5/specific gravity at 15.6°C) – 131.5). In general, the higher the API gravity value, the lighter the oil. Specific gravities of oils tend to be in the 0.73 to 1.0 range, with paraffin type oils being lighter than asphalt-base oils.[3] Despite the fact that it does not provide any significant chemical information, the parameter is easily measured and widely used. Table 1 gives a brief survey of some recent fluorescence studies of crude oils, clearly showing the dearth of chemical information usually provided. Therefore, one must be aware that when discussing petroleum fluorescence, qualitative descriptions dominate, and quantitative analysis may only apply to individual studies.

8.3. PETROLEUM OIL FLUORESCENCE

The use of fluorescence for the analysis of crude oils has been in use for the past 60 years particularly for mud logging where UV light is used to detect the presence of oil in drilling mud.[27] Fluorescence is also used in the analysis of core samples, again to identify the presence of oils.[3] The fluorescence of crude petroleum oils derives from the aromatic hydrocarbon fraction,[28] and this fluorescence emission is strongly influenced by the chemical composition (e.g. fluorophore and quencher concentrations) and physical characteristics (e.g. viscosity and optical density) of the oil. Unfortunately, crude petroleum oils encompass a very wide range of physical and chemical characteristics, making the fluorescence analysis of crude petroleum oils rather difficult.

Crude oils vary in appearance from black tars to clear liquids, indicating a complex absorption profile. This is shown experimentally in Figure 8.1, where the spectra of 22 oils of different compositions are displayed. It should also be noted that the shape of the absorption edge is very similar in each case. Examination of the electronic absorption spectra of crude oils reveals that the electronic absorption edge is similar to the "Urbach tail," in which the absorption coefficient depends exponentially on the photon energy.[17]

The absorption edge moves to the red, as crude oils get heavier (lower API gravity), and the distribution and number of chromophores increases. These absorption studies, in conjunction with fluorescence emission studies reveal that when the Urbach tail accurately describes the absorption, the dominant absorption process in the Urbach tail region, corresponds to excitation of the lowest-energy electronic transitions of the corresponding chromophores. In addition, Mullins and his co-authors suggest that the absorption tail gives a direct measure of the chromophores population distribution, with larger chromophores being present in exponentially decreasing quantities.[17]

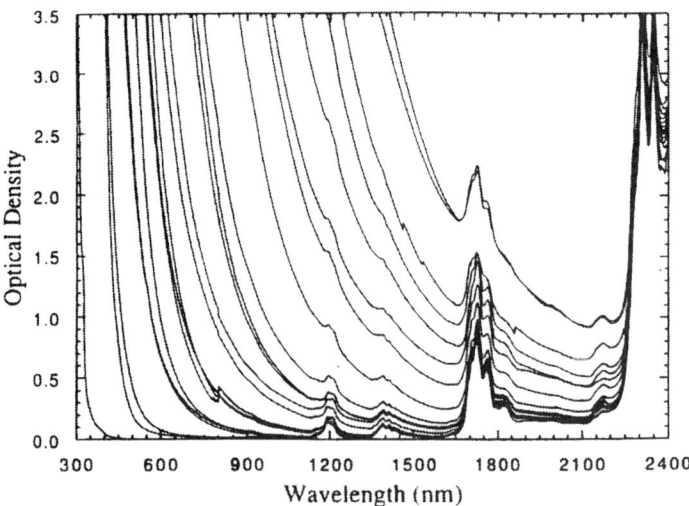

Figure 8.1. Optical absorption spectra of 22 crude oils, with the heavier oils to the right of the plot. The vibrational absorption peaks are superimposed on a highly variable, continuous, monotonic electronic absorption profile. The spectra were collected in 2 mm pathlength cells and referenced against a 2mm cell filled with CCl$_4$. *Reproduced with permission from Ref [17] © 1995, Society for Applied Spectroscopy.*

In practical terms, this means that the excitation wavelength used for fluorescence spectroscopy of crude oils needs to be selected with care to enable efficient excitation of all crude oils types. From Figure 8.1 it is clear that visible excitation (>~450 nm) may not be suitable for the lightest crude oils and/or condensates. Furthermore, different excitation wavelengths results in the excitation of different fluorophore populations, which has an impact on the fluorescence emission produced. Another point, arising from the absorption studies, is the potential for inner filter and energy transfer effects to influence fluorescence emission. The chemical complexity of crude oils ensures that in most cases, there is a high probability the fluorescence emitted by one species will be absorbed by another fluorophore resulting in energy transfer. This is particularly true for UV or blue excitation. Therefore, for the study of bulk, undiluted oils, front-surface geometries are required,[10, 21-23, 29] and in all other cases, oils need to be highly diluted to minimise inner filter effects.

8.3.1. Steady-State Emission

In general, light oils (high API gravity) tend to have relatively narrow, strong fluorescence emission bands with a smaller Stoke shift than that found for of heavier oils (lower API gravity) where the emission tends to be weaker, broader, and red-shifted. Figure 8.2 shows the emission spectra obtained from a series of five crude oils, using 337 nm excitation. Similar results were later reported by Rayner and Szabo,[20] also using 337 nm excitation. These gross changes in fluorescence emission are due to the higher concentration of fluorophores and quenchers present in the heavier oils, which leads to a higher rate of energy transfer and fluorescence quenching, producing the broader, weaker, red shifted emission.[18, 30]

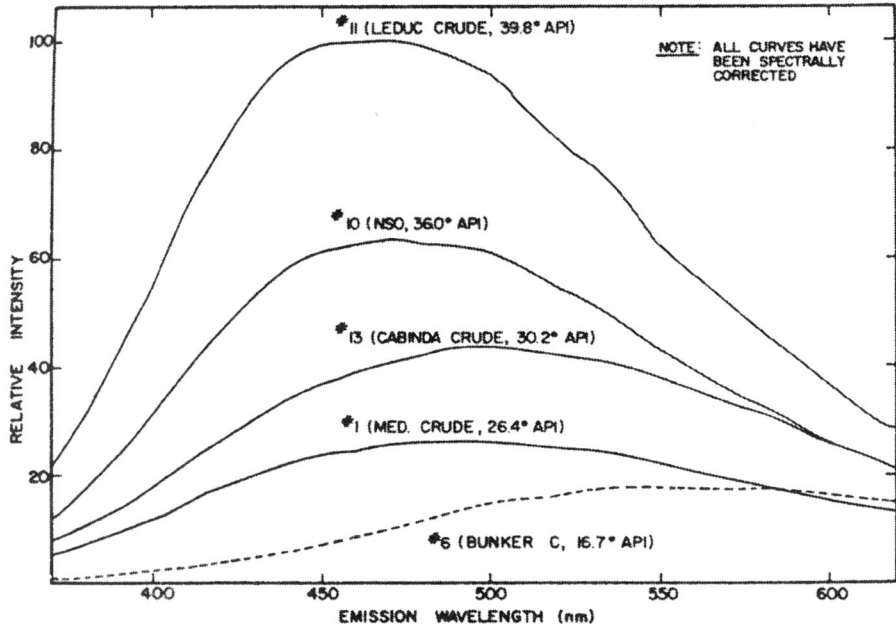

Figure 8.2. Fluorescence spectra of crude oils obtained using 337 nm excitation. *Reproduced with permission from Ref [16], © 1974, Optical Engineering.*

Energy transfer processes result in red shifted, and spectrally broadened emission while the quenching processes reduce emission intensity. The effect however diminishes with increasing excitation wavelength (Figure 8.3) until with near-IR excitation, the profile of the emission spectra of light, heavy, and diluted samples of each are very nearly identical. At these NIR excitation wavelengths, there is nearly no energy transfer. The effect can be quantified by diluting crude oils (Figure 8.4) with solvents such as benzene,[18, 25] n-heptane,[25] or cyclohexane.[29] In each case, dilution results in an increase in fluorescence intensity and a blue shift. Stern-Volmer plots constructed from highly diluted crude oils imply that the quenching and energy transfer processes depend linearly with crude oil concentration.[18, 25] Analysis of the wavelength dependence of the Stern-Volmer plots indicate that the collisional decay constants are greater at shorter emission wavelengths[25] and excitation wavelengths.[18]

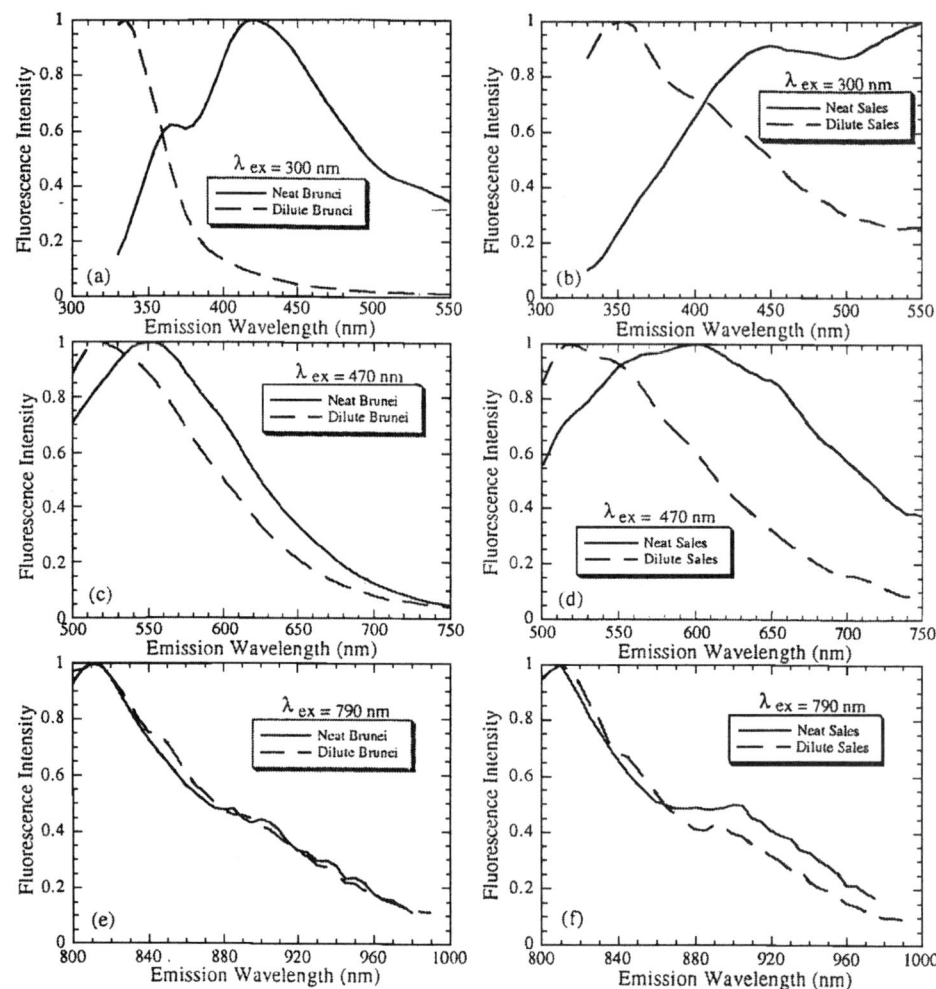

Figure 8.3. Fluorescence emission spectra for neat and dilute solutions of a light (Brunei) and a heavy (Sales) crude oil for 300-, 400-, and 790- nm excitation wavelengths. Collisional energy transfer produces large spectral differences between neat and dilute solutions for short excitation wavelengths. Collisional energy transfer decreases with increasing excitation wavelength. Determining the contribution of the nascent (dilute solution) emission to the neat spectrum allows the quantitative determination of collisional energy transfer. *Reproduced with permission from Ref [18] © 1995, Society for Applied Spectroscopy.*

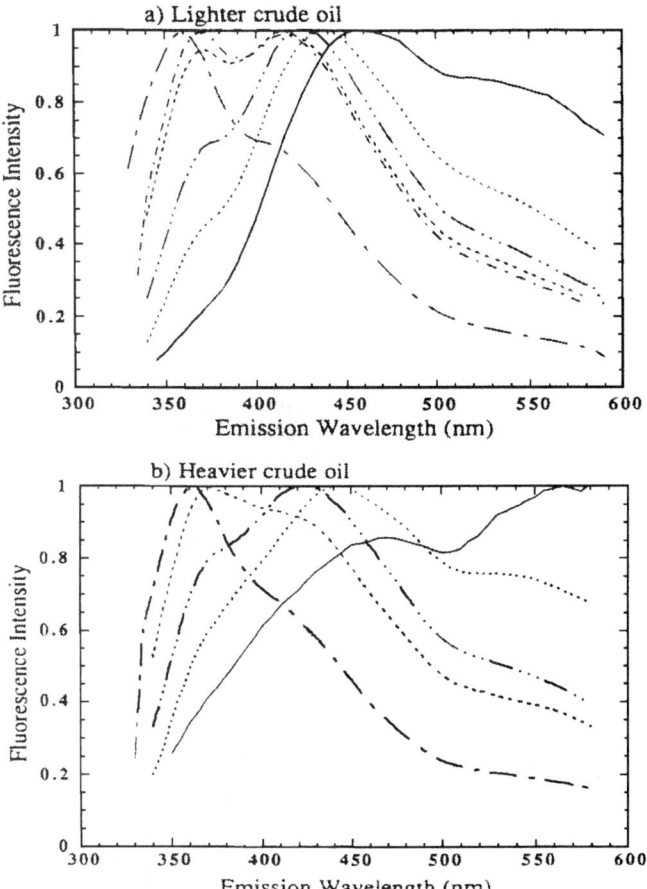

Figure 8.4. The emission spectra with 316-nm excitation for (a) North Sea, the lighter crude oil, and (b) Sales, the heavier crude oil, at different concentrations. A substantial red shift with increasing concentration results from energy transfer. Dilution factors: (a) (— · —) 1:35,000; (- - -) 1:625; (---) 1:125; (————) 1:25; (·····) 1:5; (——) neat. (b) (— · —) 1:35,000; (- - -) 1:625; (---) 1:125; (————)1:25; (·····) 1:5; (——) neat. *Reproduced with permission from Ref [25] © 1994, Society for Applied Spectroscopy.*

The quantum yields of crude oils are highly dependant on the excitation wavelength, with excitation in the visible or red being much less efficient than UV excitation (Figure 8.5). This is largely due to the reduction in optical absorption (Figure 8.1) and the increase in non-radiative decay pathways (internal conversion) with increasing excitation wavelength. The relative change in quantum yield was also demonstrated to be largely similar for a wide range of light to heavy oils, which can be accounted for by the energy dependence of internal conversion. Dilution of crude oils also increases the quantum yield by reducing the quenching rate.[30] Diluted heavy oils however, have lower quantum yields than diluted light oils due to their higher concentrations of larger (red absorbing) chromophores, which are more likely to undergo non-radiative internal conversion.

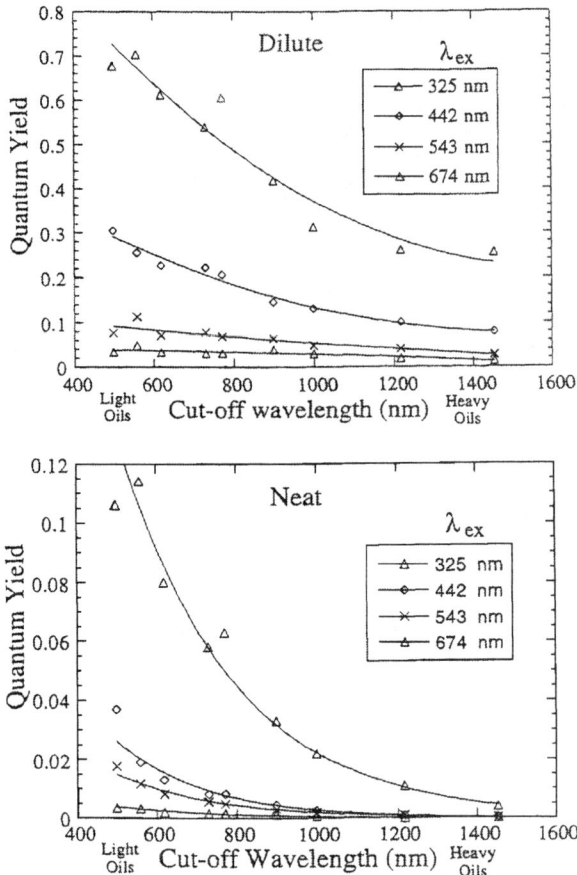

Figure 8.5. The absolute quantum yields for neat (bottom) and dilute (top) crude oils. The cutoff wavelength is an optical measure of the oil weight; heavy oils have a large cutoff wavelength. The cutoff wavelength is defined as being the point at which the optical density reaches a value of 3.0 for a 2mm optical pathlength. Quantum yields decrease greatly with increasing excitation wavelength due to internal conversion. Quenching from chromophores interactions causes neat crude oils to have much smaller quantum yields than dilute crude oils, especially for heavy oils. For the dilute case, the quantum yields of heavy oils are lower due to the greater fraction of large, aromatic chromophores. *Reproduced with permission from Ref [30] © 1995, Society for Applied Spectroscopy.*

The balance between collisional energy transfer and quenching is the dominant feature affecting crude oil fluorescence. Downare and Mullins[18] calculated (from emission spectra) and plotted the ratio of energy transfer to total emission against excitation wavelengths for three representative oils (Figure 8.6). This plot shows how influential the selection of excitation wavelength is on the photophysics of crude oil emission. This has obvious consequences for the consideration of excitation sources for petroleum fluorescence applications. For several studies in the 1970's and 80's, which employ 316 nm or 337 nm excitation sources, emission arises largely (>75%) from energy transfer, while more recent studies employing semiconductor light sources (380 nm LED or 405 nm violet laser diodes) produce emission that has a lower energy transfer contribution (75-60%). Therefore, direct comparison of experimental results between studies using different excitation sources needs to be approached with caution.

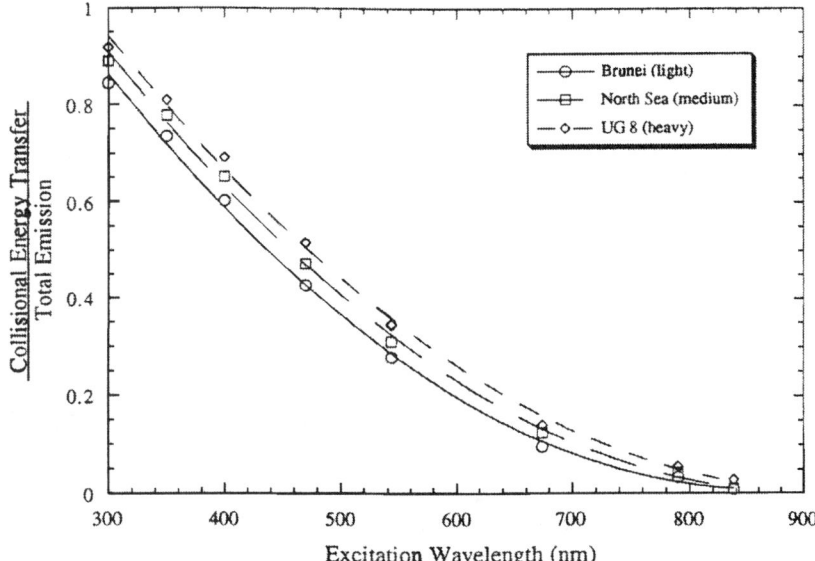

Figure 8.6. The ratio of energy transfer to total emission plotted against excitation wavelengths for three crude oils ranging from light to heavy, which exhibits several systematic trends. Collisional energy transfer varies from nearly 100% for the shortest wavelength excitation to nearly 0% for the longest wavelength excitation. All crude oils show nearly the same behaviour. Thus, for crude oils, the fraction of collisional energy transfer is not a function of chromophores concentration; all crude oils are in the high concentration limit. *Reproduced with permission from Ref [18], © 1995, Society for Applied Spectroscopy.*

The changes observed in the fluorescence emission of crude oils with increasing excitation wavelength are: a narrowing of the emission band,[18, 21] and reductions in the Stokes shift, quantum yield,[30] and fluorescence lifetime.[21] This decrease is caused by the complex interaction between energy transfer and quenching processes. At short excitation wavelengths, energy transfer processes dominate since most of the absorbing fluorophores have large bandgaps and can transfer energy to the large numbers of smaller bandgap molecules. At longer excitation wavelength, the excited fluorophores have small bandgaps and there are fewer molecules with smaller bandgaps for energy transfer, so most collisions result in quenching, with the subsequent reduction in fluorescence lifetime. Furthermore, as the bandgaps of the excited fluorophores decreases, there is an increased rate of internal conversion, which also contributes to the reduction in lifetime.[21, 26, 30, 31]

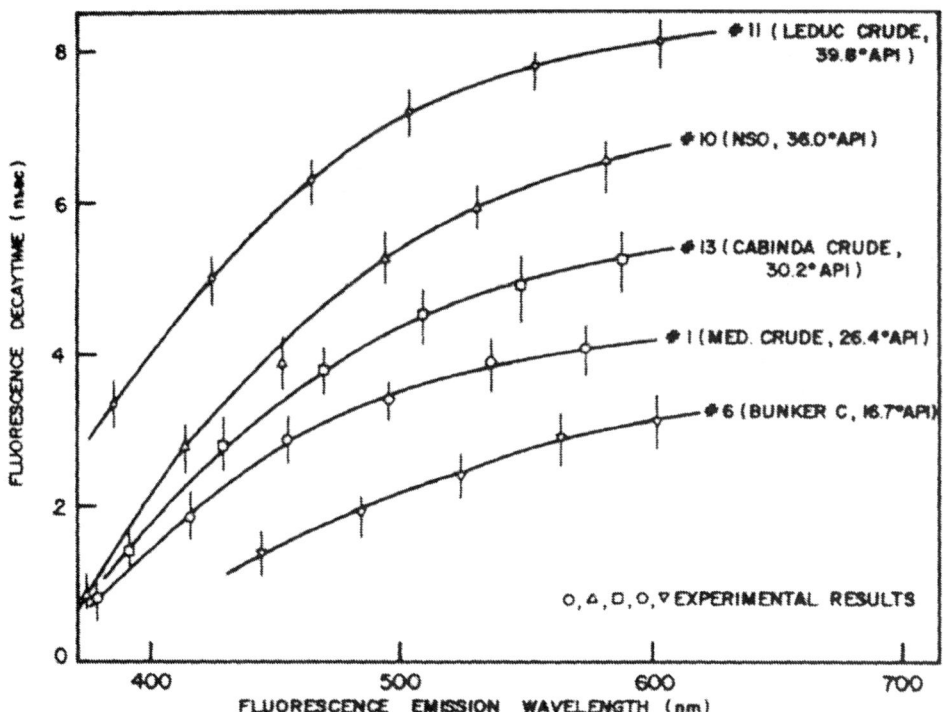

Figure 8.7. Variation of fluorescence decay time with emission wavelength for several crude oils. Data was collected with 337 nm excitation and using a single exponential fitting approach. *Reproduced with permission from Ref [16], © 1974, Optical Engineering.*

8.3.2. Time-Resolved Fluorescence

The fluorescence lifetime of crude oils is very sensitive to composition, with heavy oils having shorter lifetimes than light oils (Figure 8.7).[10, 16, 21, 22, 25] The 1974 study also showed a positive correlation between API gravity for four crude oils, and the decay time being measured at λ_{max}. The changes in fluorescence lifetime is a result of the interplay between energy transfer and quenching, and is also emission wavelength dependant, since each emission wavelength represents a different population of emitting fluorophores. The numerical value of the fluorescence lifetime is however, very dependant on the method by which it is calculated. In Figure 8.7, the lifetimes were calculated via a single exponential convolution process, which should not be used other than for the most qualitative of descriptions. Rayner and Szabo,[20] used Time Correlated Single Photon Counting (TCSPC) with a pulsed nitrogen lamp source to measure fluorescence lifetime data from a series of 13 oils, including five crude oils. They calculated lifetime values via a deconvolution method, the fitting of two decay terms, and the requirement that the residuals of the fit be randomly distributed about zero. This resulted in reproducible fits with three parameters being reported: τ_a, τ_b, and F_a/F_b, with F_a/F_b being the ratio of the pre-exponential factors for the two lifetime components.

However, no attempt was made to correlate the lifetime data with chemical composition. In 1987, Abu-Zeid et al. revisited the use of fluorescence lifetimes as a potential tool for identifying oil slicks.[11] Unfortunately, their lifetime measurements did not include any pre-exponential factors so it is not possible to assess the true fluorescence lifetimes of the oils.

The complex composition of crude oils ensures that the measured fluorescence decay curve is the sum of a distribution of individual decay curves. As such, one needs to exercise caution in discussing fluorescence lifetimes. Wang and Mullins,[25] in their detailed study of the fluorescence lifetime of crude oils used a bi-exponential fit to describe the fluorescence decay curves. More recently, we have used the intensity-averaged lifetime ($\bar{\tau}$) calculated using either bi-or tri- exponential fits as a reproducible and standardised lifetime measurement for crude oils.[10, 21-23, 32] Figure 8.8 shows the $\bar{\tau}$ dependence with emission wavelength for a series of North Sea oils. In general, the wavelength at which the maximum average lifetime is measured occurs at longer wavelength for heavy crude oils than it does for light oils.[16, 21, 22] The variation in lifetime going from 450 nm to ~600 nm emission wavelength is more pronounced for the lighter oils as is seen from Figure 8.8. At a particular point, a maximum value for the average lifetime ($\bar{\tau}_{max}$) is recorded at a wavelength $\lambda_{\tau max}$ (e.g. ~ 600 nm for N11 with 380 nm excitation), which is dependant on the chemical characteristics of each oil. Figure 8.8 also shows a noticeable decrease in the average lifetime at emission wavelengths greater than ~600 nm. This 'curved' wavelength dependence is due to the complex interplay between collisional energy transfer and quenching processes.[21, 22]

At short emission wavelengths, fluorophores have large bandgaps and can readily undergo collisional energy transfer with larger, aromatic, small bandgap molecules. The rate of energy transfer is dependant on the bandgap between the two populations and so the value of $\bar{\tau}$ relative to the maximum lifetime ($\bar{\tau}_{max}$) is shortened the most at short emission wavelengths where the bandgap is greatest. As the emission wavelength increases the energy gap between donor and acceptor decreases, and the concentration of large aromatic acceptor molecules decreases. The first factor results in an increase in the energy transfer rate but the second factor results in a decrease in the rate, as there are fewer acceptor molecules. Overall, the effect of both factors results in a net overall decrease in the rate of energy transfer leading to an increase in $\bar{\tau}$ until $\bar{\tau}_{max}$ is reached at a wavelength $\lambda_{\tau max}$. $\lambda_{\tau max}$ is red-shifted as the API gravity of the oils decreases which is evident from Figure 8.8.

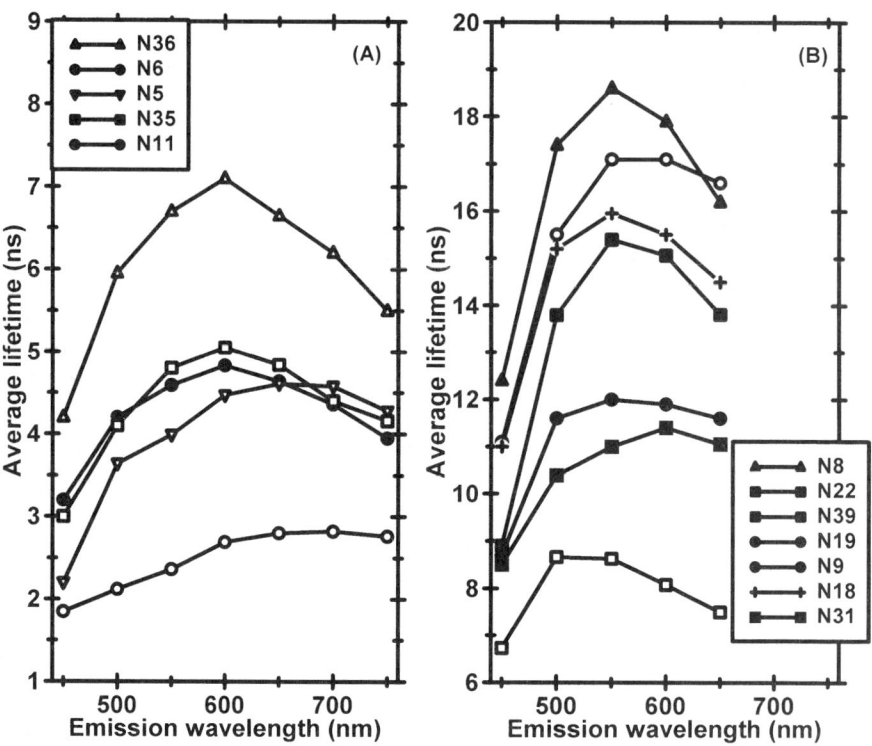

Figure 8.8. Plots of intensity averaged fluorescence ($\bar{\tau}$) lifetime vs. emission wavelength for a series of medium-heavy oils, API gravity < 40° (A), and light oils, API gravity > 35° (B). The lifetimes were measured using a 380 nm LED excitation source and TCSPC instrumentation. *Reproduced and adapted with permission from Ref [22], © 2002, Society for Applied Spectroscopy.*

At emission wavelengths longer than $\lambda_{\tau max}$, collisional quenching takes over as the dominant effect because as the emission wavelength increases, the bandgap between excited fluorophores and acceptor molecule gets smaller. Eventually the bandgap is so small that the acceptor molecules act as quenchers since they can de-excite themselves (internal conversion), and so collisional quenching dominates resulting in a decrease in $\bar{\tau}$ relative to $\bar{\tau}_{max}$. The excited fluorophores can also undergo internal conversion the rate of which is inversely proportional to the bandgap and this leads to a decrease in $\bar{\tau}$. The greater degree of lifetime change associated with lighter oils is due to the comparatively dilute concentration of fluorophores which so the average lifetime is more sensitive to collisional and internal conversion processes. With heavy oils, the concentration of fluorophores and quenchers is so high that small changes in the concentration of donor or quencher species will have very little effect on the average lifetime. The decrease in lifetime attributable to internal conversion is far outweighed by collisional quenching for the heavy oils.

8.3.3. Multidimensional Techniques

Multidimensional techniques involve the collection of multiple fluorescence parameters and offer a more detailed method for studying petroleum photophysics or providing a spectroscopic fingerprint for identification purposes. Excitation-Emission Matrix (EEM) methods are a convenient way of mapping the fluorescence space that complex oils occupy.[33, 34] They can be used in the laboratory to determine the optimum excitation-emission parameters for simpler measurements/instrumentation, or as a method for the discrimination of oil types. The main drawbacks of EEM include time-consuming data collection and the requirement for complex instrumentation. Synchronous Fluorescence Spectroscopy (SFS) has been applied to the study of petroleum-based materials since the mid-1970's.[35, 36] Constant wavelength mode SFS has been used to study crude oils,[37] motor oils,[38, 39] and asphaltenes.[40] SFS has also been applied to the study of Shale oils,[41] fluid inclusions,[42] and reservoir compartmentalization.[43] Constant energy SFS was used for the analysis of both crude oil and gasoline.[44] 3-D fluorescence spectra have been used to discriminate oils into condensate, light oil, and heavy oils.[45, 46] Smith and Sinski[47] studied the red-shift cascade,* which has also been used to investigate the degradation of aged petroleum by mycobacteria.[48] More recently, Total Synchronous Fluorescence Scan Spectroscopy (TSFS) has been used to discriminate different refined and crude petroleum liquids.[49, 50]

The compositional diversity of crude petroleum oils is evident in Figure 8.9 where the TSFS plots of nine different oils are displayed. For comparison purposes, and to account for instrument instability / sampling effects, each TSFS plot was normalised to the point of maximum fluorescence intensity. This allows for a general comparison between the different oils based on the identity of the emitting species. All the TSFS plots show a general diagonal contour trend from short λ_{ex} /large $\Delta\lambda$, to long λ_{ex} /short $\Delta\lambda$ which represents a maximum fluorescence emission in the 350 – 500 nm range for these excitation wavelengths. This diagonal trend represents the extensive impact energy transfer processes have on crude petroleum oil fluorescence. The top row of Figure 8.9 shows the TSFS plots of 3 light oils (API > 40°) with low polar concentrations (<4%). There are considerable differences in the plot topology and this is due to changes in the aromatic concentration (as measured). The measured aromatic concentration increases across the top row from 1.8% to 6.6%, and then to 18.2% for (C). Oil (A) is classed as a late maturity oil and as such, most of the larger polyaromatic species will have been broken down to alkanes and small aromatic species. This results in a very tight TSFS contour plot centred at $\lambda_{ex} = 390$ nm, $\Delta\lambda = 40$ nm, indicating a somewhat homogenous, and restricted mixture of fluorophores, with an emission maximum around 430 nm. The more diverse and wider ranging contour plot of oil (B) cannot be explained just on the basis of a ~4% increase in aromatic concentration, but also by a change in the type of aromatic species present. Since most crude oil fluorophores are aromatic, it follows that the increase in aromatic concentration causes TSFS contours to spread out over a larger parameter space. The primary process driving this is the increased rates of collisional energy transfer from small to large aromatic species.

---

* The red shift cascade is the greater degree of energy transfer, which occurs at high concentrations of crude oils. The excitation energy will continue to cascade to larger fluorophores, producing greater red shifts in the emission spectra.[25]

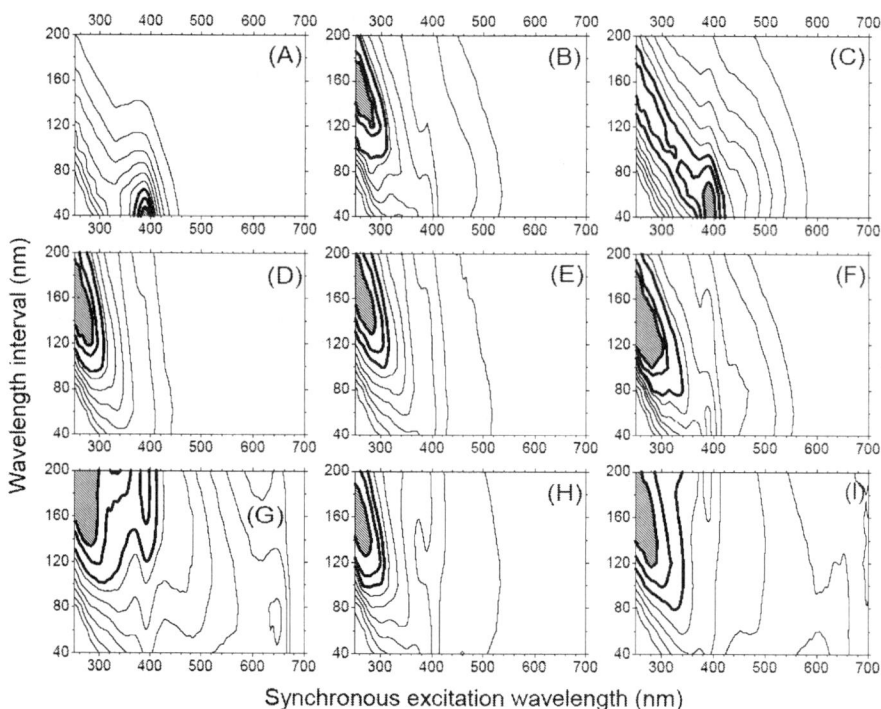

Figure 8.9: Total Synchronous Fluorescence Scan plots for 9 different crude petroleum oils recorded from 250 nm to 700 nm over a wavelength interval of 40-200 nm in a front surface sampling geometry: (A) 7703: API = 50.6. polar = 1.6%; (B) 7197: API = 45.1, polar = 1.9%; (C) 7058: API = 40.1, polar=3.5%; (D) 7062: API = 36, polar = 17.2%; (E) 7093: API = 30.9, polar = 11.5%; (F) 7633: API =24.8, polar =3.73%; (G) 7169: API = 21.6, polar=20.3%; (H) 7130: API = 15.6, polar=24.6%; (I) 7033, API = 12.8, Polar = 26.3%. *Reproduced with permission from Ref [50], © 2003, Journal of Fluorescence Spectroscopy.*

Generally as a crude oil matures, the aromatic fraction is gradually reduced and therefore it would seem possible to assess the maturity of the oils by measuring the changes in TSFS topography. Unfortunately, this is probably only applicable to oils from a single source, because the second row of Figure 8.9 shows that for oils with similar aromatic concentrations as C (D & E), the topography of the TSFS contour plots are significantly different. This is caused by a relatively higher polar concentration, which results in increased rates of collisional quenching, with the greatest effect being observed at λ_{ex} ~400 nm and $\Delta\lambda$ of <100 nm. In the TSFS plot for (F), the contours extend further out into the red because this oil has a relatively low polar concentration leading to a reduced quenching rate. The bottom row of Figure 8.9 shows the TSFS plots for some heavy oils, all of which have relatively large concentrations of polar constituents. This results in much weaker fluorescence intensity, but apart from (G), the TSFS topography does not appear to be very different from the TSFS in the preceding rows. Case (G) is a unique in that it is heavily degraded which has resulted in the formation of a much wider range of fluorophores as evidenced by the spread of high intensity contours into more of the parameter space.

The large degree of similarity between some of the TSFS plots is largely due to the fact that normalisation was done at the point of maximum fluorescence intensity and this obscures the huge differences in fluorescence intensity observed for the various samples (~200 fold between light (A) and heavy oils (I)).[50]

8.3.4. Instrumentation

The instrumentation for laboratory based fluorescence analysis of petroleum fluids is not that significantly different from that employed for routine biophysical or chemical physics measurements. In the author's laboratory at NUI-Galway, standard fluorescence steady-state instrumentation (Perkin-Elmer LS50B) and time-resolved instrumentation (TCSPC instrumentation assembled from commercial components) has been used. Apart from using front-surface excitation for all measurements, there are no significant novel requirements. [10, 21-23, 32, 50] Most of the novel instrumentation research involves the development of systems for specific applications. Examples include using fibre optics coupled with fluorescence and reflectivity measurements for quantifying the amounts of oil, water, and gas in a continuous flow,[51, 52] and fibre optic systems for the analysis of solid materials particularly from core samples,[53] and soil (or water).[54, 55] Several studies have discussed the utility of various excitation sources, such as tuneable dye lasers[56] and LED's[57, 58] for oil fluorescence applications.

8.4. QUALITATIVE AND QUANTITATIVE OIL ANALYSIS

The use of fluorescence methods for the qualitative and quantitative analysis of crude oils is fairly common, covering diverse applications from oil spill identification to the study of microscopic petroleum fluid inclusions within rocks. The correlation of different fluorescence parameters with chemical (or physical) characteristics of bulk crude oils have been used to develop qualitative or quantitative models for industrial or research applications. The selection (or indeed discovery) or the best fluorescence parameters for quantitative crude oil analysis is a complex issue and has not yet been fully resolved. One of the key challenges however, is to develop a method that is suitable for any crude oil type from any source. In this section, we survey the use of both steady-state and time-resolved fluorescence parameters for the analysis of crude oils.

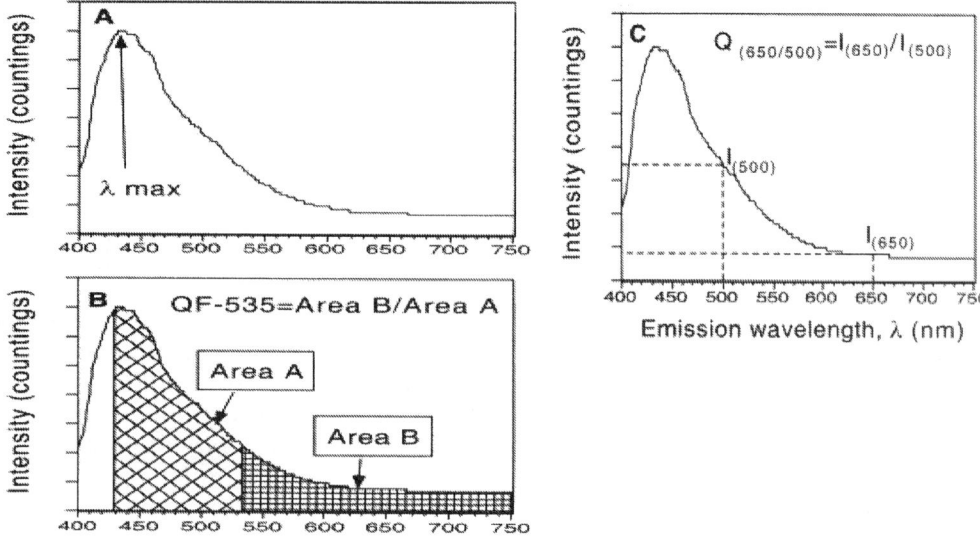

Figure 8.10. Definitions of fluorescence emission parameters. (A) The wavelength of maximum intensity, λ_{max}.; (B) The QF-535 factor; and (C) The $Q_{650/500}$ (Red-Green-Quotient) factor. *Reproduced with permission from Ref. [68], © 2001, Elsevier.*

8.4.1. Steady-State

Three of the most common steady-state parameters used within the geological science community for oil analysis are illustrated in Figure 8.10. The first and most obvious fluorescence parameter to consider is the emission wavelength and in particular the point of maximum emission (λ_{max}). Hagemann & Hollerbach,[15] correlated λ_{max} with the density (g/ccm) of a series of crude oils (Figure 8.11). The wavelength of maximum fluorescence emission (λ_{max}), using a 365 ± 30 nm excitation source, was later demonstrated to correlate with saturate and aromatic concentrations (Figure 8.12) for a series of mostly Canadian crude oils.[59] Using a 380 nm LED source and a small set of North Sea oils, it was found that the width of the emission band correlated better with API gravity than did λ_{max}.[22] However, the data set was too restricted and there was no detailed gross compositional data available.

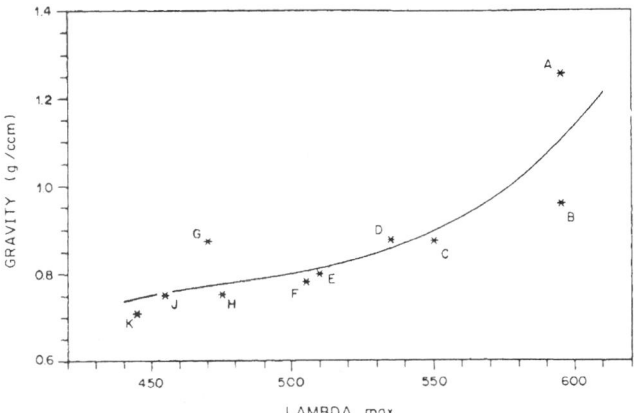

Figure 8.11. The variation in the fluorescence parameter λ_{max}, with decreasing gravity of the crude oils and condensates. (A) Papa Playa, biodegraded; (B) Georgsdorf, Germany, Heavy Oil; (C) Shell Monarch, USA, low mature; (D) Arco McCone, USA, low/high mature; (E) Arlesried, Germany, low/high mature; (F) Hassi-Messaoud, Algeria, low/high maturity; (G) Amerado USA Morton, low/high maturity; (H) east Mereenie, Australia, high mature; (J) Amerdo F.E. Weedeman, USA, condensate; (K) Bilabari, Nigeria, USA, condensate. *Reproduced with permission from Ref [15], © 1986, Pergamon Journals Ltd.*

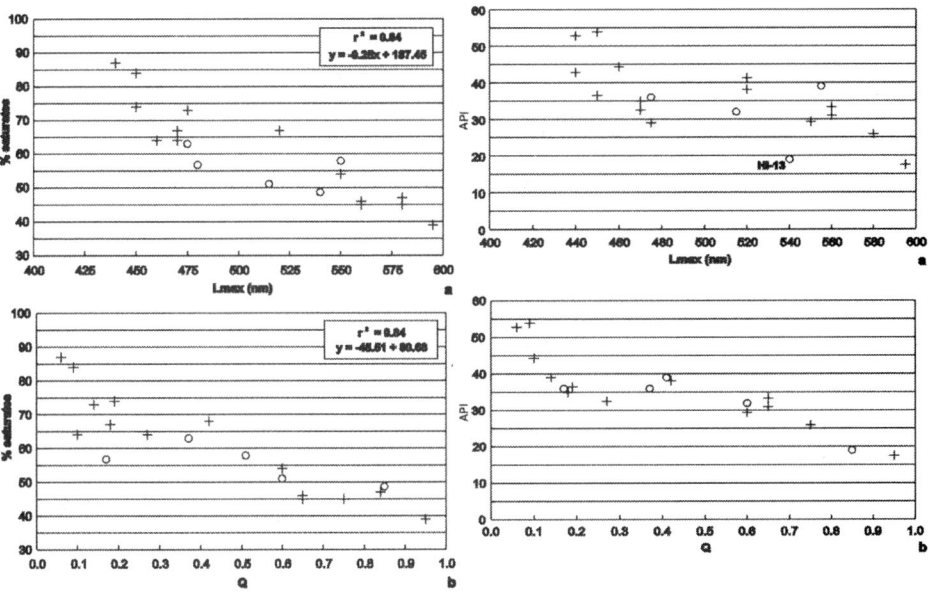

Figure 8.12. Plots of fluorescence parameters, λ_{max} (top) and Q(bottom) versus oil saturate concentration (left) and API gravity (right). The crosses represent data acquired from bulk crude oils while the circles are data obtained from Hydrocarbon bearing fluid inclusions. *Reproduced with permission from ref. 59, © 1997, Elsevier.*

A second parameter, the Red-Green ratio shows a better correlation with oil viscosity, which is linearly related to oil density (Figure 8.13). However, this was not pursued further and chromaticity diagrams based on the emission data were used to discriminate the various oil samples. More methods based on these chromaticity diagrams have been used for the analysis of hydrocarbon bearing fluid inclusions, and are discussed in the section on fluid inclusions. Unfortunately, these methods are dependant on the use of a set excitation wavelength, usually 365 nm from a Mercury arc lamp.[15, 28] This means that the results cannot be compared directly with those obtained using other sources such as a 337 nm nitrogen laser,[16] or 380 nm LED sources.[22] Another study using the Red-Green ratio demonstrated a correlation between this factor and an oil quality parameter based on the combined concentration of the aromatic, resin, and asphaltene fractions.[12] However, the method does not seem to have been used to any great extent by other researchers. This study also raises the potential for extracting hydrocarbons into epoxy resins for fluorescence analysis. However, this has to be approached with caution since no data is presented on the effects of immobilization on quenching and energy transfer rates. It would be expected that the fluorescence behavior of oils entrapped in epoxy resins would be significantly different from the free bulk oils.

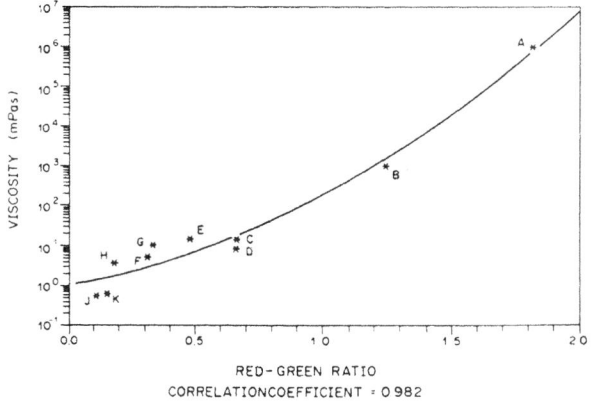

Figure 8.13. Relationship between the Red-Green-Quotient and viscosity of the crude oils and condensates. (A) Papa Playa, biodegraded; (B) Georgsdorf, Germany, Heavy Oil; (C) Shell Monarch, USA, low mature; (D) Arco McCone, USA, low/high mature; (E) Arlesried, Germany, low/high mature; (F) Hassi-Messaoud, Algeria, low/high maturity; (G) Amerado USA Morton, low/high maturity; (H) east Mereenie, Australia, high mature; (J) Amerdo F.E. Weedeman, USA, condensate; (K) Bilabari, Nigeria, USA, condensate. *Reproduced with permission from Ref [15], © 1986, Pergamon Journals Ltd.*

Another emission ratio method is the QFT-II method developed by Texaco in the 1990's for oil well logging.[60] QFT-II uses 254 nm excitation and measures fluorescence intensity at emission wavelengths of 287 and 365 nm from solvent extracted material. The method was calibrated using a set of 70 crude oils (no data given on source or chemical composition) using two different extraction solvents (hexane and *iso*-propyl-alcohol). The method is designed for on-site analysis and only yields an estimate for API gravity, and is unsuited for heavy oils (API gravity <15).

The ratio of the 535-750 nm flux to the 430-535 nm flux of fluorescence emission spectra using 365 nm excitation was also used as a parameter to characterise crude oils but no assessment of its quantitative accuracy was provided.[28, 70] In general, there are several practical disadvantages in using fluorescence intensity based measurements for oil characterisation particularly for remote sensing applications. The main problem lies in the difficulty in the accurate, reproducible measurement of absolute fluorescence emission intensity.[†] These measurements can be affected by instabilities in the excitation source, detector electronics, sample turbidity, and photobleaching.[61, 62]

8.4.2. Time-Resolved

Time-resolved fluorescence measurements are largely insensitive to the negative factors that affect steady-state measurements and are more easily referenced, making sensing applications more robust.[63] Time resolved fluorescence techniques,[64] which have also been employed for characterization of petroleum products, are not only largely free of these artefacts,[22, 25] but in addition contain information that is lost in the time-averaging process inherent in steady-state methods. In the case of crude oils, which are complex mixtures of fluorophores, time-resolved fluorescence measurements offer the best approach for fully revealing the influence of quenching and energy transfer processes on fluorescence behaviour.

In 1987, the fluorescence lifetimes measured at a range of different emission wavelengths for a selection of Kuwaiti and Australian oils were measured using a 337.1 nm excitation source.[11] Regrettably, no chemical compositional data was supplied for the oils, and the reported lifetime data is incomplete (no fractional intensities supplied). A photophysical model for the fluorescence lifetime behaviour of crude petroleum oils detailed the complex balance between energy transfer and quenching that governs the fluorescence lifetimes of crude oils was derived from dilution studies on several oils.[25] This study used a variety of different excitation sources but did not try to correlate chemical composition with fluorescence lifetime data.

More recently, we have revisited the use of intensity average fluorescence lifetime for the characterisation of crude petroleum oils and have sought to correlate fluorescence lifetime data with physical characteristics (API gravity) and compositional factors such alkane, aromatic, polar, and sulphur concentration.[21, 22] Figure 8.14 shows the plot of lifetime versus alkane concentration (calculated 3 different ways) for a series of 23 oils. The oils studied are topped oils, where the light hydrocarbon fraction was removed, and they encompassed a wide range of sources and compositions.[10, 32] Using a longer excitation wavelength (405 nm) resulted in a slightly better correlation for both corrected alkane (Figure 8.15) and polar concentration (Figure 8.16).[32] In each case, however, there is still considerable scatter about the best-fit line, preventing the development of accurate, quantitative models.

---

† Fluorescence ratio methods like the red-green quotient and the QFT-II avoid the problem of absolute emission intensity measurements.

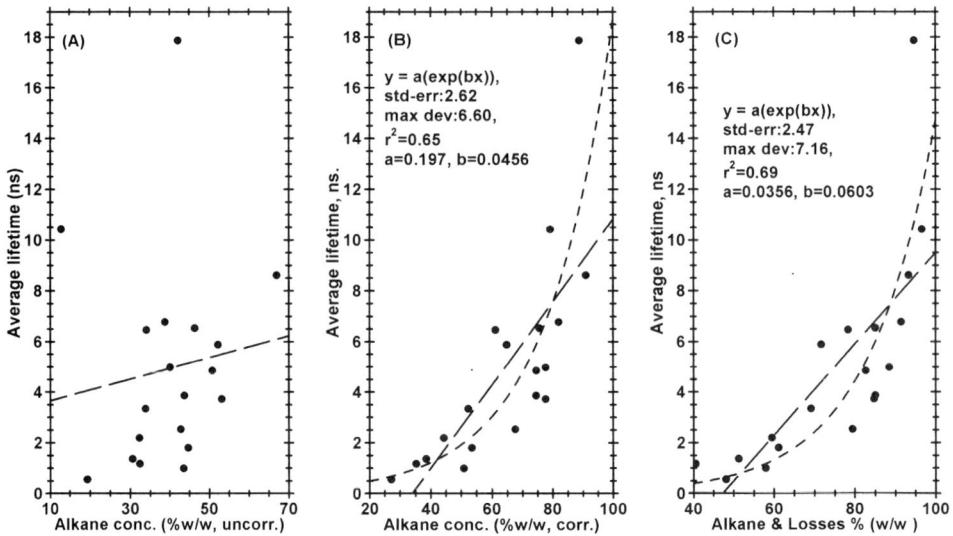

Figure 8.14. Plot of average fluorescence lifetime at an emission wavelength of 500 nm (380 nm LED excitation) versus (A) uncorrected alkane concentration, (B) corrected alkane concentration and (C) sum of uncorrected alkane concentration and column losses. *Reproduced with permission from Ref [22], © 2004, Society for applied Spectroscopy.*

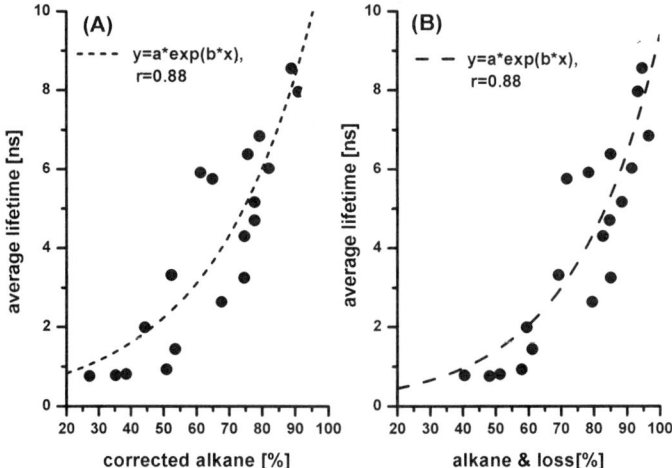

Figure 8.15. Plot of average fluorescence lifetime ($\overline{\tau}$) versus: A) corrected alkane concentration; B) sum of alkane concentration and column losses at the emission wavelength of 540 nm. The best exponential growth fit and the value of correlation coefficient *r* for each concentration are also plotted. *Reproduced with permission from Ref [32], © 2004, Society for applied Spectroscopy.*

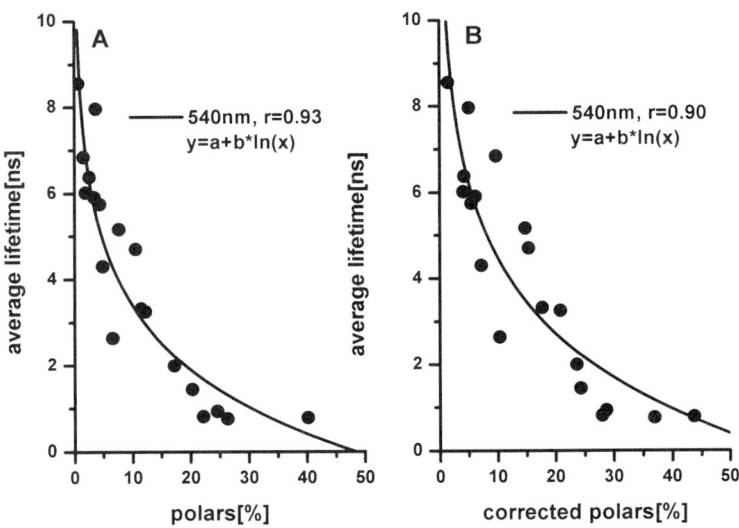

Figure 8.16. Plot of average fluorescence lifetime ($\overline{\tau}$) versus: A) measured polar concentration; B) corrected polar concentration; at the emission wavelength of 540 nm. The best logarithmic fit and the value of correlation coefficient *r* for each concentration are also plotted. *Reproduced with permission from Ref [32], © 2004, Society for applied Spectroscopy.*

The data plotted in Figures 8.15 and 8.16 highlight the difficulty in developing all inclusive fluorescence methods suitable for any oil type. The wide diversity of sources from which the test oils were derived almost certainly contributes to the large scatter in the plots. The use of simple lifetime models is another potential source of error, since they cannot fully account for the complex decay kinetics of crude oils. Further investigations are underway to try to resolve this issue and will be reported in the near future. The use of fluorescence lifetime ratios for quantitative analysis was also investigated,[21, 22] and the initial results were promising, but, these studies used a restricted set of oils from the North Sea. The application of lifetime ratios to the more diverse oil sample sets does not result in accurate correlations suitable for analytical use.[65]

8.5. APPLICATIONS: FLUID INCLUSION STUDIES

Hydrocarbon-containing fluid inclusions (HCFI) are small, generally < 10 μm in diameter, micro-cavities filled with fluid trapped during crystallization or healing of the fractures in minerals such us quartz, fluorite, or calcite.[66, 67] The fluids entrapped within HCFI are fossilised samples of oils, gases, aqueous solutions, and solids, whose composition may not have changed since the trapping event. The accurate determination of the chemical composition of entrapped fluids can provide (together with micro-thermometric studies) essential information about the crystal growth, temperature, and timing of fluid migration. This information is of significant importance to the petroleum exploration industry, particularly in regard to the study of petroleum reservoirs. The analysis of HCFI is done either by crushing bulk rock samples and extracting the entrapped fluid for analysis, or by the analysis of single fluid inclusions.[68] The oil composition data obtained from bulk fluid inclusion analysis (by crushing) suffers from a variety of problems including: sample destruction, mixing of fluids from heterogeneous fluid inclusions, and contamination from materials within the rock sample itself.[69]

Fluorescence based methods are widely used for studying HCFI and the most common identification method is by observing their fluorescence under UV illumination.[70, 71] The use of visually determined fluorescence colour, is widely used as a qualitative guide for assessing the maturity of oil in HCFI.[67] Unfortunately, the use of fluorescence colour is intrinsically prone to error, and does not yield quantitative results.[72, 73] In addition, there are issues with instrumental variation (excitation wavelength, emission filters, etc.) and reproducibility.[74, 75] However, when calibrated with oils from the same basin/reservoir, it has been used to show variation in HCFI composition (changes in API gravity) in a single fluid inclusion assemblage (Figure 8.17).[13]

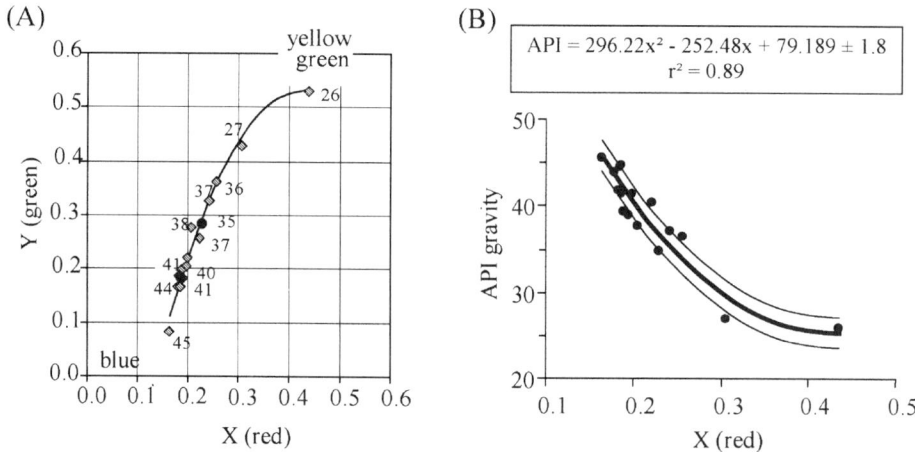

Figure 8.17. Reference oil fluorescence results: (A) – calibration oil chromaticity coordinates. Black solids indicate 3/9A-6 well oils. Solid circles are indexed by their API gravity. Note the colour evolution towards green-yellow with decreasing API gravity; (B) Oil API gravity versus red chromaticity parameters. *(Reproduced with permission from ref. [13]).*

The red-green quotient Q (Q = intensity 650 nm / intensity 500 nm), and the wavelength of maximum fluorescence emission intensity (λ_{max}), were found to correlate with API gravity and gross chemical composition (%w saturates, aromatics, polars, and asphaltenes) for a set of HCFI synthesized using a sample set of Canadian crude oils.[59] Both Q and λ_{max} were also shown to correlate well with gross chemical composition of Athabaska bitumen sub-fractions.[24]

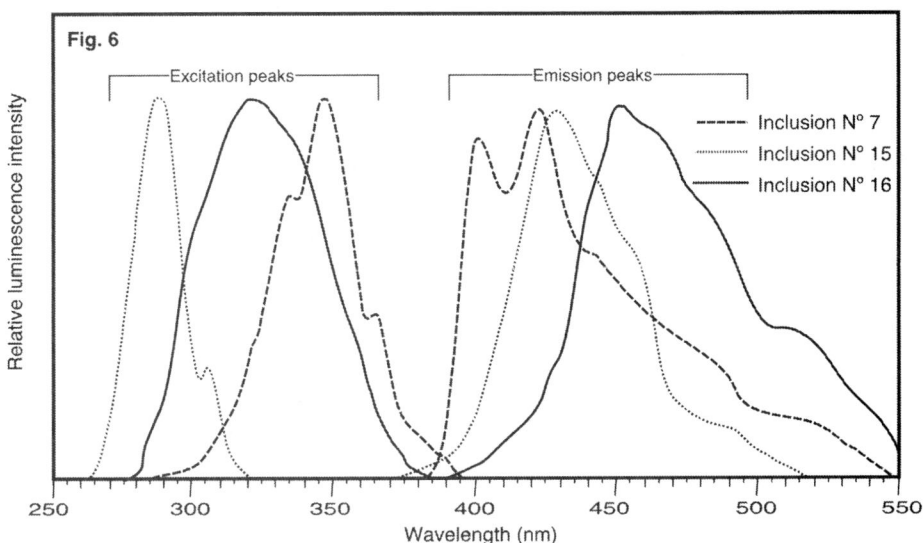

Figure 8.18: FLEEMS measurements of inclusion showing optimum excitation and optimum emission spectra for each inclusion. Excitation slit = 10 nm. Emission slit = 10 nm. Scanning speed = 60 nm per min. All spectra are normalized to same fluorescence intensity level for ease of comparison. Measuring microscope objective: Olympus D Apo UV 100 x , N.A. 1.3 oil immersion. *Reproduced with permission from Ref. [76]* © *1996,* Elsevier Science Ltd

Kihle coupled a standard steady-state fluorimeter to an epi-fluorescence microscope using fibre optics to enable the collection of excitation, emission, and synchronous excitation-emission from microscopic sample areas, such as single HCFI.[76] When the monochromators are set to zero order it enabled the collection of the emission spectrum at optimum excitation conditions, and the excitation spectrum at optimum emission conditions (Figure 8.18). This methodology was used to discriminate different HCFI on the basis of these spectra or via values extracted such as optimum excitation wavelength or the Stokes shift. He also demonstrated the collection of synchronous fluorescence excitation-emission spectra from individual fluid inclusions.[76]

However, steady-state based fluorescence measurements for HCFI studies suffer from several drawbacks (apart from those mentioned in section 4.1). The apparent fluorescence intensity and spectral distribution can adversely be influenced by the physical properties of the sample such as geometry, opacity, sample turbidity, and the scattering properties of both the actual inclusion and host mineral. The last factor is particularly important in the context of HCFI studies where there is a large variation in sample type. Photobleaching of synthetic and natural HCFI has been shown to cause changes in the fluorescence emission intensity.[61]

In 1987, McLimans outlined the potential of using fluorescence lifetime measurements for HCFI analysis.[67] He showed that fluorescence lifetime (355 nm excitation) increased on going from condensate, to very mature, to moderately mature, to immature oils. There was also a noticeable increase in lifetime for all oils with emission wavelength over the 400-600 nm range. However, no data was provided on the method for calculating fluorescence lifetime or about the chemical composition / source of the oils samples, nor was any HCFI data presented.

Recently, in NUI-Galway, fluorescence lifetime data has been acquired from a series of HCFI and correlated with data obtained from a test set of 23 crude oils.[32] The HCFI all emitted similar coloured fluorescence under 405 nm excitation, however, all were clearly discriminated on the basis of their lifetime measurements. Unfortunately, the study also showed that the fluorescence lifetimes of some HCFI oils were much longer than the lightest oils in the text set. This is probably because the oils in HCFI contain higher fractions of light hydrocarbon liquids and gases that are not present in the bulk oils sampled. This will necessitate the development of larger calibration datasets, which comprise of non-topped crude oils (or analogues) to account for the gases and volatiles entrapped in HCFI. It will be also necessary to fabricate artificial HCFI with entrapped oils of known compositions to further validate the methodology. Artificial HCFI can be synthesised in halite at low temperature,[77] and more recently from quartz at higher temperatures.[78] Unfortunately, it is not clear that it will be possible to have a well defined (known chemical composition) oil in these HCFI as there is evidence that partitioning and/or thermal degradation can occur.

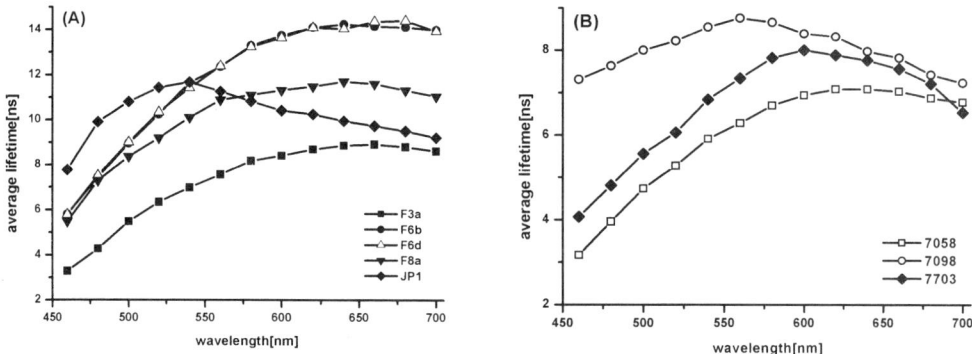

Figure 8.19: Plot of average fluorescence lifetime versus emission wavelength for: A) HCFI samples, B) several light crude oils, API gravity = 50.6° - 40.1°. *Reproduced with permission from Ref. [32] © 2004, Society for applied Spectroscopy.*

The use of lifetime based measurements for HCFI characterization, while still in its infancy does offer considerable benefits when compared to the more traditional colour, intensity, and spectral measurements. Finally, one novel application of fluorescence to inclusion studies involves using confocal microscopy to measure the volume of oil and gas within fluid inclusions by utilising the spatial resolution of modern confocal microscopes.[79]

8.6. APPLICATIONS: REMOTE SENSING/OIL SPILL IDENTIFICATION

Oil spill identification is another area where fluorescence based methods are being applied.[80, 81] The development of laser fluorosensors for oil spill identification was recently reviewed by Fingas, concentrates on ship-borne and airborne sensors using 308 or 355 nm ultraviolet excitation.[82] Studies by several groups have concentrated on the development of remote fluorosensors capable of measuring both emission spectra and fluorescence lifetimes,[11, 83, 84] of crude oils on sea/water surfaces.[14, 85, 86, 87] The photochemical weathering of crude oils on water can be monitored using fluorescence intensity, where there is a sharp decrease in intensity upon solar irradiation.[8, 88, 89] However, this decrease in intensity is probably due to light faction loss by evaporation rather than the formation of any oxygenated species. The fluorescence intensity of crude oils is extremely sensitive to the alkane concentration, and any loss will result in large changes in fluorescence intensity.[10, 32] Chemometric methods such as Principal Component Analysis (PCA) have been applied with fluorescence emission spectroscopy to demonstrate the potential discrimination of oils types and weathering effects.[90] Computational models for discriminating or matching oil types vary from simple vector models[91] to more complex Artificial Neural Networks.[92, 93] ANN methods have also been used to help separate fluorescence signals from oil and humic substances in seawater.[94]

8.7. APPLICATIONS: SUNDRY TECHNIQUES

Asphaltene aggregation was studied by front surface fluorescence spectroscopy,[95] and the use of fluorescence depolarisation techniques have been shown to track differences in asphaltene isolation methods.[96] Fluorescence has also been used to assess asphaltene structure,[97] measure asphaltene size,[98, 99] and to assess the effect of temperature.[100] Khorasani and Michelsen,[101] used PCA and fluorescence spectroscopy to derive a discrimination method for the assessment of maturity and oil generation from marine algal kerogens. The use of fluorescence to study the impact of crude oils on the environment is also becoming more important. Various fluorescence methods have been used to look at the Polycyclic Aromatic Hydrocarbon (PAH) content of crude oils and petroleum products,[102] and to characterise petroleum contaminated soils[103] and the formation of oil-mineral aggregates.[104] Fluorescence techniques have also been used to measure crude oil contamination in offshore sediments[105, 106] and the formation of oil in water emulsions is also studied by fluorescence.[107] There have been studies on estimating hydrocarbon levels in benthic organisms.[108] Fluorescence methods were also used to study the cracking kinetics of crude oils at high temperatures and pressures using a diamond anvil cell.[109] Time-resolved fluorescence spectra (plotted as contour maps) have also been proposed as a method for discriminating different oil types.[110, 111] The method, based on 250 nm laser excitation, shows potential, but as yet no quantitative reports have been shown.

8.8. CONCLUSIONS

The use of fluorescence spectroscopy in a wide variety of methodologies for the characterization and analysis of crude petroleum products is well established. Yet there are still considerable opportunities for the development of robust, accurate, and quantitative fluorescence methods for crude oil analysis. The rapid miniaturization and increasing capability of fluorescence instrumentation is further expanding the range of applications, particularly with regard to on-site and *in-situ* measurements. However, the wide range of compact, portable excitation sources now available, presents a problem to the crude oil fluorescence community. This diversity of excitation sources, each of which excites different fluorophore populations within the crude oils, makes it increasingly difficult to compare quantitatively results from different studies. This combined with the fact that most research laboratories work on widely different crude oils further fragments this economically important research field. There is therefore, a need to develop a set of crude oil standards (either natural or synthesised in the laboratory), freely available to all, with which new methods and instrumentation can be calibrated, and the results from different laboratories more easily compared.

8.9. ACKNOWLEDGMENTS

The author would like to acknowledge support from Science Foundation Ireland by way of the Grant Scheme for Investigators (Grant no. 02/IN.1/M231), and the National Centre for Biomedical Engineering Science (NUI-Galway) as part of the Higher Education Authority Programme for Research in Third Level Institutions.

8.10. REFERENCES

1   R. C. Selley, *Elements of Petroleum Geology, 2nd. Ed.*, (Academic Press, 1998).
2   J. M. Hunt, *Petroleum Geochemistry and Geology*, (W.H. Freeman and Company San Francisco, 1979).
3   F. K. North, *Petroleum Geology*, (Chapman & Hall, London, 1985).
4   O. C. Mullins, T. Daigle, C. Crowell, H. Groenzin, and N. B. Joshi, Gas-oil ratio of live crude oils determined by near-infrared spectroscopy, *Appl. Spectrosc.* **55**(2), 197-201 (2001).
5   Z. D. Wang and M. F. Fingas, Development of oil hydrocarbon fingerprinting and identification techniques, *Mar. Pollut. Bull.* **47**(9-12), 423-452 (2003).
6   American Society for the Testing of Materials, *Annual Book of ASTM Standards, Section 5, Petroleum Products and Lubricants (I-IV)*, 2003.
7   Institute of Petroleum, *Standard Methods for Analysis and Testing of Petroleum and Related Products and British Standard 2000 Parts*, (John Wiley & Sons, 2001)
8   D. E. Nicodem, C. L. B. Guedes, M. Conceição, Z. Fernandes, D. Severino, R. J. Correa, M. C. Coutinho, and J. Silva, Photochemistry of petroleum, *Prog. React. Kinect. Mec.* **26**(2-3), 219-238 (2001).
9   O. C. Mullins in: *Structure and Dynamics of Asphaltenes*, edited by O. C. Mullins and E. Y. Sheu, (Plenum Press, New York, 1998), pp. 21-77.
10   A. G. Ryder, A time-resolved fluorescence spectroscopic study of crude petroleum oils: influence of chemical composition, *Appl. Spectrosc.* **58**(5), 613-623 (2004).
11   M. E. Abu-Zeid, K. S. Bhatia, M. A. Marafi, Y. Y. Makdisi, and M. F. Amer, Measurement of fluorescence decay of crude oil: a potential technique to identify oil slicks, *Environ. Pollut.* **46**, 197-207 (1987).
12   B. Alpern, M. J. L. DeSousa, H. J. Pinheiro, and X. Zhu, Detection and evaluation of hydrocarbons in-source rocks by fluorescence microscopy, *Org. Geochem.* **20**(6), 789-795 (1993).
13   A. Blanchet, M. Pagel, F. Walgenwitz, and A. Lopez, Microspectrofluorimetric and micro thermometric evidence for variability in hydrocarbon fluid inclusions in quartz overgrowths: implications for inclusion trapping in the Alwyn North field, North Sea, *Org. Geochem.* **34**(11), 1477-1490 (2003).
14   P. Camagni, A. Colombo, C. Koechler, N. Omenetto, P. Qi, and G. Rossi, Fluorescence response of

mineral oils: spectral yield vs absorption and decay time, *Appl. Opt.* **30**(1), 26-35 (1991).

15   H. W. Hagemann and A. Hollerbach, The fluorescence behavior of crude oils with respect to their thermal maturation and degradation, *Org. Geochem.* **10**, 473-480 (1986).

16   R. M. Measures, W. R. Houston, and D. G. Stephenson, Laser induced fluorescent decay spectra – a new form of environmental signature, *Opt. Eng.* **13**(6), 494-501 (1974).

17   O. C. Mullins, S. Mitra-Kirtley, and Y. Zhu, The electronic absorption-edge of petroleum, *Appl. Spectrosc.* **46**(9), 1405-1411 (1992).

18   T. D. Downare and O. C. Mullins, Visible and near-infrared fluorescence of crude oils, *Appl. Spectrosc.* **49**(6), 754-764 (1995).

19   M. F. Quinn, S. Joubian, F. Al-Bahrani, S. Al-Aruri, and O. Alameddine, A de-convolution technique for determining the intrinsic fluorescence decay lifetimes of crude oils, *Appl. Spectrosc.* **42**(3), 406-410 (1988).

20   D. M. Rayner and A. G. Szabo, Time-resolved laser fluorosensors: A laboratory study of their potential in the remote characterization of oil, *Appl. Optics* **17**(10), 1624-1630 (1978).

21   A. G. Ryder, T. J. Glynn, M. Feely, and A. J. G. Barwise, Characterization of crude oils using fluorescence lifetime data, *Spectrochim. Acta A* **58**(5), 1025-1037 (2002).

22   A. G. Ryder, Quantitative analysis of crude oils by fluorescence lifetime and steady state measurements using 380 nm excitation, *Appl. Spectrosc.* **56**(1), 107-116 (2002).

23   A. G. Ryder, T. J. Glynn, and M. Feely, Influence of chemical composition on the fluorescence lifetimes of crude petroleum oils, *Proc SPIE – Int. Soc. Opt. Eng.* **4876**, 1188-1195 (2003).

24   L. D. Stasiuk, T. Gentzis, and P. Rahimi, Application of spectral fluorescence microscopy for the characterization of Athabasca bitumen vacuum bottoms, *Fuel.* **79**, 769-775 (2000).

25   X. Wang and O. C. Mullins, Fluorescence lifetime studies of crude oils, *Appl. Spectrosc.* **48**(8), 977-984 (1994).

26   Y. Zhu and O. C. Mullins, Temperature dependence of fluorescence of crude oils and related compounds, *Energy & Fuels.* **6**(5), 545-552 (1992).

27   M. V. Reyes, Application of fluorescence techniques for mud-logging analysis of oil drilled with oil-based muds, *SPE Formation Evaluation*, **9**(4), 300-305 (1994).

28   B. Pradier, C. Largeau, S. Derenne, L. Martinez, P. Bertrand, and Y. Pouet, Chemical basis of fluorescence alteration of crude oils and kerogens .1. Microfluorimetry of an oil and its isolated fractions - relationships with chemical-structure, *Org. Geochem.* **16**(1-3), 451-460 (1990).

29   G. Ellingsen and S. Fery-Forgues, Application de la spectroscopie de fluorescence à l'étude du pétrole: la défi de la complexité, *Rev. I. Fr. Petrol.* **53**(2), 201-216 (1998).

30   C. Y. Ralston, X. Wu, and O. C. Mullins, Quantum yields of crude oils, *Appl. Spectrosc.* **50**(12), 1563-1568 (1996).

31   D. E. Nicodem, M. F. V. Da Cunha, and C. L. B. Guedes, Time-resolved single photon counting study of the quenching of fluorescent probes by petroleum: Probing the energy distribution of the nonaliphatic components, *Appl. Spectrosc.* **54**(9), 1409-1411 (2000).

32   A. G. Ryder, M. A. Przyjalgowski, M. Feely, B. Szczupak, and T. J. Glynn, Time-resolved fluorescence microspectroscopy for characterizing crude oils in bulk and hydrocarbon bearing fluid inclusions, *Appl. Spectrosc.* **58**(9), 1106-1115 (2004).

33   A. E. Dudelzak, S. M. Babichenko, L. V. Poryvkina, and K. J. Saar, Total luminescent spectroscopy for remote laser diagnostics of natural water conditions *Appl. Optics* **30**(4), 453- (1991).

34   D. Patra and A. K. Mishra, Study of diesel fuel contamination by excitation emission matrix spectral subtraction fluorescence, *Anal. Chim. Acta* **454**(2), 209-215 (2002).

35   P. John and I. Soutar, Identification of crude oils by synchronous excitation spectrofluorimetry, *Anal. Chem.* **48**(3), 520-524 (1976).

36   S. G. Wakeham, Synchronous fluorescence spectroscopy and its application to indigenous and petroleum-derived hydrocarbons in Lacustrine sediments, *Environ. Sci. Technol.* **11**(3), 272-276 (1977).

37   J. M. Song and L. F. Wang, Study on the characteristic and significance of synchronous fluorescence spectrum of crude oil and nature gas samples, *Spectrosc. Spect. Anal.* **22**(5), 803-805 (2002).

38   D. Patra, K. L. Sireesha, and A. K. Mishra, Determination of synchronous fluorescence scan parameters for certain petroleum products, *J. Sci. Ind. Res. India* **59**(4), 300-305 (2000).

39   D. Patra and A. K. Mishra, Concentration dependent red shift: qualitative and quantitative investigation of motor oils by synchronous fluorescence scan, *Talanta* **53**(4), 783-790 (2001).

40   E. Buenrostro-Gonzalez, S. I. Andersen, J. A. Garcia-Martinez, C. Lira-Galeana, Solubility/molecular structure relationships of asphaltenes in polar and nonpolar media, *Energ. Fuel.* **16**(3), 732-741 (2002).

41   L. J. Shadle, K. S. Seshadri, and D. L. Webb, Characterization of shale oils .1. Analysis of Fischer assay oils and their aromatic fractions using advanced analytical techniques, *Fuel Process. Technol.* **37**(2), 101-120 (1994).

42   J. A. Musgrave, R. G. Carey, D. R. Janecky, and C. D. Tait, Adaption of synchronously scanned luminescence spectroscopy to organic-rich fluid inclusion microanalysis, *Rev. Sci. Instrum.* **65**(6), 1877-1882 (1994).

43   A. Permanyer, L. Douifi, A. Lahcini, J. Lamontagne, and J. Kister, FTIR and SUVF spectroscopy applied to reservoir compartmentalization: a comparative study with gas chromatography fingerprints results, *Fuel* **81**(7), 861-866 (2002).

44   L. A. Files, M. Moore, M. J. Kerkhoff, and J. D. Winefordner, Gasoline and crude-oil fingerprinting using constant energy synchronous luminescence spectrometry, *Microchem. J.* **35**(3), 305-314 (1987).

45   K. L. Yong and J. G. Lu, Common and diverse characteristics of three-dimensional fluorescence spectra of crude oils, *Spectrosc. Lett.* **33**(6), 963-970 (2000).

46   J. Lu, and K. Yong, Fluorescence quenching phenomena in three-dimension of fluorescence determination of crude oils. *Fenxi Shiyanshi* **17**(6), 28-31, (1998).

47   G. C. Smith and J. F. Sinski, The red-shift cascade: Investigations into the concentration-dependent wavelength shifts in three-dimensional fluorescence spectra of petroleum samples, *Appl. Spectrosc.* **53**(11), 1459-1469 (1999).

48   J. F. Sinski, B. S. Compton, B. S. Perkins, and M. C. Nicoson, Utilizing three-dimensional fluorescence's red-shift cascade effect to monitor mycobacterium PRY-1 degradation of aged petroleum, *Appl. Spectrosc.* **58**(1), 91-95 (2004).

49   D. Patra and A. K. Mishra, Total synchronous fluorescence scan spectra of petroleum products, *Anal. Bioanal. Chem.* **373**(4-5), 304-309 (2002).

50   A. G. Ryder, Assessing the Maturity of crude petroleum oils using total synchronous fluorescence scan spectra, *J. Fluor.* **14**(1), 99-104 (2004).

51   X. Wu, E. B. Dussan V, and O. C. Mullins, Using an optical sensor to quantify the amount of oil, water, and gas in a water-continuous flow, *Proc SPIE – Int. Soc. Opt. Eng.* **3856**, 298-307 (1999).

52   US Patent 6,704109 B2.

53   T. D. Downare, O. C. Mullins, and X. Wu, Optimization of a fluorescence detection system for the characterization of solids, *Appl. Spectrosc.*, **48**(12), 1483-1490 (1994).

54   J. Bublitz, M. Dickenhausen, M. Gratz, S. Todt, and W. Schade, Fiberoptic laser-induced fluorescence probe for the detection of environmental-pollutants, *Appl. Opt.* **34**(18), 3223-3233 (1995).

55   W. Schade and J. Bublitz, On-site laser probe for the detection of petroleum products in water and soil, *Environ. Sci. Technol.* **30**(5), 1451-1458 (1996).

56   M. L. Pascu, N. Moise, and A. Staicu, Tunable dye laser applications in environment pollution monitoring, *J. Mol. Struct.* **598**(1), 57-64 (2001).

57   S. Landgraf, Application of semiconductor light sources for investigations of photochemical reactions, *Spectrochim. Acta A* **57**(10), 2029-2048 (2001).

58   S. Landgraf, Use of ultrabright LEDs for the determination of static and time-resolved florescence information of liquid and solid crude oil samples, *J. Biochem. Bioph. Meth.*, In Press, (2004).

59   L. D. Stasiuk and L. R. Snowdon, Fluorescence micro-spectrometry of synthetic and natural hydrocarbon fluid inclusions: crude oil chemistry, density and application to petroleum migration, *Appl. Geochem.* **12**(3), 229-233 (1997).

60   P. L. Delaune, K. K. Spilker, S. A. Hanson, A. C. Wright, and R. Quagliaroli, Enhanced wellsite technique for oil detection and characterization, *SPE-56802*, in *1999 SPE annual technical conference and exhibition proceedings, v., Formation evaluation and reservoir geology*, 801-816, (1999).

61   J. R. Lakowicz, *Principles of Fluorescence Spectroscopy, 2nd. ed* (Kluwer Academic/Plenum Publishers, New York, 1999).

62   J. Pironon and B. Pradier, Ultraviolet-fluorescence alteration of hydrocarbon fluid inclusions, *Org. Geochem.* **18**(4), 501-509 (1992).

63   H. Szmacinski and J. R. Lakowicz in: *Topics in fluorescence spectroscopy: Vol. 4. Probe Design and Chemical Sensing*, edited by J. R. Lakowicz, Ed. (Plenum Press, New York, 1994), pp. 295-329.

64   D. J. S. Birch and R. E. Imhof, in: *Topics in Fluorescence Spectroscopy, Vol. 1 Techniques*, edited by J. R. Lakowicz (Plenum Press, New York and London, 1992), pp. 1-95.

65   M. A. Przyjalgowski and A. G. Ryder, unpublished results.

66   E. Roedder, *Mineral Soc. Am., Rev. Mineral.*, **12**, 1- (1984).

67   R. K. McLimans, The application of fluid inclusions to migration of oil and diagenesis in petroleum

reservoirs, *Appl. Geochem.* **2**, 585-603 (1987).

68   I. A. Munz, Petroleum inclusions in sedimentary basins: systematics, analytical methods and applications, *Lithos* **55**(1-4), 195-212 (2001).

69   D. Emery and A. G. Robinson, *Inorganic geochemistry: Applications to petroleum geology* (Blackwell Science, UK, 1993).

70   N. Guilhaumou, N. Szydlowskii, and B. Pradier, Characterization of hydrocarbon fluid inclusions by infra-red and fluorescence microspectrometry, *Mineral. Mag.* **54**, 311-324 (1990).

71   B. Alpern, M. J. Lemos de Sousa, H. J. Pinheiro, and X. Zhu, Optical morphology of hydrocarbons and oil progenitors in sedimentary rocks-relations with geochemical parameters. *Publ. Mus. Labor. miner. geol. Fac. Ciênc. Porto.* **3**, 1-21, (1992)

72   S. C. George, T. E. Ruble, A. Dutkiewicz, The use and abuse of fluorescence colours as maturity indicators of oil in inclusions from Australian petroleum systems. *APPEA Journal.* **41**(1), 505-522 (2001).

73   S. C. George, T. E. Ruble, A. Dutkiewicz, and P. J. Eadington, Assessing the maturity of oil trapped in fluid inclusions using molecular geochemistry data and visually-determined fluorescence colours, *Appl. Geochem.* **16**(4), 451-473 (2001).

74   N. H. Oxtoby, Comments on: Assessing the maturity of oil trapped in fluid inclusions using molecular geochemistry data and visually-determined fluorescence colours, *Appl. Geochem.* **17**(10), 1371-1374 (2002).

75   S. C. George, T. E. Ruble, A. Dutkiewicz, and P. J. Eadington, Reply to comment by Oxtoby on "Assessing the maturity of oil trapped in fluid inclusions using molecular geochemistry data and visually-determined fluorescence colours", *Appl. Geochem.* **17**(10), 1375-1378 (2002).

76   J. Kihle, Adaptation of fluorescence excitation-emission micro-spectroscopy for characterization of single hydrocarbon fluid inclusions, *Org. Geochem.* **23**(11-12), 1029-1042 (1995).

77   J. Pironon, Synthesis of hydrocarbon fluid inclusions at low temperature. *Am. Mineral.* **75**, 226–229 (1990).

78   S. Teinturier and J. Pironon, Experimental growth of quartz in petroleum environment. part I: procedures and fluid trapping, *Geochim. Cosmochim. Ac.* **68**(11), 2495-2507 (2004).

79   J. Pironon, M. Canals, J. Dubessy, F. Walgenwitz, and C. Laplace-Builhe, Volumetric reconstruction of individual oil inclusions by confocal scanning laser microscopy, *Eur. J. Mineral.* **10**(6), 1143-1150 (1998).

80   C. E. Brown, R. D. Nelson, M. F. Fingas, and J. V. Mullin, Laser fluorosensor overflights of the Santa Barbara oil seeps, *Spill Sci. Technol. B.* **3**(4), 227-230 (1996).

81   C. E. Brown and M. F. Fingas, Review of the development of laser fluorosensors for oil spill application, *Mar. Pollut. Bull.* **47**(9-12), 477-484 (2003).

82   P. Lambert, M. Goldthorp, B. Fieldhouse, Z. Wang, M. Fingas, L. Pearson, and E. Collazzi, Field fluorometers as dispersed oil-in-water monitors, *J. Hazard. Mater.* **102**(1), 57-79 (2003).

83   M. F. Quinn, A. S. Al-Otaibi, A. Abdullah, P. S. Sethi, F. Al-Bahrani, and O. Alameddine, Determination of intrinsic fluorescence lifetime parameters of crude oils using a laser fluorosensor with a streak camera detection system, *Instrum. Sci. Technol.* **23**(3), 201-215 (1995).

84   S. D. Alaruri, M. Rasas, O. Alamedine, S. Jubian, F. Al-Bahrani, and M. Quinn, Remote characterization of crude and refined oils using a laser fluorosensor system, *Opt. Eng.* **34**(1), 214-221 (1995).

85   D. M. Rayner, M. Lee, and A. G. Szabo, Effect of sea-state on performance of laser fluorosensors, *Appl. Optics* **17**(17), 2730-2733 (1978).

86   J. S. Knoll, Visible fluorescence from ultraviolet excited crude oil, *Appl. Optics* **24**(14), 2121-2123 (1985).

87   T. Hengstermann and R. Reuter, Lidar fluorosensing of mineral oil spills on the sea surface, *Appl. Optics*, **29**(22), 3218-3227 (1990).

88   D. E. Nicodem, C. L. B. Guedes, and R. J. Correa, Photochemistry of petroleum I. Systematic study of a Brazilian intermediate crude oil, *Mar. Chem.* **63**(1-2), 93-104 (1998).

89   A. Boukir, M. Guiliano, L. Asia, A. El Hallaoui, G. Mille, A fraction to fraction study of photo-oxidation of BAL 150 crude oil asphaltenes, *Analusis* **26**(9), 358-364 (1998).

90   J. Li, S. Fuller, J. Cattle, C. Pang Way, and D. B. Hibbert, Matching fluorescence spectra of oil spills with spectra from suspect sources, *Anal. Chim. Acta* **514**(1), 51-56 (2004).

91   T. J. Killeen, D. Eastwood and M. Schulz Hendrick, Oil-matching by using a simple vector model for fluorescence spectra, *Talanta* **28**(1), 1-6 (1981).

92   J. M. Andrews and S. H. Lieberman, Neural-Network approach to qualitative identification of fuels and

oils from laser-induced fluorescence-spectra, *Anal. Chim. Acta* **285**(1-2), 237-246 (1994).

93   L. M. He, L. L. Kear-Padilla, S. H. Lieberman, and J. M. Andrews, Rapid in situ determination of total oil concentration in water using ultraviolet fluorescence and light scattering coupled with artificial neural networks, *Anal. Chim. Acta* **478**(2), 245-258 (2003).

94   T. A. Dolenko, V. V. Fadeev, I. V. Gerdova, S. A. Dolenko, and R. Reuter, Fluorescence diagnostics of oil pollution in coastal marine waters by use of artificial neural networks, *Appl. Optics* **41**(24), 5155-5166 (2002).

95   F. C. Albuquerque, D. E. Nicodem, K. Rajagopal, Investigation of asphaltene association by front-face fluorescence spectroscopy, *Appl. Spectrosc.* **57**(7), 805-810 (2003).

96   S. I. Andersen, A. Keul, and E. Stenby, Variation in composition of subfractions of petroleum asphaltenes, *Petrol. Sci. Technol.* **15**(7-8), 611-645 (1997).

97   H. Groenzin, and O. C. Mullins, Asphaltene molecular size and structure, *J. Phys. Chem. A* **103**(50), 11237-11245 (1999).

98   H. Groenzin and O. C. Mullins, Molecular size and structure of asphaltenes from various sources, *Energy & Fuels* **14**(3), 677-684 (2000).

99   H. Groenzin, O. C. Mullins, S. Eser, J. Mathews, M. G. Yang, and D. Jones, Molecular size of asphaltene solubility fractions, *Energy & Fuels* **17**(2), 498-503 (2003).

100   L. Buch, H. Groenzin, E. Buenrostro-Gonzalez, S.I. Andersen, C. Lira-Galeana, and O. C. Mullins, Molecular size of asphaltene fractions obtained from residuum hydrotreatment, *Fuel* **82**(9), 1075-1084 (2003).

101   G. K. Khorasani and J. K. Michelsen, Four-dimensional fluorescence imaging of oil generation: development of a new fluorescence imaging technique, *Org. Geochem.* **22**(1), 211-223 (1995).

102   J. R. Kershaw and J. C. Fetzer, The room-temperature fluorescence analysis of polycyclic aromatic-compounds in petroleum and related materials, *Polycycl. Aromat. Comp.* **7**(4), 253-268 (1995).

103   H. G. Lohmannsroben and T. Roch, In situ laser-induced fluorescence (LIF) analysis of petroleum product-contaminated soil samples, *J. Environ. Monitor.* **2**(1), 17-22 (2000).

104   P. E. Kepkay, J. B. C. Bugden, K. Lee, and P. Stoffyn-Egli, Application of ultraviolet fluorescence spectroscopy to monitor oil–mineral aggregate formation, *Spill Sci. Technol. B.* **8**(1), 101-108 (2002).

105   B. T. Hargrave and G. A. Phillips, Estimates of oil in aquatic sediments by fluorescence spectroscopy, *Environ. Pollut.* **8**(3), 193-215 (1975).

106   S. S. Al-Lihaibi and L. Al-Omran, Petroleum hydrocarbons in offshore sediments from the gulf, *Mar. Pollut. Bull.* **32**(1), 65-69 (1996).

107   P. Lianos, J. Lang, J. Sturm, and R. Zana, Fluorescence-probe study of oil-in-water microemulsions. 3. further investigations involving other surfactants and oil mixtures. *J. Phys. Chem.* **88**(4), 819-822 (1984).

108   M. Picer, Simple spectrofluorometry methods for estimating petroleum hydrocarbons levels in various sea benthic organisms, *Chemosphere* **37**(4), 607-617 (1998).

109   W. L. Huang and G. A. Otten, Cracking kinetics of crude oil and alkanes determined by diamond anvil cell-fluorescence spectroscopy pyrolysis: technique development and preliminary results, *Org. Geochem.* **32**(6), 817-830 (2001).

110   E. Hegazi, A. Hamdan, and J. Mastromarino, New approach for spectral characterization of crude oil using time-resolved fluorescence spectra, *Appl. Spectrosc.* **55**(2), 202-207 (2001).

111   E. Hegazi and A. Hamdan, Estimation of crude oil grade using time-resolved fluorescence spectra, *Talanta* **56**(6), 989-995 (2002).

NOVEL INSIGHTS INTO PROTEIN STRUCTURE AND DYNAMICS UTILIZING THE RED EDGE EXCITATION SHIFT APPROACH

H. Raghuraman, Devaki A. Kelkar, and Amitabha Chattopadhyay*

ABSTRACT

A shift in the wavelength of maximum fluorescence emission toward higher wavelengths, caused by a corresponding shift in the excitation wavelength toward the red edge of the absorption band, is termed the red edge excitation shift (REES). This effect is mostly observed with polar fluorophores in motionally restricted media such as viscous solutions or condensed phases where the dipolar relaxation time for the solvent shell around a fluorophore is comparable to or longer than its fluorescence lifetime. REES arises from slow rates of solvent relaxation (reorientation) around an excited state fluorophore which depends on the motional restriction imposed on the solvent molecules in the immediate vicinity of the fluorophore. Utilizing this approach, it becomes possible to probe the mobility parameters of the environment itself (which is represented by the relaxing solvent molecules) using the fluorophore merely as a reporter group. Further, since the ubiquitous solvent for biological systems is water, the information obtained in such cases will come from the otherwise 'optically silent' water molecules. This makes REES extremely useful since hydration plays a crucial modulatory role in the formation and maintenance of organized molecular assemblies such as folded proteins in aqueous solutions and biological membranes. The application of REES as a powerful tool to monitor the organization and dynamics of a variety of soluble, cytoskeletal, and membrane-bound proteins is discussed.

---

* H. Raghuraman, Devaki A. Kelkar, and Amitabha Chattopadhyay, Centre for Cellular and Molecular Biology, Hyderabad 500 007, India. E-mail: amit@ccmb.res.in.

9.1. INTRODUCTION

Proteins are highly ordered, dynamic, complex biological macromolecules involved in many cellular functions in their native state. In order to understand the structure-function relationship in biological macromolecules such as proteins, it is necessary to appreciate the dynamics of the constituent molecules as well as that of the system as a whole. Water plays a crucial role in determining the structure and the dynamics, and in turn the functionality of proteins.[1-11] Protein-water interactions are therefore vital to biological functions.[12] It is estimated that a threshold level of hydration (less than 0.4 grams of water per gram of protein) is required to fully activate the dynamics and function of globular proteins.[13-15] The most direct evidence for the importance of water in protein structure and function is that in the absence of water, proteins cannot diffuse and become non-functional. Lack of motion and function have been observed when proteins are transferred to organic solvents[16] and dehydration studies show that at least a monolayer of water molecules is required for the protein to be fully functional.[17] These studies not only prove that the presence of water is essential for the organization and function of proteins, but also that other solvents cannot serve as substitutes. In addition, it has become increasingly evident that water molecules mediate lipid-protein interactions[18-20] and hence the function of membrane proteins.[21-23]

Although detailed and precise structural information of proteins, particularly soluble proteins, can be obtained from x-ray crystallographic diffraction data, such information is necessarily static. Interestingly, global and local dynamics exhibited by proteins and specific regions in them play important roles in their function.[24, 25] Further, a detailed crystallographic database is still not available in the case of membrane proteins due to the inherent difficulty in crystallizing them.[26, 27] This is apparent when one considers that the number of membrane proteins with known x-ray crystal structures is still very small and represents only ~0.2% of all solved protein structures.[27-31] Spectroscopic techniques which provide both structural and dynamic information therefore become very useful for analyses of proteins. Fluorescence spectroscopy represents one such approach and is widely used in the analysis of protein structure, dynamics and function. The advantages of using fluorescence techniques are intrinsic sensitivity, suitable time scale, non-invasive nature, and minimum perturbation.[32, 33] This review is focussed on the application of a relatively novel approach, the red edge excitation shift (REES), as a powerful tool to monitor the organization and dynamics of proteins and peptides.

REES represents a powerful approach which can be used to directly monitor the environment and dynamics around a fluorophore in a complex biological system.[34-39] A shift in the wavelength of maximum fluorescence emission toward higher wavelengths, caused by a shift in the excitation wavelength toward the red edge of the absorption band, is termed REES. This effect is mostly observed with polar fluorophores in motionally restricted media such as viscous solutions or condensed phases where the dipolar relaxation time for the solvent shell around a fluorophore is comparable to or longer than its fluorescence lifetime. REES arises from slow rates of solvent relaxation (reorientation) around an excited state fluorophore which depends on the motional restriction imposed on the solvent molecules in the immediate vicinity of the fluorophore. Utilizing this approach, it becomes possible to probe the mobility parameters of the environment itself (which is represented by the relaxing solvent molecules) using the fluorophore merely as a reporter group. Further, since the ubiquitous solvent for biological systems is water, the information obtained in such cases will come from the otherwise 'optically silent' water molecules. This makes REES extremely useful since hydration plays a crucial modulatory role in a large number of important cellular events.[40] The application of REES to elucidate the organization and dynamics of membranes[36, 38] and membrane-bound peptides and proteins has been reviewed recently.[39]

9.2. RED EDGE EXCITATION SHIFT (REES)

In general, fluorescence emission is governed by Kasha's rule which states that fluorescence normally occurs from the zero vibrational level of the first excited electronic state of a molecule.[41, 42] It is obvious from this rule that fluorescence emission should be independent of excitation wavelength. In fact, such a lack of dependence of fluorescence emission parameters on excitation wavelength is often taken as a criterion for purity and homogeneity of a molecule. Thus, for a fluorophore in a bulk non-viscous solvent, the fluorescence decay rates and the wavelength of maximum emission are usually independent of the excitation wavelength.

However, this generalization breaks down in case of polar fluorophores in motionally restricted media such as very viscous solutions or condensed phases, that is, when the mobility of the surrounding matrix relative to the fluorophore is considerably reduced. This situation arises because of the importance of the solvent shell and its dynamics around the fluorophore during the process of absorption of a photon and its subsequent emission as fluorescence. Under such conditions, when the excitation wavelength is gradually shifted to the red edge of the absorption band, the maximum of fluorescence emission exhibits a concomitant shift toward higher wavelengths. Such a shift in the wavelength of maximum emission toward higher wavelengths, caused by a corresponding shift in the excitation wavelength toward the red edge of the absorption band, is termed the red edge excitation shift (REES).[35-39] Since REES is observed only under conditions of restricted mobility, it serves as a reliable indicator of the dynamics of fluorophore environment.

The genesis of REES lies in the change in fluorophore-solvent interactions in the ground and excited states brought about by a change in the dipole moment of the fluorophore upon excitation, and the rate at which solvent molecules reorient around the excited state fluorophore. For a polar fluorophore, there exists a statistical distribution of solvation states based on their dipolar interactions with the solvent molecules both in the ground and excited states. Since the dipole moment of a molecule changes upon excitation, the solvent dipoles have to reorient around this new excited state dipole of the fluorophore, so as to attain an energetically favorable orientation. This readjustment of the dipolar interaction of the solvent molecules with the fluorophore essentially consists of two components: first, the redistribution of electrons in the surrounding solvent molecules due to the altered dipole moment of the excited state fluorophore, and then, the physical reorientation of the solvent molecules around the excited state fluorophore. The former process is almost instantaneous, *i.e.*, electron redistribution in solvent molecules occurs at about the same time scale as the process of excitation of the fluorophore itself (10^{-15} sec). The reorientation of the solvent dipoles, however, requires a net physical displacement. It is therefore a much slower process and is dependent on the restriction to their mobility offered by the surrounding matrix. More precisely, for a polar fluorophore in a bulk non-viscous solvent, this reorientation occurs at a time scale of the order of 10^{-12} sec, so that all the solvent molecules completely reorient around the excited state dipole of the fluorophore well within its excited state lifetime, which is typically of the order of 10^{-9} sec. Hence, irrespective of the excitation wavelength used, emission is observed *only* from the solvent-relaxed state. However, if the same fluorophore is now placed in a viscous medium, this reorientation process is slowed down to 10^{-9} sec or longer. Under these conditions, excitation by progressively lower energy quanta, *i.e.*, excitation wavelength being gradually shifted toward the red edge of the absorption band, selectively excites those fluorophores which interact more strongly with the solvent molecules in the excited state. These are the fluorophores around which the solvent molecules are oriented in such a way as to be more similar to that found in the solvent-relaxed state. Thus, the necessary condition for giving rise to REES is that a different average population is excited at each excitation wavelength and, more importantly, that the difference is maintained in the time scale of fluorescence lifetime. As discussed above, this requires that the dipolar relaxation time for the solvent shell be comparable to or longer than the fluorescence lifetime, so that fluorescence occurs from various partially relaxed states. This implies a reduced mobility of the surrounding matrix with respect to the fluorophore.

The essential criteria for the observation of REES can therefore be summarized as follows: (i) The fluorophore should normally be polar so as to be able to suitably orient the neighboring solvent molecules in the ground state (molecules such as bianthryl which are nonpolar in the ground state and yet could be polar in the excited state due to intramolecular charge transfer reaction are exceptions to this[43-45]); (ii) The solvent molecules surrounding the fluorophore should be polar; (iii) The solvent reorientation time around the excited state dipole of the fluorophore should be comparable to or longer than the fluorescence lifetime; and (iv) There should be a relatively large change in the dipole moment of the fluorophore upon excitation. The observed spectral shifts thus depend both on the properties of the fluorophore itself (*i.e.*, the vectorial difference between the dipole moments in the ground and excited states), and also on properties of the environment interacting with it (which is a function of the solvent reorientation time). It has previously been shown for tryptophan, the most commonly found intrinsic fluorophore in proteins, that a dipole moment change of ~6 D upon excitation[46, 47] is enough to give rise to significant red edge effects. A recent comprehensive review on REES is provided by Demchenko.[37]

9.3. INTRINSIC FLUORESCENCE OF PROTEINS AND PEPTIDES: TRYPTOPHAN AS THE FLUOROPHORE OF CHOICE

The aromatic amino acids tryptophan, tyrosine and phenylalanine are capable of contributing to the intrinsic fluorescence of proteins. When all three residues are present in a protein (termed as class B protein),[48] pure emission from tryptophan can be obtained only by photoselective excitation at wavelengths above 295 nm.[32] Although tyrosine and phenylalanine are natural fluorophores in proteins, tryptophan is the most extensively used amino acid for fluorescence analysis of proteins. In a protein containing all three naturally fluorescent amino acids, observation of tyrosine and phenylalanine fluorescence is often complicated because of interference by tryptophan due to resonance energy transfer.[32, 49]   The application of tyrosine and phenylalanine fluorescence is therefore mostly limited to tryptophan-free proteins (however, a recent study reports an exception to this[50]). More importantly, tyrosine fluorescence is insensitive to environmental factors such as polarity and does not exhibit appreciable solvatochromism in sharp contrast to tryptophan fluorescence.[51] This is a clear disadvantage for a fluorescent reporter group in biological applications. Fluorescence of phenylalanine is weak and seldom used in protein studies.[49]   Hence, the term 'natural protein fluorescence' is almost always associated with tryptophan fluorescence.[52]

Tryptophan residues serve as intrinsic, site-specific fluorescent probes for protein structure and dynamics[32] and are generally present at about 1 mol% in proteins.[49] The low tryptophan content of proteins is a favorable feature of protein structure since a protein may typically possess few tryptophan residues which facilitate interpretation of fluorescence data and avoid complications due to inter-tryptophan interactions. The well documented sensitivity of tryptophan fluorescence to environmental factors such as polarity and mobility makes tryptophan fluorescence a valuable tool in studies of protein structure and dynamics by providing specific and sensitive information of protein structure and its interactions.[32, 49, 53, 54] The presence of tryptophan residues as intrinsic fluorophores in most peptides and proteins makes them an obvious choice for fluorescence spectroscopic analysis as apparent from the fact that ~300 papers utilizing tryptophan fluorescence in proteins are published per year.[55]

The interesting spectral properties of tryptophan are attributed to a number of factors. The tryptophan residue has a large indole side chain that consists of two fused aromatic rings (see Figure 9.1). Tryptophan has two overlapping $S_o \rightarrow S_1$ electronic transitions denoted as ^1L_a and ^1L_b which are almost perpendicular to each other.[32] Both $S_o \rightarrow {}^1L_a$ and $S_o \rightarrow {}^1L_b$ transitions occur in the 260-300 nm range. In nonpolar solvents, ^1L_a has higher energy than ^1L_b. In polar solvents, however, the energy level of ^1L_a is lowered making it the lowest energy state. This inversion is believed to occur because ^1L_a transition has a higher dipole moment (as it is directed through the –NH group of the

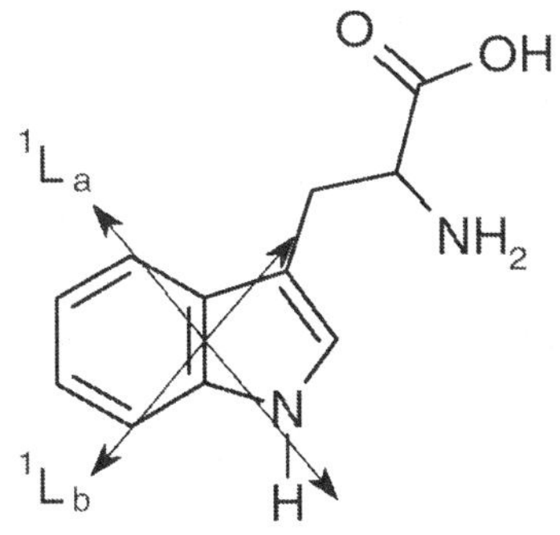

Figure 9.1. Chemical structure of tryptophan showing the transition moment directions for the ^1L_a and ^1L_b transitions. The ^1L_a is the fluorescing state in most proteins and has higher dipole moment (as it is directed through the ring –NH group), and can have dipole-dipole interactions with polar solvent molecules. In nonploar solvents, ^1L_a has higher energy than ^1L_b. However, the energy level of ^1L_a is lower in polar solvents due to favorable dipole-dipole interactions making it the lowest energy state. Adapted and modified from ref. 32.

indole ring), and can have dipole-dipole interactions with polar solvent molecules. Irrespective of whether ^1L_a or ^1L_b is the lowest S_1 state, equilibration between these two states is believed to be very fast (of the order of 10^{-12} sec), so that emission only from the lower S_1 state is observed.[56] It is generally believed that ^1L_a is the fluorescing state in all proteins with the possible exception of Trp-48 of azurin.[55]   In molecular terms, tryptophan is a unique amino acid since it is capable of both hydrophobic and polar interactions. This is due to the fact that while tryptophan has the polar –NH group which is capable of forming hydrogen bonds, it also has the largest nonpolar accessible surface area among the naturally occurring amino acids.[57] Due to its aromaticity, the tryptophan residue is capable of π–π interactions and weakly polar interactions.[58, 59]   This amphipathic character of tryptophan gives rise to its unique hydrogen bonding property and ability to function through long range electrostatic interaction.[60] The amphipathic character of tryptophan also explains its interfacial localization in membranes[61, 62] which is characterized by unique motional and dielectric characteristics different from the bulk aqueous phase and the more isotropic hydrocarbon-like deeper regions of the membrane (see later). More importantly, apart from the structural and spectral properties, the role of tryptophan residues in maintaining the structure and function of both soluble[63-65] and membrane[60, 66-72] proteins has attracted considerable attention.   The importance of tryptophan residues is exemplified by the fact that any perturbation to tryptophan residues (substitution, deletion, chemical modification, or photodamage) often results in reduction or loss of protein functionality.

9.4. APPLICATION OF REES IN THE ORGANIZATION AND DYNAMICS OF PROTEINS

9.4.1.   Soluble Proteins

The local dynamics of the protein matrix around a given amino acid residue can be examined from the rate at which the matrix responds to (relaxes around) the newly created excited state dipole moment of the fluorophore. In other words, the magnitude of REES could be utilized to estimate the relative rigidity of the region of the protein surrounding the fluorophore. The first protein for which REES was documented using its intrinsic fluorescence was human serum albumin.[73] Pioneering work in the application of REES to elucidate protein organization and dynamics was carried out by Demchenko and co-workers (recently reviewed in ref. 37). In one of their early studies, it was shown that the fluorescence emission spectra of 2-(p-toluidinylnaphthalene)-6-sulfonate (TNS) bound to proteins such as β-lactoglobulin, β-casein, and bovine and human serum albumins depend on the excitation wavelength used, giving rise to REES of the order of 10 nm in all cases.[74] The fact that this effect was actually a result of the slow rate of solvent dipolar relaxation was confirmed when a similar effect was observed in case of the same fluorophore in glucose glass and in glycerol at 1 °C, but not in liquid solutions.

Based on an in-depth study of a variety of single tryptophan proteins, it was observed that the presence of red edge effects in such proteins is influenced by the position of emission maximum of tryptophan fluorescence when excited at the absorption maximum.[75] It was found that proteins (such as azurin, parvalbumin, and ribonucleases

T_1 and C_2) with maximum of emission at very short wavelengths (*i.e.*, between 307 and 323 nm) do not show red edge effects. This is attributed to the fact that the tryptophan environments in this class of proteins are nonpolar and the dipole-orientational broadening of the spectra is insufficient to create a large enough distribution of differentially solvated substrates that could be photoselected in order to give rise to REES. However, a recent study provides an exception to this generalization in which a single tryptophan mutant of triosephosphate isomerase from the parasite *Plasmodium falciparum*, with an emission maximum of 321 nm, has been shown to exhibit REES.[76] Interestingly, REES was also not observed for proteins (such as myelin basic protein, β-casein, and monomeric melittin in water) with long wavelength emission maximum (beyond 341 nm). The tryptophan residue in these cases is exposed to aqueous environment undergoing much faster solvent relaxation as compared to the fluorescence lifetimes of the concerned tryptophans. It is for this reason that tryptophans in denatured proteins do not exhibit REES, e.g., tubulin[77] (see later). However, considerable REES effects were observed for many proteins (such as tetrameric form of melittin, human serum albumin, albumin complexed with sodium dodecyl sulfate) with intermediate maxima of emission (between 325 and 341 nm). For these proteins, there is enough range of dipole-orientational broadening and at the same time, the rate of solvent reorientation is slow enough to give rise to REES. It is interesting to note that monomeric melittin in aqueous solution does not exhibit REES whereas the tetrameric form of melittin exhibits a REES of 6 nm indicating that the tryptophan residues are in a motionally restricted environment in the aggregated form of melittin.[78] However, monomeric melittin has been shown to exhibit significant REES upon membrane binding (see section 4.2.3).

While human serum albumin shows REES in the absence of detergents,[75] the emission maximum of bovine serum albumin is independent of the excitation wavelength. However, upon binding to detergents, there is an increase in solvent restriction around the tryptophan residues in both human and bovine serum albumins, as evident by the enhanced red edge effects.[75, 79] Interestingly, the presence of slow solvent relaxation at ambient temperature has been demonstrated utilizing REES of the single tryptophan protein Bj2S, a seed albumin from *Brassica juncea*.[80] The REES approach has been used in a number of cases to monitor the tryptophan environment and dynamics in proteins such as human α_1-acid glycoprotein,[81] bothropstoxin-I from the venom of *Bothrops jararacussu*,[82] ascorbate oxidase,[83] smooth muscle myosin light chain kinase,[84] skeletal myosin rod,[85] the human leukocyte antigen complex,[86] and the pore-forming α-toxin from *Staphylococcus aureus*.[87] In addition, it has been shown that the environment of tryptophans in cytoskeletal proteins such as tubulin[77] and spectrin[88, 89] are motionally restricted and display REES. Importantly, observation of REES in multitryptophan proteins considerably rules out the possibility of homotransfer among tryptophans.[90]

In addition to the application of REES to monitor protein conformation in solution, it has also been applied to more complex systems such as intact eye lens and cornea. About 35% of the lens (by weight) is made up of three closely related proteins called crystallins. The magnitude of REES has been effectively utilized as a parameter to study the photophysical and chemical properties of isolated intact eye lenses and to probe the change in the organization of the lens upon photodamage.[91] The use of REES has also been extended to relate the properties of the lens with those of its constituent proteins and their homo- and heteroaggregates[92] and of proteoglycans and crystallins from intact cornea.[93, 94]

9.4.1.1. REES as a Tool to Explore Residual Local Structure in Denatured Proteins

As mentioned earlier, tryptophans in denatured proteins generally do not exhibit REES due to fast solvent relaxation in the denatured state. For example, tryptophans in the cytoskeletal protein tubulin show REES of 7 nm. However, no REES is observed when the protein is denatured in 8 M urea.[77] In a recent study on the cytoskeletal protein erythroid spectrin, the tryptophans show REES indicating the localization in environments which are motionally restricted due to slow solvent relaxation.[88, 89] Interestingly, spectrin display REES even when denatured in 8 M urea.[88] This surprising result is in contrast to earlier studies where it was shown that the emission maximum of tryptophans in denatured proteins do not exhibit excitation wavelength dependence. This is because the tryptophans are exposed to water when denatured and therefore do not offer any restriction to the solvent (water) dipoles around them in the excited state.[75] The observation of REES in denatured spectrin indicates that the tryptophans are shielded from bulk solvent even when denatured and indicates local residual structure in the denatured protein. The tryptophan microenvironment in spectrin is, therefore, characterized by unique structural and dynamic features that are maintained to a significant extent even when denatured with urea. This is further supported by analysis of fluorescence quenching data using acrylamide as quencher.[88] Such residual structure in an unfolded protein is thought to reside predominantly in hydrophobic clusters, where residues like tryptophan stabilize these networks through cooperative long range nonnative interactions.[95] Although residual structure in denatured proteins has been studied before,[96, 97] this example constitutes the first demonstration of slow solvent relaxation in a completely denatured protein. In addition, the influence of ionic strength induced conformational change of spectrin on the solvent-restricted environment of the tryptophan residues has been recently addressed.[89]

9.4.2. Membrane Peptides and Proteins

9.4.2.1. Membrane Interface: An Appropriate System for the REES Approach

Biological membranes are complex assemblies of lipids and proteins that allow cellular compartmentalization and act as the interface through which cells communicate with each other and with the external milieu. Organized molecular assemblies such as membranes can be considered as large cooperative units with characteristics very different from the individual structural units that constitute them. A direct consequence of such highly organized systems is the restriction imposed on the mobility of their constituent structural units. It is well known that interiors of biological membranes are viscous, with the effective viscosity comparable to that of light oil.[98-100] In addition, membranes exhibit considerable degree of anisotropy along the axis perpendicular to the bilayer. While the center of the bilayer is nearly isotropic, the upper portion, only a few angstroms away toward the membrane surface, is highly ordered.[38, 101-105] Properties such as polarity, fluidity, segmental motion, ability to form hydrogen bonds and extent of solvent penetration vary in a depth-dependent manner in the membrane. The interfacial region in membranes is the most important region so far as the dynamics and function of the membrane is concerned. The membrane interface is characterized by unique motional and dielectric characteristics distinct from both the bulk aqueous phase and the more isotropic hydrocarbon-like interior of the membrane.[101-104, 106] It is a chemically heterogeneous region composed of lipid headgroup, water and portions of the acyl chain.[107] Overall, the interfacial region of the membrane accounts for 50% of the thermal thickness of the bilayer.[106] This specific region of the membrane is known to exhibit slow solvent (water) relaxation and participate in intermolecular charge interactions[108] and hydrogen bonding through the polar headgroup.[109-111] These structural features, which slow down the rate of solvent reorientation, have previously been recognized as typical features of environments giving rise to significant red edge effects.[38, 112] It is therefore the membrane interface which is most likely to display REES.[105, 113-115] The application of REES to study the organization and dynamics of membranes has been reviewed recently.[36, 38]

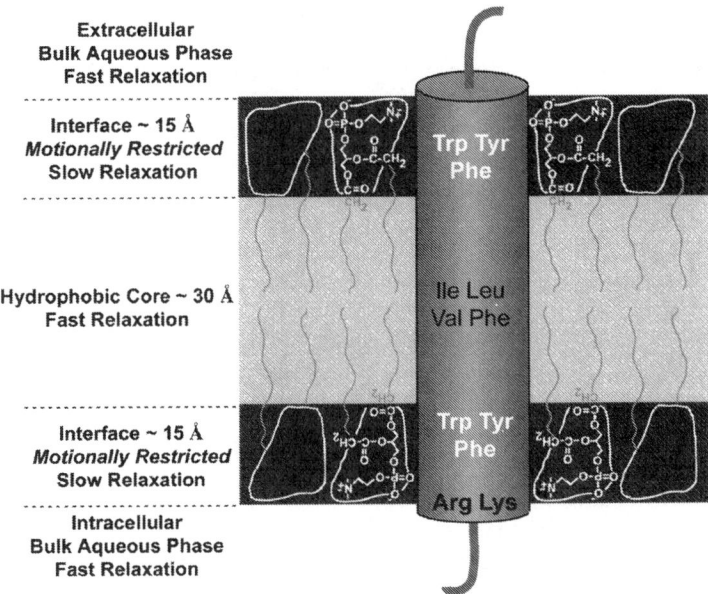

Figure 9.2. A schematic representation of the membrane showing the various regions of the bilayer displaying motional anisotropy. The membrane lipids shown have two hydrophobic tails with a phosphatidylcholine (PC) head group. The preferred locations of various amino acids present in a membrane-spanning transmembrane domain are also shown. It should be noted that the tryptophan residues are localized in the membrane interface region, a region characterized by unique organization, dynamics, hydration and functionality. Adapted from ref. 39.

9.4.2.2. Tryptophan and the Membrane Interface

The biological membrane provides a unique environment to membrane-spanning proteins and peptides influencing their structure and function. Membrane-spanning proteins have distinct stretches of hydrophobic amino acids that form the membrane-spanning domain and are reported to have a significantly higher tryptophan content than soluble proteins.[116] In addition, it has been observed that tryptophan residues in integral membrane proteins and peptides are not uniformly distributed and that they tend to be localized toward the membrane interface, possibly because they are involved in hydrogen bonding[117] with the lipid carbonyl groups or interfacial water molecules (see Figure 9.2). For instance, crystal structures of membrane proteins such as the potassium channel,[118] bacteriorhodopsin,[119] maltoporin,[120] and others have shown that most tryptophans are located in a saddle-like 'aromatic belt' around the membrane interfacial region. Statistical studies of sequence databases and available crystal structures of integral membrane proteins also show preferential clustering of tryptophan residues at the membrane interface.[121-123] Furthermore, for transmembrane peptides and proteins, tryptophan has been found to be an efficient anchor at the membrane interface[116, 124] and defines the hydrophobic length of transmembrane helices.[125] Importantly, the role of tryptophan residues in maintaining the structure and function of membrane proteins is exemplified by the fact that substitution or deletion of tryptophans often results in reduction or loss of protein functionality.[60, 66, 69]

The exact location and orientation of tryptophan residues at the membrane interface is, however, not clear. Some experiments suggest that tryptophan residues have a preference for the lipid headgroup side of the interface but others suggest that the preference is for the fatty acyl chain side.[126-129] Nevertheless, the preferential location of tryptophan residues at the membrane interface is thought to be due to the aromaticity of the indole moiety and the overall amphipathic nature of tryptophan.[126] The tryptophan rich aromatic belt at the membrane interface in transmembrane helices is thought to stabilize the helix with respect to the membrane environment.[122] The tryptophan residue has a large indole side chain that consists of two fused rings. In fact, the tryptophan side chain has the largest volume of all the amino acid side chains,[130] with a volume of 228 Å^3 which is comparable to the volume of a phosphatidylcholine headgroup,[131] *i.e.* 319 Å^3. Wimley and White have shown from partitioning of model peptides to membrane interfaces that the experimentally determined interfacial hydrophobicity of tryptophan is the highest among the naturally occurring amino acid residues thus accounting for its specific interfacial localization in membrane-bound peptides and proteins.[61] The amphipathic character of tryptophan gives rise to its hydrogen bonding ability which could account for its orientation in membrane proteins and its function through long-range electrostatic interactions.[60, 132] This amphipathic character of tryptophan also explains its interfacial localization in membranes due to its tendency to be solubilized in this region of the membrane, besides favorable electrostatic interactions and hydrogen bonding.

9.4.2.3. Membrane Peptides

The membrane environment and the relative position within the membrane have a profound influence on the dynamics of amino acid residues of membrane-spanning helices. Both natural and synthetic membrane peptides represent convenient molecules to explore the specific role of membrane and peptide structural properties on the orientation, incorporation, stability and function of such peptides in the membrane milieu.[133-139]

Tryptophan octyl ester (TOE) has been recognized as an important model for membrane-bound tryptophan residues.[62, 140-142] The fluorescence characteristics of TOE incorporated into model membranes and membrane-mimetic systems have been shown to be similar to that of membrane-bound tryptophans. Consistent with the preferred interfacial location of the tryptophan residue and the solvent-restricted environment of the interface, TOE has been found to exhibit REES. Further, the extent of REES was found to be dependent on pH indicating more motional freedom on deprotonation at higher pH.[62]

Melittin, the major toxic component in the venom of the European honey bee, *Apis mellifera*, was one of the earliest membrane peptides studied utilizing REES. It is an amphipathic cationic hemolytic peptide with a single, functionally important tryptophan residue.[143] The amphipathic nature is characteristic of many membrane-bound peptides and putative transmembrane helices of membrane proteins.[143, 144] This has made melittin very popular as a model to study lipid-protein interactions in membranes.[143, 145] Results from studies utilizing REES of the sole tryptophan residue showed that when bound to zwitterionic membranes, the microenvironment of the functionally active tryptophan of melittin is motionally restricted, consistent with the interfacial location of the tryptophan residue.[146] Interestingly, further results employing REES indicate that the microenvironment of the tryptophan residue in melittin is modulated when bound to negatively charged membranes, and this could be related to the functional difference in the lytic activity of the peptide observed in these two cases.[147] REES is therefore sensitive to changes in dynamics of hydration caused by varying electrostatic interactions. This has been recently confirmed using melittin bound to micelles of various charge types.[148] In addition, the REES approach has recently been successfully applied to monitor the effect of membrane cholesterol on the organization and dynamics of melittin.[149]

Reverse micelles serve as appropriate membrane-mimetic host assembly for the characterization of membrane-active peptides and proteins utilizing REES.[150] This is due to the fact that reverse micelles offer the unique advantage of monitoring dynamics of embedded molecules with varying degrees of hydration which is difficult to achieve with complex systems such as membranes. It has thus been shown that the magnitude of REES for melittin bound to reverse micelles is sensitive to the change in water content of the system.[151] This constitutes the first report directly demonstrating that REES is sensitive to the changing dynamic hydration profile of an amphiphilic peptide. In another study, the magnitude of REES of membrane-bound melittin has been shown to be dependent on the extent of lipid chain unsaturation in the membrane. It is interesting to note that the magnitude of REES obtained with melittin in membranes containing more than two double bonds is considerably higher (up to 19 nm)[152] than what is usually reported for membrane-bound tryptophan residues,[62, 146, 147, 153-156] although higher REES has been reported in a few cases (see below). In addition, the environment of tryptophan residues of several tryptophan-rich antimicrobial peptides in the membrane-bound form has been explored utilizing REES.[157]

The phenomenon of REES, in conjunction with time-resolved fluorescence spectroscopic parameters such as wavelength-dependent fluorescence lifetimes and time-resolved emission spectra (TRES) was utilized to probe the localization and dynamics of the functionally important tryptophan residues in the gramicidin channel.[153, 156] Gramicidin belongs to a family of prototypical cation-selective channel formers, which are naturally fluorescent due to the presence of four tryptophan residues.[158] These interfacially localized tryptophans are known to play a crucial role in the organization and function of the channel.[60, 66] The results from these studies point out the motional restriction experienced by the tryptophans at the peptide-lipid interface of the gramicidin channel.[153, 156] This is consistent with other studies[60, 66] in which such a restriction is thought to be imposed due to hydrogen bonding between the indole rings of the tryptophan residues in the channel conformation and the neighboring lipid carbonyls. The significance of such organization in terms of functioning of the channel is brought out by the fact that substitution, photodamage, or chemical modification of these tryptophans are known to give rise to channels with altered conformation and reduced conductivity.[60, 66, 159-163] More importantly, REES and related fluorescence approaches have been used to distinguish between the channel and non-channel conformations of gramicidin in membranes.[156] REES of gramicidin is therefore sensitive to the conformation adopted by the peptide in the membrane. This provides a convenient spectroscopic handle to monitor the functional status of this important ion channel peptide. In addition, REES has been used to monitor the effects of structural transition of the host assembly and hydration on the organization and dynamics of gramicidin.[164, 165] Further, REES of gramicidin has been found to be dependent on the lipid composition[166] and phase state of the membrane.[167]

In the absence of crystal structures, fluorescence approaches to study the topography and membrane interactions of specific peptide fragments have proved to be very useful. REES has been used to probe the membrane interaction of synthetic peptides corresponding to the functionally important regions of membrane binding proteins.[168, 169] The 579-601 fragment of the ectodomain of the HIV-1 gp-41 protein which is essential for its activity, was shown to be incorporated in model membranes.[168] Upon red edge excitation, there was a substantial red shift in fluorescence emission indicating a motionally restricted environment for the single tryptophan of this peptide. The REES result is somewhat unusual not only due to the rather large magnitude of REES (18 nm) but also due to the observation of REES for a peptide with the emission maximum at 348 nm when excited at the absorption maximum. This is in contrast with the previously accepted notion that REES can only be found for tryptophan residues which emit between 325 and 341 nm.[75] Large magnitude of REES (28 nm) has also been observed for the reporter tryptophan in membrane-bound PT-(1-46)F4W synthetic peptide which corresponds to the ω-loop region of the human prothrombin γ-carboxyglutamic acid domain.[169] In both these cases, the peptides are found to partition to the interfacial region of the membrane. The observed REES would therefore be a direct consequence of the motionally restricted membrane interface.[38] In yet another study, the organization and topography of the membrane-binding C2 domain of factor VIII have been monitored using REES in membrane-mimetic environments.[170]

The interaction of the peptide corresponding to the γM4 transmembrane domain of the nicotinic acetylcholine receptor with membranes has been investigated.[171] Studies using REES showed that the helix is oriented in such a way that the N-terminal tryptophan (Trp-453) is located in a motionally restricted environment at the membrane interface, confirming the previous observation that γM4 forms a transmembrane helix.[172] The application of REES has been extended in a novel way to determine the topology and membrane localization of tryptophan residues in the colicin E1 channel peptide by measuring REES of single tryptophan containing channel peptides in the soluble aqueous state, and when bound to membranes.[154] From the difference in REES values of the channel peptides in the membrane-bound and soluble state, the topology of the peptide was mapped out.

9.4.2.4. Membrane Proteins

An early application of REES to study membrane protein organization consisted of studying the microconformational heterogeneity of the membrane binding domain of cytochrome b_5 by comparing the information obtained from the native protein and its mutant which has a single tryptophan residue in this domain.[155] Both these proteins show a red shift in the emission spectrum when excited at the long wavelength edge of the excitation spectrum indicating thereby that the tryptophan residue(s) in both cases are localized in a region of motional constraint in the membrane. In another study, the tryptophans of the pore-forming *Staphylococcus aureus* α-toxin were shown to exhibit REES.[87] However, no significant difference was observed between REES displayed by the soluble and membrane-bound forms of the toxin indicating no drastic change in the tryptophan environment upon membrane binding. The tryptophans in *E. coli* porin Omp-F, a pore-forming channel protein, also exhibit REES.[173]

Membrane-active proteins that exist in soluble and membrane-bound forms undergo considerable conformational change in the transition from the soluble to the membrane-bound state. One such membrane-active protein is the mitochondrial creatine kinase which shows a large increase in REES upon membrane binding.[174] The soluble form exhibits REES of 6 nm which increases to a REES of 19 nm in the presence of membranes which is indicative of a restricted environment for the tryptophans in the membrane-bound form. Since the tryptophan residues are localized in the protein interior,[175] the observed motional restriction has been attributed to conformational changes of the protein induced by lipid binding.

9.4.3.   Extrinsic Fluorophores

The analysis of fluorescence from multitryptophan proteins is often complicated due to the complexity of fluorescence processes in such systems, and the heterogeneity in fluorescence parameters (such as quantum yield and lifetime) due to environmental sensitivity of individual tryptophans.[32] This problem can be avoided by the use of extrinsic fluorophores which often display improved spectral properties. Such extrinsic fluorescent probes are widely used to study the dynamics of proteins and membranes.[74, 105, 113, 114, 176] The advantage of this approach is that one has a choice of the fluorescent label to be used and, therefore, specific probes with appropriate characteristics can be designed for specific applications. Many of these probes display high sensitivity to the polarity of the local environment. In addition, many extrinsic fluorescent probes are weakly fluorescent or nonfluorescent in water but fluoresce strongly when bound to proteins and membranes making the contribution from the unbound probe negligible.

Fluorescent probes such as 6-(*p*-toluidinyl)-2-naphthalenesulfonic acid (TNS) and 8-anilinonaphthalene-1-sulfonic acid (ANS) are representative examples of this type of probes. These probes are considered as indicators of binding site polarity, and the shifts in fluorescence spectra on binding are commonly correlated with solvent polarity scales. For example, by comparing data obtained for TNS in solvents and bound to macromolecules, it has been found that there is a slower relaxation around TNS bound to the heme site of the apomyoglobin.[176] As mentioned earlier, in one of the early studies of REES, it was shown that TNS bound to proteins such as β-lactoglobulin, β-casein, and bovine and human serum albumins exhibits REES of the order of 10 nm.[74] In addition, the complexes of TNS with tetrameric melittin[177] and $α_1$-acid glycoprotein[178] demonstrated that the TNS-binding site is rigid on the nanosecond time scale in these proteins giving rise to REES.

The hydrophobic fluorescent probe 6-propionyl-2-dimethylaminonaphthalene (PRODAN) shows polarity-sensitive fluorescence.[179] The dipole moment of PRODAN changes by ~5-8 D upon excitation.[180, 181] A change in dipole moment of this magnitude, along with its hydrogen bonding capability,[181] makes PRODAN a suitable probe for monitoring REES effects to characterize hydrophobic binding sites in proteins. For example, PRODAN binds erythroid spectrin with a high affinity.[182] The organization and dynamics of the PRODAN binding site in erythroid spectrin have been monitored recently.[88] The observation of REES for spectrin-bound PRODAN suggests that PRODAN is in an environment where its mobility is considerably reduced. Since PRODAN binds to the hydrophobic site in spectrin, this result shows that this region offers considerable restriction to the reorientational motion of the solvent dipoles around the excited state fluorophore.[88] In addition, pronounced REES was reported for complexes of PRODAN with model lipoproteins.[183] Interestingly, a derivative of PRODAN, 2'-(*N,N*-dimenthyl)amino-6-naphthopyl-4-*trans*-cyclohexanoic acid (DANCA), has been used to determine the polarity of the heme-binding pocket in apomyoglobin utilizing the excitation wavelength dependence of the emission spectrum.[184]

In an attempt to understand the dynamics of ligand-binding pockets in proteins, REES studies have also been carried out utilizing naturally fluorescent ligands. For instance, bilirubin is a natural ligand for the human and bovine serum albumins. It has been shown to exhibit REES when complexed with albumins in the presence of micellar concentrations of Triton X-100 which confer rigidity to the binding pocket of bilirubin in albumins.[79] In another study, REES of the antimitotic drug colchicine and its derivatives was detected when bound to tubulin using colchicine fluorescence in order to understand structure-function relationship in the tubulin-colchicine complex.[77] In addition, REES of flavin adenine dinucleotide (FAD) bound to lipoamide dehydrogenase from *Azotobacter vinelandii* has also been reported.[185] Taken together, these results show that REES can be conveniently used to monitor the organization and dynamics of ligand-binding pockets of proteins.

A widely used extrinsic probe is the 7-nitrobenz-2-oxa-1,3-diazol-4-yl (NBD) group which reacts with amino or thiol groups to form stable, highly fluorescent compounds.[186] In this way, various NBD-labeled lipids and proteins have been generated and these have proved to be useful for monitoring structure and dynamics of proteins, peptides, and membranes. NBD-labeled lipids are extensively used as fluorescent analogues of native lipids in biological and model membranes to study a variety of processes. The NBD moiety possesses some of the most desirable properties for serving as an excellent probe for spectroscopic applications. It is very weakly fluorescent in water and is relatively photostable. Upon transfer to a hydrophobic medium, it fluoresces brightly in the visible range and exhibits a high degree of environmental sensitivity.[113, 187-190] More importantly, it has earlier been shown using solvatochromic and quantum chemical approaches that the dipole moment of the NBD group changes by ~4 D upon excitation,[189] an important criterion for a fluorophore to exhibit REES effects.[35, 38] The NBD group therefore has been used extensively to explore the organization and dynamics in membranes and membrane-mimetic media utilizing the REES approach.[113, 114, 191-197]

Apolipoprotein C-II (apoC-II) is an exchangeable 79-residue apolipoprotein found in blood plasma and is essential for the activation of lipoprotein lipase and therefore in the hydrolysis of triacylglycerols and the prevention of hypertriglyceridaemia.[198] By covalently labeling the peptide fragment of apolipoprotein C-II (apoC-II$_{19-39}$) with NBD, it was found that NBD-apoC-II$_{19-39}$ exhibits an increased REES of 25 nm when bound to egg yolk phosphatidylcholine membranes when compared to the corresponding REES in aqueous solution (9 nm).[199] This observation, along with results from time-resolved fluorescence measurements, points out that the NBD-labeled peptide is located near the polar-interfacial region of the bilayer where it experiences a heterogeneous environment. In addition, the newly developed approach of site-directed fluorescence labeling (SDFL)[200, 201] at a single site on the target protein, could be extremely useful to monitor the environmental dynamics of proteins in a site-specific manner (see later), particularly for multitryptophan containing proteins.

9.5. CONCLUSION AND FUTURE PERSPECTIVES

The lack of a suitable fluorophore often makes it difficult to monitor the dynamics in a region of interest in proteins. Advances in molecular biological techniques, however, have made it possible to incorporate (or substitute) endogeneous labels such as tryptophan in regions of choice in soluble[202] as well as integral membrane[203] proteins. A major limitation in working with multitryptophan proteins is that the analysis of fluorescence data is often complicated due to the complexity of fluorescence processes in such systems and lack of specific information.[32, 204] Site-specific incorporation of extrinsic probes, accomplished by using unnatural amino acid mutagenesis,[205-207] could help avoid this complication. The incorporation of tryptophan analogues has been shown to be useful in this regard. Successful biosynthetic incorporation of tryptophan analogues has been accomplished with various derivatives of azatryptophan, hydroxytryptophan, and fluorotryptophan.[208] The main advantage of these analogues is that the spectral properties are distinct from that of tryptophan. For example, the absorption and fluorescence emission spectra of 7-azatryptophan are significantly red shifted (by 10 nm and 50 nm respectively) from those of tryptophan.[209] Consequently, it is possible to selectively excite these tryptophan analogues in proteins, in the presence of a 'sea' of tryptophans of other proteins in their native environment. More importantly, it has been shown that the incorporation of these analogues causes minimal perturbation to protein structure, stability, and function.[210-212] Interestingly, 7-azatryptophan has been shown to display REES and is sensitive to slow solvent relaxation at low temperatures,[213] which makes it an ideal candidate to explore site-specific environmental rigidity in proteins.

As mentioned earlier, water plays a crucial role in the formation and maintenance of organized molecular assemblies such as proteins and membranes in a cellular environment. Knowledge of dynamics of hydration at the molecular level is therefore of considerable importance in understanding the cellular structure and function.[3, 12, 15, 21-23,] [40, 214-218] REES is based on the change in fluorophore-solvent interactions in the ground and excited states brought about by a change in the dipole moment of the fluorophore upon excitation, and the rate at which solvent molecules reorient around the excited state fluorophore. The unique feature about REES is that while all other fluorescence techniques such as fluorescence quenching, energy transfer and polarization measurements yield information about the fluorophore (either intrinsic or extrinsic) itself, REES provides information about the relative rates of solvent (water in biological systems) relaxation dynamics, which is not possible to obtain by other techniques. Since the dynamics of hydration is directly associated with the functionality of proteins, REES in combination with the molecular biological approaches for incorporating novel fluorophores at specific sites, could prove to be a novel and extremely powerful tool to explore the organization and dynamics of both soluble and membrane proteins and other organized molecular assemblies.

9.6. ACKNOWLEDGEMENTS

Work in A. C.'s laboratory was supported by the Council of Scientific and Industrial Research, Department of Science and Technology, Government of India, and the Third World Academy of Sciences. H. R. and D. A. K. thank the Council of Scientific and Industrial Research for the award of Senior Research Fellowships. Some of the work described in this article was carried out by former and present members of A. C.'s research group whose contribution is gratefully acknowledged. We thank members of our laboratory for critically reading the manuscript.

9.7. REFERENCES

1.   R. Pethig, Dielectric studies of protein hydration, in: *Protein-Solvent Interactions,* edited by R. B. Gregory (Marcel Dekker, New York, 1995) pp. 265-288.
2.   M. M. Teeter, Water-protein interactions: theory and experiment, *Annu. Rev. Biophys. Biophys. Chem.* **20,** 577-600 (1991).
3.   P. W. Fenimore, H. Frauenfelder, B. H. McMahon, and F. G. Park, Slaving: solvent fluctuations dominate protein dynamics and function, *Proc. Natl. Acad. Sci. U.S.A.* **99,** 16047-16051 (2002).
4.   K. Ogata, and S. J. Wodak, Conserved water molecules in MHC class-I molecules and their putative structural and functional roles, *Protein Eng.* **15,** 697-705 (2002).
5.   M. Fasano, M. Orsale, S. Melino, E. Nicolai, F. Forlani, N. Rosato, D. Cicero, S. Pagani, and M. Paci, Surface changes and role of buried water molecules during the sulfane sulfur transfer in rhodanese from *Azotobacter vinelandii:* a fluorescence quenching and nuclear magnetic relaxation dispersion spectroscopic study, *Biochemistry* **42,** 8550-8557 (2003).
6.   L. A. Munishkina, J. Henriques, V. N. Uversky, and A. L. Fink, Role of protein-water interactions and electrostatics in α-synuclein fibril formation, *Biochemistry* **43,** 3289-3300 (2004).
7.   Y. Levy, and J. N. Onuchic, Water and proteins: a love-hate relationship, *Proc. Natl. Acad. Sci. U.S.A.* **101,** 3325-3326 (2004).
8.   S. K. Pal, J. Peon, and A. H. Zewail, Biological water at the protein surface: dynamical solvation probed directly with femtosecond resolution, *Proc. Natl. Acad. Sci. U.S.A.* **99,** 1763-1768 (2002).
9.   S. N. Timasheff, Protein hydration, thermodynamic binding, and preferential hydration, *Biochemistry* **41,** 13473-13482 (2002).
10.   S. Melchionna, G. Briganti, P. Londei, and P. Cammarano, Water induced effects on the thermal response of a protein, *Phys. Rev. Lett.* **92,** 158101 (2004).

11. F. Xu, and T. A. Cross, Water: foldase activity in catalyzing polypeptide conformational rearrangements, *Proc. Natl. Acad. Sci. U.S.A.* **96,** 9057-9061 (1999).

12. C. Mattos, Protein-water interactions in a dynamic world, *Trends Biochem. Sci.* **27,** 203-208 (2002).

13. A. R. Bizzarri, and S. Cannistraro, Molecular dynamics of water at the protein-solvent interface, *J. Phys. Chem. B* **106,** 6617-6633 (2002).

14. G. P. Singh, F. Parak, S. Hunklinger, and K. Dransfeld, Role of adsorbed water in the dynamics of metmyoglobin, *Phys. Rev. Lett.* **47,** 685-688 (1981).

15. S. H. Koenig, K. Hallenga, and M. Shporer, Protein-water interaction studied by solvent ^{1}H, ^{2}H, and ^{17}O magnetic relaxation, *Proc. Natl. Acad. Sci. U.S.A.* **72,** 2667-2671 (1975).

16. C. Mattos, and D. Ringe, Proteins in organic solvents, *Curr. Opin. Struct. Biol.* **11,** 761-764 (2001).

17. G. Careri, Collective effects in hydrated proteins, in: *Hydration Processes in Biology: Theoretical and Experimental Approaches,* edited by M. C. Bellisent-Funel, (IOS Press, Amsterdam, 1999) pp. 143-155.

18. C. Ho, and C. D. Stubbs, Hydration at the membrane protein-lipid interface, *Biophys. J.* **63,** 897-902 (1992).

19. L. Essen, R. Siegert, W. D. Lehmann, and D. Oesterhelt, Lipid patches in membrane protein oligomers: crystal structure of the bacteriorhodopsin-lipid complex, *Proc. Natl. Acad. Sci. U.S.A.* **95,** 11673-11678 (1998).

20. K. E. McAuley, P. K. Fyfe, J. P. Ridge, N. W. Isaacs, R. J. Cogdell, and M. R. Jones, Structural details of an interaction between cardiolipin and an integral membrane protein, *Proc. Natl. Acad. Sci. U.S.A.* **96,** 14706-14711 (1999).

21. R. Sankararamakrishnan, and M. S. P. Sansom, Water-mediated conformational transitions in nicotinic receptor M2 helix bundles: a molecular dynamics study, *FEBS Lett.* **377,** 377-382 (1995).

22. T. Okada, Y. Fujiyoshi, M. Silow, J. Navarro, E. M. Landau, and Y. Shichida, Function role of internal water molecules in rhodopsin revealed by x-ray crystallography, *Proc. Natl. Acad. Sci. U.S.A.* **99,** 5982-5987 (2002).

23. T. Kouyama, T. Nishikawa, T. Tokuhisa, and H. Okumura, Crystal structure of the L intermediate of bacteriorhodopsin: evidence for vertical translocation of a water molecule during the proton pumping cycle, *J. Mol. Biol.* **335,** 531-546 (2004).

24. M. B. Jackson, Advances in ion channel structure, *Trends Neurosci.* **24,** 291 (2004).

25. P. M. Hwang, R. E. Bishop, and L. E. Kay, The integral membrane enzyme PagP alternates between two dynamically distinct states, *Proc. Natl. Acad. Sci. U.S.A.* **101,** 9618-9623 (2004).

26. J. Torres, T. J. Stevens, and M. Samsó, Membrane proteins: the 'wild west' of structural biology, *Trends Biochem. Sci.* **28,** 137-144 (2003).

27. S. H. White, The progress of membrane protein structure determination, *Protein Sci.* **13,** 1948-1949 (2004).

28. J. U. Bowie, Stabilizing membrane proteins, *Curr. Opin. Struct. Biol.* **11,** 397-402 (2001).

29. H. M. Berman, J. Westbrook, Z. Feng, G. Gilliland, T. N. Bhat, H. Weissig, I. N. Shindyalov, and P. E. Bourne, The protein data bank, *Nucleic Acids Res.* **28,** 235-242 (2000).

30. P. C. Preusch, J. C. Norvell, J. C. Cassatt, and M. Cassman, Progress away from 'no crystals, no grant', *Nat. Struct. Biol.* **5,** 12-14 (1998).

31. P. L. Yeagle, and A. G. Lee, Membrane protein structure, *Biochim. Biophys. Acta* **1565,** 143 (2002).

32. M. R. Eftink, Fluorescence techniques for studying protein structure, in: *Methods of Biochemical Analysis, Vol. 35,* edited by C. H. Suelter (John Wiley, New York, 1991), pp. 127-205.

33. A. Chattopadhyay, and H. Raghuraman, Application of fluorescence spectroscopy to membrane protein structure and dynamics, *Curr. Sci.* **87,** 175-180 (2004).

34. A. P. Demchenko, Site-selective excitation: a new dimension in protein and membrane spectroscopy, *Trends Biochem. Sci.* **13,** 374-377 (1988).

35. S. Mukherjee, and A. Chattopadhyay, Wavelength-selective fluorescence as a novel tool to study organization and dynamics in complex biological systems, *J. Fluoresc.* **5,** 237-246 (1995).

36. A. Chattopadhyay, Application of the wavelength-selective fluorescence approach to monitor membrane organization and dynamics, in: *Fluorescence Spectroscopy, Imaging and Probes,* edited by R. Kraayenhof, A. J. W. G. Visser, and H. C. Gerritsen (Springer-Verlag, Heidelberg, 2002) pp. 211-224.

37. A. P. Demchenko, The red-edge effects: 30 years of exploration, *Luminescence* **17,** 19-42 (2002).

38. A. Chattopadhyay, 2003, Exploring membrane organization and dynamics by the wavelength-selective fluorescence approach, *Chem. Phys. Lipids* **122,** 3-17 (2003).

39. H. Raghuraman, D. A. Kelkar, and A. Chattopadhyay, Novel insights into membrane protein structure and dynamics utilizing the wavelength-selective fluorescence approach, *Proc. Ind. Natl. Sci. Acad. A* **69,** 25-35 (2003).

40. P. Mentré, (Ed.), Water in the cell, *Cell. Mol. Biol.* **47,** 709-970 (2001).

41. J. B. Birks, *Photophysics of Aromatic Molecules* (Wiley-Interscience, London, 1970).

42. K. K. Rohatgi-Mukherjee, *Fundamentals of Photochemistry* (Wiley Eastern, New Delhi, 1978).

43. A. P. Demchenko, and A. I. Sytnik, Solvent reorganizational red-edge effect in intramolecular electron transfer, *Proc. Natl. Acad. Sci. U.S.A.* **88,** 9311-9314 (1991).

44. A. P. Demchenko, A new generation of fluorescence probes exhibiting charge-transfer reactions, *Proc. SPIE Int. Soc. Opt. Eng.* **2137,** 588-599 (1994).

45. B. Valeur, *Molecular Fluorescence: Principles and Applications,* (Wiley-VCH, Weinheim, 2002) pp. 63-65.

46. D. W. Pierce, and S. G. Boxer, Stark effect spectroscopy of tryptophan, *Biophys. J.* **68,** 1583-1591 (1995).

47. P. R. Callis, ^1L_a and ^1L_b transitions of tryptophan: applications of theory and experimental observations to fluorescence of proteins, *Methods Enzymol.* **278,** 113-150 (1997).

48. G. Weber, Fluorescence-polarization spectrum and electronic-energy transfer in proteins, *Biochem. J.* **75,** 345-352 (1960).

49. J. R. Lakowicz, *Principles of Fluorescence Spectroscopy* (Kluwer-Plenum, New York, 1999).

50. K. Ruan, J. Li, R. Liang, C. Xu, Y. Yu, R. Lange, and C. Balny, A rare protein fluorescence behavior where the emission is dominated by tyrosine: case of the 33-kDa protein from spinach photosystem II, *Biochem. Biophys. Res. Commun.* **293,** 593-597 (2002).

51. J. B. A. Ross, W. R. Laws, K. W. Rousslang, and H. R. Wyssbrod, Tyrosine fluorescence and phosphorescence from proteins and polypeptides, in: *Topics in Fluorescence Spectroscopy, Vol. 3, Biochemical Applications,* edited by J. R. Lakowicz (Plenum Press, New York, 1992), pp. 1-63.

52. A. S. Ladokhin, Fluorescence spectroscopy in peptide and protein analysis, in: *Encyclopedia of Analytical Chemistry,* edited by R. A. Meyers (John Wiley, New York, 2000) pp. 5762-5779.

53. E. P. Kirby, and R. F. Steiner, Influence of solvent and temperature upon the fluorescence of indole derivatives, *J. Phys. Chem.* **74,** 4480-4490 (1970).

54. J. M. Beechem, and L. Brand, Time-resolved fluorescence of proteins, *Annu. Rev. Biochem.* **54,** 43-71 (1985).

55. J. T. Vivian, and P. R. Callis, Mechanisms of tryptophan fluorescence shifts in proteins, *Biophys. J.* **80,** 2093-2109 (2001).

56. A. J. Ruggiero, D. C. Todd, and G. R. Fleming, Subpicosecond fluorescence anisotropy studies of tryptophan in water, *J. Am. Chem. Soc.* **112,** 1003-1014 (1990).

57. C. Chothia, The nature of the accessible and buried surfaces in proteins, *J. Mol. Biol.* **105,** 1-14 (1976).

58. S. K. Burley and G. A. Petsko, Aromatic-aromatic interaction: a mechanism of protein structure stabilization, *Science* **229,** 23-28 (1985).

59. S. K. Burley and G. A. Petsko, Weakly polar interactions in proteins, *Adv. Protein Chem.* **39,** 125-189 (1988).

60. V. Fonseca, P. Daumas, L. Ranjalahy-Rasoloarijao, F. Heitz, R. Lazaro, Y. Trudelle, and O. S. Andersen, Gramicidin channels that have no tryptophan residues, *Biochemistry* **31,** 5340-5350 (1992).

61. W. C. Wimley, and S. H. White, Experimentally determined hydrophobicity scale for proteins at membrane interfaces, *Nat. Struct. Biol.* **3,** 842-848 (1996).

62. A. Chattopadhyay, S. Mukherjee, R. Rukmini, S. S. Rawat, and S. Sudha, Ionization, partitioning, and dynamics of tryptophan octyl ester: implications for membrane-bound tryptophan residues, *Biophys. J.* **73,** 839-849 (1997).

63. T. Nagy, P. Simpson, M. P. Williamson, G. P. Hazlewood, H. J. Gilbert, and L. Orosz, All three surface tryptophans in type IIa cellulose binding domains play a pivotal role in binding both soluble and insoluble ligands, *FEBS Lett.* **429,** 312-316 (1998).

64. O. K. Gasymov, A. R. Abduragimov, T. N. Yusifov, and B. J. Glasgow, Binding studies of tear lipocalin: the role of the conserved tryptophan in maintaining structure, stability and ligand affinity, *Biochim. Biophys. Acta* **1433,** 307-320 (1999).

65. A.-X. Song, L.-Z. Li, T. Yu, S.-M. Chen, and Z.-X. Huang, Role of tryptophan 121 in the soluble Cu_A domain of cytochrome c oxidase: structure and electron transfer studies, *Protein Eng.* **16,** 435-441 (2003).

66. M. D. Becker, D. V. Greathouse, R. E. Koeppe, and O. S. Andersen, Amino acid sequence modulation of gramicidin channel function: effects of tryptophan-to-phenylalanine substitutions on the single-channel conductance and duration, *Biochemistry* **30,** 8830-8839 (1991).

67. J. Kolena, S. Scsukova, M. Tatara, J. Vranova, and M. Jezova, Involvement of tryptophan in the structural alteration of the rat ovarian LH/hCG receptor, *Exp. Clin. Endocrinol. Diabetes* **105,** 304-307 (1997).

68. E. H. Clark, J. M. East, and A. G. Lee, The role of tryptophan residues in an integral membrane protein: diacylglycerol kinase, *Biochemistry* **42,** 11065-11073 (2003).

69. A. S. Miller, and J. J. Falke, Side chains at the membrane-water interface modulate the signaling state of a transmembrane receptor, *Biochemistry* **43,** 1763-1770 (2004).

70. H.-S. Won, S.-H. Park, H. E. Kim, B. Hyun, M. Kim, B. J. Lee, and B.-J. Lee, Effects of a tryptophanyl substitution on the structure and antimicrobial activity of C-terminally truncated gaegurin 4, *Eur. J. Biochem.* **269**, 4367-4374 (2002).

71. Y. Tang, F. Zaitseva, R. A. Lamb, and L. H. Pinto, The gate of the influenza virus M2 proton channel is formed by a single tryptophan residue, *J. Biol. Chem.* **277**, 39880-39886 (2002).

72. Q. Hong, I. Gutiérrez-Aguirres, A. Barlic, P. Malovrh, K. Kristan, Z. Podlesek, P. Macek, D. Turk, J. M. González-Mañas, J. H. Lakey, and G. Anderluh, Two-step binding by equinatoxin II, a pore-forming toxin from the sea anemone, involves an exposed aromatic cluster and a flexible loop, *J. Biol. Chem.* **277**, 41916-41924 (2002).

73. A. P. Demchenko, Dependence of human serum albumin fluorescence spectrum on excitation wavelength, *Ukr. Biochim. Zh.* **53**, 22-27 (1981).

74. A. P. Demchenko, On the nanosecond mobility in proteins. Edge excitation fluorescence red shift of protein-bound 2-(*p*-toluidinylnaphthalene)-6-sulfonate, *Biophys. Chem.* **15**, 101-109 (1982).

75. A. P. Demchenko, Red-edge-excitation fluorescence spectroscopy of single-tryptophan proteins, *Eur. Biophys. J.* **16**, 121-129 (1988).

76. P. Pattanaik, G. Ravindra, C. Sengupta, K. Maithal, P. Balaram, and H. Balaram, Unusual fluorescence of W168 in *Plasmodium falciparum* triosephosphate isomerase, probed by single-tryptophan mutants, *Eur. J. Biochem.* **270**, 745-756 (2003).

77. S. Guha, S. S. Rawat, A. Chattopadhyay, and B. Bhattacharyya, Tubulin conformation and dynamics: a red edge excitation shift study, *Biochemistry* **35**, 13426-13433 (1996).

78. A. P. Demchenko, and A. S. Ladokhin, Temperature-dependent shift of fluorescence spectra without conformational changes in protein; studies of dipole relaxation in the melittin molecule, *Biochim. Biophys. Acta* **955**, 352-360 (1988).

79. S. K. Patra, and M. K. Pal, Red edge excitation shift emission spectroscopic investigation of serum albumins and serum albumin-bilirubin complexes, *Spectrochim. Acta A* **53**, 1609-1614 (1997).

80. M. Ghose, S. Mandal, D. Roy, R. K. Mandal, and G. Basu, Dielectric relaxation in a single tryptophan protein, *FEBS Lett.* **509**, 337-340 (2001).

81. J. R. Albani, Correlation between dynamics, structure and spectral properties of human α1-acid glycoprotein (orosomucoid): a fluorescence approach, *Spectrochim. Acta A* **54**, 175-183 (1998).

82. A. H. C. de Oliveira, J. R. Giglio, S. H. Andrião-Escarso, and R. J. Ward, The effect of resonance energy homotransfer on the intrinsic tryptophan fluorescence emission of the bothropstoxin-I dimer, *Biochem. Biophys. Res. Commun.* **284**, 1011-1015 (2001).

83. A. Di Venere, G. Mei, G. Gilardi, N. Rosato, F. De Matteis, R. McKay, E. Gratton, and A. Finazzi Agrò, Resolution of the heterogeneous fluorescence in multi-tryptophan proteins: ascorbate oxidase, *Eur. J. Biochem.* **257**, 337-343 (1998).

84. M. Chabbert, E. Piémont, F. G. Prendergast, and H. Lami, Fluorescence of a tryptophan bearing peptide from smooth muscle myosin light chain kinase upon binding to two closely related calmodulins, *Arch. Biochem. Biophys.* **322**, 429-436 (1995).

85. Y.-C. Chang, and R. D. Ludescher, Tryptophan photophysics in rabbit skeletal myosin rod, *Biophys. Chem.* **49**, 113-126 (1994).

86. D. M. Gakamsky, E. Haas, P. Robbins, J. L. Strominger, and I. Pecht, Selective steady-state and time-resolved fluorescence spectroscopy of an HLA-A2-peptide complex, *Immunol. Lett.* **44**, 195-201 (1995).

87. S. M. Raja, S. S. Rawat, A. Chattopadhyay, and A. K. Lala, Localization and environment of tryptophans in soluble and membrane-bound states of a pore-forming toxin from *Staphylococcus aureus*, *Biophys. J.* **76**, 1469-1479 (1999).

88. A. Chattopadhyay, S. S. Rawat, D. A. Kelkar, S. Ray, and A. Chakrabarti, Organization and dynamics of tryptophan residues in erythroid spectrin: novel structural features of denatured spectrin revealed by the wavelength-selective fluorescence approach, *Protein Sci.* **12**, 2389-2403 (2003).

89. D. A. Kelkar, A. Chattopadhyay, A. Chakrabarti, and M. Bhattacharyya, Effect of ionic strength on the organization and dynamics of tryptophan residues in erythroid spectrin: a fluorescence approach, (submitted for publication).

90. S. Ercelen, D. Kazan, A. Erarslan, and A. P. Demchenko, On the excited-state energy transfer between tryptophan residues in proteins: the case of penicillin acylase, *Biophys. Chem.* **90**, 203-217 (2001).

91. C. M. Rao, S. C. Rao, and P. B. Rao, Red edge excitation effect in intact eye lens, *Photochem. Photobiol.* **50**, 399-402 (1989).

92. S. C. Rao, and C. M. Rao, Red edge excitation shifts of crystallins and intact lenses. A study of segmental mobility and inter-protein interactions, *FEBS Lett.* **337**, 269-273 (1994).

93. L. Uma, Y. Sharma, and D. Balasubramanian, Conformation, stability and interactions of corneal keratan sulfate proteoglycan, *Biochim. Biophys. Acta* **1294**, 8-14 (1996).

94. L. Uma, D. Balasubramanian, and Y. Sharma, *In situ* fluorescence spectroscopic studies on bovine cornea, *Photochem. Photobiol.* **59**, 557-561 (1994).
95. J. Klein-Seetharaman, M. Oikawa, S. B. Grimshaw, J. Wirmer, E. Duchardt, T. Ueda, T. Imoto, L. J. Smith, C. M. Dobson, and H. Schwalbe, Long-range interactions within a nonnative protein, *Science* **295**, 1719-1722 (2002).
96. D. Neri, M. Billeter, G. Wider, and K. Wüthrich, NMR determination of residual structure in a urea-denatured protein, the 434-repressor, *Science* **257**, 1559-1563 (1992).
97. N. S. Bhavesh, S. C. Panchal, and R. V. Hosur, An efficient high-throughput resonance assignment procedure for structural genomics and protein folding research by NMR, *Biochemistry* **40**, 14727-14735 (2001).
98. R. A. Cone, Rotational diffusion of rhodopsin in the visual receptor membrane, *Nat. New Biol.* **236**, 39-43 (1972).
99. M. M. Poo, and R. A. Cone, Lateral diffusion of rhodopsin in the photoreceptor membrane, *Nature* **247**, 438-441 (1974).
100. M. Edidin, Lipids on the frontier: a century of cell-membrane bilayers, *Nat. Rev. Mol. Cell Biol.* **4**, 414-418 (2003).
101. J. Seelig, Deuterium magnetic resonance: theory and application to lipid membranes, *Quart. Rev. Biophys.* **10**, 353-418 (1977).
102. R. G. Ashcroft, H. G. L. Coster, and J. R. Smith, The molecular organisation of bimolecular lipid membranes: The dielectric structure of the hydrophilic/hydrophobic interface, *Biochim. Biophys. Acta* **643**, 191-204 (1981).
103. E. Perochon, A. Lopez, and J. F. Tocanne, Polarity of lipid bilayers: a fluorescence investigation, *Biochemistry* **31**, 7672-7282 (1992).
104. R. M. Venable, Y. Zhang, B. J. Hardy, and R. W. Pastor, Molecular dynamics simulations of a lipid bilayer and of hexadecane: an investigation of membrane fluidity, *Science* **262**, 223-226 (1993).
105. A. Chattopadhyay, and S. Mukherjee, Depth-dependent solvent relaxation in membranes: wavelength-selective fluorescence as a membrane dipstick, *Langmuir* **15**, 2142-2148 (1999).
106. S. H. White, and W. C. Wimley, Peptides in lipid bilayers: structural and thermodynamic basis for partitioning and folding. *Curr. Opin. Struct. Biol.* **4**, 79-86 (1994).
107. M. C. Wiener, and S. H. White, Structure of a fluid dioleoylphosphatidylcholine bilayer determined by joint refinement of x-ray and neutron diffraction data. III. Complete structure, *Biophys. J.* **61**, 434-447 (1992).
108. P. L. Yeagle, in: *The Membranes of Cells* (Academic Press, Orlando, Florida, 1987), pp. 89-91.
109. J. M. Boggs, Lipid intermolecular hydrogen bonding: influence on structural organization and membrane function, *Biochim. Biophys. Acta* **906**, 353-404 (1987).
110. R. B. Gennis, in: *Biomembranes: Molecular Structure and Function* (Springer-Verlag, New York, 1989), pp. 47-48.
111. T. B. Shin, R. Leventis, and J. R. Silvius, Partitioning of fluorescent phospholipid probes between different bilayer environments. Estimation of the free energy of interlipid hydrogen bonding, *Biochemistry* **30**, 7491-7497 (1991).
112. J. Sykora, P. Kapusta, V. Fidler, and M. Hof, On what time scale does solvent relaxation in phospholipid bilayers happen?, *Langmuir* **18**, 571-574 (2002).
113. A. Chattopadhyay, and S. Mukherjee, Fluorophore environments in membrane-bound probes: a red edge excitation shift study, *Biochemistry* **32**, 3804-3811 (1993).
114. A. Chattopadhyay, and S. Mukherjee, Red edge excitation shift of a deeply embedded membrane probe: implications in water penetration in the bilayer, *J. Phys. Chem. B* **103**, 8180-8185 (1999).
115. M. Hof, Solvent relaxation in biomembranes, in: *Applied Fluorescence in Chemistry, Biology and Medicine*, edited by W. Rettig, B. Strehmel, S. Schrader, and H. Seifert (Springer-Verlag, Heidelberg, 1999) pp. 439-456.
116. M. Schiffer, C. H. Chang, and F. J. Stevens, The functions of tryptophan residues in membrane proteins, *Protein Eng.* **5**, 213-214 (1992).
117. J. A. Ippolito, R. S. Alexander, D. W. Christianson, Hydrogen bond stereochemistry in protein structure and function, *J. Mol. Biol.* **215**, 457-471 (1990).
118. D. A. Doyle, J. M. Cabral, R. A. Pfuetzner, A. Kuo, J. M. Gulbis, S. L. Cohen, B. T. Chait, and R. MacKinnon, The structure of the potassium channel: molecular basis of K+ conduction and selectivity, *Science* **280**, 69-77 (1998).
119. H. Luecke, B. Schobert, H.-T. Richter, J.-P. Cartailler, and J. K. Lanyi, Structural changes in bacteriorhodopsin during ion transport at 2 angstrom resolution, *Science* **286**, 255-261 (1999).
120. T. Schirmer, T. A. Keller, Y. F. Wang, and J. P. Rosenbusch, Structural basis for sugar translocation through maltoporin channels at 3.1 Å resolution, *Science* **267**, 512-514 (1995).

121. R. A. F. Reithmeier, Characterization and modeling of membrane proteins using sequence analysis, *Curr. Opin. Struct. Biol.* **5**, 491-500 (1995).

122. C. Landolt-Marticorena, K. A. Williams, C. M. Deber, and R. A. F. Reithmeier, Non-random distribution of amino acids in the transmembrane segments of human type I single span membrane proteins, *J. Mol. Biol.* **229**, 602-608 (1993).

123. M. B. Ulmschneider, and M. S. P. Sansom, Amino acid distributions in integral membrane protein structures, *Biochim. Biophys. Acta* **1512**, 1-14 (2001).

124. M. R. R. de Planque, D. V. Greathouse, R. E. Koeppe, H. Schafer, D. Marsh, and J. A. Killian, Influence of lipid/peptide hydrophobic mismatch on the thickness of diacylphosphatidylcholine bilayers. A ^{2}H NMR and ESR study using designed transmembrane alpha-helical peptides and gramicidin A, *Biochemistry* **37**, 9333-9345 (1998).

125. J. A. Demmers, E. van Duijn, J. Haverkamp, D. V. Greathouse, R. E. Koeppe, A. J. R. Heck, and J. A. Killian, Interfacial positioning and stability of transmembrane peptides in lipid bilayers studied by combining hydrogen/deuterium exchange and mass spectrometry, *J. Biol. Chem.* **276**, 34501-34508 (2001).

126. W.-M. Yau, W. C. Wimley, K. Gawrisch, and S. H. White, The preference of tryptophan for membrane interfaces, *Biochemistry* **37**, 14713-14718 (1998).

127. S. Persson, J. A. Killian, and G. Lindblom, Molecular ordering of interfacially localized tryptophan analogs in ester- and ether-lipid bilayers studied by ^{2}H-NMR, *Biophys. J.* **75**, 1365-1371 (1998).

128. R. E. Jacobs, and S. H. White, The nature of the hydrophobic binding of small peptides at the bilayer interface: implications for the insertion of transbilayer helices, *Biochemistry* **28**, 3421-3437 (1989).

129. J. W. Brown, and W. H. Huestis, Structure and orientation of a bilayer-bound model tripeptide: a proton NMR study, *J. Phys. Chem.* **97**, 2967-2973 (1993).

130. C. Chothia, Structural invariants in protein folding, *Nature* **254**, 304-308 (1975).

131. H. I. Petrache, S. E. Feller, and J. F. Nagle, Determination of component volumes of lipid bilayers from simulations, *Biophys. J.* **72**, 2237-2242 (1997).

132. O. S. Andersen, D. V. Greathouse, L. L. Providence, M. D. Becker, and R. E. Koeppe, Importance of tryptophan dipoles for protein function: 5-fluorination of tryptophans in gramicidin A channels, *J. Am. Chem. Soc.* **120**, 5142-5146 (1998).

133. J. A. Lundbaek, A. M. Maer, and O. S. Andersen, Lipid bilayer electrostatic energy, curvature stress, and assembly of gramicidin channels, *Biochemistry* **36**, 5695-5701 (1997).

134. R. J. Webb, J. M. East, R. P. Sharma, and A. G. Lee, Hydrophobic mismatch and the incorporation of peptides into lipid bilayers: a possible mechanism for retention in the Golgi, *Biochemistry* **37**, 673-679 (1998).

135. M. R. R. de Planque, J. A. W. Kruijtzer, R. M. J. Liskamp, D. Marsh, D. V. Greathouse, R. E. Koeppe, B. de Kruijff, and J. A. Killian, Different membrane anchoring positions of tryptophan and lysine in synthetic transmembrane alpha-helical peptides, *J. Biol. Chem.* **274**, 20839-20846 (1999).

136. J. Ren, S. Lew, J. Wang, and E. London, Control of the transmembrane orientation and interhelical interactions within membranes by hydrophobic helix length, *Biochemistry* **38**, 5905-5912 (1999).

137. K. Hristova, C. E. Dempsey, and S. H. White, Structure, location, and lipid perturbations of melittin at the membrane interface, *Biophys. J.* **80**, 801-811 (2001).

138. A. N. Ridder, W. van de Hoef, J. Stam, A. Kuhn, B. de Kruijff, and J. A. Killian, Importance of hydrophobic matching for spontaneous insertion of a single-spanning membrane protein, *Biochemistry* **41**, 4946-4952 (2002).

139. R. F. M. de Almeida, L. M. S. Loura, M. Prieto, A. Watts, A. Fedorov, and F. J. Barrantes, Cholesterol modulates the organization of the γM4 transmembrane domain of the muscle nicotinic acetylcholine receptor, *Biophys. J.* **86**, 2261-2272 (2004).

140. A. S. Ladokhin, and P. W. Holloway, Fluorescence of membrane-bound tryptophan octyl ester: a model for studying intrinsic fluorescence of protein-membrane interactions, *Biophys. J.* **69**, 506-517 (1995).

141. B. de Foresta, J. Gallay, J. Sopkova, P. Champeil, and M. Vincent, Tryptophan octyl ester in detergent micelles of dodecylmaltoside: fluorescence properties and quenching by brominated detergent analogs, *Biophys. J.* **77**, 3071-3084 (1999).

142. B. Sengupta, and P. K. Sengupta, Influence of reverse micellar environments on the fluorescence emission properties of tryptophan octyl ester, *Biochem. Biophys. Res. Commun.* **277**, 13-19 (2000).

143. C. F. Dempsey, The actions of melittin on membranes, *Biochim. Biophys. Acta* **1031**, 143-161 (1990).

144. Y. Shai, Molecular recognition between membrane-spanning polypeptides, *Trends Biochem. Sci.* **20**, 460-464 (1995).

145. G. Saberwal, and R. Nagaraj, Cell-lytic and antibacterial peptides that act by perturbing the barrier function of membranes: facets of their conformational features, structure-function correlation and membrane-perturbing abilities, *Biochim. Biophys. Acta* **1197**, 109-131 (1994).

146. A. Chattopadhyay, and R. Rukmini, 1993, Restricted mobility of the sole tryptophan in membrane-bound melittin, *FEBS Lett.* **335**, 341-344 (1993).
147. A. K. Ghosh, R. Rukmini, and A. Chattopadhyay, Modulation of tryptophan environment in membrane-bound melittin by negatively charged phospholipids: implications in membrane organization and function, *Biochemistry* **36**, 14291-14305 (1997).
148. H. Raghuraman, and A. Chattopadhyay, Effect of micellar charge on the conformation and dynamics of melittin, *Eur. Biophys. J.* (2004), Apr 8, 2004 [Epub ahead of print], PMID: 15071759.
149. H. Raghuraman, and A. Chattopadhyay, Interaction of melittin with membrane cholesterol: a fluorescence approach, *Biophys. J.* (2004), in press.
150. A. Guz, and Z. Wasylewski, Red-edge excitation fluorescence spectroscopy of proteins in reversed micelles, *J. Protein Chem.* **13**, 393-399 (1994).
151. H. Raghuraman, and A. Chattopadhyay, Organization and dynamics of melittin in environments of graded hydration: a fluorescence approach, *Langmuir* **19**, 10332-10341 (2003).
152. H. Raghuraman, and A. Chattopadhyay, Influence of lipid chain unsaturation on membrane-bound melittin: a fluorescence approach, *Biochim. Biophys. Acta* (2004), in press.
153. S. Mukherjee, and A. Chattopadhyay, Motionally restricted tryptophan environments at the peptide-lipid interface of gramicidin channels, *Biochemistry* **33**, 5089-5097 (1994).
154. M. C. Tory, and A. R. Merrill, Determination of membrane protein topology by red-edge excitation shift analysis: application to the membrane-bound colicin E1 channel peptide, *Biochim. Biophys. Acta* **1564**, 435-448 (2002).
155. A. S. Ladokhin, L. Wang, A. W. Steggles, and P. W. Holloway, Fluorescence study of a mutant cytochrome b5 with a single tryptophan in the membrane-binding domain, *Biochemistry* **30**, 10200-10206 (1991).
156. S. S. Rawat, D. A. Kelkar, and A. Chattopadhyay, Monitoring gramicidin conformations in membranes: a fluorescence approach, *Biophys. J.* (2004), in press.
157. D. J. Schibli, R. F. Epand, H. J. Vogel, and R. M. Epand, Tryptophan-rich antimicrobial peptides: comparative properties and membrane interactions, *Biochem. Cell. Biol.* **80**, 667-677 (2002).
158. R. E. Koeppe, and O. S. Andersen, Engineering the gramicidin channel, *Annu. Rev. Biophys. Biomol. Struct.* **25**, 231-258 (1996).
159. F. Heitz, G. Spach, and Y. Trudelle, Single channels of 9, 11, 13, 15-destryptophyl-phenylalanyl-gramicidin A, *Biophys. J.* **39**, 87-89 (1982).
160. D. Jones, E. Hayon, and D. Busath, Tryptophan photophysics is responsible for gramicidin-channel inactivation by ultraviolet light, *Biochim. Biophys. Acta* **861**, 62-66 (1986).
161. M. Strässle, G. Stark, M. Wilhelm, P. Daumas, F. Heitz, and R. Lazaro, Radiolysis and photolysis of ion channels formed by analogues of gramicidin A with a varying number of tryptophan residues, *Biochim. Biophys. Acta* **980**, 305-314 (1986).
162. K. U. Prasad, T. L. Trapane, D. Busath, G. Szabo, and D. W. Urry, Synthesis and charaterization of (1-13C) Phe[9] gramicidin A. Effects of side chain variation, *Int. J. Pept. Protein Res.* **22**, 341-347 (1983).
163. P. Daumas, F. Heitz, L. Ranjalahy-Rasoloarijao, and R. Lazaro, Gramicidin A analogs: influence of the substitution of the tryptophan by naphthylalanines, *Biochimie*, **71**, 77-81(1989).
164. S. S. Rawat, D. A. Kelkar, and A. Chattopadhyay (unpublished observations).
165. D. A. Kelkar, and A. Chattopadhyay, Effects of varying hydration on the organization and dynamics of an ion channel: a fluorescence approach (manuscript in preparation).
166. S. S. Rawat, and A. Chattopadhyay, Ion channel modulation by sterols: inactivation of gramicidin channels by cholesterol domains, *Prog. Biophys. Mol. Biol.* **65** (Suppl 1), 110 (1996).
167. D. A. Kelkar, and A. Chattopadhyay (unpublished observations).
168. N. C. Santos, M. Prieto, and M. A. R. B. Castanho, Interaction of the major epitope region of HIV protein gp41 with membrane model systems. A fluorescence spectroscopy study, *Biochemistry* **37**, 8764-8775 (1998).
169. L. A. Falls, B. C. Furie, M. Jacobs, B. Furie, and A. C. Rigby, The ω-loop region of the human prothrombin γ-carboxyglutamic acid domain penetrates anionic phospholipid membranes, *J. Biol. Chem.* **276**, 23895-23902 (2001).
170. S. Veeraraghavan, J. D. Baleja, and G. E. Gilbert, Structure and topography of the membrane-binding C2 domain of factor VIII in the presence of dodecylphosphocholine micelles, *Biochem. J.* **332**, 549-555 (1998).
171. R. F. M. de Almeida, L. M. S. Loura, F. J. Barrantes, and M. Prieto, Interaction of the γM4 transmembrane segment of the acetylcholine receptor with cholesterol-rich phases, *Biophys. J.* **80**, 545a (2001).

172. F. J. Barrantes, S. A. Antonilli, M. P. Blanton, and M. Prieto, Topography of nicotinic acetylcholine receptor membrane-embedded domains, *J. Biol. Chem.* **275**, 37333-37339 (2000).
173. B. R. Pattnaik, S. Ghosh, and M. R. Rajeswari, Selective excitation of tryptophans in OmpF: a fluorescence emission study, *Biochem. Mol. Biol. Int.* **42**, 173-181 (1997).
174. T. Granjon, M.-J. Vacheron, C. Vial, and R. Buchet, Mitochondrial creatine kinase binding to phospholipids decreases fluidity of membranes and promotes new lipid-induced β structures as monitored by red edge excitation shift, laurdan fluorescence, and FTIR, *Biochemistry* **40**, 6016-6026 (2001).
175. K. Fritz-Wolf, T. Schnyder, T. Wallimann, and W. Kabsch, Structure of mitochondrial creatine kinase, *Nature* **381**, 341-345 (1996).
176. J. R. Lakowicz, and S. Keating-Nakamoto, Red-edge excitation of fluorescence and dynamic properties of proteins and membranes, *Biochemistry* **23**, 3013-3021 (1984).
177. A. P. Demchenko, Fluorescence molecular relaxation studies of protein dynamics. The probe binding site of melittin is rigid on the nanosecond time scale, *FEBS Lett.* **182**, 99-102 (1985).
178. J. R. Albani, Motions studies of the human α1-acid glycoprotein (orosomucoid) followed by red-edge excitation spectra and polarization of 2-*p*-toluidinylnaphthalene-6-sulfonate (TNS) and of tryptophan residues, *Biophys. Chem.* **44**, 129-137 (1992).
179. G. Weber, and F. J. Farris, Synthesis and spectral properties of a hydrophobic fluorescent probe: 6-propionyl-2-(dimethylamino)naphthalene, *Biochemistry* **18**, 3075-3078 (1979).
180. A. Balter, W. Nowak, W. Pawelkiewicz, and A. Kowalczyk, Some remarks on the interpretation of the spectral properties of prodan, *Chem. Phys. Lett.* **143**, 565-570 (1988).
181. A. Samanta, and R. W. Fessenden, Excited state dipole moment of PRODAN as determined from transient dielectric loss measurements, *J. Phys. Chem. A* **104**, 8972-8975 (2000).
182. A. Chakrabarti, Fluorescence of spectrin-bound prodan. *Biochem. Biophys. Res. Commun.* **226**, 495-497 (1996).
183. J. B. Massey, H. S. She, and H. J. Pownall, Interfacial properties of model membranes and plasma lipoproteins containing ether lipids, *Biochemistry* **24**, 6973-6978 (1985).
184. R. B. Macgregor, and G. Weber, Estimation of the polarity of the protein interior by optical spectroscopy, *Nature* **319**, 70-73 (1986).
185. P. I. H. Bastiaens, A. van Hoek, W. J. H. van Berkel, A. de Kok, and A. J. W. G. Visser, Molecular relaxation spectroscopy of flavin adenine dinucleotide in wild type and mutant lipoamide dehydrogenase from *Azotobacter vinelandii*, *Biochemistry* **31**, 7061-7068 (1992).
186. A. Chattopadhyay, Chemistry and biology of *N*-(7-nitrobenz-2-oxa-1,3,-diazol-4-yl)-labeled lipids: fluorescent probes of biological and model membranes, *Chem. Phys. Lipids* **53**, 1-15 (1990).
187. A. Chattopadhyay, and E. London, Spectroscopic and ionization properties of *N*-(7-nitrobenz-2-oxa-1,3-diazol-4-yl)-labeled lipids in model membranes, *Biochim. Biophys. Acta* **938**, 24-34 (1988).
188. S. Lin, and W. S. Struve, Time-resolved fluorescence of nitrobenzoxadiazole-aminohexanoic acid: effect of intermolecular hydrogen-bonding on non-radiative decay, *Photochem. Photobiol.* **54**, 361-365 (1991).
189. S. Mukherjee, A. Chattopadhyay, A. Samanta, and T. Soujanya, Dipole moment change of NBD group upon excitation studied using solvatochromic and quantum chemical approaches: implications in membrane research, *J. Phys. Chem.* **98**, 2809-2892 (1994).
190. S. Fery-Forgues, J.-P. Fayet, and A. Lopez, Drastic changes in the fluorescence properties of NBD probes with the polarity of the medium: involvement of a TICT state?, *J. Photochem. Photobiol. A* **70**, 229-243 (1993).
191. S. S. Rawat, S. Mukherjee, and A. Chattopadhyay, Micellar organization and dynamics: a wavelength-selective fluorescence approach, *J. Phys. Chem. B* **101**, 1922-1929 (1997).
192. S. S. Rawat, and A. Chattopadhyay, Structural transitions in the micellar assembly: a fluorescence study, *J. Fluoresc.* **9**, 233-244 (1999).
193. A. Chattopadhyay, S. Mukherjee, and H. Raghuraman, Reverse micellar organization and dynamics: a wavelength-selective fluorescence approach, *J. Phys. Chem. B* **106**, 13002-13009 (2002).
194. S. Mukherjee, H. Raghuraman, S. Dasgupta, and A. Chattopadhyay, Organization and dynamics of *N*-(7-nitrobenz-2-oxa-1,3,-diazol-4-yl)-labeled lipids: a fluorescence approach, *Chem. Phys. Lipids* **127**, 91-101 (2004).
195. H. Raghuraman, S. K. Pradhan, and A. Chattopadhyay, Effect of urea on the organization and dynamics of Triton X-100 micelles: a fluorescence approach, *J. Phys. Chem. B* **108**, 2489-2496 (2004).
196. D. A. Kelkar, and A. Chattopadhyay, Depth-dependent solvent relaxation in reverse micelles: a fluorescence approach, *J. Phys. Chem. B* (2004), in press.
197. V. Tsukanova, D. W. Grainger, and C. Salesse, Monolayer behavior of NBD-labeled phospholipids at the air/water interface, *Langmuir* **18**, 5539-5550 (2002).

198. L. R. McLean, and R. L. Jackson, Interaction of lipoprotein lipase and apolipoprotein C-II with sonicated vesicles of 1,2-ditetradecylphosphatidylcholine: comparison of binding constants, *Biochemistry* **24**, 4196-4201 (1985).

199. C. E. MacPhee, G. J. Howlett, W. H. Sawyer, and A. H. A. Clayton, Helix-helix association of a lipid-bound amphipathic α-helix derived from apolipoprotein C-II, *Biochemistry* **38**, 10878-10884 (1999).

200. A. J. Heuck, and A. E. Johnson, Pore-forming protein structure analysis in membranes using multiple independent fluorescence techniques, *Cell Biochem. Biophys.* **36**, 89-101 (2002).

201. A. H. A. Clayton, and W. H. Sawyer, Site-specific tryptophan fluorescence spectroscopy as a probe of membrane peptide structure and dynamics, *Eur. Biophys. J.* **31**, 9-13 (2002).

202. V. Gopal, H.-W. Ma, M. K. Kumaran, and D. Chatterji, A point mutation at the junction of domain 2.3/2.4 of transcription factor sigma 70 abrogates productive transcription and restores its expected mobility on a denaturing gel, *J. Mol. Biol.* **242**, 9-22 (1994).

203. M. E. Menezes, P. D. Roepe, and H. R. Kaback, Design of a membrane transport protein for fluorescence spectroscopy, *Proc. Natl. Acad. Sci. U.S.A.* **87**, 1638-1642 (1990).

204. A. Chattopadhyay, and M. G. McNamee, Average membrane penetration depth of tryptophan residues of the nicotinic acetylcholine receptor by the parallax method, *Biochemistry* **30**, 7159-7164 (1991).

205. V. W. Cornish, D. R. Benson, C. A. Altenbach, K. Hideg, W. L. Hubbell, and P. G. Schultz, Site-specific incorporation of biophysical probes into proteins, *Proc. Natl. Acad. Sci. U.S.A.* **91**, 2910-2914 (1994).

206. M. W. Nowak, P. C. Kearney, J. R. Sampson, M. E. Saks, C. G. Labarca, S. K. Silverman, W. Zhong, J. Thorson, J. N. Abelson, N. Davidson, P. G. Schultz, D. A. Dougherty, and H. A. Lester, Nicotinic receptor binding site probed with unnatural amino acid incorporation in intact cells, *Science* **268**, 439-442 (1995).

207. B. D. Cohen, T. B. McAnaney, E. S. Park, Y. N. Jan, S. G. Boxer, and L. Y. Jan, Probing protein electrostatics with a synthetic fluorescent amino acid, *Science* **296**, 1700-1703 (2002).

208. J. B. A. Ross, A. G. Szabo, and C. W. V. Hogue, Enhancement of protein spectra with tryptophan analogs: fluorescence spectroscopy of protein-protein and protein-nucleic acid interactions, *Methods Enzymol.* **278**, 151-190 (1997).

209. J. Guharay, and P. K. Sengupta, Characterization of the fluorescence emission properties of 7-azatryptophan in reverse micellar environments, *Biochem. Biophys. Res. Commun.* **219**, 388-392 (1996).

210. C.-Y. Wong, and M. R. Eftink, Biosynthetic incorporation of tryptophan analogues into staphylococcal nuclease: effect of 5-hydroxytryptophan and 7-azatryptophan on structure and stability, *Protein Sci.* **6**, 689-697 (1997).

211. D. Mendel, J. A. Ellman, Z. Chang, D. L. Veenstra, P. A. Kollman, and P. G. Schultz, Probing protein stability with unnatural amino acids, *Science* **256**, 1798-1802 (1992).

212. J. Broos, F. ter Veld, and G. T. Robillard, Membrane protein-ligand interactions in *Escherichia coli* vesicles and living cells monitored via a biosynthetically incorporated tryptophan analogue, *Biochemistry* **38**, 9798-9803 (1999).

213. B. Sengupta, J. Guharay, A. Chakraborty, and P. K. Sengupta, Low temperature luminescence behaviours of 7-azatryptophan, 5-hydroxytryptophan and their chromophoric moieties, *Spectrochim. Acta A Mol. Biomol. Spectrosc.* **58**, 2005-2012 (2002).

214. J. H. Crowe, and L. M. Crowe, Effect of dehydration on membranes and membrane stabilization at low water activities, *Biol. Membr.* **5**, 57-103 (1984).

215. R. P. Rand, and V. A. Parsegian, Hydration forces between phospholipid bilayers, *Biochim. Biophys. Acta* **988**, 351-376 (1989).

216. C. Ho, M. B. Kelly, and C. D. Stubbs, The effect of phospholipid unsaturation and alcohol perturbation at the protein/lipid interface probed using fluorophore lifetime heterogeneity, *Biochim. Biophys. Acta* **1193**, 307-315 (1994).

217. W. B. Fischer, S. Sonar, T. Marti, H. G. Khorana, and K. J. Rothschild, Detection of a water molecule in the active-site of bacteriorhodopsin: hydrogen bonding changes during the primary photoreaction, *Biochemistry* **33**, 12757-12762 (1994).

218. H. Kandori, Y. Yamazaki, J. Sasaki, R. Needleman, J. K. Lanyi, and A. Maeda, Water-mediated proton transfer in proteins: an FTIR study of bacteriorhodopsin, *J. Am. Chem. Soc.* **117**, 2118-2119 (1995).

RNA FOLDING AND RNA-PROTEIN BINDING ANALYZED BY FLUORESCENCE ANISOTROPY AND RESONANCE ENERGY TRANSFER

Gerald M. Wilson*

10.1. INTRODUCTION

Ribonucleic acids (RNA) perform a host of functions in all living things. The vast majority of these roles are associated with diverse aspects of gene expression, including but not limited to delivery of genetic information between the genome (usually DNA) and protein synthetic machinery, coordination of processing events on nascent RNA molecules, and delivery of amino acids to translating ribosomes, which themselves are largely composed of RNA. More recent concepts in RNA metabolism include *cis*-regulation of gene expression by riboswitches[1] and *trans*-regulation through small, non-coding RNAs[2]. Over the past two decades, a large body of work has described catalytic functions for many RNA molecules, from small structured RNAs capable of self-cleavage to the peptidyltransfer functions of ribosomes[3, 4]. Finally, the heterogeneous nature of RNA structure and its amenability to reiterative selection procedures has prompted significant biotechnological interest in RNA aptamers. These are generally short RNA sequences that exhibit high affinity and selectivity for specific molecular targets, and show considerable promise as tools for diagnostic sensing and modulation of biomolecular function[5, 6].

A unique feature of RNA that lends itself to functional complexity in biological systems is that RNA molecules, while encoded by double-stranded DNA templates, are themselves synthesized without a complimentary strand. The plethora of hydrogen bond potential presented by single-stranded RNA molecules thus provides myriad possibilities for conformational variation through intra- and inter-molecular hybridization[7]. Furthermore, many biological functions of RNA are mediated through interactions with sequence-specific RNA-binding proteins. In this review, I describe steady-state fluorescence-based strategies for quantitative analyses of RNA folding and RNA-protein binding events, along with the merits and limitations of these techniques relative to alternative

*Gerald M. Wilson, Department of Biochemistry and Molecular Biology and Center for Fluorescence Spectroscopy, University of Maryland School of Medicine, 108 N. Greene St., Baltimore, MD 21201; Tel: (410)706-8904; Fax: (410)706-8297; e-mail: gwils001@umaryland.edu.

methods. Finally, I summarize some recent progress in the study of mechanisms direct-
ing cytoplasmic mRNA decay that has been facilitated by fluorescence approaches, and
briefly consider some new frontiers that will permit analyses of RNA structure and function at previously unattainable levels of complexity and subtlety.

10.2. METHODOLOGY

10.2.1. Site-specific Labeling of RNA Substrates with Fluorophores

In general, assessment of RNA folding or protein-binding events by fluorescence
spectroscopy requires the conjugation of one or more fluorescent dyes to the substrate
RNA molecule. For many cases, this task has been simplified by recent improvements in
the quality and cost-effectiveness of solid-phase polyribonucleotide synthesis, and the
concomitant proliferation of commercial sources offering such services. We have obtained custom RNA substrates up to 80 nucleotides in length from Dharmacon Research
(Lafayette, CO), and selected shorter oligoribonucleotides from Integrated DNA Technologies (Coralville, IA) and others. Both of the abovementioned suppliers offer a number of options for 5'-fluorophore conjugation to synthetic RNA substrates, including fluorescein (Fl), cyanine 3 (Cy3), cyanine 5 (Cy5), and rhodamine (TAMRA). Dharmacon

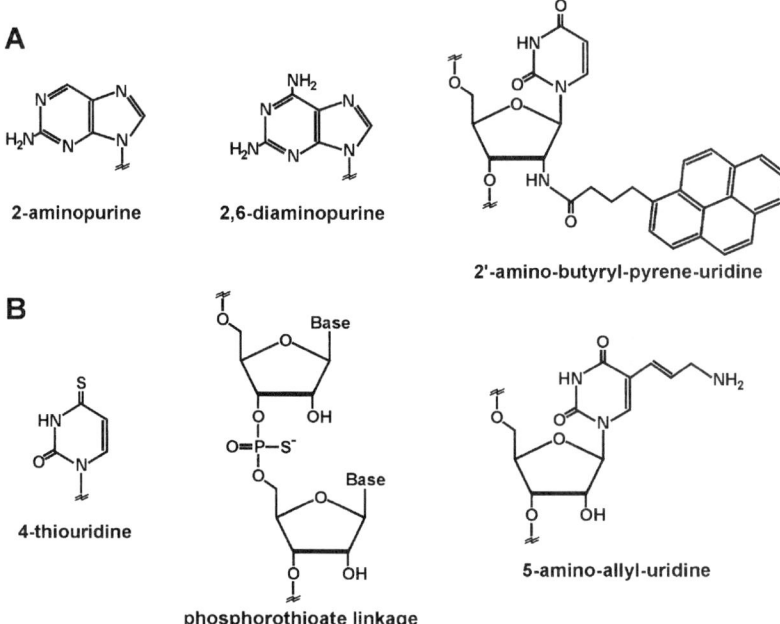

Figure 10.1. (A) Selected fluorescent bases and nucleotides available for incorporation by solid-phase RNA
synthesis, and (B) bases and nucleotides permitting conjugation of extrinsic fluorophores at internal sites within
synthetic oligoribonucleotides.

further offers each of these fluorophores as 3'-conjugates. At present, incorporation of fluorescent dyes at internal sites on RNA substrates is largely limited to the ultraviolet-excited probes 2-aminopurine and 2,6-diaminopurine (Figure 10.1A). Recently, Dharmacon also began offering site-specific incorporation of a uridine analogue containing a 2'-amide-linked pyrene.

Incorporation of different fluorophores at either terminus, internally, or in the context of larger (> 80 nucleotides) RNA substrates generally require more elaborate synthetic strategies. For oligonucleotides up to 80 nucleotides, incorporation of 5'- or 3'-terminal amino or thiol groups during solid-phase synthesis readily permits subsequent linkage to fluorophores in the form of N-hydroxysuccinymidyl (NHS) esters (for amino groups) or maleimide/iodoacetimide conjugates (for thiol groups). A host of suitable fluorescent dyes with excitation/emission wavelengths spanning both the ultraviolet and visible spectra are available from Molecular Probes (Eugene, OR). Similarly, inclusion of functional groups at specific internal locations within custom-synthesized RNA substrates provides additional opportunities for conjugation of extrinsic fluorophores. For example, 4-thio-uridine has been used for base-specific linkage of thiol-reactive probes[8]. Substituting a phosphorothioate group for the generic phosphodiester linkage at a specific point in the RNA polymer may permit similar fluorescent dyes to be targeted to the RNA backbone. Dharmacon also offers to site-specifically incorporate 5-amino-allyl-uridine for conjugation of NHS ester-linked fluorophores (Figure 10.1B).

We prepare fluorescent-labeled RNA substrates larger than 80 nucleotides by one of two principal strategies. The first applies when a single fluorescent dye is required at either the 5'- or 3'-terminus. In this case, the applicable RNA is first synthesized by *in vitro* transcription from a double-stranded DNA template. 5'-labeling is performed by removal of the 5'-triphosphate with alkaline phosphatase, then incorporation of a thiophosphate group on the 5'-OH using adenosine 5'-[γ-thio]triphosphate (ATPγS) and T4 polynucleotide kinase[9]. A fluorophore is then linked to the thiophosphate group as a maleimide conjugate. In our hands, this method typically achieves 20-30% labeling efficiency. Attachment of fluorescent dyes to RNA 3'-termini is achieved by oxidation of

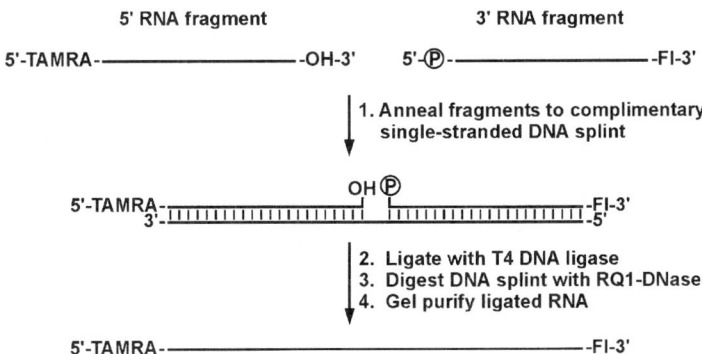

Figure 10.2. Tandem linkage of RNA fragments using single-stranded DNA splints to generate an extended, double-labeled RNA substrate.

the 2',3'-vicinal diol with periodate, then reaction of the resulting dialdehyde with hydra-zine-coupled fluorophores. This method achieves labeling efficiencies approaching 100%, and has been reviewed at length elsewhere[10]. The second strategy applies to extended substrates requiring internal or multiple fluorescent dyes linked to a common RNA molecule. In these cases, RNA substrates are generally synthesized in two or three segments, involving solid-phase synthesis and/or *in vitro* transcription. Following incorporation of appropriate fluorophores or functional groups into each segment using the strategies described above, the fragments are assembled by DNA splint-directed ligation and purified (Figure 10.2). A variant of this technique that permits selective incorporation of 4-thiouridine at junction sites has also been described[11].

10.2.2. Assessment of RNA-protein Interactions by Fluorescence Anisotropy

10.2.2.1. Comparison with Alternative Strategies

Many biological functions of RNA, particularly those pertaining to the regulation or utilization of genetic information, involve interactions with cellular proteins[12-15]. Discrimination of binding mechanisms and elucidation of cognate thermodynamic and kinetic parameters require quantitative assay systems to monitor RNA:protein binding events. While we and others have employed gel mobility shift assays (GMSAs) for this purpose[16-18], we find that complex dissociation during electrophoresis often leads to underestimation of solution binding affinity. This limitation is particularly significant for highly dynamic binding events[19]. Similarly, nitrocellulose filter-binding strategies to separate protein-bound from -unbound RNA substrates require washing steps to minimize non-specific RNA retention[20], during which the equilibrium is modified by lowering the effective concentration of each reagent. In our hands, filter-binding assays also show a high degree of retention for some RNA substrates, even in the absence of protein. Finally, both GMSAs and filter-binding assays suffer from the limitation that reaction components are not measured in free, aqueous solution; in GMSAs, the products are resolved by prolonged passage through a semi-solid (usually polyacrylamide) matrix, while filter-binding assays retain products in the solid phase. An additional method for monitoring RNA-protein equilibria that has recently gained prominence is surface plasmon resonance (SPR), also referred to as BIACORE analysis[21]. However, use of this assay is complicated by three principal limitations. First, one reagent must be affixed to the assay tube, essentially placing the ligand in the solid phase, rather than fully solvated in solution. Heterogeneity in substrate linkage with this surface can generate very complex populations of binding activities, which have been described in some cases by fractal analyses[22]. Second, SPR measurements are kinetic, so equilibrium constants must be derived from the quotient of the on- and off-rates. While this quotient yields the equilibrium constant in ideal cases, the matter becomes more complex for multi-phasic reactions, since multiple, interrelated on- and off-rates must be extracted. The accuracy of the on- and off-rates is also likely influenced by the limitations on ligand diffusion imposed by linking the molecule to a solid support; namely, that the tethered molecule would experience retarded rotational and translational motions. Finally, data analysis must also account for discontinuous fluid dynamics through the assay tube, since frictional interactions between the liquid and solid phases may cause fluids to flow more slowly near the walls of a tube as compared to its center. Since apparent association/dissociation rates will be influenced by the rate of substrate presentation, this factor should also be considered when extracting equilibrium binding data from SPR experiments.

In light of these considerations, we have employed fluorescence anisotropy to evaluate RNA-protein interactions. Using this system, protein binding to a fluorescent RNA substrate may be detected by the change in fluorescence anisotropy resulting from increases in molecular volume and/or decreases in RNA flexibility[19, 23, 24]. For quantitative evaluation of macromolecular binding, this system offers several advantages over other techniques. First, fluorescence measurements may be taken under true equilibrium conditions, since reaction products are not fractionated (unlike GMSAs or filter-binding). Second, since all reaction components are in solution, this detection method is not complicated by heterogeneity in ligand presentation or solvation (unlike SPR). Third, anisotropy may be measured under both steady-state and pre-steady-state conditions, thus permitting independent assessment of thermodynamic and kinetic reaction parameters (unlike all methods listed above). Fourth, binding assays are logistically simplified by abrogating the need for radioactive compounds to track reaction products (unlike GMSAs or filter-binding). Finally, the mathematical relationships describing fluorescence anisotropy across mixed populations of fluorescent molecules have been well described[25, 26], thus permitting more complex mechanisms of macromolecular interaction such as sequential or cooperative binding events to be quantitatively considered and compared.

10.2.2.2. Theoretical and Practical Considerations

Several of the RNA-binding factors under investigation in our laboratory utilize sequential binding events to form oligomeric structures on RNA substrates (Figure 10.3). Accordingly, we have used this generic framework to develop algorithms for elucidation

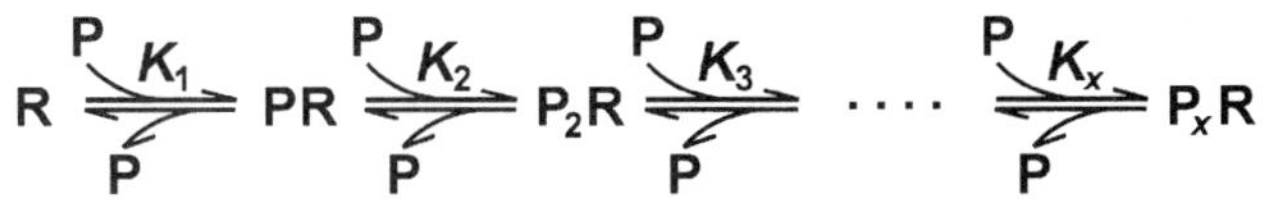

$$R \underset{P}{\overset{P \quad K_1}{\rightleftharpoons}} PR \underset{P}{\overset{P \quad K_2}{\rightleftharpoons}} P_2R \underset{P}{\overset{P \quad K_3}{\rightleftharpoons}} \cdots \underset{P}{\overset{P \quad K_x}{\rightleftharpoons}} P_xR$$

Figure 3. Sequential binding model for assembly of an oligomer of x protein molecules (P) on a common RNA substrate (R). Thermodynamics of individual binding steps are described by distinct association equilibrium constants (K).

of equilibrium binding constants from data generated by fluorescence anisotropy measurements. In practice, a wide variety of RNA-protein binding events may be considered using subsets of this general model. The steady-state concentration of each RNA:protein complex is related to the concentrations of free RNA ([R]), free protein ([P]), and relevant association equilibrium constants by:

$$[PR] = [R][P]K_1$$
$$[P_2R] = [R][P]^2 K_1 K_2$$
$$\downarrow$$
$$[P_xR] = [R][P]^x \cdot \prod_{i=1}^{x} K_i$$

Scheme 10.1

Under conditions of constant fluorescence quantum yield, the total measured anisotropy (A_t) of a mixture of fluorescent species labeled with a common fluorophore may be interpreted by Eq. (1)[25, 27, 28].

$$A_t = \sum_i A_i f_i \qquad (1)$$

Here, A_i represents the intrinsic anisotropy of each fluorescent species and f_i its fractional concentration. Applying this function to binding reactions containing fluorophore-coupled RNA substrates and unlabeled proteins, the relevant fractional concentrations of fluorescent reaction components are given by Eq. (2), where $[R]_{tot}$ is the total concentration of RNA substrate in the binding reaction.

$$f_R = \frac{[R]}{[R]_{tot}} ; f_{PR} = \frac{[PR]}{[R]_{tot}} ; \equiv ; f_{PxR} = \frac{[P_xR]}{[R]_{tot}} \qquad (2)$$

Substitution into Eq. (1) yields:

$$A_t = \frac{1}{[R]_{tot}} \left(A_R[R] + A_{PR}[PR] + \cdots + A_{PxR}[P_xR] \right) \qquad (3)$$

Incorporating Scheme 10.1 and the conservation of mass given in Eq. (4) thus yields the general relationship between A_t and free protein concentration [P] given in Eq. (5).

$$[R]_{tot} = [R] + [PR] + \cdots + [P_xR] \qquad (4)$$

$$A_t = \frac{A_R + A_{PR}[P]K_1 + \cdots + A_{PxR}[P]^x \prod_{i=1}^{x} K_i}{1 + [P]K_1 + \cdots + [P]^x \prod_{i=1}^{x} K_i} \qquad (5)$$

For practical application of this algorithm, $[R]_{tot}$ must be limiting, and in general should be at least 5-fold lower than any K_d (= $1/K$) value. Under these circumstances, the total concentration of protein in the reaction system ($[P]_{tot}$) closely approximates the free protein concentration ([P]), thus permitting solution of reaction parameters by nonlinear regression of A_t versus $[P]_{tot}$ data sets. While many commercially available software packages adequately resolve such algorithms, we have found the PRISM package (GraphPad, San Diego, CA) particularly useful for these analyses, based on both ease of equation customization and ample assessment of uncertainties in regression solutions.

The simplest type of RNA-protein interaction is the reversible, binary association of a single protein molecule with an RNA substrate described by a single equilibrium constant (ie: $x = 1$). By this model, Eq. (5) thus resolves to:

$$A_t = \frac{A_R + A_{PR} K[P]}{1 + K[P]} \qquad (6)$$

The intrinsic anisotropy of the unbound RNA substrate (A_R) may be measured directly from binding reactions lacking added protein, while the remaining constants (A_{PR}, K) are resolved by nonlinear regression of A_t versus [P] ($\approx [P]_{tot}$) data sets. An example of this model is given by the association of the heat shock protein, Hsp70, with an AU-rich RNA substrate (Section 3).

Solutions of A_t versus [P] involving two distinct binding steps ($x = 2$) are described by Eq. (7).

$$A_t = \frac{A_R + A_{PR} K_1[P] + A_{P2R} K_1 K_2[P]^2}{1 + K_1[P] + K_1 K_2[P]^2} \qquad (7)$$

Interaction of the mRNA-destabilizing factor, AUF1, with selected AU-rich RNA substrates is well described by this model (Section 3). While A_R is measured independently of protein as described for Eq. (6), the remaining constants (A_{PR}, A_{P2R}, K_1, and K_2) can generally be well resolved from A_t versus [P] provided saturation is approached at high protein concentrations (for approximation of A_{P2R}), and association binding constants differ by a factor of at least 5, with $K_1 > K_2$. When K_2 approaches or is greater than K_1, resolution of a concise solution is hampered by difficulties in establishing the intrinsic anisotropy of the intermediate complex (A_{PR}), largely owing to compensatory influences of A_{PR} and K_2 on regression convergence. However, this limitation can be overcome if an independent measure of either constant is available, as may be obtained by use of selected protein or RNA mutants under some circumstances[19]. Similarly, application of the general model described by Eq. (5) when $x > 2$ is complicated by the large number of reaction parameters requiring resolution. In such cases, values for selected A_{PxR} or K_x constants must generally be obtained in independent experiments using mutant components or reaction conditions which limit x. Subsequently, the remaining reaction parameters may be resolved by global analyses of A_t versus [P] data sets.

In some cases, resolution of independent association constants for multi-step binding equilibria may be overly odious or even unnecessary when relative differences in binding activities are under investigation. This is particularly apparent in cooperative protein binding mechanisms, where intrinsic anisotropy values are difficult to assess for intermediate protein:RNA complexes, largely due to their low fractional concentrations. Under these circumstances, we have considered reversible interactions between multiple protein molecules with a fluorescent RNA substrate using the general scheme:

$$xP + \text{Fl-RNA} \ \Phi \ P_x\cdot\text{Fl-RNA}$$

Scheme 10.2

Under conditions where RNA concentration is limiting (ie: $[P]_{free} \approx [P]_{total}$), A_t remains dependent on total protein concentration, but may be resolved by a variant of the Hill model[29]:

$$A_t = \frac{A_R + A_{complex} K[P]^h}{1 + K[P]^h} \qquad (8)$$

In this model, K represents an aggregate equilibrium constant, h is the Hill co-efficient, and A_R and $A_{complex}$ are the intrinsic anisotropy values of the free and maximally protein-associated fluorescent RNA substrates, respectively. Adapting a transformation of the Hill model[30] to A_t *versus* $[P]$ data sets returns an additional parameter, $[P]_{1/2}$, which approximates the concentration of protein yielding half-maximal binding saturation:

$$A_t = A_R + \left(A_{complex} - A_R\right) \times \left[\frac{\left([P]/[P]_{1/2}\right)^h}{1 + \left([P]/[P]_{1/2}\right)^h}\right] \qquad (9)$$

Association of Hsp70 with polyuridylate RNA substrates is well described by this model (Section 3).

To this point, all binding algorithms described have assumed that RNA concentrations are limiting, thus permitting the approximation $[P]_{free} \approx [P]_{total}$. With readily available instrumentation and high quantum yield fluorophores, this condition generally holds for binding events where $K_d > 1$ nM. However, in some circumstances, the affinity of selected RNA-binding proteins for cognate substrates may exceed this limit[31]. In such situations, reversible, binary binding events may be resolved in terms of the total concentrations of RNA and protein in the system[26]:

$$A_t = A_R + (A_{PR} - A_R) \times$$
$$\left[\frac{1 + K[R]_{tot} + K[P]_{tot} - \sqrt{\{(1 + K[R]_{tot} + K[P]_{tot})^2 - 4[R]_{tot}[P]_{tot} K^2\}}}{2K[R]_{tot}}\right] \qquad (10)$$

A second condition of the aforementioned binding algorithms is that all reaction products exhibit similar fluorescence quantum yields. This condition ensures that each fluorescent species (ie: R *versus* PR) makes an equivalent molar contribution to total fluorescence emission, hence permitting their contributions to A_t to be considered by simple additivity as defined in Eq. (1). However, interaction of some RNA-binding proteins with fluorescent RNA substrates may alter quantum yield, possibly through direct contact with the fluorophore, by altering fluorophore solvent accessibility, or possibly through changes in local RNA structure, ionic strength, or pH. In our experience, the simplest solutions in these instances are to either change the fluorophore (eg: Fl to Cy3 or TAMRA), move the fluorescent dye to a different location on the RNA substrate (eg: 3'-linkage in place of 5'), or add intervening nucleotides or carbon spacers to increase the distance between the fluorophore and the protein-binding site. Failing this, quantitative solutions are still attainable for reversible, binary binding equilibria by correction of the fractional contributions of free and bound RNA to A_t based on their relative quantum yields, given by Q_R and Q_{PR}, respectively[26, 32]:

$$\frac{[PR]}{[R]} = \frac{A_t - A_R}{A_{PR} - A_t} \times \frac{Q_R}{Q_{PR}} = \frac{A_c - A_R}{A_{PR} - A_c} \qquad (11)$$

Rearrangement thus provides a solution for the quantum yield-corrected anisotropy (A_c) of each binding reaction:

$$A_c = \frac{A_R + [(A_t - A_R)/(A_{PR} - A_t)](Q_R/Q_{PR})(A_{PR})}{1 + [(A_t - A_R)/(A_{PR} - A_t)](Q_R/Q_{PR})} \qquad (12)$$

This parameter may then be employed in A_c versus [P] data sets to resolve equilibrium binding constants as described above.

10.2.2.3. Instrumentation for Measurement of Fluorescence Anisotropy

Measurement of fluorescence anisotropy requires a steady-state spectrofluorometer containing excitation and emission polarizers. Selection of excitation and emission wavelengths may be achieved using monochromators or optical filters. For each binding reaction, fluorescence emission must be measured using polarizers fixed in both parallel (I_{VV}: vertical excitation polarizer, vertical emission polarizer) and perpendicular (I_{VH}: vertical excitation polarizer, horizontal emission polarizer) orientations. In the conventional "L" spectrofluorometer format, this requires two separate measurements, while "T" format instruments, which contain two independent photodetectors, permit both parameters to be measured simultaneously. Anisotropy is then calculated as:

$$A_t = \frac{I_{VV} - G \cdot I_{VH}}{I_{VV} + 2G \cdot I_{VH}} \qquad (13)$$

G is a correction factor that compensates for differences in the detection efficiency of vertically- versus horizontally-polarized light, and is calculated using steady-state fluorescence as $G = I_{HV}/I_{HH}$[25].

In our hands, routine measurement of fluorescence anisotropy has been greatly simplified using the Beacon 2000 Variable Temperature Fluorescence Polarization System, manufactured by Panvera (Madison, WI). Utility of this instrument is far more limited in scope than conventional spectrofluorometers, although several features make the Beacon an attractive and relatively inexpensive tool for anisotropy measurements. First, the Beacon is supplied with fluorescein excitation (λ_{max} = 485 nm) and emission (λ_{max} = 535 nm) filters, which are optimized to permit very sensitive detection of fluorescein-conjugated biomolecules. Using Fl-tagged RNA substrates, we can reliably measure fluorescence anisotropy at concentrations approaching 10^{-10} M. However, additional filters must be purchased separately for detection of fluorescence at other wavelengths. Using the 6 mm × 50 mm sample holder with disposable glass tubes, we measure RNA-protein binding reactions in a total volume of 100 µl, at temperatures ranging from 6° to 65°C. The Beacon measures fluorescence in all required polarizer orientations in a single operation, and returns measurements of total fluorescence intensity and anisotropy, with blank subtraction performed automatically if selected. The high sensitivity and small sample size permitted by this instrument greatly facilitate generation of the large data sets necessary for resolution of complex RNA-protein binding mechanisms.

10.2.3. Assessment of RNA Folding by FRET

The conformational heterogeneity and flexibility of RNA molecules significantly contribute to the often complex nature of their function. For example, the catalytic activities of ribozymes are intimately dependent on adoption of correct three-dimensional RNA structures[33-35], and protein-binding events may depend on presentation of RNA substrates in specific conformations[36-40]. Complex structures including pseudoknots contribute to the regulation of ribosomal frameshifting[41-43]. Furthermore, changes in system temperature, pH, cation population, and protein binding events can all potentially modulate the stability and/or dynamics of RNA folding[30, 39, 44-48].

Fluorescence resonance energy transfer (FRET) is emerging as a powerful tool in the elucidation of RNA folding mechanisms[49-51], due in part to the increasing ease with which fluorophores may be site-specifically conjugated within complex RNA molecules. We routinely evaluate folding of small (< 40 bases) RNA substrates by selectively labeling their 3'-termini with Fl and 5'-termini with Cy3 (Figure 10.4). The utility of this

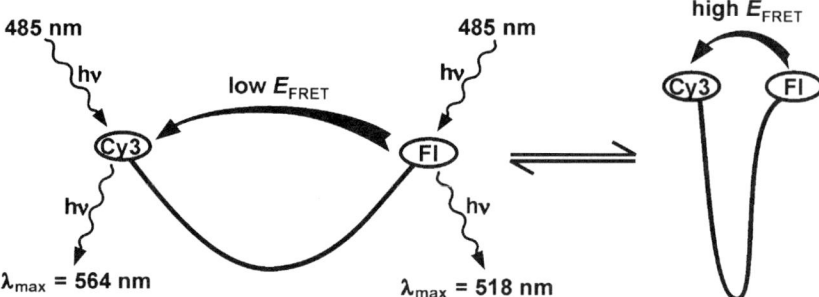

Figure 10.4. Measurement of RNA folding by FRET between conjugated 3'-Fl and 5'-Cy3 moieties. Changes in the distance between the termini in the unfolded (left) *versus* folded (right) states are detected by differences in FRET efficiency.

method is underscored by the relationship between the efficiency of FRET (E_{FRET}) and the scalar distance (r) between a fluorescent donor and acceptor:

$$E_{FRET} = R_0^6/(R_0^6 + r^6), \quad (14)$$

R_0 is the Förster distance for the donor-acceptor pair, defined as the distance at which FRET efficiency is 50%[25, 49]. In the case of the Fl-Cy3 fluorophore pair conjugated to single-stranded DNA, R_0 has been calculated as 55.7 Å[52]. We have also employed Fl-Cy5 as a donor-acceptor pair ($R_0 = 47$ Å) for measurements involving shorter RNA substrates[53], although many other options exist given the plethora of fluorophores available for RNA conjugation (Section 2).

We typically calculate E_{FRET} for double-labeled RNA substrates by measuring the decrease in fluorescence emission of the FRET donor in the presence of the acceptor[25, 50, 54]:

$$E_{FRET} = 1 - (F_{DA}/F_D) \quad (15)$$

F_{DA} is the blank-corrected fluorescence of the donor in the presence of the acceptor, measured using a double-labeled RNA substrate ($\lambda_{ex} = 490$ nm and $\lambda_{em} = 518$ nm for fluorescein donors), while F_D is the fluorescence of the donor in the absence of the acceptor, measured using an RNA labeled only with Fl. Since quantum emission from donor moieties is not decreased by FRET in RNA substrates lacking acceptors, it is important that the labeling efficiency of the acceptor be very high (> 90%) for optimal resolution of E_{FRET}. This factor is particularly significant if estimates of inter-fluorophore distance are to be generated using Eq. (14). It has been our experience that custom oligoribonucleotides synthesized by commercial suppliers are more efficiently labeled at their 5'-termini, rather than using 3'-linkages. As such, we typically direct suppliers to conjugate FRET acceptor dyes to the 5'-end of synthetic RNA substrates. Alternatively, we have achieved efficient 3'-linkage through periodate oxidation and hydrazide coupling[29], although this must be performed in the absence of other tethered fluorophores to prevent oxidative damage to the dye, thus requiring ligation across single-stranded DNA splints (Figure 10.2) to generate the double-labeled RNA substrate.

In our lab, FRET has proven useful for measuring the thermal stability of folded RNA structures, cation-dependence of RNA folding, and local conformational changes induced by protein binding events. Selected examples of these experiments and some of the molecular details that they have revealed are outlined in Section 3.

10.3. ELUCIDATION OF MECHANISMS CONTRIBUTING TO REGULATION OF CYTOPLASMIC mRNA TURNOVER BY FLUORESCENCE SPECTROSCOPY

In eukaryotes, gene expression is a highly regulated process exhibiting control at many levels to ensure that gene products are maintained within levels appropriate for cellular growth and function. A critical determinant governing the synthetic rates of proteins are the concentrations of cytoplasmic mRNAs encoding them. As with any biological system, the steady-state level of a cytoplasmic mRNA is dependent on its rates of

both synthesis and degradation. The production rate of a cytoplasmic mRNA is a cumu-
lative function of transcription, pre-mRNA processing, and nucleo-cytoplasmic transport,
each of which may be subject to independent regulatory control. Cytoplasmic mRNA
turnover is also tightly regulated, with mammalian mRNAs displaying a spectrum of decay rates spanning up to two orders of magnitude. Generally, determinants of both con-
stitutive and inducible mRNA turnover rates are present as *cis*-acting sequences within
individual mRNAs[55, 56].

AU-rich elements (AREs) constitute a varied family of RNA sequences localized to
the 3'-untranslated regions (3'UTRs) of many labile mRNAs[24, 57]. The ability of AREs to
modulate mRNA decay rates is mediated by association of cytoplasmic *trans*-acting factors [14]. Some proteins, like AUF1[58-60], tristetraprolin[61, 62], and KSRP[63], promote rapid
decay of ARE-containing transcripts, while some others, including members of the Hu
family of RNA-binding proteins, prevent mRNA degradation[64, 65]. Given the heterogeneity in size and sequence of AREs from different mammalian mRNAs, together with the
plethora of cytoplasmic factors competing for these *cis*-acting elements, two of our principal research foci have been to characterize the substrate preferences of selected ARE-
binding factors, and evaluate the structural and functional consequences of these *cis-trans*
interactions. The following subsections describe some findings contributing to our understanding of this regulatory system, and the involvement of fluorescence spectroscopic
techniques in these studies.

10.3.1. Evaluation of *Trans*-factor Binding Mechanisms and Affinity by Fluores-
cence Anisotropy

AUF1 was first identified as an activity capable of accelerating the decay of ARE-
containing mRNAs in a cell-free system[66]. Subsequent purification and cloning revealed
that AUF1 is expressed as a family of four protein isoforms through alternative splicing
of a common pre-mRNA[67, 68]. The isoforms are denoted by their apparent molecular
weights as p37^{AUF1}, p40^{AUF1}, p42^{AUF1}, and p45^{AUF1}, and all possess some degree of ARE-
binding activity[68]. The p42^{AUF1} and p45^{AUF1} isoforms are exclusively nuclear in most cell
types, while p37^{AUF1} and p40^{AUF1} are typically found in both nuclear and cytoplasmic
compartments[67,69]. In numerous biochemical and cell biological systems, the expression
and ARE-binding activity of p37^{AUF1} and p40^{AUF1} are closely associated with the rapid
turnover of ARE-containing mRNAs[58-60, 70-72].

Our current model suggests that AUF1 binding to an ARE substrate functions as a
targeting system to recruit subsequent components of the cytoplasmic mRNA decay machinery to the mRNA[73]. However, the RNA sequence and/or structural determinants that
promote AUF1 binding to one ARE over another remain unclear. To gain insight into
this question, we characterized the association of recombinant AUF1 proteins with a
number of model RNA substrates. By GMSA, p37^{AUF1} binding to the core ARE from
tumor necrosis factor α (TNFα) mRNA forms two protein:RNA complexes in a concentration-dependent manner[19]. Hydrodynamic studies and chemical cross-linking indicated
that p37^{AUF1} is a dimer in solution, but forms protein tetramers on the TNFα ARE substrate, and larger oligomeric structures in the presence of longer ARE sequences[19, 74].
Using fluorescence anisotropy-based assays, we demonstrated that p37^{AUF1} binding to the
TNFα ARE substrate was consistent with the sequential association of protein dimers

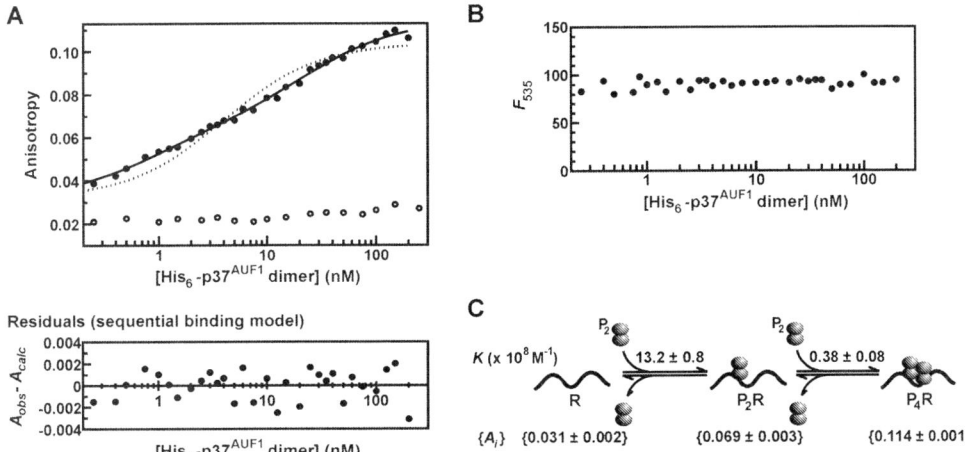

Figure 10.5. Assessment of p37^{AUF1} binding to RNA substrates by fluorescence anisotropy. (A) Anisotropy increases as a function of protein concentration for recombinant His$_6$-p37^{AUF1} binding to a fluorescent substrate containing the TNFα ARE sequence (solid circles) but not to an unrelated β-globin RNA fragment (open circles). Binding to the ARE substrate is well resolved by a two-stage binding algorithm defined by Eq. (7) (solid line), while the single-site binding model of Eq. (6) is clearly inappropriate (dotted line). (B) Total fluorescence emission from the Fl-TNFα ARE substrate does not vary with protein concentration, indicating no significant change in probe quantum yield and validating use of algorithms derived from Eq. (1). (C) Association binding (K) and intrinsic anisotropy (A_i) constants for p37^{AUF1} association with the ARE substrate. Data are from Ref. 29.

(Figure 10.5). Furthermore, this method allowed us to determine the equilibrium association constants describing both stages of AUF1 tetramer assembly on this substrate. Binding is not explicitly specific for the ARE sequence, since p37^{AUF1} also binds to polyuridylate RNA sequences with high affinity[19, 29]. Additional studies verified that p40^{AUF1} binds the TNFα ARE by a comparable mechanism, but that the affinity of p40^{AUF1} for this substrate is regulated by phosphorylation at two distinct sites[75]. Finally, the ability of AUF1 proteins to bind RNA is regulated by the structural presentation of each target site (described in Section 3.2), thus adding a new dimension to the complexity of RNA substrate selectivity by this *trans*-acting factor.

The inducible 70 kDa heat shock protein, Hsp70, has also been implicated as an ARE-binding factor and was identified in a common cytoplasmic complex with AUF1 by co-immunoprecipitation[76, 77]. Recombinant Hsp70 binding to the TNFα ARE was consistent with 1:1 stoichiometry by GMSA and fluorescence anisotropy-based assays, however, this protein formed cooperative, multimeric structures on polyuridylate substrates[78]. Solution of A_t *versus* [Hsp70] data sets for substrates containing a 32-nucleotide polyuridylate sequence resolved a Hill coefficient of 1.7 ± 0.1 (Figure 10.6). Unlike AUF1, however, association of Hsp70 with ARE substrates was not significantly influenced by conformational changes in the RNA. Together, these studies illustrated the mechanistic heterogeneity of ARE recognition by different ARE-binding proteins, and demonstrated that the selectivity of some proteins (ie: AUF1) but not others (ie: Hsp70) could be influenced by local higher order structures involving the RNA target site.

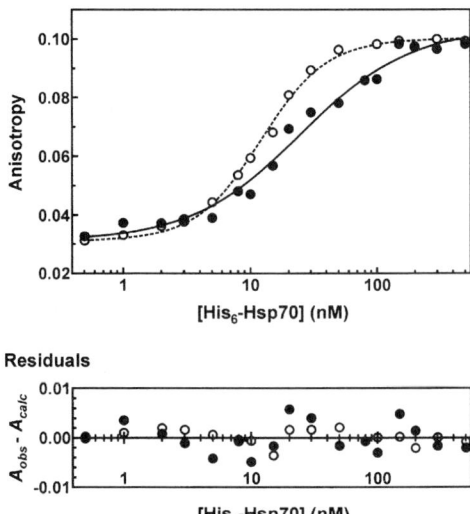

Figure 10.6. Association of recombinant Hsp70 with fluorescein-labeled RNA substrates containing the TNFα ARE resolved by Eq. (6) (solid circles, solid line) giving $K = 4.0 \pm 0.4 \times 10^7$ M^{-1} ($K_d = 25$ nM), or a 32-base polyuridylate sequence resolved by Eq. (9) (open circles, dashed line) giving $[Hsp70]_{1/2} = 12.2 \pm 0.4$ nM. Data are from Ref. 78.

10.3.2. Higher Order Structures Involving the TNFα ARE Regulate AUF1 Binding

The first indications that the TNFα ARE was capable of adopting a higher order RNA structure were based on an increase in the intrinsic anisotropy of fluorescein-labeled RNA substrates containing this element in the presence of Mg^{2+} [29], concomitant with inhibition of p37^{AUF1} binding activity. Both inhibition of AUF1 binding and restriction of segmental RNA motion by Mg^{2+} were dependent on the ARE sequence, since neither effect was observed with the polyuridylate substrate. Subsequent experiments verified that the cation-induced structural change in the ARE substrate was an intramolecular event[29]. To more rigorously assess the mechanism and consequences of the Mg^{2+}-induced change in ARE structure, we synthesized selected RNA substrates with 3'-fluorescein and 5'-Cy3 moieties for conformational analyses by FRET[39]. Addition of Mg^{2+} to samples containing the double-labeled ARE substrate resulted in a large increase in E_{FRET} based on decreased emission from the fluorescein donor in the presence of the acceptor (Figure 10.7), indicating that the RNA termini are positioned closer together in solution in the presence of the cation. Similar to the measurements of segmental motion by fluorescence anisotropy, these cation-dependent changes in RNA folding were largely dependent on the ARE sequence, since the structure of a double-labeled polyuridylate substrate was only modestly affected by Mg^{2+}.

Thermal denaturation experiments indicated that the TNFα ARE is capable of forming a weak condensed structure in the absence of Mg^{2+}, but that this folding event is stabilized in the presence of the cation. The potential of different cations to stabilize the

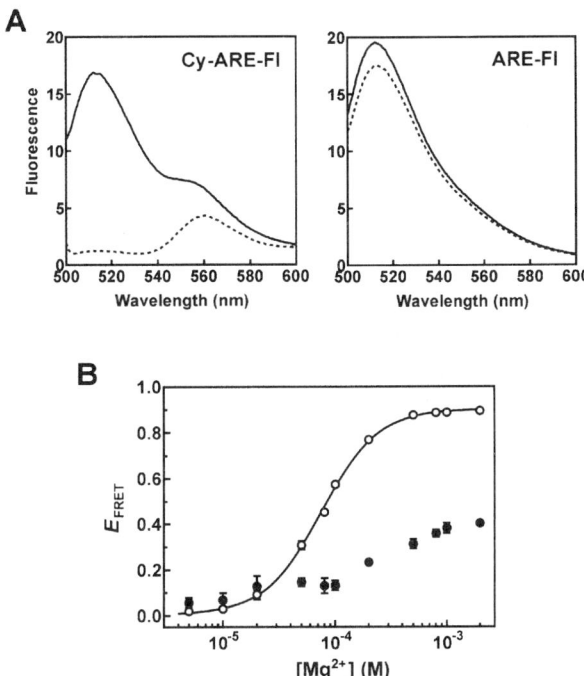

Figure 10.7. Assessment of RNA folding by FRET. (A) Emission spectra (λ_{ex} = 490 nm) of TNFα ARE RNA substrates labeled with FRET donor-acceptor pairs (Cy-ARE-Fl) or the donor alone (ARE-Fl) measured in the absence (solid line) or presence (dashed line) of 1 mM MgCl$_2$. (B) Mg^{2+} dependence of RNA folding measured for RNA substrates containing the TNFα ARE (open circles) or the 32-base polyuridylate sequence (solid circles). The cooperative nature of ARE folding with respect to Mg^{2+} was resolved using Eq. (9) (solid line), where h = 1.7 ± 0.1 and [Mg^{2+}]$_{1/2}$ = 75 ± 2 µM. Data are from Ref. 39.

folded ARE structure was then assayed to help define a mechanistic basis for this effect. In all cases, adoption of the condensed RNA conformation was cooperative with respect to the cation ($h > 1$), but the ion concentrations necessary for stabilizing this structure varied widely. For example, an inorganic trivalent cation (Co(NH$_3$)$_6$$^{3+}$) stabilized the folded ARE at concentrations 1000-fold lower than inorganic, divalent cations (Mg^{2+}, Ca^{2+}, Mn^{2+}), and 1 million-fold lower than monovalent cations (Na$^+$, K$^+$). In addition, a cation where the positive charges are tightly packed (Co(NH$_3$)$_6$$^{3+}$) was 13-fold more effective at stabilizing the folded ARE than a comparably charged organic cation (spermidine^{3+}), where the positive charges are distributed across 11 Å. Based on the preference of the folded ARE for highly charged, condensed cations, we concluded that cations likely stabilize the folded ARE structure by targeted counterion neutralization at regions of high negative charge density[39]. To our knowledge, this represented the first indication that AREs may form higher-order RNA structures, and indicated that these structural transitions may have a significant impact on binding of *trans*-acting factors.

10.3.3. AUF1 Binding Modulates Local RNA Conformation

The FRET system was also used to determine whether association of AUF1 proteins with RNA substrates could modify local RNA structure. In the absence of Mg^{2+}, $p37^{AUF1}$ binding to the TNFα ARE resulted in a decrease in the scalar distance between the 5'- and 3'-termini, suggesting the adoption of a more condensed RNA structure. The folded structure appeared to be distinct from that stabilized by cations, however, since the FRET efficiency indicated a distance of 48-51 Å between the RNA termini, significantly larger than the 38 Å upper limit calculated for the cation-conjugated structure. Further evidence for this distinction was provided by measuring the protein-dependence of FRET in the presence of Mg^{2+}. At low protein concentrations, the RNA was tightly condensed, consistent with the cation-stabilized structure. However, as the protein concentration was increased, the distance between the RNA termini also increased, again resolving to the 48-51 Å range characteristic of the AUF1:ARE complex in the absence of Mg^{2+}. This

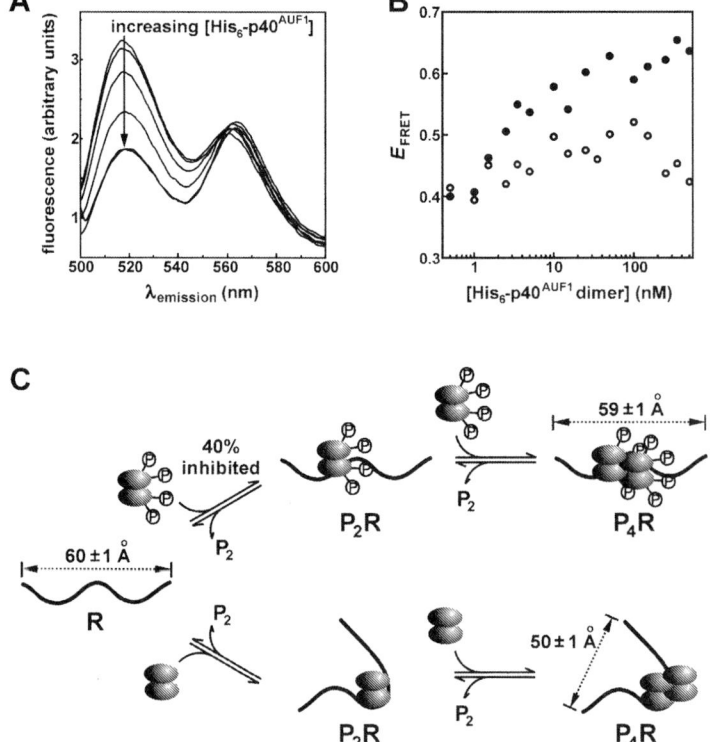

Figure 10.8. (A) Emission spectra (λ_{ex} = 490 nm) of the 5'-Cy3/3'-Fl labeled TNFα ARE RNA substrate under conditions of increasing recombinant $p40^{AUF1}$ concentration. (B) FRET efficiency as a function of $p40^{AUF1}$ concentration for the unphosphorylated protein (solid circles) compared with $p40^{AUF1}$ phosphorylated on Ser83 and Ser87 (open circles). (C) Schematic showing the assembly of $p40^{AUF1}$ tetramers on the TNFα ARE RNA substrate, and the roles of phosphorylation on the affinity and structural consequences of this process. Data are from Ref. 75.

result indicated that, at high protein concentrations, the structural influence of AUF1 binding can override cation-stabilized condensation of ARE conformation, and supported our previous hypothesis that oligomerization of AUF1 on the TNFα ARE converges to a common complex regardless of the presence of Mg^{2+} [29]. Finally, AUF1 association with a polyuridylate substrate also induced a change in RNA structure, similar to that observed with the ARE substrate. Together, these data indicated that while ion-stabilized ARE structural changes appear to be RNA sequence-dependent, AUF1-induced RNA folding is an RNA sequence-independent consequence of protein binding [39].

A physiological role for AUF1-induced RNA folding was suggested by parallel studies involving $p40^{AUF1}$. In the monocytic leukemia cell line THP-1, activation of the protein kinase C pathway with phorbol esters induced rapid but transient expression of ARE-containing mRNAs encoding the cytokine interleukin-1β and the inflammatory mediator TNFα, in part through a 6- to 12-fold increase in the stability of each transcript [79]. Concomitant with inhibition of mRNA turnover, changes were observed in the distribution of cytoplasmic ARE-binding complexes containing AUF1. Purification of AUF1 from polysome complexes revealed that stabilization of the ARE-containing transcripts was accompanied by loss of phosphate from Ser83 and Ser87 of $p40^{AUF1}$. Fluorescence anisotropy and FRET-based experiments using recombinant $p40^{AUF1}$ phosphorylated at these sites revealed that local remodeling of ARE structure is inhibited by $p40^{AUF1}$ phosphorylation (Figure 10.8). Interestingly, phosphorylation of both sites is required to inhibit AUF1-induced changes in RNA conformation; selective modification at Ser83 or Ser87 individually did not prevent adoption of a folded state [75]. Taken together, these data suggest that ARE-directed mRNA decay may be regulated in part through reversible phosphorylation of $p40^{AUF1}$ by modification of local RNA structure flanking the AUF1 binding site, possibly by selectively promoting or inhibiting subsequent factor recruitment.

10.4. FUTURE DIRECTIONS

In our hands, fluorescence anisotropy and FRET-based techniques have made vital contributions to our ongoing studies of ARE-directed mRNA turnover, and have permitted the perusal of mechanistic questions that would not have been otherwise feasible. On a broader scale, owing largely to the ease with which fluorescent-labeled RNA substrates may now be procured, many other labs are now applying similar techniques to a wide range of questions directed at understanding the intricate relationships between RNA structure and function.

Given the demonstrated utility of fluorescence-based assay systems in the study of RNA metabolism, coupled with the wealth of instrumentation and analysis tools available to investigators today, emerging studies will greatly benefit from the current state of the art in fluorescence technology. For example, time resolved fluorescence permits highly detailed analyses of RNA folding events by allowing explicit detection of localized conformational states. This strategy was applied to the cap region of the iron response element to reveal base-specific transitions between stacked and unstacked conformations [80]. In addition, time resolved FRET is a powerful method for assessing the distribution of distances between two points across a population of molecules. This methodology has recently been employed to detect RNA residues critical for ribozyme folding [81] and dynamic features of an essential subdomain of the hepatitis C virus internal ribosome entry site[82]. Finally, the emerging development and application of single molecule fluorescence spectroscopy is permitting conformational events involving individual RNA molecules to be visualized in real time. For example, folding and catalytic events have been visualized for individual ribozyme molecules[83, 84], providing heretofore unobservable perspectives of structural and dynamic processes within these RNAs, and opening a new frontier for quantitative assessment of macromolecular functions without statistical thermodynamics.

10.5. ACKNOWLEDGEMENTS

Funding for our projects is provided by NCI Grant R01 CA102428 from the National Institutes of Health and a Scientist Development Grant from the American Heart Association. Additional funding for the Center for Fluorescence Spectroscopy is provided by NCRR Grant P41 RR08119 from the National Institutes of Health.

10.6. REFERENCES

1. A. G. Vitreschak, D. A. Rodionov, A. A. Mironov, and M. S. Gelfand, Riboswitches: the oldest mechanism for regulation of gene expression? *Trends Genet.* **20**, 44-50 (2004).
2. G. Storz, J. A. Opdyke, and A. Zhang, Controlling mRNA stability and translation with small, noncoding RNAs. *Curr. Opin. Microbiol.* **7**, 140-144 (2004).
3. T. A. Steitz and P. B. Moore, RNA, the first macromolecular catalyst: The ribosome is a ribozyme. *Trends Biochem. Sci.* **28**, 411-418 (2003).
4. D. M. J. Lilley, The origins of RNA catalysis in ribozymes. *Trends Biochem. Sci.* **28**, 495-501 (2003).
5. M. Rajendran and A. D. Ellington, In vitro selection of molecular beacons. *Nucleic Acids Res.* **31**, 5700-5713 (2003).
6. T. Hermann and D. J. Patel, Adaptive recognition by nucleic acid aptamers. *Science* **287**, 820-825 (2000).
7. J. A. Doudna, A molecular contortionist. *Nature* **388**, 830 (1997).
8. A. E. Johnson, H. J. Adkins, E. A. Matthews, and C. R. Cantor, Distance moved by transfer RNA during translocation from the A site to the P site on the ribosome. *J. Mol. Biol.* **156**, 113-140 (1982).
9. J. Czworkowski, O. W. Odom, and B. Hardesty, Fluorescence study of the topology of messenger RNA bound to the 30S ribosomal subunit of *Escherichia coli. Biochemistry* **30**, 4821-4830 (1991).
10. P. Z. Qin and A. M. Pyle, Site-specific labeling of RNA with fluorophores and other structural probes. *Methods* **18**, 60-70 (1999).
11. Y.-T. Yu, Construction of 4-thiouridine site-specifically substituted RNAs for cross-linking studies. *Methods* **18**, 13-21 (1999).
12. C. G. Burd and G. Dreyfuss, Conserved structures and diversity of functions of RNA-binding proteins. *Science* **265**, 615-621 (1994).
13. J. E. G. McCarthy and H. Kollmus, Cytoplasmic mRNA-protein interactions in eukaryotic gene expression. *Trends Biochem. Sci.* **20**, (1995).
14. G. M. Wilson and G. Brewer, The search for trans-acting factors controlling messenger RNA decay. *Prog. Nucleic Acids Res. Mol. Biol.* **62**, 257-291 (1999).
15. L. E. Maquat, Nonsense-mediated mRNA decay: splicing, translation and mRNP dynamics. *Nature Rev. Mol. Cell Biol.* **5**, 89-99 (2004).
16. G. M. Wilson and G. Brewer, Identification and characterization of proteins binding A+U-rich elements. *Methods* **17**, 74-83 (1999).
17. W.-J. Ma, S. Cheng, C. Campbell, A. Wright, and H. Furneaux, Cloning and characterization of HuR, a ubiquitously expressed elav-like protein. *J. Biol. Chem.* **271**, 8144-8151 (1996).

18. C. T. DeMaria and G. Brewer, AUF1 binding affinity to A+U-rich elements correlates with rapid mRNA degradation. *J. Biol. Chem.* **271**, 12179-12184 (1996).
19. G. M. Wilson, Y. Sun, H. Lu, and G. Brewer, Assembly of AUF1 oligomers on U-rich RNA targets by sequential dimer association. *J. Biol. Chem.* **274**, 33374-33381 (1999).
20. I. Wong and T. M. Lohman, A double-filter method for nitrocellulose-filter binding: application to protein-nucleic acid interactions. *Proc. Natl. Acad. Sci. USA* **90**, 5428-5432 (1993).
21. P. S. Katsamba, S. Park, and I. A. Laird-Offringa, Kinetic studies of RNA-protein interactions using surface plasmon resonance. *Methods* **26**, 95-104 (2002).
22. A. Sadana, A kinetic study of analyte-receptor binding and dissociation, and dissociation alone, for biosensor applications: a fractal analysis. *Anal. Biochem.* **291**, 34-47 (2001).
23. W. J. Checovich, R. E. Bolger, and T. Burke, Fluorescence polarization - a new tool for cell and molecular biology. *Nature* **373**, 254-256 (1995).
24. G. Brewer, Characterization of c-*myc* 3' to 5' mRNA decay activities in an *in vitro* system. *J. Biol. Chem.* **273**, 34770-34774 (1998).
25. J. R. Lakowicz, *Principles of Fluorescence Spectroscopy*, Kluwer Academic/Plenum, New York, NY (1999).
26. J. R. Lundblad, M. Laurance, and R. H. Goodman, Fluorescence polarization analysis of protein-DNA and protein-protein interactions. *Mol. Endocrinol.* **10**, 607-612 (1996).
27. D. M. Jameson and W. H. Sawyer, Fluorescence anisotropy applied to biomolecular interactions. *Methods Enzymol.* **246**, 283-300 (1995).
28. G. Weber, Polarization of the fluorescence of macromolecules 2. Fluorescent conjugates of ovalbumin and bovine serum albumin. *Biochem. J.* **51**, 155-167 (1952).
29. G. M. Wilson, K. Sutphen, K. Chuang, and G. Brewer, Folding of A+U-rich RNA elements modulates AUF1 binding: Potential roles in regulation of mRNA turnover. *J. Biol. Chem.* **276**, 8695-8704 (2001).
30. S. L. Heilman-Miller, D. Thirumalai, and S. A. Woodson, Role of counterion condensation in folding of the *Tetrahymena* ribozyme. I. Equilibrium stabilization by cations. *J. Mol. Biol.* **306**, 1157-1166 (2001).
31. P. S. Katsamba, D. G. Myszka, and I. A. Laird-Offringa, Two functionally distinct steps mediate high affinity binding of U1A protein to U1 hairpin II RNA. *J. Biol. Chem.* **276**, 21476-21481 (2001).
32. W. B. Dandliker, M.-L. Hsu, J. Levin, and B. R. Rao, Equilibrium and kinetic inhibition assays based upon fluorescence polarization. *Methods Enzymol.* **74**, 3-28 (1981).
33. R. Russell, I. S. Millett, M. W. Tate, L. W. Kwok, B. Nakatani, S. M. Gruner, S. G. Mochrie, V. Pande, S. Doniach, D. Herschlag, and L. Pollack, Rapid compaction during RNA folding. *Proc. Natl. Acad. Sci. USA* **99**, 4266-4271 (2002).
34. J. Pan, M. L. Deras, and S. A. Woodson, Fast folding of a ribozyme by stabilizing core interactions: evidence of multiple folding pathways in RNA. *J. Mol. Biol.* **296** , 133-144 (2000).
35. D. Gilley and E. H. Blackburn, The telomerase RNA pseudoknot is critical for the stable assembly of a catalytically active ribonucleoprotein. *Proc. Natl. Acad. Sci. USA* **96**, 6621-6625 (1999).
36. K. J. Addess, J. P. Basilion, R. D. Klausner, T. A. Rouault, and A. Pardi, Structure and dynamics of the iron responsive element RNA: Implications for binding of the RNA by iron regulatory binding proteins. *J. Mol. Biol.* **274**, 72-83 (1997).
37. Y. Ke, J. Wu, E. A. Leibold, W. E. Walden, and E. C. Theil, Loops and bulge/loops in iron-responsive element isoforms influence iron regulatory protein binding. *J. Biol. Chem.* **273**, 23637-23640 (1998).
38. L. B. Blyn, L. M. Risen, R. H. Griffey, and D. E. Draper, The RNA-binding domain of ribosomal protein L11 recognizes an rRNA tertiary structure stabilized by both thiostrepton and magnesium ion. *Nucleic Acids Res.* **28**, 1778-1784 (2000).
39. G. M. Wilson, K. Sutphen, M. Moutafis, S. Sinha, and G. Brewer, Structural remodeling of an A+U-rich RNA element by cation or AUF1 binding. *J. Biol. Chem.* **276**, 38400-38409 (2001).
40. P. Bouvet, F. H. T. Allain, L. D. Finger, T. Dieckmann, and J. Feigon, Recognition of pre-formed and flexible elements of an RNA stem-loop by nucleolin. *J. Mol. Biol.* **309**, 763-775 (2001).
41. C. A. Theimer and D. P. Giedroc, Equilibrium unfolding pathway of an H-type RNA pseudoknot which promotes programmed-1 ribosomal frameshifting. *J. Mol. Biol.* **289**, 1283-1299 (1999).
42. G. M. Wilson and G. Brewer, Slip-sliding the frame: programmed -1 frameshifting on eukaryotic transcripts. *Genome Res.* **9**, 393-394 (1999).

43. E. P. Plant, K. L. Muldoon Jacobs, J. W. Harger, A. Meskauskas, J. L. Jacobs, J. L. Baxter, A. N. Petrov, and J. D. Dinman, The 9-Å solution: How mRNA pseudoknots promote efficient programmed -1 frameshifting. *RNA* **9**, 168-174 (2003).
44. P. E. Cole, S. K. Yang, and D. M. Crothers, Conformational changes of transfer ribonucleic acid. Equilibrium phase diagrams. *Biochemistry* **11**, 4358-4368 (1972).
45. J. Flinders and T. Dieckmann, A pH controlled conformational switch in the cleavage site of the VS ribozyme substrate RNA. *J. Mol. Biol.* **308**, 665-679 (2001).
46. T. C. Gluick, R. B. Gerstner, and D. E. Draper, Effects of Mg^{2+}, K^+, and H^+ on an equilibrium between alternative conformations of an RNA pseudoknot. *J. Mol. Biol.* **270**, 451-463 (1997).
47. L. G. Laing, T. C. Gluick, and D. E. Draper, Stabilization of RNA structure by Mg ions: Specific and non-specific effects. *J. Mol. Biol.* **237**, 577-587 (1994).
48. S. Bernacchi, S. Stoylov, E. Piemont, D. Ficheux, B. P. Roques, J. L. Darlix, and Y. Mely, HIV-1 nucleocapsid protein activates transient melting of least stable parts of the secondary structure of TAR and its complementary sequence. *J. Mol. Biol.* **317**, 385-399 (2002).
49. R. M. Clegg, Fluorescence resonance energy transfer and nucleic acids. *Methods Enzymol.* **211**, 353-388 (1992).
50. D. Klostermeier and D. P. Millar, RNA conformation and folding studied with fluorescence resonance energy transfer. *Methods* **23**, 240-254 (2001).
51. D. M. J. Lilley and T. J. Wilson, Fluorescence resonance energy transfer as a structural tool for nucleic acids. *Curr. Opin. Chem. Biol.* **4**, 507-517 (2000).
52. D. G. Norman, R. J. Grainger, D. Uhrín, and D. M. J. Lilley, Location of cyanine-3 on double-stranded DNA: Importance for fluorescence resonance energy transfer studies. *Biochemistry* **39**, 6317-6324 (2000).
53. B. Y. Brewer, J. Malicka, P. J. Blackshear, and G. M. Wilson, RNA sequence elements required for high affinity binding by the zinc finger domain of tristetraprolin: Conformational changes coupled to the bipartite nature of AU-rich mRNA-destabilizing motifs. *J. Biol. Chem.* **279**, 27870-27877 (2004).
54. P. Wu and L. Brand, Resonance energy transfer: Methods and applications. *Anal. Biochem.* **218**, 1-13 (1994).
55. J. Guhaniyogi and G. Brewer, Regulation of mRNA stability in mammalian cells. *Gene* **265**, 11-23 (2001).
56. C. J. Wilusz, M. Wormington, and S. W. Peltz, The cap to tail guide to mRNA turnover. *Nature Rev. Mol. Cell Biol.* **2**, 237-246 (2001).
57. C.-Y. A. Chen and A.-B. Shyu, AU-rich elements: characterization and importance in mRNA degradation. *Trends Biochem. Sci.* **20**, 465-470 (1995).
58. P. Loflin, C.-Y. A. Chen, and A.-B. Shyu, Unraveling a cytoplasmic role for hnRNP D in the *in vivo* mRNA destablization directed by the AU-rich element. *Genes Dev.* **13**, 1884-1897 (1999).
59. A. Lapucci, M. Donnini, L. Papucci, E. Witort, A. Tempestini, A. Bevilacqua, A. Niolin, G. Brewer, N. Schiavone, and S. Capaccioli, AUF1 is a *bcl-2* A+U-rich element-binding protein involved in *bcl-2* mRNA destabilization during apoptosis. *J. Biol. Chem.* **277**, 16139-16146 (2002).
60. B. Sarkar, Q. Xi, C. He, and R. J. Schneider, Selective degradation of AU-rich mRNAs promoted by the p37 AUF1 protein isoform. *Mol. Cell. Biol.* **23**, 6685-6693 (2003).
61. W. S. Lai, E. Carballo, J. M. Thorn, E. A. Kennington, and P. J. Blackshear, Interactions of CCCH zinc finger proteins with mRNA: Binding of tristetraprolin-related zinc-finger proteins to AU-rich elements and destabilization of mRNA. *J. Biol. Chem.* **275**, 17827-17837 (2000).
62. W. S. Lai, E. Carballo, J. R. Strum, E. A. Kennington, R. S. Phillips, and P. J. Blackshear, Evidence that tristetraprolin binds to AU-rich elements and promotes the deadenylation and destabilization of tumor necrosis factor alpha mRNA. *Mol. Cell. Biol.* **19**, 4311-4323 (1999).
63. C.-Y. Chen, R. Gherzi, S.-E. Ong, E. L. Chan, R. Raijmakers, G. J. M. Pruijn, G. Stoecklin, C. Moroni, M. Mann, and M. Karin, AU binding proteins recruit the exosome to degrade ARE-containing mRNAs. *Cell* **107**, 451-464 (2001).
64. S. S. Y. Peng, C.-Y. A. Chen, N. Xu, and A.-B. Shyu, RNA stabilization by the AU-rich element binding protein, HuR, an ELAV protein. *EMBO J.* **17**, 3461-3470 (1998).
65. X. C. Fan and J. A. Steitz, Overexpression of HuR, a nuclear-cytoplasmic shuttling protein, increases the *in vivo* stability of ARE-containing mRNAs. *EMBO J.* **17**, 3448-3460 (1998).
66. G. Brewer, An A+U-rich element RNA-binding factor regulates c-*myc* mRNA stability in vitro. *Mol. Cell. Biol.* **11**, 2460-2466 (1991).

67. W. Zhang, B. J. Wagner, K. Ehrenman, A. W. Schaefer, C. T. DeMaria, D. Crater, K. DeHaven, L. Long, and G. Brewer, Purification, characterization, and cDNA cloning of an AU-rich element RNA-binding protein, AUF1. *Mol. Cell. Biol.* **13**, 7652-7665 (1993).

68. B. J. Wagner, C. T. DeMaria, Y. Sun, G. M. Wilson, and G. Brewer, Structure and genomic organization of the human AUF1 gene : alternative pre-RNA splicing generates four protein isoforms. *Genomics* **48**, 195-202 (1998).

69. Y. Arao, R. Kuriyama, F. Kayama, and S. Kato, A nuclear matrix-associated factor, SAF-B, interacts with specific isoforms of AUF1/hnRNP D. *Arch. Biochem. Biophys.* **380**, 228-236 (2000).

70. J. S. Buzby, S. Lee, P. van Winkle, C. T. DeMaria, G. Brewer, and M. S. Cairo, Increased granulocyte-macrophage colony-stimulating factor mRNA instability in cord versus adult mononuclear cells is translation-dependent and associated with increased levels of A+U-rich element binding factor. *Blood* **88**, 2889-2897 (1996).

71. A. Pende, K. D. Tremmel, C. T. DeMaria, B. C. Blaxall, W. A. Minobe, J. A. Sherman, J. D. Bisognano, M. R. Bristow, G. Brewer, and J. D. Port, Regulation of the mRNA-binding protein AUF1 by activation of the β-adrenergic receptor signal transduction pathway. *J. Biol. Chem.* **271**, 8493-8501 (1996).

72. O. I. Sirenko, A. K. Lofquist, C. T. DeMaria, J. S. Morris, G. Brewer, and J. S. Haskill, Adhesion-dependent regulation of an A+U-rich element-binding activity associated with AUF1. *Mol. Cell. Biol.* **17**, 3898-3906 (1997).

73. G. M. Wilson and G. Brewer, Regulation of mRNA stability by AUF1. In *RNA Binding Proteins: New Concepts in Gene Regulation* ed. by K. Sandberg and S. E. Molroney, pp 101-117. Kluwer Academic Publishers, Norwell, MA (2002).

74. C. T. DeMaria, Y. Sun, L. Long, B. J. Wagner, and G. Brewer, Structural determinants in AUF1 required for high affinity binding to A+U-rich elements. *J. Biol. Chem.* **272**, 27635-27643 (1997).

75. G. M. Wilson, J. Lu, K. Sutphen, Y. Suarez, S. Sinha, B. Brewer, E. C. Villanueva-Feliciano, R. M. Ylsa, S. Charles, and G. Brewer, Phosphorylation of p40^AUF1 regulates binding to A+U-rich mRNA-destabilizing elements and protein-induced changes in ribonucleoprotein structure. *J. Biol. Chem.* **278**, 33039-33048 (2003).

76. T. Henics, E. Nagy, H. J. Oh, C. Csermely, A. von Gabain, and J. R. Subjeck, Mammalian Hsp70 and Hsp110 proteins bind to RNA motifs involved in mRNA stability. *J. Biol. Chem.* **274**, 17318-17324 (1999).

77. G. Laroia, R. Cuesta, G. Brewer, and R. J. Schneider, Control of mRNA decay by heat shock-ubiquitin-proteosome pathway. *Science* **284**, 499-502 (1999).

78. G. M. Wilson, K. Sutphen, S. Bolikal, K. Chuang, and G. Brewer, Thermodynamics and kinetics of Hsp70 association with A+U-rich mRNA-destabilizing sequences. *J. Biol. Chem.* **276**, 44450-44456 (2001).

79. G. M. Wilson, J. Lu, K. Sutphen, Y. Sun, Y. Huynh, and G. Brewer, Regulation of A+U-rich element-directed mRNA turnover involving reversible phosphorylation of AUF1. *J. Biol. Chem.* **278**, 33029-33038 (2003).

80. K. B. Hall and D. J. Williams, Dynamics of the IRE RNA hairpin loop probed by 2-aminopurine fluorescence and stochastic dynamics simulations. *RNA* **10**, 34-47 (2004).

81. D. Klostermeier and D. P. Millar, Energetics of hydrogen bond networks in RNA: Hydrogen bonds surrounding G+1 and U42 are the major determinants for the tertiary structure stability of the hairpin ribozyme. *Biochemistry* **41**, 14095-14102 (2002).

82. S. E. Melcher, T. J. Wilson, and D. M. J. Lilley, The dynamic nature of the four-way junction of the hepatitis C virus IRES. *RNA* **9**, 809-820 (2003).

83. X. Zhuang, L. E. Bartley, H. P. Babcock, R. Russell, T. Ha, D. Herschlag, and R. Cuesta, A single-molecule study of RNA catalysis and folding. *Science* **288**, 2048-2051 (2000).

84. Z. Xie, N. Srividya, T. R. Sosnick, T. Pan, and N. F. Scherer, Single-molecule studies highlight conformational heterogeneity in the early folding steps of a large ribozyme. *Proc. Natl. Acad. Sci. USA* **101**, 534-539 (2004).

TIME-RESOLVED EVANESCENT WAVE-INDUCED FLUORESCENCE ANISOTROPY MEASUREMENTS

Trevor A. Smith,[*] Michelle L. Gee and Colin A. Scholes

11.1. ABSTRACT

The technique of time-resolved evanescent wave-induced fluorescence spectroscopy (TR-EWIFS) has been extended to incorporate fluorescence anisotropy measurements. We report on the application of these time-resolved evanescent wave-induced fluorescence anisotropy measurements (EW-TRAMs) as a source of information concerning the motion and conformational changes of macromolecules that can occur near a solid/liquid interface. We have applied EW-TRAMs to studies of the adsorption of proteins and polymers onto silica from solution. We also discuss the implications and potential complications of the polarisation properties of the standing evanescent field and how these properties can be used to advantage.

11.2. INTRODUCTION

The study of macromolecular interactions and dynamics of adsorbed species at an interface is of great importance to the understanding of a variety of phenomena in chemistry, physics and biotechnology.[1, 2] The interfacial properties of polymers and polyelectrolytes are the key to lubrication, stability and rheology in a vast range of colloidal systems.[3] Understanding the adhesion of dyes and polymer films to solid substrates is of critical importance in the optimisation of paints and dyed materials, and in the preparation of device structures such as polymer-based displays. Likewise, the behaviour of proteins at an interface impacts in many fields.[4] For example, the strong tendency to adsorption exhibited by proteins is exploited in areas such as food science

---

[*] School of Chemistry, University of Melbourne, Parkville, 3010, Victoria, Australia

and the cosmetics industry, where proteins are used as stabilisers and emulsifiers. In biocompatibility, protein adsorption is thought to be the first step in biofouling, in which bioadhesion to a surface can have an adverse affect on a wide range of biomedical devices. Filtration, transport and storage systems, as well as ship hulls are also prone to the destructive effects of biofouling.

Understanding the extent, specificity and dynamics of polymer and protein interactions upon the introduction of a macromolecule to an interface, plays a critical role in the development of these fields. Since polymer and protein structure is integral to the performance, function and properties of the macromolecule at the interface, any insight that can be acquired regarding structural/conformational changes the macromolecule undergoes as a result of its interaction with the solid substrate, is highly sought after. There is, therefore, an increasing interest in the development of powerful methodologies to enhance and extend surface characterisation methods to study polymer and protein surface activity. Studies involving the use of indirect experimental methods such as the measurement of dynamic surface tension and surface pressure have provided some information on macromolecular adsorption.[5-8] However, with these techniques, information on macromolecule conformation must be inferred. Photon correlation spectroscopy (PCS)[9-11] and specular neutron or X-ray reflectivity, (SNR),[12-14] and (SXR)[15] have also been used to study the conformation of adsorbed macromolecules such as proteins. The use of non-linear optical effects (in particular second harmonic and sum-frequency generation) at surfaces and interfaces is another means of achieving a high degree of surface specificity.[16-20] These methods can provide useful information regarding orientation and order of small molecular species at an interface but can be difficult to interpret in macromolecular systems. Another attractive way to probe interfacial phenomena is through the utilisation of total internal reflection(TIR)-based spectroscopic methods in which the penetrating evanescent field is used to photo-excite an absorbing species positioned within the penetrating field.[21] Such techniques can be used at the interface between many phases (e.g. liquid/solid, liquid/liquid, solid/solid etc.). In what follows, we discuss the use of time-resolved fluorescence evanescent wave techniques that we have used successfully in studies of the conformation of macromolecules at a silica/solution or silica/polymer film interface.

The basic principles behind evanescent wave spectroscopy are illustrated schematically in Figure 11.1. When light transmitted in a medium of refractive index, n_1, encounters a medium of lower refractive index, n_2, two process can occur: when the angle of incidence is less than the critical angle, θ_c, defined by Snell's law as $\theta_c = \sin^{-1}(n_2/n_1)$, refraction occurs. For angles of incidence greater than θ_c total internal reflection (TIR) occurs. Maxwell's equations of electromagnetic radiation predict that, upon TIR, a small amount of the light field penetrates the lower refractive index medium in the form of a standing wave known as an evanescent wave.[22]

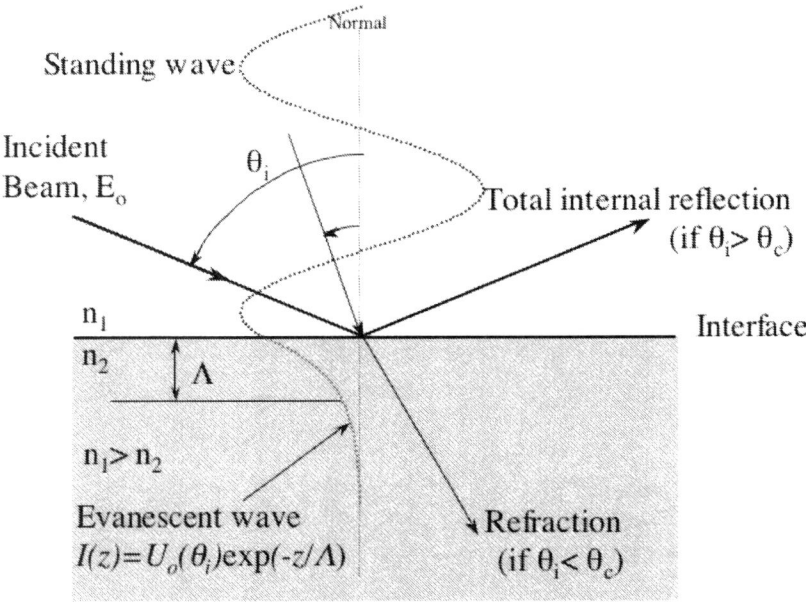

Figure 11.1. Schematic representation of the Total Internal Reflection (TIR) principle and the resulting standing evanescent wave. Λ is the penetration depth of the evanescent field (defined in text) which is related to the angle of incidence of the excitation light, θ_i, relative to the normal. z is the distance away from the interface.

The intensity profile within the evanescent field is related to the square of the E-field and decays exponentially with distance, z, from the interface through Eq. (1):

$$I(z) = U_0(\theta)\exp\left(\frac{-z}{\Lambda}\right) \tag{1}$$

where $U_0(\theta)$ $(= E_o^2)$ is the intensity of the light at the interface, and the penetration depth, Λ, is defined as the distance into medium 2, normal to the interface, at which the light intensity of the evanescent wave has decreased to $1/e$ of its interface value. Λ is related to the wavelength of the excitation radiation, λ, the angle of incidence, θ_i, the refractive index of medium 1, n_1, and the critical angle for the pair of media, θ_c, through Eq. (2):

$$\Lambda = \frac{\lambda}{4\pi n_1} \bullet \frac{1}{\sqrt{\sin^2\theta_i - \sin^2\theta_c}} \tag{2}$$

If a light absorbing species is located within close proximity to an interface formed between two media of differing refractive indices, the intensity of the totally internally reflected beam will be reduced due to extraction of the light field intensity through absorption of the evanescent field penetrating the medium 2. The evanescent field can thus be used at a specific wavelength to probe species within the interfacial region as a function of distance from the interface by varying the depth of penetration of the evanescent field into medium 2 (usually over the order of tens of nanometres) by altering the incident angle or the wavelength of the excitation light, or through changes in the refractive index of one of the media. The disadvantage of absorption-based TIR techniques[23-27] is that they rely on either spectral changes or are otherwise restricted to intensity information that is related to the concentration and extinction coefficient of the absorbing species within the evanescent field. Additionally, limited attenuation of the transmitted beam results in limited sensitivity, usually requiring multiple reflection geometries to achieve adequate signal intensities.

TIR techniques can be extended to the fluorescence domain, which affords several advantages over absorption-based TIR methods. If the species absorbing the evanescent wave intensity is also fluorescent, the evanescent field can induce fluorescence in the interfacial region, the intensity, I_f, of which is described by:

$$I_f \propto \int_0^\infty \phi_f(z)c(z)U_0 \exp\left(-\frac{z}{\Lambda}\right)dz \qquad (3)$$

where ϕ_f is the fluorescence quantum efficiency of the fluorophore, $c(z)$ is the fluorophore concentration profile in the z direction and U_0 is the fluorescence intensity at z=0. The inherently higher sensitivity of fluorescence-based methods over transmission-based techniques permits the evanescent wave-induced fluorescence (EWIF) approach to employ a single reflection geometry with which to probe the interfacial region, although a multiple internal reflection fluorescence geometry has been used in the early work.[28]

This combination of the evanescent wave principle with conventional fluorescence spectroscopy[29, 30] takes advantage of the inherent multidimensionality and sensitivity of fluorescence-based techniques to provide a reasonably surface-specific method of characterising processes in the interfacial region. The process of emission yields information that is intrinsically related to the nature of the fluorophore and its immediate microenvironment through variations in viscosity, polarity, pH, molecular motion etc.[31] Hence, in studying macromolecules at an interface, the emission resulting from photo-excitation by the evanescent field of a fluorescent species inherent to, or physically or chemically attached to, the macromolecule, is detected. Additionally, by making use of the dependence of the evanescent field penetration depth on the angle of incidence (Eq. 2), fluorescence can be measured as a function of the distance from the interface thus giving spatial information regarding the chromophore's distribution at the interface. EWIFS techniques can, in principle, give information about the density distribution of adsorbed polyelectrolytes by monitoring the intensity of induced fluorescence as a function of Λ. EWIF spectroscopy (EWIFS) has been used to characterise the interfacial properties of many species, ranging from small dye molecules to macromolecules such as proteins, polymers, enzymes and DNA.[32-36] The surface specificity of the evanescent wave has also been successfully used in many biosensing applications.[32] TIR-based fluorescence imaging is now also gaining great popularity[37] and instrumentation for such applications is now commercially available through the major microscope companies.[38-40]

EWIFS has also been combined with fluorescence correlation spectroscopy to monitor the association/dissociation rates, local diffusion coefficients and absolute densities of fluorescent molecules at the interface of a solution and a planar substrate.[41, 42] Atomic force microscopy has also been combined with EWIFS for the study of biological systems.[43, 44]

In practice, steady-state EWIF measurements share some similar limitations to absorption-based TIR methods. Principally, if there are no spectral changes upon adsorption, then the steady state fluorescence signal can only contain intensity information that will be proportional to both the extent of light absorption (through concentration, excitation pathlength and extinction coefficient) and the fluorescence quantum yield. In much steady state EWIFS work to date, it has been assumed that the fluorescence quantum efficiencies are known and invariant with distance from, and normal to the interface. Under these interpretation constraints, the EWIF intensity measurements are related directly to the surface excess of an adsorbed species. However, some uncertainty exists in the area of EWIFS concerning the effects of the surface itself on the photophysical properties of the fluorophore, and more specifically, the fluorescence quantum yield and fluorescence decay time,[22] as discussed below. It must be considered that even a 'simple' interface will almost certainly contain more than one type of fluorescent centre[32] thus providing a heterogeneous environment for molecular species.[45] Emission characteristics, such as fluorescence quantum yield, will therefore be dependent on the precise location of individual fluorophores within this environment. Thus the interpretation of steady-state measurements is restricted.[45]

The incorporation of time resolution into EWIFS, i.e. TR-EWIFS (time resolved evanescent wave induced fluorescence), has the potential to resolve some of the limitations of steady state measurements due to any changes in the photophysical properties of the fluorophore being reflected in the fluorescence decay profile.[22] Furthermore, TR-EWIFS has the capability to detect any variation of fluorescence quantum yield as a function of distance normal to the interface[22, 45] through varying the angle of incidence of the excitation light in order to vary the penetration depth, Λ, of the evanescent wave (Eq. 2), and monitoring the time-resolved fluorescence decay profiles from a fluorophore within an adsorbed layer. TR-EWIFS has been used to great effect by a number of authors, employing both time- and frequency domain methods on a variety of time-scales and on a range of experimental systems.[22, 29, 31, 32, 45-64] Time-resolved fluorescence imaging by evanescent wave excitation has also been demonstrated.[65] Despite the advantages of time-resolved EWIF measurements over steady-state methods, we will show that even TR-EWIFS has its limitations.

Variable angle ("depth profiling") TR-EWIFS can be further extended to evanescent wave-induced time-resolved fluorescence *anisotropy* measurements, EW-TRAMs; a technique well suited to the study of fluorescence depolarising processes whose dynamics occur on the nanosecond and sub-nanosecond time scales such as chromophore/segmental motion, energy transfer and energy migration, etc. For a species in bulk solution, the fluorescence decay profile from a fluorophore is monitored, following excitation with vertically polarised light, through polarisers set parallel and perpendicular to the polarisation of the excitation light, i.e. $I_{VV}(t)$ and $I_{VH}(t)$ respectively. The time-resolved fluorescence anisotropy function, r(t), (Eq. 4) contains information on the motion of the fluorophores on the time scale of the lifetime of the excited state:

$$r(t) = \frac{I_{VV}(t) - I_{VH}(t)}{I_{VV}(t) + 2I_{VH}(t)} \tag{4}$$

If the photo-excitation is achieved via a penetrating evanescent field, it is possible to study chromophore motion within the interfacial region, which can provide information relating to the binding of an adsorbate at an interface. Furthermore, such binding can be monitored as a function of distance from the interface simply by varying the penetration depth, Λ, of the field (Eq. 2).

In this paper we discuss the use of TR-EWIF measurements in studies of the adsorption of Bovine Serum Albumin (BSA) from aqueous solution onto a silica surface, *in situ*, and as a function of distance normal to the solid/solution interface. We have measured the time-resolved fluorescence decay behaviour of BSA using 1-anilinonaphthalene-8-sulfonic acid (ANS) which is a fluorescent probe whose fluorescence decay kinetics depend on the probe's microenvironment such that, in a polar microenvironment, the fluorescence is rapidly and efficiently quenched. We also discuss the influences on the fluorescence decay characteristics of fluorophores near an interface that can pose problems in the interpretation of fluorescence decays collected by TR-EWIF measurements.

We also introduce the technique of variable angle TR-EWIFS coupled with fluorescence anisotropy measurements and show that these measurements can be used as a source of additional information, in particular regarding the molecular motion and conformation changes of a macromolecule at an interface. There is vast potential for EW-TRAMs to provide spatial information on the fluorescence depolarising processes near an interface, however, to date this has been under exploited. To illustrate the potential of EW-TRAMs, we have applied EW-TRAMs to studies of the adsorption of proteins (e.g. the BSA/ANS system) and polymers onto silica from solution. The implications and potential complications of the polarisation properties of the standing evanescent field in EW-TRAMs are discussed, along with the experimental considerations and instrumental corrections that are needed to account for these complications in order to maximise the information available from such measurements. This includes preliminary results of the excitation polarisation dependence on the time-resolved fluorescence anisotropy profiles of a fluorescent probe incorporated in a polymer film.

11.3. TIME-RESOLVED FLUORESCENCE MEASUREMENTS NEAR AN INTEREFACE

The fluorescence quantum yield and excited state lifetime of a fluorophore near a dielectric interface are known to differ from those of the same fluorophores in bulk solution. The fluorescence quantum yield is defined as:

$$\phi_f = \frac{k_r}{k_r + \sum k_{nr}} \tag{5}$$

where k_r is the radiative rate constant describing fluorescence, and Σk_{nr} is the sum of non-radiative rate constants for deactivation of the excited state encompassing vibrational deactivation, intersystem crossing, internal conversion etc.

Clearly, changes in either k_r or k_{nr} will alter ϕ_f. If k_r dominates over Σk_{nr} then changes in k_r will not alter ϕ_f appreciably since these will cancel from numerator and denominator, but changes in the rates of the non-radiative processes can influence ϕ_f. On the other hand, if Σk_{nr} is much greater than k_r, variations in Σk_{nr} or k_r can impose a quite marked influence on the fluorescence quantum yield. Since steady-state fluorescence intensity measurements depend on the absorptivity of the fluorophore, its concentration and fluorescence quantum yield, as well as instrumental collection efficiency considerations (as a function of incident and observation angles),[66] changes in apparent emission intensity can be deceptive. The fluorescence lifetime, τ_f, is defined as the inverse of the sum of the rate constants of all deactivating pathways of the excited state, ($\tau_f = 1/(k_r + \Sigma k_{nr})$) and is therefore related to the fluorescence quantum yield ($\phi_f = k_r \tau_f$). The fluorescence quantum yield is proportional to the integrated area beneath the fluorescence decay profile, and changes in ϕ_f should be reflected through proportional changes in τ_f. Time-resolved fluorescence measurements are therefore less susceptible to the effects influencing steady-state intensity measurements. An excellent review discussing time-resolved techniques in the study of fluorophores at an interface has been published elsewhere.[63]

Both the radiative and non-radiative rate constants can be affected by the presence of a dielectric medium interface. Adsorption of a fluorophore to an interface is commonly accompanied by a decrease (or increase) in the fluorescence quantum yield. There are several reports of increased emission quantum yield of dyes near an interface, as exposed by a lengthening of fluorescence decay curves as the penetration depth is reduced.[22, 45, 46] This is particularly well illustrated by the example of malachite green which has a very low fluorescence quantum yield in bulk aqueous solution but exhibits significantly enhanced emission when adsorbed at an aqueous/quartz interface.[46] Such increases in quantum yields are often attributed solely to changes in k_{nr}, induced by the inhibition of vibrations of the flexible bonds of the fluorophore through interactions between the fluorophore and the interface. k_r is often assumed to remain unchanged.

However, the radiative rate can also be affected by the presence of a dielectric interface, due in part because of interfacial electromagnetic boundary conditions imposed on the radiating field of a dipole. This is primarily because the interface reflects the radiating field produced by the fluorophores. If the reflection returns to the fluorophore in phase, emission will be enhanced and the apparent decay rate will increase. The stronger the reflection, the greater the potential for modification to the rate by this mechanism. Strong variations in fluorescence lifetime and quantum yield as a function of distance from dielectric interfaces have been reported for reflective metallic and non-metallic interfaces.[66-74] However, as discussed by elsewhere[31] this effect is often expected to be minor for the fluorophore/silica systems of the type under consideration here.

Clearly time-resolved measurements present several advantages over the steady-state counterparts, especially in evanescent wave-induced fluorescence experiments. Time-resolved fluorescence measurements have the added ability to determine the existence of multiple emitting species through multiple discrete fluorescence decay times. Even a 'simple' interfacial region may well comprise a heterogeneous environment for molecular species.[32, 45] Emission characteristics, such as fluorescence quantum yield and hence fluorescence decay time, will therefore be dependent on the precise location of individual fluorophores within this environment. In the extreme, fluorescence decay profiles recorded from a fluorophore near an interface may be best modelled on the basis of a distribution of fluorescence decay times, but these distributions can still be used to identify different environments experienced by the fluorophore.

11.4. FLUORESCENCE ANISOTROPY NEAR AN INTERFACE

Whilst fluorescence anisotropy methods are well established for studying species in bulk solution, the situation is far more complex following evanescent wave excitation of interfacial species. Firstly, the intensity of emission from an oriented dipole at a dielectric interface is found to be dependent on the orientation of the electric dipole moment of the fluorophore with respect to the plane of the interface[31, 67] (and hence the polarisation of the incident light). The angular distribution of a fluorophore's emission can be represented by a radial distribution function.[74, 75] In addition to being dependent on the distance of the fluorophore from the interface, the intensity of emission from an oriented dipole at a dielectric interface is also dependent on the detection angle relative to the interface,[66, 70, 73, 76] as shown in Figure 11.2 for a dipole oriented parallel to the plane of the interface.

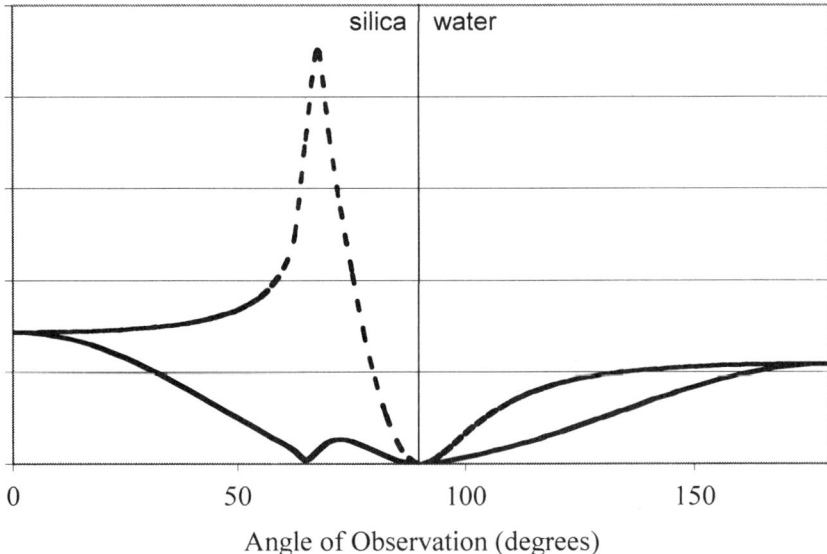

Figure 11.2. Polarisation-dependent angular distribution of emission from a dipole oriented parallel to a silica/water interface as a function of observation angle. Dipoles located on the water side of the interface, solid line: *p*-polarisation, dashed line: *s*-polarisation.

The greatest difference between s and p polarised excitation is seen to occur at observation angles equal to the critical angle of the system, and is least when emission collection is normal to the interface. The apparent fluorescence quantum yield and decay time are thus highly susceptible to the polarisation biases inherent in fluorescence anisotropy measurements.

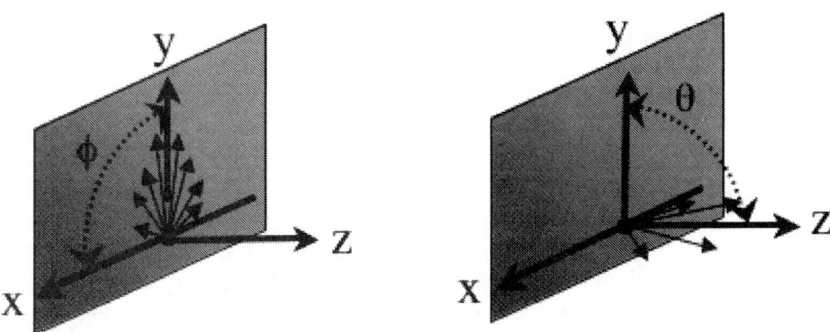

Figure 11.3. Schematic diagram illustrating the frame of reference and the in- and out- of plane rotations of an emission dipole positioned at an interface. Depolarisation within the plane of the interface is designated, ϕ, and out of the plane, θ

Furthermore, while propagating light can be polarised with respect to two orientations, within the evanescent (standing) wave, a component of the E-field exists in each of the three directions. While the penetration depth Λ is polarisation independent[77] the energy density at $z=0$, E_0, depends on the polarisation of the incident light. For unpolarised excitation, the expressions for the electric field amplitudes, E_0, and the intensity of light, U_0 $(= E_0^2)$, in the three directions (as defined in Figure 11.3) at the interface in the rarer medium are, respectively:[21, 49]

$$E_{0,y}(\theta_i) = \frac{2\cos\theta_i}{\left(1 - n_{21}^2\right)^{1/2}} \qquad U_{0,y}(\theta_i) = \frac{4\cos^2\theta_i}{\left(1 - n_{21}^2\right)} \qquad (6)$$

$$E_{0,x}(\theta_i) = \frac{2\cos\theta_i\left[\sin^2\theta_i - n_{21}^2\right]^{1/2}}{\left(1 - n_{21}^2\right)^{1/2}\left[\left(1 + n_{21}^2\right)\sin^2\theta_i - n_{21}^2\right]^{1/2}} \qquad U_{0,x}(\theta_i) = \frac{4\cos^2\theta_i\left[\sin^2\theta_i - n_{21}^2\right]}{\left(1 - n_{21}^2\right)\left[\left(1 + n_{21}^2\right)\sin^2\theta_i - n_{21}^2\right]} \qquad (7)$$

$$E_{0,z}(\theta_i) = \frac{2\sin\theta_i\cos\theta_i}{\left(1 - n_{21}^2\right)^{1/2}\left[\left(1 + n_{21}^2\right)\sin^2\theta_i - n_{21}^2\right]^{1/2}} \qquad U_{0,z}(\theta_i) = \frac{4\sin^2\theta_i\cos^2\theta_i}{\left(1 - n_{21}^2\right)\left[\left(1 + n_{21}^2\right)\sin^2\theta_i - n_{21}^2\right]} \qquad (8)$$

In Equations (6-10), $n_{21} = \sin\theta_c = n_2/n_1$. For an interface that is oriented in the vertical direction, vertically polarised excitation corresponds to s polarisation and horizontally polarised excitation corresponds to p polarisation. For s polarised excitation the intensity of light at the interface, $U_0^s(\theta_i)$ is given by $U_{0,y}(\theta_i)$ whilst for p polarised excitation, $U_0^p(\theta_i)$ is given by $U_{0,x}(\theta_i)+U_{0,z}(\theta_i)$. Hence:

$$U_0^s\left(\theta_i\right)=\frac{4\cos^2\theta_i}{\left(1-n_{21}^2\right)} \tag{9}$$

$$U_0^p\left(\theta_i\right)=\frac{4\cos^2\theta_i\left[2\sin^2\theta_i-n_{21}^2\right]}{\left[n_{21}^4\cos^2\theta_i+\sin^2\theta_i-n_{21}^2\right]} \tag{10}$$

In other words, when the evanescent wave is generated from s polarised light, molecules with some vector component of their absorption transition dipoles oriented parallel to the polarisation of the incident light (i.e. in the y direction) are excited preferentially, whereas for p polarised incident light, dipoles oriented both parallel (with a component in the x direction) and perpendicular to the interface (z direction) can be excited simultaneously.[21] Emission from both dipole orientations are then detected simultaneously and with different efficiency depending on the detection frame of reference, as discussed above.

These features of polarised evanescent wave-induced fluorescence measurements might be considered as undesired complicating factors. However, in principle they can potentially be exploited to provide a means by which the orientation of fluorophores in an interfacial region can be deduced through differences in molecular motion in two orthogonal planes,[78, 79] as illustrated schematically in Figure 11.3.

In a typical experiment, duplicate measurements at a particular angle of incidence (penetration depth) are carried out in which s (vertically) and p (horizontally) polarised excitation light is used and emission corresponding to all four permutations of the excitation and emission polarisation orientations (I_{VV}, I_{VH}, I_{HV} and I_{HH}) is collected. The anisotropy functions defined in Eqs. (11) and (12)[78] (c.f. the corresponding equation for the bulk, Eq. 4) are then calculated. (Note I_{XY}, X corresponds to the polarisation of the excitation radiation and Y corresponds to the orientation of the emission polarisation).

In-plane:
$$r_\phi=\frac{I_{YY}-I_{YX}}{I_{YY}+I_{YX}} \tag{11}$$

Out-of-plane:
$$r_\theta=\frac{I_{ZY}-\frac{1}{2}\left(I_{YY}+I_{YX}\right)}{I_{ZY}+I_{YY}+I_{YX}} \tag{12}$$

A consequence of these equations is that fluorescence anisotropy can be measured in- and out-of-the plane of the interface following excitation by both s and p polarised incident light, thus providing information relating to possible differences in the modes of emission depolarising processes (such as molecular rotation) through differences between the fluorescence anisotropy. In light of the discussion above concerning the advantages of time-resolved, compared with steady-state, fluorescence measurements in evanescent wave applications, and the well documented value of time-resolved fluorescence anisotropy measurements (TRAMs), EW-TRAMs are expected to prove a valuable source of information. Changes in EW-TRAMs behaviour can be illustrative even without rigorous analytical interpretation, but if the TRAMs profiles can be interpreted analytically, a great deal of additional information can be gained. Clearly, there is vast potential for TRAMS coupled with variable angle, evanescent wave excitation to provide spatial information on the fluorescence depolarising processes near an interface. However, while a limited number of steady state EWIF anisotropy measurements have been published,[49, 78, 80-83] remarkably few studies taking advantage of the time-resolved aspects appear to have been reported to date.[79, 84-89]

11.5. INSTRUMENTATION/INSTRUMENTATION DEVELOPMENT

The majority of the limited number of TR-EWIFS instruments reported have principally used the time-correlated single photon counting (TCSPC)[90, 91] detection technique, although frequency domain detection[62] and the use of an image intensifier/optical multichannel analyser system has also been used.[65, 92] TCSPC is a widely used tool, with exceptional sensitivity, dynamic range, and flexibility and is ideally suited to TREWIFS and hence TCSPC was used here, as described below.

The EWIFS apparatus was constructed in-house and based on the hemicylindrical prism designs reported previously.[32, 49, 56, 59] This arrangement is especially useful for depth profiling experiments since, in this configuration, the incident and totally internally reflected light entering and exiting the prism normal to the surface of the cylindrical part of the prism, regardless of the angle of incidence relative to the interfacial region, thus minimising complications due to refraction effects and spurious reflections within the prism. In most reports employing this configuration, the hemi-cylindrical prism is mounted on a rotation stage to facilitate the alteration of the angle of incidence relative to the interface. Details of our original experimental rig adopting this configuration (Figure 11.4a) have been reported previously.[89]

For isotropic solutions of fluorescent species the contribution of molecular reorientation to the fluorescence decay can be eliminated by detecting the emission through a polariser aligned at $54.7°$ relative to the excitation polarisation. For evanescent wave excitation on a macroscopic scale, this "magic angle" may not be $54.7°$ and may be a function of the angle of incidence and the refractive indices of the two media constituting the interface. Wirth et al.[78, 79] have derived a mathematical expression for the magic angle condition under evanescent wave conditions. Evanescent wave excitation polarised at the appropriate magic angle combined with emission polarised at $45°$ in the X,Y plane collected normal to the interface, eliminates contributions from molecular

(a)

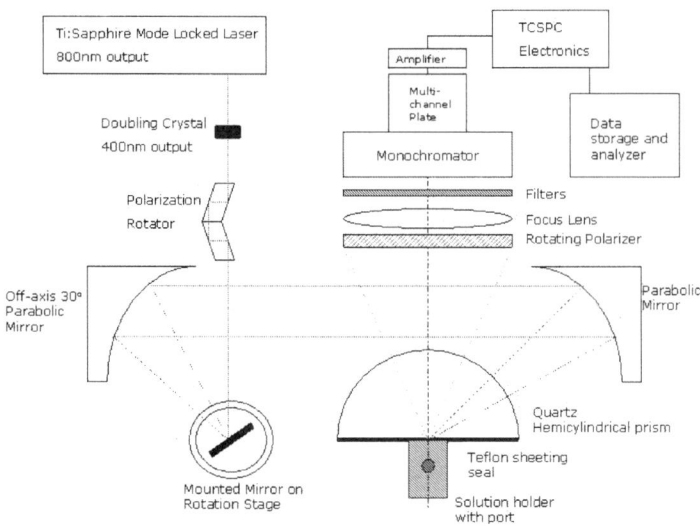

(b)

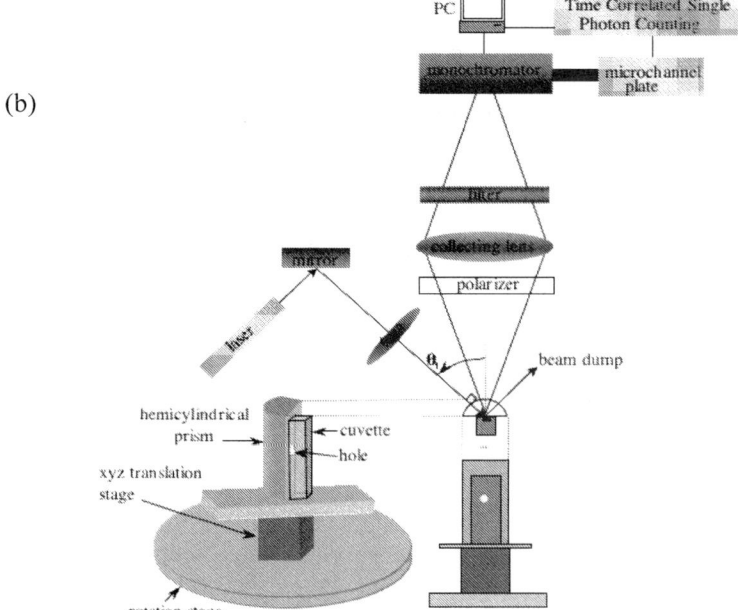

Figure 11.4. Schematic representation of the optical arrangement used for TR-EWIFS and EW-TRAMs (a) original configuration (b) modified arrangement that facilitates collection of emission normal to the interface regardless of incident angle.

reorientation to the fluorescence emission regardless of orientation distribution of ground states with respect to the incident angle.

The angular distribution of fluorescence from liquids following evanescent wave excitation[66] represented by a radial distribution function,[74, 75] such as that Figure 11.2, adds weight to the premise that collection of the emission should be along the Z-axis, i.e. normal to the interface, in order to minimise the polarisation dependencies discussed above. To optimise the detection of polarised emission, we have modified our experimental configuration as illustrated in Figure 11.4b. This modified EWIFS configuration ensures that the emission is collected normal to the interface for all angles of incidence. The excitation source used for these measurements was the frequency doubled (~400 nm) output of a femtosecond Kerr lens mode-locked titanium:sapphire laser (Coherent, Mira 900F-Innova 400) operating at ~ 800 nm and a pulse repetition rate of 76 MHz. The repetition rate of the mode-locked laser pulse train was reduced with a home-built pulse-picker based on a TeO_2 Bragg cell (Gooch & Housego, UK) and synchronised driver electronics (CAMAC, CD5000). The polarisation of the excitation beam is changed manually using a silica double-rhomb polarisation rotator (Halbo) to allow polarisation-dependent emission decay profiles to be recorded for all permutations of excitation and emission polarisation pairs, i.e. $I_{VV}(t)$, $I_{VH}(t)$, $I_{HV}(t)$ and $I_{HH}(t)$. The vertically, s, or horizontally, p, polarised beam is incident on a plane mirror mounted on a rotation stage arranged so that the rotation axis is at the focal point of a $30°$ off-axis parabolic reflector (Edmund Optics). A second identical off-axis parabolic reflector re-converges the excitation beam to a point on the sample. Through this configuration we can vary the angle of incidence over $15°$ while maintaining the emission collection at $90°$ to the interface (Figure 11.4b), and simultaneously maintaining the condition that the incident and totally internally reflected light enters and exits the prism normal to the exterior surface.

The emission was collected through a rotatable, large aperture emission polarisation analyser (King) mounted as the first optical component in order to minimise polarisation effects within the lens and other optical components, and imaged with a wide-angle telephoto camera lens (Nikon). For fluorescence decay measurements, the *excitation* polarisation was oriented at the appropriate magic angle relative to the vertical frame of reference and emission was analysed at $45°$, as discussed by Wirth.[93] For emission anisotropy measurements the rotatable emission polarisation analyser was aligned at $0°$ and $90°$ relative to the polarised excitation to measure the two fluorescence decays for a given excitation polarisation by toggling (at 30 second intervals) the memory group of the multichannel analyser (MCA). The other pair of fluorescence decays corresponding to the orthogonal excitation polarisation was collected sequentially in the same manner.

The collected emission was passed through suitable filters and onto the slit of a 1/4-metre monochromator (CVI Digikrom CM110). The spectrally- and polarisation-analysed fluorescence was detected by a microchannel plate photomultiplier (Eldy, St. Petersburg, Russia, model EM1-132) and recorded using conventional TCSPC electronics described elsewhere.[94] Data were transferred from the MCA to a computer where the experimentally obtained $I_{ExcEm}(t)$ decays were used to generate the time resolved anisotropy, $r(t)$, functions (Eqs. 4, 11 and 12). An instrumental correction factor, G, discussed below, was included to take into account the bias of the detector system (in particular due to the monochromator) to the polarisation. Fluorescence decay data were usually collected to at least 20,000 counts at the channel of maximum intensity in order to generate time-resolved fluorescence anisotropy decay curves of adequate signal to noise ratio.

11.5.1. Materials and cleaning methods

All of the chemical reagents were used without any further purification. All experiments were carried out at room temperature (approx. 20° C). Prior to each experiment, the hemicylindrical prism was cleaned by first being soaked in acetone (AR), followed by washing in a warm ammonia/peroxide solution, then rinsed in copious amounts of Milli-Q water ($18M\Omega$/cm).

11.5.2. BSA/ANS Solutions:

Bovine serum albumin (BSA) and 1-anilinonaphthalene-8-sulfonic acid (ANS) were purchased from Sigma Chemical Co. The solvent used for the BSA/ANS work was a 0.01 M acetate buffer solution, giving a pH of 5. For TR-EWIF measurements and EW-TRAMs, the BSA/ANS complex was allowed to adsorb to the silica surface from buffered solutions (BSA concentration ~100 ppm). It is assumed that adsorption was complete after 3 hours contact, based on the observation of negligible variation in the fluorescence decays after this time. TR-EWIFS experiments were carried out as a function of angle of incidence to allow depth profiling of the interfacial region. Two sets of experimental conditions were used, referred to as Type A and Type B. Type A measurements were performed on BSA/ANS adsorbed from solution, maintaining equilibrium with BSA-ANS in the bulk. A set of fluorescence decay profiles was obtained for a variety of angles of incidence ranging from 69° to 89° with 4° increments. Thus, the penetration depth ranged from 100 nm to 40 nm (Eq. 2). At the highest angle of incidence employed, the depth of the penetration (i.e. Λ=40 nm) is longer than the thickness of the interfacial region. Therefore, in these measurements, the fluorescence intensity has a contribution from the surface region, in addition to a significant proportion of emission from any free BSA/ANS complex remaining in bulk solution. The Type B measurements were performed after the bulk solution of BSA/ANS had been replaced by thorough flushing of the cell with acetate buffer solution to remove any non-adsorbed or loosely adsorbed BSA. This procedure ensured that the silica surface was always in contact with aqueous solution. Even though the evanescent field penetrates beyond the interfacial region, the bulk buffer solution will not contribute to the fluorescence intensity. Hence, in type B measurements, the fluorescence decays are comprised solely of the signal from the surface-adsorbed form of the BSA/ANS complex. EW-TRAMs were also carried out as a function of penetration depth of the evanescent wave.

The majority of this work employed the apparatus shown in Figure 11.4(a).[89] To characterise the anisotropy of non-adsorbed BSA, TRAMs were also performed in bulk solution with the sample solution contained in a standard silica quartz cuvette and the emission collected in the usual right angle geometry. A full description of the instrumentation used for the bulk measurements is given elsewhere.[95]

11.5.3. Fluorophore Doped Polymer Films

Poly(acrylic acid) (PAA) was synthesised in our laboratory previously. PAA films doped with acridine (L.Light and Co.) were spin-coated from water or acetone onto glass microscope coverslips (Menzelglaser), from a ~6w/w% polymer solution, providing a film thickness of ~100-200nm. The fluorophore doped polymer films on the glass substrate were clamped to the prism's face with index-matching fluid between the film and the silica prism. The refractive index for the polymer film was assumed to be close to that of the solvent, since visual inspection indicated that the critical angle of the silica/polymer interface was approximately the same as for the glass-water and glass-acetone systems i.e. in the absence of polymer. Fluorescence of acridine was detected at the wavelength of peak emission intensity, 470nm for water cast films, and 520nm for films cast from acetone. Incident angles were varied over a range of five degrees greater than the relevant critical angle.

A correction term, (or "G-factor", Eq. 13) is usually required in calculating the fluorescence anisotropy function to account for any polarisation bias in the detection equipment, in particular due to the emission monochromator.[96] In bulk solution this can be measured in a number of ways[91] including the use of excitation light polarised parallel to the line of detection (i.e. horizontally polarised excitation in a right angle geometry). Another method is to use a rapidly rotating fluorophore so emission should be completely depolarised thus providing equal emission intensity in both polarisation analyser orientations. Alternatively, a value for the G-factor can be deduced from anisotropy measurements of a fluorophore with well characterised emission polarisation properties if it can be completely immobilised.

$$r(t) = \frac{I_{VV}(t) - GI_{VH}(t)}{I_{VV}(t) + 2GI_{VH}(t)} \quad \text{with} \quad G = \frac{I_{HV}}{I_{HH}} \tag{13}$$

EW-TRAMs also requires correction for polarisation bias, however, determining the correction factor is more complex than in the bulk solution case. To obtain the G-factor ratio for EW-TRAMs, p-polarised excitation light must be incident on the interface under critical angle conditions. Absorption transition dipoles oriented in the z-direction are then preferentially excited and, under the assumption that the emission depolarisation is the same in both the x and y-orientations, emission intensity in the two detection planes should be equal. Any discrepancy can then be attributed to detection equipment bias which can be evaluated using the above relationship (Eq. 13). In practice, determining the G-factor for EW excitation has proven challenging, but efforts to date using a range of techniques indicate that the G-factor correction term required to account for the polarisation effects is, surprisingly and fortuitously, very close to unity for our apparatus. It may be possible that the effects induced by the individual dipoles, which give rise to the structure in the radial distribution function, are averaged out due to the wide excitation and collection zone of our system, thus reducing the overall effect. This postulation is currently under further investigation.[97] An overall G-factor, accounting for the polarisation bias of both these effects and the detection system was used in the following discussion.

11.6. RESULTS AND DISCUSSION

11.6.1. Adsorption of BSA/ANS to silica

ANS is an extraordinarily sensitive fluorescent probe for determining the microscopic viscosity and polarity of chemical and biological systems. It is well known that under aqueous conditions, ANS is virtually non-fluorescent (the quantum yield, ϕ_f=0.004) due to an efficient fluorescence quenching mechanism.[98] A decrease in polarity of the ANS environment leads to both an increase in the fluorescence lifetime (and a concomitant increase in fluorescence quantum yield, ϕ_f=0.98) and a slightly blue-shifted fluorescence maximum.[98] It is well documented that neither unbound ANS nor BSA alone fluoresce significantly in buffered aqueous solution under the excitation and emission wavelengths used here (λ_{exc}~390 nm, λ_{em}~490 nm). When no BSA is present, no hydrophobic regions exist into which the ANS can partition, and the emission of ANS is quenched very efficiently.

The adsorption of BSA onto silica has been investigated in detail previously by EWIFS methods using either the intrinsic fluorescence of BSA[52] following direct excitation of the amino acids, especially tryptophan, (around 344 nm in bulk solution and ~333 nm in close proximity to surfaces),[98] or a variety of fluorescence probes including ANS.[47, 50, 84, 98-100] The behaviour of ANS itself near interfaces has also been investigated.[51, 101] The work to date has indicated that upon adsorption to silica, BSA undergoes some degree of conformational change.[52, 100] Hlady et al.[50, 98, 99] present data that indicate the presence of some sort of surface aggregates at high BSA surface concentrations, in agreement with the existence of two-dimensional aggregates of adsorbed BSA concluded by others. These authors also interpret a change in the distance between energy donors and acceptors as a result of conformational changes (partial unfolding of adsorbed BSA) upon adsorption.[99] Fukumura and Hayashi[84] also reported that BSA undergoes conformational change upon adsorption to a silica surface. In addition, Hlady and Andrade[98] reported that the conformation change involves embedding of the tryptophan group of BSA within the more hydrophobic regions of the coil, while the ANS probe groups experience a more polar environment (hence fluorescence quenching) through increased exposure to more hydrophilic regions.

Since there is negligible spectral shift of ANS emission profile upon complexation with BSA or upon the adsorption of the BSA/ANS complex to the silica surface, time-resolved methods present clear advantages over steady-state methods. The capacity of variable angle TR-EWIF measurements to yield information on the conformation of a protein at an interface has been illustrated elsewhere,[89] in which TR-EWIF decay measurements were recorded from the silica-adsorbed BSA/ANS complex as a function of excitation angle relative to the interface (and hence penetration depth). The fluorescence decay profiles obtained by EW excitation were shown to be significantly shorter than those of the BSA/ANS complex in bulk solution, and shortened further as the penetration depth was reduced. The decay curves collected under Type B conditions showed that the fluorescence decay of ANS in the BSA/ANS complex is most rapid when the penetration depth, Λ, is smallest. This implies that the BSA unfolds upon adsorption to the silica thus exposing the BSA-bound ANS groups to an increasingly more polar microenvironment the closer the ANS is to the silica surface, and significantly more polar than that experienced by the ANS in the BSA/ANS complex in bulk solution. The BSA unfolds to the greatest extent at the silica surface so as to optimise its interaction with silica, and further away from the silica surface, an adsorbed BSA molecule is increasingly more coiled and so retains increasingly more hydrophobic sites where the ANS resides. The time-resolved fluorescence decay results illustrate the presence of a hydrophobicity gradient in the adsorbed layer normal to the surface, and are also consistent with the existence of a conformational change of BSA upon adsorption to the silica interface.

Fluorescence anisotropy measurements, while also suffering from drawbacks such as poorer signal quality than TR-EWIF measurements and added complexity of analysis, have the potential to provide additional qualitative and quantitative information on dynamic processes that can contribute to a loss of emission polarisation including energy transfer, migration and fluorophore/segmental rotation motion. TRAMs were conducted on complexed BSA/ANS in bulk solution (BSA concentration 100 ppm), as shown in Figure 11.5 as the curve labelled "bulk".

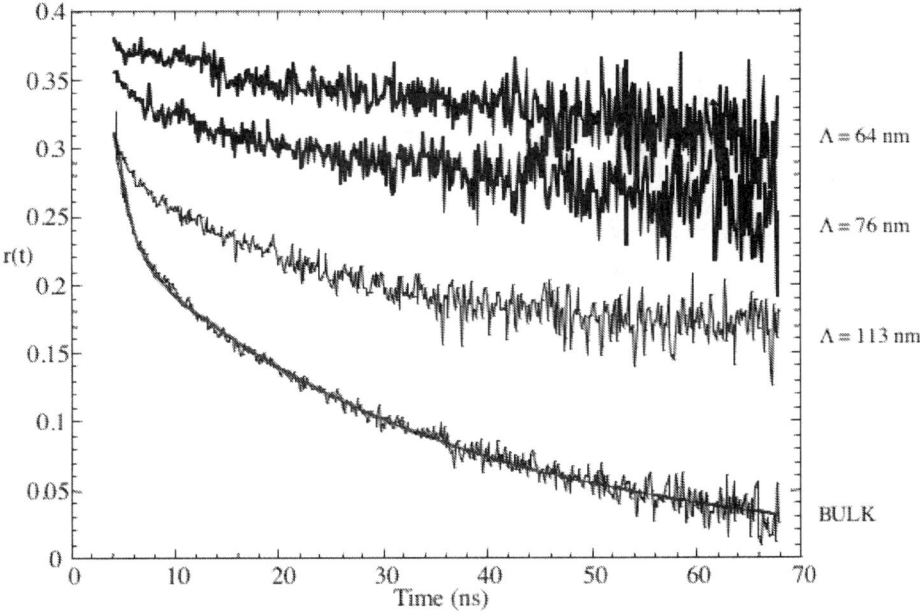

Figure 11.5: Time resolved fluorescence decay curves of the BSA-ANS complex as a function of penetration depth, Λ, of the evanescent wave normal to the interface. Two data sets are presented, corresponding to Type A and to Type B experimental protocols, as described in the Experimental Section. The penetration depths range from around 40 nm to around 100 nm, in 12nm increments within each data set. Note that the fluorescence emission decays most rapidly the smaller the penetration depth, as discussed in the text.

The fluorescence anisotropy decay observed from BSA/ANS in bulk solution can be analysed using Eq. (14) to model a simple picture of ANS, assuming it exhibits a single fluorescence decay time, it is free to rotate through a limited angular range (within a cone of semi-angle θ) and it is attached to a BSA molecule that undergoes isotropic rotational diffusion.

$$r(t) = r_0[\beta_1 \exp(-t / \tau_{c1}) + \beta_2 \exp(-t / \tau_{c2})] \qquad (14)$$

The two rotational correlation times, τ_{C1} and τ_{C2}, extracted resulting from using Eq. (14) were found to be ~2 ns and ~40 ns respectively. The 40 ns correlation time is assumed to correspond to the overall molecular motion of the BSA/ANS coil. The shorter component (τ_{C1} ~ 2ns) is tentatively attributed to a segmental motion of the BSA peptide chain with complexed ANS chromophores. The overall rotational motion of the BSA/ANS is therefore comprised of rapid rotational motion of the probe superimposed upon that of the entire protein molecule.

EW-TRAMs should aid in identifying whether the changes observed in the fluorescence decay profiles of BSA/ANS as a function of penetration depth are in fact related to partial denaturation of the protein upon adsorption, thus exposing the ANS to more aqueous environments, or to some other effect reducing the fluorescence decay kinetics. EW-TRAMs on the surface-adsorbed BSA/ANS complex were conducted as a function of penetration depth, under the Type B conditions using just vertically polarised excitation. The resulting time-resolved fluorescence anisotropy profiles from these experiments are also shown in Figure 11.5. It is clear that as the depth of penetration of the evanescent wave is reduced, the fluorescence anisotropy decays far more slowly than is the case in bulk solution. This indicates that as a result of protein adsorption, the motion of the parts of the protein to which the ANS is bound is more restricted, as might be expected. This suggests that, not only is the BSA coil as a whole undergoing restricted rotation, but also the more rapid segmental motion of the ANS-bound segments of the protein (the 2 ns component observed in the bulk measurements) is also severely hindered. One mechanism by which these observations may be explained is an unravelling of the BSA coil at the silica surface leading to an increase in the number of points of surface attachment. Any ANS bound to portions of the BSA that experience this uncoiling and adsorption to the surface will consequently exhibit restricted rotation due to the inhibited coil and chain segment motion.

In addition to the two modes of isotropic (segmental and coil) rotation of the BSA assumed in the discussion above, other factors may influence the fluorescence anisotropy observed from a fluorophore in close proximity to an interface. Certain adsorbed species might exhibit spatial alignment relative to the surface leading to anisotropy of the absorption and emission dipole moments.[78, 93] Furthermore, the dynamics of rotational motion may not be the same in planes parallel and perpendicular to the interface. Assuming a vertically oriented interface (e.g. Figure 11.3), both *s* and *p* polarised evanescent wave excitation (as discussed above) can be used to investigate rotational motion in two planes i.e. in- and out- of the plane of the interface. Preliminary results from EW-TRAMs for the BSA/ANS system with a penetration depth of ~64 nm are presented in Figure 11.6.[87]

There is a clear difference between the decay of the anisotropy profiles corresponding to the in- and out- of-plane motion of the fluorophore (Eqs. (11) and (12) respectively) indicating that the rate of reorientation of the emission dipoles is more rapid within the plane of the interface than it is in the plane orthogonal to it. While these data are not of the highest signal/noise standard, the potential of the EW-TRAMs method is nonetheless illustrated. This example, of an excitation polarisation-dependent EW-TRAM, illustrates the potential of this approach to spatially resolving two-dimensional motion of fluorophores near an interface. A more striking example of this is given in the next Section.

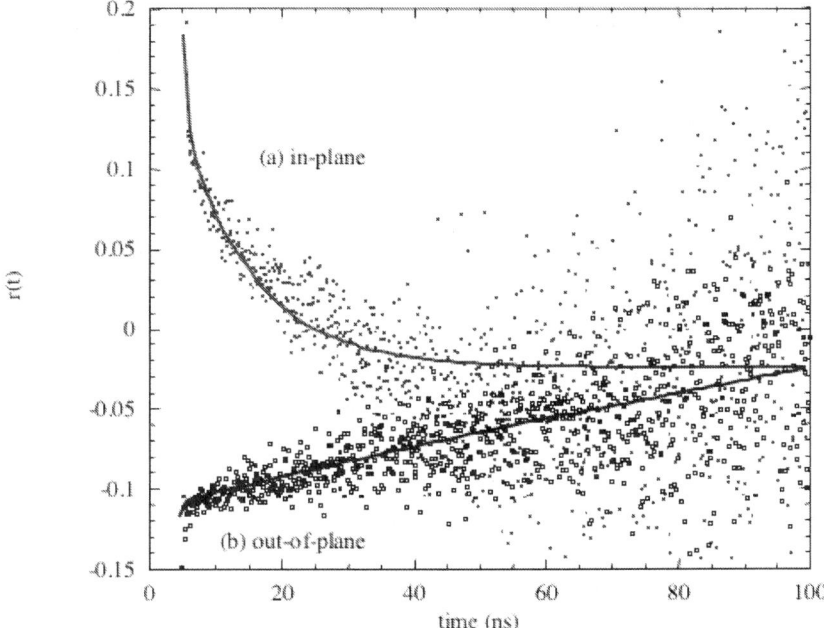

Figure 11.6. Excitation polarisation dependent, EW-TRAM profiles recorded for BSA-ANS adsorbed onto silica, illustrating differences in the rate of in- and out- of plane motion. The penetration depth is ~64 nm.

11.6.2. Acridine in poly(acrylic acid) films

We have also used EW-TRAMs to study films of poly(acrylic acid) (PAA), cast from either water or acetone onto silica, employing a low concentration of acridine ($\sim 5 \times 10^{-4}$ M) as a fluorescent probe. This work has exposed some intriguing time-dependent fluorescence polarisation behaviour, confirming the potential of the TR-EWIFs technique to probe differences in molecular motion in- and out-of the plane of the interface. Figure 11.7 shows the time-dependent fluorescence anisotropy function from acridine/PAA produced by evanescent wave excitation of the acridine when the excitation light was polarised vertically (i.e. in the plane of the interface) and horizontally (out of the plane of the interface). The presence of time-dependence of the r(t) function indicates the occurrence of some dynamic emission depolarisation process, which could include energy migration between acridine molecules or more likely, considering the low concentration of fluorophores used, fluorophore rotation. The time-scale of the observed time-dependent r(t) indicates that the fluorophores are not held rigidly within the polymer film but rather exhibit considerable motion. This suggests that the polymer film consists of low viscosity microenvironments or solvent-containing pockets. Furthermore, isotropic fluorophore rotation would be expected to result in a reasonably simple decay of the r(t) function, and the behaviour in- and out-of the plane of the interface would be expected to be similar. Instead, the time-dependence observed for the r(t) function is unexpectedly complex, showing an unusual but marked decay and rise behaviour. In addition, we also observe dramatically different behaviour, reported by the fluorophores corresponding to different modes of motion in the two planes relative to the interface, and the behaviour exhibited by the water-cast films is different to that of the acetone-cast films.

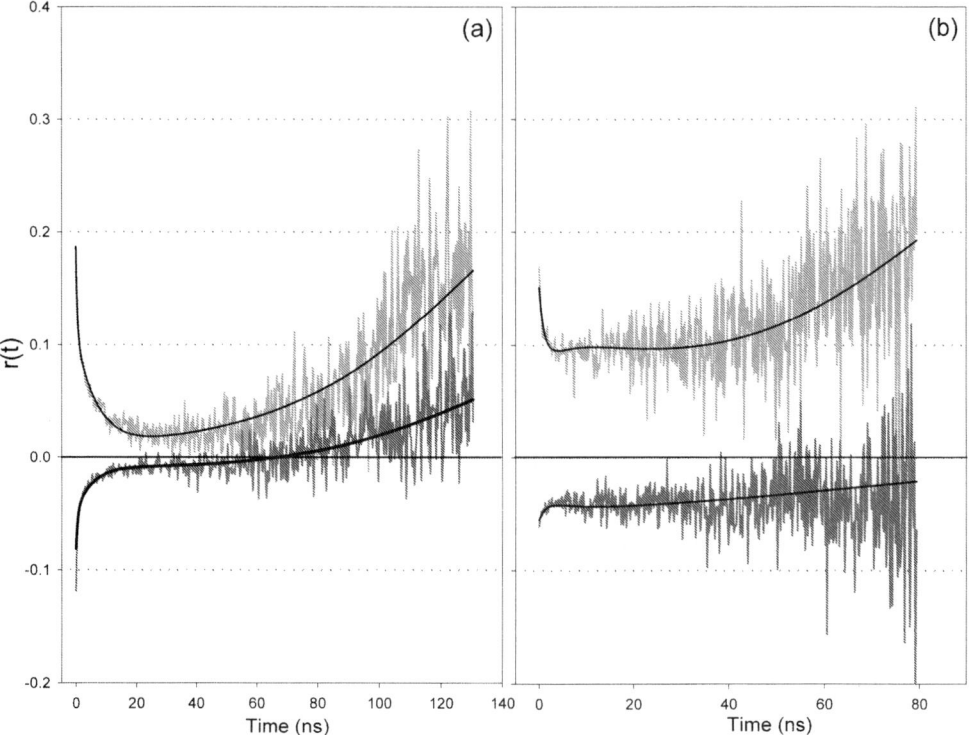

Figure 11.7. Excitation polarisation dependent, EW-TRAM profiles recorded for acridine/PAA films cast on silica from (a) water and (b) acetone.

Similar dip and rise behaviour in time-resolved fluorescence anisotropy profiles has been observed from a range of chemical and biological systems,[95, 102-107] and has been interpreted in terms of a temporally-dependent variation of the contribution towards the total r(t) intensity of fluorophores exhibiting multiple fluorescence decay and rotational correlation times. In the case of the fluorophore experiencing two distinct environments, an analytical form for r(t) can be derived:

$$r(t) = f_1(t)r_1(t) + f_2(t)r_2(t) \qquad (15)$$

where $f_1(t)$ is the fraction of fluorescence at time t due to the fluorophore experiencing environment 1. $r_1(t)$ describes the fluorescence anisotropy decay corresponding to environment 1. $f_2(t)$ and $r_2(t)$ are the corresponding quantities for environment 2. The fraction term for environment 1 is given by Eq. (16) and an analogous expression describes the fraction term for environment 2:

$$f_1(t) = \frac{x_1 \sum_{j=1}^{N} \alpha_{j1} \exp(-t/\tau_{j1})}{x_1 \sum_{j=1}^{N} \alpha_{j1} \exp(-t/\tau_{j1}) + x_2 \sum_{j=1}^{N} \alpha_{j2} \exp(-t/\tau_{j2})} \qquad (16)$$

where τ_{j1}, α_{j1}, and τ_{j2}, α_{j2} are the corresponding fluorescence decay times and amplitudes for the probe in the two environments. Immediately after excitation (t=0), $f_1(t)$ and $f_2(t)$ are the actual fractions of the probe molecules in each environment, x_1 and x_2, respectively, assuming that the extinction coefficient and radiative rate constant are the same for both populations. $r_1(t)$ and $r_2(t)$ in Eq. (15) are given by single exponential functions: $r_1(t) = r_{0_1}\exp(-t/\tau_{c_1})$, $r_2(t) = r_{0_2}\exp(-t/\tau_{c_2})$ where r_{0_1} and r_{0_2} are the initial anisotropy values for the fluorophore; τ_{c_1} and τ_{c_2} are the rotational correlation times of the two modes of rotation. It is apparent from Eq. (15) that if the larger fluorescence lifetime component is associated with a slower rotation time, then at long times after excitation, r(t) can also show a growth in intensity due to the increasing contribution of this species to r(t) relative to the shorter lived, faster rotating free dye molecules.
In the case of the acridine/PAA films, the fluorescence decay profiles cannot be analysed as sums of discrete exponential terms and it has been necessary to invoke the use of "lifetime distributions",[32, 45, 46] suggesting heterogeneity of the environment of the polymer film near the interface. The decay time distributions recovered indicate at least two decay time bands. For the water-cast film one decay time distribution band centred around 9 ns is tentatively assigned to acridine in an acidic aqueous environment, and another centred around 23 ns is assigned to be acridine associated more strongly with the polymer. For the acetone-cast film a short decay time band (~<2 ns - acridine within acetone microenvironments) and a broad distribution (~12 ns) are observed. The broad nature of this distribution, compared to that of the water-based film, indicates a very large variation in polymer-acetone interactions.

While a full interpretation of the complex TRAMs behaviour is still under consideration, the observed behaviour may be interpretable, in broad terms, on the model described by Eq. (15). If we assume that the fluorophore experiences pockets of solvent (with the shorter fluorescence decay times) that allow rapid rotation of the acridine, as well as environments with more restricted motion (and longer fluorescence decay times) associated with the polymer segmental motion, then the motion (due to short fluorescence lifetime) dominates the r(t) intensity on short time scales, while at medium to longer time scales the polymer segment motion contribution dominates. The interpretation of the difference in motion "in-plane" and "out-of-plane" motion inferred by the data in Figure 11.7 is currently under consideration, but the observed behaviour indicates that a very slow process is also contributing to the overall emission depolarisation, and that this process has different kinetics in the difference planes relative to the interface.

11.7. CONCLUSIONS

We have illustrated the potential of time-resolved evanescent wave-induced fluorescence anisotropy measurements to provide information regarding the orientation and motion of fluorophores near an interface. Changes in the fluorescence anisotropy of ANS as a function of distance from the interface have been used to determine changes in the conformation of bovine serum albumin on adsorption to a silica surface. We have also used acridine as a fluorescent probe of the local environment in poly(acrylic acid) films on silica. The excitation polarisation dependence on EW-TRAMs enables motion in- and out-of the plane of the intercept to be differentiated, providing a powerful experimental technique for interfacial studies.

11.8. ACKNOWLEDGEMENTS

TAS would like to acknowledge Prof. David Phillips, Dr. Garry Rumbles and Dr. Ben Crystall for introducing him to the field of evanescent wave-induced fluorescence spectroscopy. The authors also thank Prof. Ken Ghiggino for providing access to the laser equipment. We also acknowledge the contributions of Mr. Levie Lensun and Dr. Fiona Scholes in the early stages of this work, and the Australian Research Council for financial support in the form of a Discovery Project (MLG) and QEII Fellowship (TAS).

11.9. REFERENCES

1.   R. M. Cornelius and J. L. Brash, Adsorption from plasma and buffer of single- and two-chain high molecular weight kininogen to glass and sulphonated polyurethane surfaces *Biomaterials* **20**, 341-250 (1999).
2.   B. D. Ratner, in *Comprehensive Polymer Science. The Synthesis, Characterization Reactions & Applications of Polymers*, Edited by S. L. Aggarwal, (Pergamon Press, Oxford, 1989), pp. 201-247.
3.   E. D. Goddard and B. Vincent (Editors), *Polymer Adsorption and Dispersion Stability, ACS Symposium Series*, vol. 240, (Washington D.C.: American Chemical Society, 1984).
4.   S. Magdassi, *Surface Activity Of Proteins: Chemical And Physicochemical Modifications* (Marcel Dekker, Inc., New York, 1996).
5.   E. Tornberg, The application of the drop volume technique to measurements of the adsorption of proteins at interfaces *J. Coll. Interface Sci.* **64**, 391-402 (1978).
6.   B. C. Tripp, J. J. Magda, and J. D. Andrade, Adsorption of globular proteins at the air/water interface as measured via dynamic surface tension: Concentration dependence, mass-transfer considerations and adsorption kinetics *J. Colloid Interface Sci.* **173**, 16-27 (1995).

7. J. Wang and J. McGuire, Surface tension kinetics of the wild type and four synthetic stability mutants of T4 phage lysozyme at the air/water interface *J. Colloid Interface Sci.* **185**, 317-323 (1997).

8. C. Ybert and J. M. di Meglio, Study of protein adsorption by dynamic surface tension measurements: Diffusive regime *Langmuir* **14**, 471-475 (1998).

9. D. G. Dalgleish, The conformations of proteins on solid/water interfaces - caseins and phosvitin on polystyrene lattices *Colloids and Surfaces* **46**, 141-155 (1990).

10. E. Dickenson, E. W. Robson, and G. Stainsby, Colloid stability of casein-coated polystyrene particles *J. Chem. Soc. Faraday Trans. I* **79**, 2937-2952 (1983).

11. J. Leaver, and D.S. Horne, Chymosin-catalysed hydrolysis of glycosylated and nonglycosylated bovine kappa-casein adsorbed on latex particles *J. Colloid Interface Sci.* **181**, 220-224 (1996).

12. T. J. Lu, T. J. Su, P. N. Thirtle, J. K. Thomas, A. R. Rennie, and R. Cubitt, The denaturation of lysozyme layers adsorbed at the hydrophobic solid/liquid surface studied by neutron reflection *J. Colloid Interface Sci.* **206**, 212-223 (1998).

13. T. J. Su, J. R. Lu, R. K. Thomas, Z. F. Cui, and J. Penfold, The conformational structure of bovine serum albumin layers adsorbed at the silica-water interface *J. Phys. Chem. B* **102**, 8100-8108 (1998).

14. T. J. Su, J. R. Lu, R. K. Thomas, and Z. F. Cui, Effect of ph on the adsorption of bovine serum albumin at the silica/water interface studied by neutron reflection *J. Phys. Chem. B* **103**, 3727-3736 (1999).

15. A. Liebmann-Vinson, L. M. Lander, M. D. Foster, W. J. Brittain, E. A. Vogler, C. F. Majkrzak, and S. Satija, A neutron reflectometry study of human serum albumin adsorption in situ *Langmuir* **12**, 2256-2262 (1996).

16. C. D. Bain, Sum-frequency vibrational spectroscopy of the solid-liquid interface *J. Chem. Soc. Faraday Trans.* **91**, 1281-1296 (1995).

17. H. J. Paul and R. M. Corn, Second-harmonic generation measurements of electrostatic biopolymer-surfactant coadsorption at the water/1,2-dichloroethane interface *J. Phys. Chem. B.* **101**, 4494-4497 (1997).

18. D. Zhang, J. Gutow, and K. B. Eisenthal, Vibrational spectra, orientations, and phase transitions in long-chain amphiphiles at the air/water interface: Probing the head and tail groups by sum frequency generation *J. Phys. Chem.* **98**, 13729-13734 (1994).

19. X. Zhao, S. Subrahmanyan, and K. B. Eisenthal, Orientational fluctuations and phase·transition of long chain molecules at the air/water interface *Phys. Rev. Lett.* **67**, 2025-2028 (1991).

20. T. Kikteva, D. Star, and G. W. Leach, Optical second harmonic generation study of malachite green orientation and order at the fused-silica/air interface *J. Phys. Chem. B.* **104**, 2860-2867 (2000).

21. N. J. Harrick, *Internal reflection spectroscopy.* (Wiley Interscience, New York,1967).

22. G. Rumbles, A. J. Brown, and D. Phillips, Time-resolved evanescent wave induced fluorescence spectroscopy part 1. - Deviations in the fluorescence lifetime of tetrasulphonated aluminium phthalocyanine at a fused silica/methanol interface *J. Chem. Soc. Faraday Trans.* **87**, 825-830 (1991).

23. D. J. Neivandt, M. L. Gee, C. P. Tripp, and M. L. Hair, Coadsorption of poly(styrenesulfonate) and cetyltrimethylammonium bromide on silica investigated by attenuated total reflection techniques *Langmuir* **13**, 2519-2526 (1997).

24. D. J. Neivandt and M. L. Gee, Variable angle of incidence evanescent wave spectroscopy of the adsorption of quaternarized poly(vinylpyridine) on silica *Langmuir* **11**, 1291-1296 (1995).

25. M. Trau, F. Grieser, T. W. Healy, and L. R. White, Evanescent wave spectroscopy: Application to the study of the spatial distribution of charged groups on an adsorbed polyelectrolyte at the silica/water interface *J. Chem. Soc. Faraday Trans.* **90**, 1251-1259 (1994).

26. A. Pirinia and C. S. P. Sung, Adsorption studies of poly(n-vinylcarbazole) on sapphire by a fibre-optic UV-attenuated reflection technique *Macromolecules* **24**, 6104-6109 (1991).

27. A. Tronin and J. K. Blasie, Variable acquisition angle total internal reflection fluorescence: A new technique for orientation distribution studies of ultrathin films *Langmuir* **17**, 3696-3703 (2001).

28. N. J. Harrick and G. I. Loeb, Multiple internal reflection fluorescence spectrometry *Anal. Chem.* **45**, 687-691 (1973).

29. H. Masuhara, N. Mataga, S. Tazuke, T. Murao, and I. Yamazaki, Time-resolved total internal reflection fluorescence spectroscopy of polymer films *Chem. Phys. Lett.* **100**, 415-419 (1983).

30. D. Ausserre, H. Hervet, and F. Rondelez, Concentration dependence of the interfacial depletion layer thickness for polymer solutions in contact with nonadsorbing walls *Macromolecules* **19**, 85-88 (1986).

31. C. D. Byrne, A. J. de Mello, and W. L. Barnes, Variable-angle time-resolved evanescent wave-induced fluorescence spectroscopy (VATR-EWIFS): A technique for concentration profiling fluorophores at dielectric interfaces *J. Phys. Chem. B.* **102**, 10326-10333 (1998).

32. A. J. de Mello, J. A. Elliott, and G. Rumbles, Evanescent wave-induced fluorescence study of rhodamine 101 at dielectric interfaces *J. Chem. Soc. Far. Trans.* **93**, 4723-4731 (1997).

33. L. W. Liebmann, J. A. Robinson, and K. G. Mann, A dual beam total internal reflection fluorescence spectrometer for dynamic depth resolved measurements of biochemical liquid-solid interface binding reactions in opaque solvents *Rev. Sci. Instrum.* **62**, 2083-2092 (1991).

34. H. Yao, H. Ikeda, and N. Kitamura, Surface-induced J aggregation of pseudoisocyanine dye at a glass/solution interface studied by total-internal reflection fluorescence spectroscopy *J. Phys. Chem.* **102**, 7691-7694 (1998).

35. D. Parsons, R. Harrop, and E. G. Mahers, The kinetics of particle and polymer adsorption by total internal reflection fluorescence *Coll. & Surf.* **64**, 151-160 (1992).

36. H. Watarai and F. Funaki, Total internal reflection fluorescence measurements of protonation equilibria of rhodamine B and octadecylrhodamine B at a toluene/water interface *Langmuir* **12**, 6717-6720 (1996).

37. K. Stock, R. Sailer, W. S. L. Strauss, R. Pavesi, M. Lyttek, H. Emmert, and H. Schneckenburger, in *Fluorescence microscopy and fluorescent probes*, edited by A. Kotyk (Espero Publishing, Prague, 1999), pp. 67-79.

38. Olympus; http://www.olympusmicro.com/primer/techniques/fluorescence/tirf/tirfhome.html

39. Nikon; http://www.ave.nikon.co.jp/inst/Biomedical/tirf/

40. Zeiss; http://www.zeiss.com/4125681F004CA025/?Open

41. T. E. Starr and N. L. Thompson, Total internal reflection with fluorescence correlation spectroscopy: Combined surface reaction and solution diffusion *Biophys. J.* **80**, 1575-1584 (2001).

42. T. E. Starr and N. L. Thompson, Local diffusion and concentrations of IgG near planar membranes: Measurement by total internal reflection with fluorescence correlation spectroscopy *J. Phys. Chem. B.* **106**, 2365-2371 (2002).

43. A. B. Mathur, G. A. Truskey, and W. M. Reichert, Total internal reflection microscopy and atomic force microscopy (TIRFM-AFM) to study stress transduction mechanisms in endothelial cells *Crit. Rev. Biomed. Eng.* **28**, 197-202 (2000).

44. A. B. Mathur, G. A. Truskey, and W. M. Reichert, Atomic force and total internal reflection fluorescence microscopy for the study of force transmission in endothelial cells *Biophys. J.* **78**, 1725-1735 (2000).

45. A. J. de Mello, B. Crystall, and G. Rumbles, Evanescent wave spectroscopic studies of surface enhanced fluorescence quantum efficiencies *J. Coll. Int. Sci.* **169**, 161-167 (1995).

46. M. A. Bell, B. Crystall, G. Rumbles, G. Porter, and D. R. Klug, The influence of a solid/liquid interface on the fluorescence kinetics of the triphenylmethane dye malachite green *Chem. Phys. Lett.* **221**, 15-22 (1994).

47. B. Crystall, G. Rumbles, and T. A. Smith, Time resolved evanescent wave induced fluorescence measurements of surface adsorbed bovine serum albumin *J. Coll. Interface Sci.* **155**, 247-250 (1993).

48. C. K. Parmar, G. Rumbles, and C. J. Winscom, Aggregation of azamethine dyes on hydrated glass surfaces: An evanescent wave-induced fluorescence study *Phys. Chem. Chem. Phys.* **4**, 1766-1775 (2002).

49. G. Rumbles, D. Bloor, A. J. Brown, A. J. de Mello, B. Crystall, D. Phillips, and T. A. Smith, in *Microchemistry: Spectroscopy and Chemistry in Small Domains*, Edited by H. Masuhara, F. C. De Schryver, N. Kitamura, and N. Tamai, (North-Holland Delta Series, Elsevier Science B.V., London, 1994), pp. 269-286.

50. P. Suci and V. Hlady, Fluorescence lifetime components of Texas Red-labelled bovine serum albumin: Comparison of bulk and adsorbed states *Coll. & Surf.* **51**, 89-104 (1990).

51. K. Bessho, T. Uchida, A. Yamauchi, T. Shioya, and N. Teramae, Microenvironments of 8-anilino-1-naphthalenesulfonate at the heptane-water interface: Time-resolved total internal reflection fluorescence spectroscopy *Chem. Phys. Lett.* **264**, 381-386 (1997).

52. M. R. Rainbow, S. Atherton, and R. C. Eberhart, Fluorescence lifetime measurements using total internal reflection fluorimetry: Evidence for a conformational change in albumin adsorbed to quartz *J. Biomed. Mat. Res.* **21**, 539-555 (1987).

53. R. A. Dryfe, Z. Ding, R. G. Wellington, P. F. Brevet, A. M. Kuznetzov, and H. H. Girault, Time-resolved laser-induced fluorescence study of photoinduced electron transfer at the water/1,2-dichloroethane interface *J. Phys. Chem. A.* **101**, (1997).

54. S. Ishizaka, S. Habuchi, H.-B. Kim, and N. Kitamura, Excitation energy transfer from sulforhodamine 101 to acid blue 1 at a liquid/liquid interface: Experimental approach to estimate interfacial roughness *Anal. Chem.* **71**, 3382-3389 (1999).

55. S. Hamai, N. Tamai, and M. Yanagimachi, in *Microchemistry: Spectroscopy and Chemistry in Small Domains*, Edited by H. Masuhara, F. C. De Schryver, N. Kitamura, and N. Tamai, (North-Holland Delta Series, Elsevier Science B.V., London, 1994), pp. 335-348.

56. M. Toriumi and M. Yanagimachi, in *Microchemistry: Spectroscopy and Chemistry in Small Domains*, Edited by H. Masuhara, F. C. De Schryver, N. Kitamura, and N. Tamai, (North-Holland Delta Series, Elsevier Science B.V., London, 1994), pp. 257-268.

57. Y. Taniguchi, M. Mitsuyi, N. Tamai, I. Yamazaki, and H. Masuhara, Time- and depth-resolved fluorescence spectra of layered organic films prepared by vacuum deposition *J. Coll. Int. Sci.* **104**, 596-598 (1985).

58. H. Masuhara, S. Tazuke, N. Tamai, and I. Yamazaki, Time-resolved total internal reflection fluorescence spectroscopy for surface photophysics studies *J. Phys. Chem.* **90**, 5830-5835 (1986).

59. S. Hamai, N. Tamai, and H. Masuhara, Excimer formation of pyrene in a solid/polymer solution interface layer. A time-resolved total internal reflection fluorescence study *J. Phys. Chem.* **99**, 4980-4985 (1995).

60. M. Yanagimachi, N. Tamai, and H. Masuhara, Excite-state proton transfer of 1-naphthol in liquid-solid interface layers. Picosecond time-resolved total internal reflection fluorescence study *Chem. Phys. Lett.* **201**, 115-119 (1993).

61. A. Itaya, A. Kurahashi, H. Masuhara, N. Tamai, and I. Yamazaki, Dynamic fluorescence microprobe method utilizing total internal reflection phenomenon *Chem. Lett.*1079-1082 (1987).

62. J. S. Lundgren, E. J. Bekos, R. Wang, and F. V. Bright, Phase-resolved evanescent wave induced fluorescence. An in situ tool for studying heterogeneous interfaces *Anal. Chem.* **66**, 2433-2440 (1994).

63. F. V. Bright, Nanosecond and sub-nanosecond time-resolved fluorescence spectroscopy at interfaces *Appl. Spectrosc. Rev.* **32**, 1-43 (1997).

64. T. Yamashita, T. Uchida, T. Fukushima, and N. Teramae, Solvation dynamics of fluorophores with an anthroyloxy group at the heptyl/water interface as studies by time-resolved total internal reflection fluorescence spectroscopy *J. Phys. Chem. B.* **107**, 4786-4792 (2003).

65. H. Schneckenburger, K. Stock, J. Eickholz, W. S. L. Strauss, M. Lyttek, and R. Sailer, in *Proc. SPIE, LaserMicroscopy*, Edited by K. König, H. J. Tanke, and H. Schneckenburger, (SPIE, 2000), pp. 36-42.

66. E.-H. Lee, R. E. Benner, and R. K. Chang, Angular distribution of fluorescence from liquids and monodispersed spheres by evanescent wave excitation *Appl. Opt.* **18**, 862-868 (1979).

67. R. R. Chance, A. Prock, and R. Silbey, Molecular fluorescence and energy transfer near interfaces *Adv. Chem. Phys.* **37**, 1-65 (1978).

68. M. Urbakh and J. Klafter, Dipole-dipole interactions near interfaces *J. Phys. Chem.* **97**, 3344-3349 (1993).

69. M. Urbakh and J. Klafter, Dipole relaxation near boundaries *J. Phys. Chem.* **96**, 3480-3485 (1992).

70. W. Lukosz, Light emission by magnetic and electric dipoles close to a plane dielectric interface. III. Radiation patterns of dipoles with arbitrary orientation. *J. Opt. Soc. Am.* **69**, 1495-1503 (1979).

71. W. Lukosz, Theory of optical-environment-dependent spontaneous-emission rates for emitters in thin layers *Phys. Rev. B.* **22**, 3030-3038 (1980).

72. W. Lukosz and R. E. Kunz, Light emission by magnetic dipoles close to a plane interface. I. Total radiated power *J. Opt. Soc. Am.* **67**, 1607-1614 (1977).

73. W. Lukosz and R. E. Kunz, Light emission by magnetic dipoles close to a plane interface. II. Radiation patterns of perpendicular oriented dipoles *J. Opt. Soc. Am.* **67**, 1615 (1977).

74. E. H. Hellen and D. Axelrod, Fluorescence emission at dielectric and metal-film interfaces *J. Opt. Soc. Am. B.* **4**, 337-350 (1987).

75. T. P. Burghardt and N. L. Thompson, Effect of planar dielectric interfaces on fluorescence emission and detection. Evanescent excitation with high-aperture collection. *Biophys. J.* **46**, 729-737 (1984).

76. D. Axelrod, E. H. Hellen, and R. M. Fulbright, in *Topics in fluorescence spectroscopy*, Edited by J. R. Lakowicz, (Plenum Press, London, 1992), pp. 289-343

77. J. Edwards, D. Ausserre, H. Hervet, and F. Rondelez, Quantitative studies of evanescent wave intensity profiles using optical fluorescence *Appl. Opt.* **28**, 1881-1884 (1989).

78. M. J. Wirth and J. D. Burbage, Adsorbate reorientation at a water/(octadecylsilyl) silica interface *Anal. Chem.* **63**, 1311-1317 (1991).

79. D. A. Piasecki and M. J. Wirth, Spectroscopic probing of the interfacial roughness of sodium dodecyl sulfate adsorbed to a hydrocarbon surface *Langmuir* **10**, 1913-1918 (1994).

80. N. L. Thompson, H. M. McConnell, and T. P. Burghardt, Order in supported phospholipid monolayers detected by the dichroism of fluorescence excited with polarized evanescent illumination *Biophys. J.* **46**, 739-747 (1984).

81. S. E. Sund, J. A. Swanson, and D. Axelrod, Cell membrane orientation visuallized by polarized total internal reflection fluorescence *Biophys. J.* **77**, 2266-2283 (1999).

82. P. A. Anfinrud, D. E. Hart, and W. S. Struve, Time-correlated photon-counting probe of singlet excitation transport and restricted rotation in Langmuir-Blodgett monolayers *J. Phys. Chem.* **92**, 4067-4073 (1988).

83. J. G. E. M. Fraaije, J. M. Kleijn, M. van der Graaf, and J. C. Dijt, Orientation of adsorbed cytochrome c as a function of the electrical potential of the interface studied by total internal reflection fluorescence *Biophys. J.* **57**, 965-975 (1990).

84. H. Fukumura and K. Hayashi, Time-resolved fluorescence anisotropy of labelled plasma proteins adsorbed on polymer surfaces *J. Coll. Interface Sci.* **135**, 435-442 (1990).

85. C. Czeslik, C. Royer, T. Hazlett, and W. Mantulin, Reorientational dynamics of enzymes adsorbed on quartz: A temperature-dependent time-resolved TIRF anisotropy study *Biophys. J.* **84**, 2533-2541 (2003).

86. S. Ishizaka, K. Nakatani, S. Habuchi, and N. Kitamura, Total internal reflection fluorescence dynamic anisotropy of sulforhodamine 101 at a liquid/liquid interface: Rotational reorientation times and interfacial structures *Anal. Chem.* **71**, 419-426 (1999).

87. M. L. Gee, L. Lensun, T. A. Smith, and C. A. Scholes, Time-resolved evanescent wave induced fluorescence anisotropy for the determination of molecular conformational changes of proteins at an interface *Eur. Biophys. J.* **33**, 130-139 (2004).

88. M. L. Gee and T. A. Smith, Shedding light on the conformation of proteins and other macromolecules at an interface *Aust. J. Chem.* **56**, 1005-1012 (2003).

89. L. Lensun, T. A. Smith, and M. L. Gee, The partial denaturation of silica-adsorbed bovine serum albumin determined by time-resolved evanescent wave-induced fluorescence spectroscopy *Langmuir* **18**, 9924-9931 (2002).

90. J. R. Lakowicz (Ed.), *Topics in Fluorescence Spectroscopy*, vol. **2**. (Plenum Press, New York, 1991)

91. D. V. O'Connor and D. Phillips, *Time Correlated Single Photon Counting.* (Academic Press, London, 1984)

92. H. Schneckenburger, R. Sailer, K. Stock, M. Lyttek, and W. S. L. Strauss, Total internal reflection fluorescence lifetime imaging (TIR-FLIM) of living cells, presented at Multidimensional Microscopy 2001, *3rd Asia-Pacific International Symposium on Confocal Microscopy and Related Technologies*, Melbourne, Australia, (2001).

93. M. J. Wirth, Magic angle lifetime measurements in evanescent wave fluorometry *App. Spect.* **47**, 651-653 (1993).

94. K. P. Ghiggino and T. A. Smith, Dynamics of energy migration and trapping in photoirradiated polymers *Prog. React. Kin.* **18**, 375-436 (1993).

95. T. A. Smith, M. Irwanto, D. J. Haines, K. P. Ghiggino, and D. P. Millar, Time-resolved fluorescence anisotropy measurements of the adsorption of rhodamine-B and a labelled polyelectrolyte onto colloidal silica *Coll. Polym. Sci.* **276**, 1032-1037 (1998).

96. J. R. Lakowicz, *Principles of Fluorescence Spectroscopy* 2nd ed. (Plenum, New York, 1999)

97. C. A. Scholes, *unpublished results* (2004).

98. V. Hlady and J. D. Andrade, Fluorescence emission from adsorbed bovine serum albumin and albumin-bound 1-anilinonaphthalene-8-sulfonate studied by TIRF *Coll. & Surf.* **32**, 359-369 (1988).

99. V. Hlady and J. D. Andrade, A TIRF titration study of 1-anilinonaphthalene-8-sulfonate binding to silica-adsorbed bovine serum albumin *Coll. & Surf.* **42**, 85-96 (1989).

100. T. P. Burghardt and D. Axelrod, Total internal reflection fluorescence study of energy transfer in surface-adsorbed and dissolved bovine serum albumin *Biochemistry* **22**, 979-985 (1983).

101. V. Hlady, C. Gölander, and J. D. Andrade, Hydrophobicity gradient on silica surfaces: A study using total internal reflection fluorescence spectroscopy *Coll. & Surf.* **33**, 185-190 (1988).

102. L. Brand, J. R. Knitson, L. Davenport, J. M. Beecham, R. E. Dale, D. G. Walbridge, and A. A. Kowalczyk, Time-resolved fluorescence spectroscopy: Some applications of associative behavior to studies of proteins and membranes, in *Spectroscopy and the Dynamics of Molecular Biological Systems*, Edited by P. M. Bayley and R. E. Dale, (Academic Press, London, 1985).

103. R. D. Ludescher, L. Peting, S. Hudson, and B. Hudson, Time-resolved fluorescence anisotropy for systems with lifetime and dynamic heterogeneity *Biophys. Chem.* **28**, 59-75 (1987).

104. C. R. Guest, R. A. Hochstrasser, C. G. Dupuy, D. J. Allen, S. J. Benkovic, and D. P. Millar, Interaction of DNA with the Kelnow fragment of DNA polymerase I studied by time-resolved fluorescence spectroscopy *Biochemistry* **30**, 8759-8770 (1991).

105. T. E. Carver, R. A. Hochstrasser, and D. P. Millar, Proofreading DNA: Recognition of aberrant DNA termini by the Klenow fragment of DNA polymerase I *Proc. Natl. Acad. Sci. USA* **91**, 10670-10674 (1994).

106. C. K. Chee, K. P. Ghiggino, T. A. Smith, S. Rimmer, I. Soutar, and L. Swanson, Time-resolved fluorescence studies of the interactions between the thermoresponsive polymer host, poly(n-isopropylacrylamide), and a hydrophobic guest, pyrene *Polymer* **42**, 2235-2240 (2001).

107. E. K. L. Yeow, K. P. Ghiggino, N. Joost, H. Reek, M. J. Crossley, A. W. Bosman, A. P. H. J. Schenning, and E. W. Meijer, The dynamics of electronic energy transfer in novel multiporphyrin functionalized dendrimers: A time-resolved fluorescence anisotropy study *J. Phys. Chem. B* **104**, 2596-2606 (2000).

APPLICATION OF FLUORESCENCE TO UNDERSTAND THE INTERACTION OF PEPTIDES WITH BINARY LIPID MEMBRANES

Rodrigo F. M. de Almeida, Luís M. S. Loura, and Manuel Prieto[*]

12.1. INTRODUCTION

The fluid mosaic model of biological membranes (Singer and Nicolson, 1972) emphasizes membrane fluidity and free lateral diffusion of membrane components. This led to the generalized idea of biomembranes as solutions of proteins embedded in bilayers of randomly distributed phospholipids. However, over the past few decades, evidence has accumulated suggesting that the lipid distribution on the bilayer is nonrandom, both in model systems and in biological membranes (Edidin, 1998). In fact, it can exhibit ordered structures with length scales ranging from micrometers (visualized by microscopy; Korlach et al., 1999) to nanometers (mostly indirect evidence, see Mouritsen and Jørgensen, 1997, for a review).

It has been long realized that model systems of membranes prepared with lipid mixtures with different main transition temperatures (T_m) can exhibit lateral phase coexistence of gel/fluid phases. Depending on the ideality of the mixture, gel-gel (e.g., 1,2-dilauroyl-sn-glycero-3-phosphocholine (DLPC)/1,2-distearoyl-sn-glycero-3-phospho-choline (DSPC); Mabrey and Sturtevant, 1976) and even gel-gel and fluid-fluid (e.g., 1,2-dielaydoyl-sn-glycero-3-phosphocholine (DEPC)/1,2-dipalmitoyl-sn-glycero-3-phospho-ethanolamine (DPPE); Wu and McConnell, 1975) phase separation can occur. The fact that phase separation occurs for mixtures of lipids coexisting in cell membranes under conditions close to physiological (e.g., Marsh, 1990) made the detection and characterization of this kind of lateral heterogeneity of considerable interest in the biophysical and biochemical communities (e.g., Edidin, 2003). The composition-temperature phase diagrams, at constant pressure (and ionic strength, etc.), are a convenient way to represent this type of behavior for each pair of phospholipids. Considering lamellar and quasi-lamellar phases, namely the gel and fluid, the variety of phase diagrams (isomorphous, eutectic, peritectic, monotectic and eutectic) is similar to

[*] Rodrigo F. M. de Almeida, Luís M. S. Loura, and Manuel Prieto, Centro de Química-Física Molecular, Instituto Superior Técnico, 1049-001 Lisboa, Portugal. Luís M. S. Loura, Departamento de Química and Centro de Química, Universidade de Évora, 7000-671 Évora, Portugal.

other areas of Chemistry, as metallurgy (a good collection of phase diagrams for binary lipid mixtures is given by Marsh (1990)). Even in the absence of macroscopic phase separation in the fluid phase, a considerable degree of lateral heterogeneity can persist above the liquidus line. This was predicted in Monte-Carlo simulations (Mouritsen and Jørgensen, 1994) and experimentally confirmed by us by means of resonance energy transfer (FRET) (de Almeida *et al.*, 2002) for the highly non-ideal mixture of DLPC/DSPC.

In the case of binary lipid mixtures, the difference in T_m between the two components can arise from differences in the acyl chain length (*e.g.*, the already mentioned DLPC/DSPC) and/or degree of unsaturation (*e.g.*, DEPC/DSPC; Wu and McConnel, 1975) and different head groups. For this last case, let us consider the phase diagram for the mixture 1,2-dimyristoyl-*sn*-glycero-3-phosphocholine (DMPC)/ 1,2-dimyristoyl-*sn*-glycero-3-phosphate (DMPA) determined from differential scanning calorimetry (DSC) experiments by Graham *et al.* (1985). At low phosphatidic acid (PA) contents, the thermograms have a complex shape, as the pre-transition of the pure phosphatidylcholine (PC) species is not abolished. The phase diagram for the mixture is shown on Figure 12.1. The diagram is peritectic, with gel-gel and gel-fluid coexistence regions.

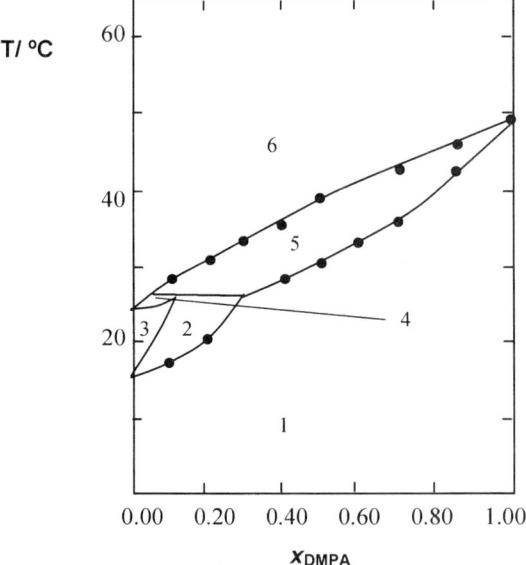

Figure 12.1. Phase diagram for the lipid mixture DMPC/DMPA. The different regions of the phase diagram are: 1) gel phase; 2) coexistence of gel rippled/gel phases; 3) gel rippled phase; 4) coexistence of gel rippled/fluid phases; 5) coexistence of fluid/gel phases; 6) fluid phase. Adapted from *Graham et al.* (1985).

The study of the interaction of model peptides with a zwitterionic lipid bilayer was fundamental to establish the hydrophilic/hydrophobic energetics of membrane proteins insertion (Wimley and White, 1996). Nonetheless, the importance of the complex composition of natural membranes has long been realized, and the presence of a single additional component, *e.g.*, the use of a binary zwitterionic/anionic phospholipid mixture, is sufficient to allow for a multiplicity of interactions. Phase separation can be induced in a binary lipid mixture with composition and temperature where otherwise only one phase would be present, namely through addition of a positively charged molecule to a mixture of a zwitterionic and an anionic phospholipids at a certain pH. In the case of the mixture 1,2-dipalmitoyl-*sn*-glycero-3-fosfo-*rac*-glycerol (DPPG) with DMPC or DSPC phase separation can be induced by addition of the cationic peptide mellitin (Lafleur *et al.*, 1989). In sum, the following possibilities have to be considered: (*i*) phase behavior of the binary lipid mixture in the absence of peptide (knowledge of the temperature/composition phase diagram); (*ii*) influence of the peptide on the phase behavior (shifts in the phase diagram, creation of new regions/types of phase separation); and (*iii*) influence of the peptide on the phase separation topology. In fact, the formation of lipid domains is thought to be a key process in several biological functions (Welti and Glaser, 1994; Simons and Ikonen, 1997) and the clarification of the relationship between lipid domains and the binding and functional properties of membrane-associated proteins is an emerging area in membrane research (Johnson and Cornell, 1999; Hurley and Meyer, 2001).

Another group of lipid mixtures that have been intensely used are phospholipid (mainly PC)/cholesterol (chol) binary systems. These mixtures' properties and interaction with peptides and proteins is of importance due to the high abundance of chol in mammalian plasma membranes. From the studies in model systems important conclusions have been inferred about the role of sterols in biomembranes.

The well-known effects of chol on the bilayer properties (see, *e.g.*, Bloom e Mouritsen, 1995; Needham e Nunn, 1990) have been rationalized considering that in the presence of high amounts of chol in a PC bilayer, the membrane is in a liquid ordered (lo) phase (using the nomenclature introduced by Ipsen *et al.*, 1987), with properties midway between the gel and the fluid. This designation highlights the facts that the translational diffusion is closer to the fluid phase (the diffusion coefficient in the lo phase is only 2 to 3 times lower than for the pure fluid phase of the PC), but the acyl chains are in a much more ordered configuration. In this nomenclature, the gel and fluid phases are designated by solid ordered (so) and liquid disordered (ld), respectively. The phase diagram is monotectic and for intermediate chol concentrations phase coexistence occurs: so and lo, below the monotectic temperature (which is close to T_m) and ld with lo, above the monotectic temperature. The latter corresponds to fluid-fluid phase separation, which is thought to be of biological relevance, namely, to the raft phenomenon (*e.g.*, Brown e London (2000)). It should be mentioned that, for PC/chol systems, the most studied being DPPC/chol, there are several phase diagrams reported, that differ considerably among them (*e.g.*, Vist and Davis, 1990; Lentz *et al.*, 1980; McMullen and McElhaney, 1995). The discrepancies are probably related to the similarity between the two phases, which makes differentiation between them difficult (London and Brown, 2000). Nevertheless, some of those discrepancies have been rationalized (de Almeida *et al.*, 2003) and ld/lo phase separation is a suitable model for the examples presented in the present review.

Although not the most intensely studied, the mixture 1-palmitoyl-2-oleoyl-*sn*-glycero-3-phosphocholne (POPC)/chol is particularly relevant. In fact, POPC is a 1-saturated, 2-unsaturated PC, a common motif found in naturally occurring phospholipids, being the major lipid component in PC isolated from several natural sources (Marsh, 1990). It has a low T_m (Marsh, 1990), but it has the ability to form a lo phase in the presence of chol both below (Thewalt and Bloom, 1992) and above (Mateo *et al.*, 1995) T_m, *i.e.*, so/lo and ld/lo phase coexistence, respectively. It is used in several studies in ternary (or higher) model systems (*e.g.*, Silvius, 1992; Milhiet *et al.*, 2001; Dietrich *et al.*, 2001; de Almeida *et al.*, 2003). The partial phase diagram POPC/chol is shown in Figure 12.2, where a broad ld/lo coexistence region can be observed. The phase diagram is similar to one previously published (Mateo *et al.*, 1995), except that for the lower temperatures the phase coexistence both begins and ends at slightly higher chol mole fractions (x_{chol}). This diagram was determined from fluorescence measurements, namely the lifetime and anisotropy of the fluorescent membrane probes 1,6-diphenyl-1,3,5-hexatriene (DPH) and *trans*-parinaric acid (*t*-PnA).

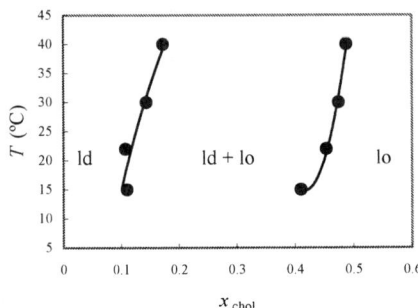

Figure 12.2. POPC/chol phase diagram. The experimental points were determined from fluorescence anisotropy and lifetimes of the lipophilic probes DPH and *t*-PnA. Reprinted from de Almeida *et al.* (2003) with permission. Copyright 2003 Biophysical Society.

After a recent review that surveys literature studies on the interaction of peptides with binary phospholipids membranes (Loura *et al.*, 2003), in the present article, we also address in a more general way lipid-protein interactions with not only phospholipid/phospholipid mixtures, but also with phospholipid/sterol mixtures, and studies in single lipid vesicles can be also addressed when considered particularly relevant. Instead of a comprehensive literature review, we illustrate the several fluorescence-based methodologies, highlighting the relevant information that can be extracted from their application, with studies carried out in our laboratory.

12.2. QUANTIFYING THE EXTENT OF INTERACTION OF THE PEPTIDE WITH THE MEMBRANE

The determination of the partition coefficient, K_p, is usually the first step in the study of the interaction of a peptide with model membranes, and should precede structural and dynamic studies. It allows quantifying the fraction of membrane bound peptide for a given lipid concentration, and it gives a measure of the "strength" of the peptide-membrane interaction, permitting, *e.g.*, quantitative comparison between different lipid species and/or phases. It is defined by (for an alternative definition and interconversion see Santos *et al.*, 2003; Loura *et al.*, 2003)

$$K_p = \frac{n_{S,L}/V_L}{n_{S,W}/V_W} \tag{1}$$

where V_i are the volumes of the phases, and $n_{S,i}$ are the moles of solute present in each phase (i = W, aqueous phase; i = L, lipid phase). An equivalent equation is obtained by expressing the membrane-bound peptide mole fraction, x_L, as a function of the lipid concentration, [L]:

$$x_L = \frac{K_p \bar{V}_L [L]}{K_p \bar{V}_L [L] + 1} \tag{2}$$

where $\bar{V}_L$ is the molar volume of the lipid phase.

For most cases, the partition coefficient of a molecule between a lipid and an aqueous phase can be obtained by fluorescence spectroscopy as long as there is a difference in a fluorescence parameter of the partitioning molecule (*e.g.*, quantum yield, fluorescence anisotropy or fluorescence lifetime) when in aqueous solution and after incorporation in the membrane (or, less commonly for peptides, if the incorporation of the molecule in the membrane leads to a change on a fluorescence property of a membrane probe). Fluorescence emission intensity, I, and steady-state anisotropy, $<r>$ (Section 12.5), can both be used to calculate the partition coefficient of a fluorescent molecule between lipid and aqueous phases:

$$I = \frac{I_W + I_L K_p \bar{V}_L [L]}{1 + K_p \bar{V}_L [L]} \tag{3}$$

$$<r> = \frac{<r>_W + <r>_L K_p \bar{V}_L [L] \varepsilon_L \Phi_L /(\varepsilon_L \Phi_L)}{1 + K_p \bar{V}_L [L] \varepsilon_L \Phi_L /(\varepsilon_L \Phi_L)} \tag{4}$$

where $\bar{V}_L$ is the lipid molar volume, ε_i is the molar absorption coefficient at the excitation wavelength and Φ_i is the fluorescence quantum yield of the peptide in phase i. In Eq. (3), the lifetime-weighted quantum yield, $<\tau>$, readily obtained from the fluorescence decay lifetimes τ_i and respective normalized amplitudes a_i from (*e.g.*, Engelborghs, 2001)

$$<\tau> = \sum_i a_i \tau_i \tag{5}$$

can be used instead of I through the relationship:

$$\langle \tau \rangle = \frac{\langle \tau \rangle_W + \langle \tau \rangle_L K_p \bar{V}_L [L]}{1 + K_p \bar{V}_L [L]} \tag{6}$$

(note that both $<\tau>$ and I are proportional to the fluorescence quantum yield). In fact, the use of data coming from time-resolved experiments is less prone to artifacts such as light scattering or inner filter effects, and is preferred to the use of steady-state intensities. Anisotropy measurements are also strongly affected by light scattering, which can be critical for the highest lipid concentrations such as shown in, *e.g.*, Figure 12.8 of Castanho and Prieto (1992). This eventually is the greatest restriction to the otherwise very sensitive fluorescence technique, because for a correct recovery of the lipid-phase parameter I_L, r_L or $<\tau>_L$ (and also K_p, due the strong correlation of the two parameters), a quasi-plateau region on the plots should be obtained. Bimolecular photophysical interactions involving peptide molecules, such as self-quenching and energy homotransfer are eventual complicating factors, mainly for situation of overcharged membranes.

The spectroscopic determinations are usually carried out by titration, *i.e.*, addition of successive amounts of lipid to the solution keeping the solute concentration constant (except for the dilution effect). However, tryptophan (Trp) and tyrosine (Tyr) are prone to photobleaching, and, in such case, preparation of separate samples with constant solute concentrations and different lipid amounts should be considered. This procedure was used to determine the partition coefficient of the peptide comprising the residues 579-601 of gp41 ectodomain from the human immunodeficiency virus type-1 (HIV-1) between water and lipid vesicles of POPC/1,2-dimyristoyl-*sn*-glycero-3-phosphoglycerol (DMPG) 8:2 (Santos *et al.*, 1998). As shown in Figure 12.3, the fluorescence intensity of the Trp residue increases with lipid concentration. In the top panel, steady-state intensities were used, and in the bottom panel, the lifetime-weighted quantum yield approach was employed. The trend of the data is analogous, and in fact the recovered partition coefficients are of the same order of magnitude. Clearly, the time-resolved data are of better quality than the steady-state data, as mentioned above, when high lipid concentrations have to be reached. The relative error associated to the recovered K_p value is reduced from 57% (steady-state data) to 32% (time-resolved data).

In case that K_p is not too high, the molar fraction of solute in water, $x_W = 1 - x_L$, can be significant. If the solute fluoresces in water (*e.g.*, Trp), the experimental spectrum, $I(\lambda)_{L+W}$, is the sum of the fractions in water, $I(\lambda)_W$, and in the membrane, $I(\lambda)_L$. The latter one can be obtained from Eq. (7):

$$I(\lambda)_L = C \cdot \left(I(\lambda)_{L+W} - x_W \frac{1}{1 + <\tau>_L / <\tau>_W} I(\lambda)_W \right) \tag{7}$$

Similarly to the fluorescence spectra, the total anisotropy decay is often difficult to analyze. In the most common case of complex decay (two or more components) for peptides, the number of needed fitting parameters would be too large. However, even in the case that the dynamic information contained in the initial part of the decay cannot be recovered, the limiting anisotropy of the bound species, $r_{\infty,L}$, which allows an easy determination of the order parameter of the system (Section 12.5), is readily obtained:

$$r_\infty^L = \left(1 + \frac{x_W \langle \tau \rangle_W}{x_L \langle \tau \rangle_L} \right) \cdot r_\infty \tag{8}$$

where r_∞ is the experimentally determined value in the presence of lipid at time $\rightarrow \infty$.

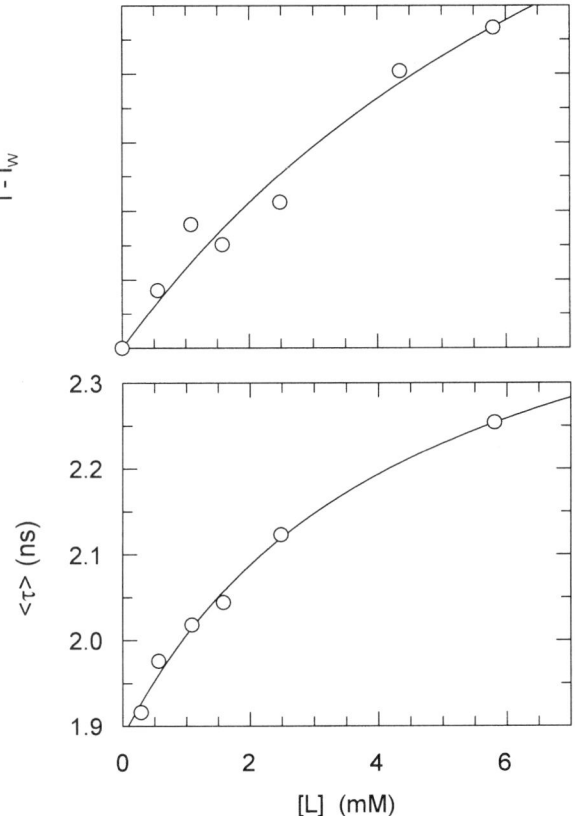

Figure 12.3. Determination of the partition constant (K_p) of the peptide comprising the residues 579-601 of gp41 ectodomain of HIV-1 between the aqueous phase and phospholipid vesicles. Peptide concentration was constant during each set of experiments. Increase in fluorescence intensity ($I - I_W$) obtained for different concentrations of POPC/DMPG 80:20 small unilamellar vesicles (SUV) (top), and fitting lines calculated using Eq. (3). Lifetime weighted quantum yield $<\tau>$ obtained for different concentrations of POPC/DMPG 80:20 SUV (bottom), and fitting lines calculated using Eq. (6). Reprinted with permission from Santos *et al.* (1998). Copyright 1998 American Chemical Society.

As an illustration of the use of these latter equations, we shall consider the interaction of the positively charged peptide α-MSH, a hormone known for its role in regulating skin pigmentation in vertebrates (Eberle, 1988) with both DMPC (zwitterionic)/DMPG (anionic) and DMPC/DMPA (anionic) (both 3:1 mixtures), which was studied using complementary techniques (DSC, infrared and ultraviolet absorption spectroscopy, and steady-state and time resolved fluorescence; Contreras *et al.*, 2001). Figure 12.4 shows the emission spectra at 20°C recovered using Eq. (7), whereas Table 12.1 shows the recovered blue-shifts relative to buffer, as well as the recovered limiting anisotropies $r_{\infty,L}$, both at 20°C and 37°C. Interestingly, a larger blue-shift is observed for DMPC/DMPG in the fluid phase relative to the gel (pointing to a deeper location of the

Trp residue in the fluid bilayer), whereas the opposite is observed for DMPC/DMPA. On the other hand, high values (for a non-transmembrane helix) of $r_{\infty,L}$ are obtained, meaning that the peptide is strongly adsorbed at the interface and highly immobilized.

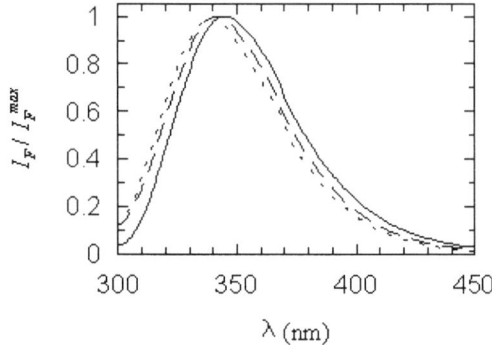

Figure 12.4. Corrected fluorescence emission spectra (λ_{exc} = 290 nm) of α-MSH at 20°C in aqueous solution (———), incorporated in gel phase large unilamellar vesicles (LUV) of DMPC/DMPG (3:1) (— —), and DMPC/DMPA (3:1) (- - - -). The spectra were corrected using Eq. (7) as described in the text. Reprinted from Contreras *et al.* (2001) with permission. Copyright 2001 Biophysical Society.

Table 12.1. Lipid/water partition coefficients (K_p), and [9]Trp photophysical parameters of α-MSH in different systems: $\langle\tau\rangle$ - lifetime-weighted quantum yield; $\Delta\lambda$ - membrane /water spectral-shift; r_{∞} - limiting anisotropy.

System	T (°C)	$K_p^{(lipid/water)}/10^2$	$\langle\tau\rangle_{W\ or\ L}$ (ns)	$\Delta\lambda$ (nm)	$r_{\infty}^{W\ or\ L}$
Buffer	20	-----	2.21 ± 0.02	-----	0
(W)	37	-----	1.51 ± 0.03	-----	0
DMPC/DMPG (3:1)	20	2.8 ± 0.2	3.22 ± 0.05	3.0	0.125 ± 0.006
(L)	37	4.0 ± 1.4	2.44 ± 0.16	5.0	0.090 ± 0.015
DMPC/DMPA (3:1)	20	6.0 ± 1.8	2.60 ± 0.05	6.0	0.127 ± 0.021
(L)	37	3.5 ± 1.0	2.25 ± 0.11	5.0	0.114 ± 0.017

Reprinted from Contreras *et al.* (2001) with permission. Copyright 2001 Biophysical Society.

In addition to water/lipid partition, one can define the partition coefficient between the two lipid phases that eventually coexist in binary lipid mixtures, identically to Eq. (1), and the same methodologies as described above can be used for its determination (in that case, values of K_p close to unity denote essentially random distribution of peptide between the two lipid phases, and K_p significantly different from unity denotes strong preference for one particular phase). However, this type of study is most uncommon for peptides, possibly because most of these have the additional complexity of significant partition to the aqueous phase. Commonly, a single aqueous phase/lipid bilayer K_p value is determined, which must be seen as a complex average between the not necessarily identical K_p values for partition between the aqueous medium and each lipid phase. Even on the frame of this simplification, useful information can be obtained. For example, returning to the aforementioned study of the interaction of α-MSH with DMPC/DMPG and DMPC/DMPA 3:1 mixtures, Table 12.1 shows the recovered values of K_p and $<\tau>_L$. The K_p values reflect the recovered blue-shifts discussed above, *i.e.*, for DMPC/DMPG a higher K_p is obtained at 37°C than at 20°C, whereas the opposite is observed for DMPC/DMPA. This interesting effect is probably related to the difference in the phase separation properties of the two lipid mixtures. At both temperatures, the DMPC/DMPG mixture is in a single phase, which is a gel phase at 20°C and a fluid phase at 37°C. A higher K_p is measured in the fluid phase, which is the usual result in one-component vesicles (Ito *et al.*, 1993). This value is similar to that measured for the DMPC/DMPA mixture at the same temperature (which is also in a single fluid phase). The interestingly increased value at 20°C for this mixture (for which there is gel/gel phase coexistence, Figure 12.1) is probably due to stronger binding of the positively charged peptide, induced by the domains enriched in anionic phospholipid.

12.3. DETERMINING THE TRANSVERSE LOCATION OF THE PEPTIDE'S FLUOROPHORE

The problem of in-depth location of the peptide using fluorescence techniques has deserved a significant attention in the fluorescence literature, and essentially addresses the location of the fluorescent amino acids Trp or Tyr. Although by now it is known that Trp is located essentially at the membrane interface, the so-called Trp anchors (de Planque *et al.*, 1999), the problem is still very relevant. The most common methodologies for obtaining topographical information are described.

12.3.1. Quenching by Lipophilic Probes

Quenching by lipophilic probes is by far the most used and perhaps the most direct methodology to obtain this type of information. Lipophilic probes which are known to be located at specific positions inside the membrane are used, and these are usually derivatized lipids or fatty acids with nitroxy-labels at specific positions along the chain (Chattopadhyay and London, 1987), or brominated lipids (Simon *et al.*, 2003). Although these are polar moieties, their location along the chain and the insertion of the chain in the phospholipid palisade structure ensure different depths in the membrane.

The data to be obtained are the Stern-Volmer plots for the fluorophore using different quenchers (a pair of quenchers is the minimum). The amount of quenching is related to the "effective quencher concentration", and this one depends on the relative positions of both quencher and fluorophore. Stern-Volmer formalisms and quenching kinetics (static or collisional mechanisms), will be described in more detail in Sub-sections 12.6.1 and 12.6.3; here a brief comment about quantitative studies on the transverse location is presented.

Studies on the quantification of the position (*e.g.*, distance to the interface), were pioneered by Chattopadhyay and London (1987) (the so-called "parallax method"). This assumes the absence of diffusion (pure static mechanism), and a unique depth inside the bilayer is considered for both fluorophore and quencher. A two-dimensional space is used to apply a quenching sphere-of-action model. Although this is not verified in reality, as both static and dynamic contributions are present (Castanho *et al.*, 1996), and a distribution exists for both the fluorophore and quencher, this methodology has been very useful for obtaining information about the location. The model was refined by the same authors (Abrams and London, 1992), and improvements were later introduced by Ladokhin (1997). This author considers a probability of quenching which is distance dependent, and more importantly an empirical Gaussian distribution for the distance distributions. This allows the recovery of a Gaussian profile for the fluorophore, assuming distributions with equal width for all the quenchers. The present state of the art of the analysis was presented by Fernandes *et al.* (2002). These authors obtained the quencher profile distribution by Brownian dynamics and recovered both the average location and width of the fluorophore distribution.

However, in case that only more immediate information is intended, direct inspection of a Stern-Volmer plot is enough to give a mapping of the in-depth position. For this purpose there is no need to obtain time-resolved data, *i.e.*, there is no need to discriminate the eventual static and dynamic contributions. Briefly, in the situation of identical depth of probe and quencher, the Stern-Volmer constant is expectedly higher, because the "effective" or "local" quencher concentration at each membrane depth, $[Q]$, is related to the overall concentration in the membrane $[Q]_L$, via $[Q]=\beta [Q]_L$, where the factor β is introduced to take into account the non-homogeneous distribution of quenchers relative to the fluorophore. It should be stressed that the overall concentration of quencher in the membrane (Castanho *et al.*, 1996), should be determined taking into account the partition coefficient of the quencher (a good reference for nitroxide-labeled fatty acids is Blatt *et al.*, 1984), as well as the lipid molar volume (Marsh, 1990).

In the literature there are very good examples of quantitative approaches such as the parallax model, and thus in this review some details about the direct comparison of Stern-Volmer constants, K_{SV}, will be described. An example of this methodology to the study of membrane embedded protein segments is a structural study about those of the nicotinic acetylcholine receptor (AChR) reconstituted in asolectin vesicles (Barrantes *et al.*, 2000). For one of them, αM1, there is a cysteine (Cys) in position 222, which according to the predicted topology would be located at the centre of the membrane, because the peptide sequence is 201-<u>IPLYFVVNVIIPC(222)LLFSFLTGLVFYLP</u>TDSGEK-242, where the underlined sequence denotes the hydrophobic stretch of the peptide. The Cys residue was derivatized with pyrene, and a differential quenching study was carried out using 5-, 7- and 12-spin labeled fatty acids. The Stern-Volmer constants are 6.59, 1.46, and 1.14 M^{-1} respectively, clearly pointing out to a surface location of the pyrene derivative, at variance with the reported topography. The existence of three proline (Pro) residues, one of which (Pro^{221}) is adjacent to the labeled Cys, is probably the ruling factor by introducing torsions and kinks on the α-helix.

Another interesting application of the differential quenching methodology is the determination of the relative positions of the potassium channel inactivating peptide Shaker B (ShB peptide) and the non-inactivating peptide mutant (ShB-L7E) (Poveda *et al.*, 2003). The inactivating one closes the channel via the so-called "ball" mechanism, and there is in the protein a relevant negative surface potential in addition to a hydrophobic pocket. Anionic phospholipid membranes were used for carrying out this study. This peptide has no Trp residues, but the Tyr one can be used for the quenching study. The spin labeled-fatty acids at positions 5 (5-NS) and 16 (16-NS), which are known to be located at 12 Å and 3 Å, respectively, from the bilayer centre (Chattopadhyay and London, 1987) were used.

As described above, the direct comparison of the plots can be used to obtain qualitative information about the transverse location. Interestingly, the quenching by 5-NS was small and similar for both peptides, and no definitive conclusion could be obtained about differences on the membrane topography of the two peptides. At variance, when using 16-NS a significantly higher efficiency was observed for ShB as compared to the ShB-L7E. In this way, a clearly shallower position for the latter peptide can be concluded. Interestingly, with the exception of the quenching of ShB-L7E by 16-NS, all the plots are clearly non-linear. This could eventually be due to the existence of a significant peptide fraction in water, which was non-accessible to the lipophilic quenchers used. However, from the independent determination of the partition coefficient of the peptides, it was concluded that only a very small amount (2 % for ShB and 5 % for ShB-L7E) was in this situation. In this way, the downward curvature is probably related to some complex accessibility of the quenchers to the membrane-bound peptide, as compared to the one in solution. This is a situation where a data analysis in the context of the parallax model is difficult, but even without this quantified information relevant conclusions are obtained.

However, in some cases the downward curvature can be successfully rationalized taking into account the fraction of fluorophore in water: in a study of a fragment of the gp41 ectodomain (HIV-1) in DMPG SUV (Santos *et al.*, 1998), a curvature was observed on the Trp quenching profile by 5-NS (Figure 12.5).

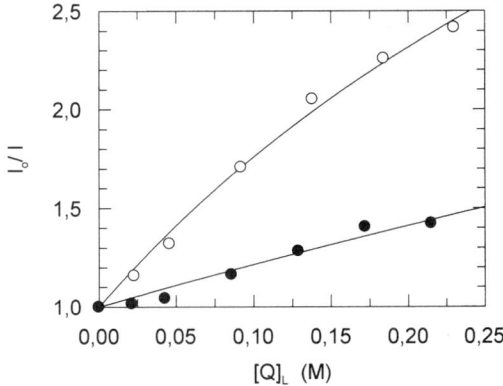

Figure 12.5. Fluorescence quenching of a fragment of the gp41 ectodomain in DMPG SUV 5.8 mM. Variation of I_0/I with the increase of the effective concentration of 5NS (O) and 16NS ( ), and fitting lines obtained using Eq. (9). Reprinted with permission from Santos *et al.* (1998). Copyright 1998 American Chemical Society.

If a typical Lehrer (1971) formalism is considered with a fraction of fluorophores accessible to the quencher contributing with a fraction f_B to the total emission, a modified Stern-Volmer plot is obtained,

$$\frac{I_o}{I} = \frac{1 + K_{SV}[Q]_L}{(1 + K_{SV}[Q]_L)(1 - f_B) + f_B} \tag{9}$$

where I_o is the fluorescence intensity in the absence of quencher, and I the fluorescence intensity in the presence of a certain quencher concentration. Data for the quenching by 5-NS could be described by Eq. (9) with $K_{SV} = 10.8 \pm 1.4$ M^{-1} and $f_B = 0.83 \pm 0.04$. This later value is in close agreement with the fraction of fluorescence intensity emitted by the peptide incorporated in the membrane at the lipid concentration used ($f_L = 0.87$), as determined from the experimental values of the partition coefficient. Therefore, all the fluorophore population in the membrane is accessible to the quencher. Data for quenching by 16-NS was also described by Eq. (9) with $K_{SV} = 2.7 \pm 0.1$ M^{-1} and the same f_B value. In this way, a surface location for this peptide can be concluded.

Sometimes, and in case that the quencher concentrations are high, static quenching contributions can be operative such as observed for the fragment 1-24 of the peptide hormone adrenocorticotropin, ACTH(1-24), in interaction with SUV of DMPC (83%) and DMPG (17%) (Moreno and Prieto, 1993). The upward curvature is described by a sphere-of-action model with a radius of 13.7 Å, which, as expected, is close to the sum of the molecular radii of Trp and quencher.

12.3.2. Quenching by Aqueous Probes

In addition to the previously described utilization of lipophilic probes, aqueous probes are also used on the study of peptide transverse location in membranes. Both acrylamide and iodide are widely used, and in most cases the Stern-Volmer constants for the peptide in interaction with the membrane is compared to the value obtained in water. In case that the peptide is shielded, i.e., internalized and not adsorbed at the membrane surface, the K_{SV} value decreases. This approach is not as informative as the methodology previously described based on lipophilic quenchers located at a graded series of depths in the membrane.

One interesting example of acrylamide quenching, involves again the peptide ACTH (1-24) introduced in the previous sub-section and a related peptide. In this case, the quenching data, shown in Figure 12.6, presents a downward curvature in both cases. This data could be described by a model (Eq. 10) taking into account the peptide fluorescence arising from the fraction in water (f_w), as well as two other fractions in the lipid, one accessible (f_{La}) and another not accessible to the quencher (f_{Ln}):

$$\frac{I_o}{I} = 1 \Big/ \left(\frac{f_W}{1 + K_{SV}^W[Q]} + \frac{f_{La}}{1 + K_{SV}^L[Q]} + f_{Lb} \right) \tag{10}$$

The value of K_{SV}^W is known from an independent study in aqueous solution, and f_w is easily calculated from the peptide partition coefficient since the lipid concentration is known. In this way, K_{SV}^L, f_{La} and f_{Ln} can be obtained from the fitting. It should be stressed that a model with just two populations (in water and in the membrane) could not describe the data. This experiment clearly shows that there is some internalization of Trp, but no precise information about the transverse location is obtained. This is only obtained via the nitroxide fatty acid quenching study, which showed that in fact the peptide is located near the surface although shielded from acrylamide.

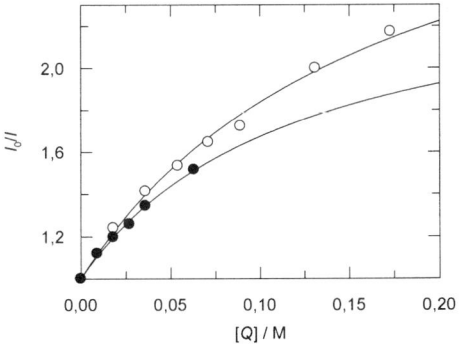

Figure 12.6. ACTH(1-24) (closed circles) and related peptide (open circles) fluorescence quenching by acrylamide in the presence of 2mM DMPC/DMPG (83:17) SUV. The lines are the fits of Eq. (10) to those data series. Adapted from Moreno and Prieto (1993).

12.3.3. Resonance Energy Transfer

The energy transfer efficiency is explicitly dependent on the inverse sixth power of the distance between donor and acceptor. This very important methodology will be dealt with in more detail later in this work, when addressing the problem of peptide aggregation in the membrane, but here its application on the determination of the peptide in-depth location in the membrane will be presented. This approach is similar to the above described methodology using lipophilic quenchers, *e.g.*, acceptors located at a graded series of depths can be used to report on proximity, but in this case it is not a contact (either collisional or static), but a dipolar interaction. Trp and Tyr are usually used as donors, as their fluorescence emission occurs at short wavelengths.

An example of this type of application is the location of Trp in the hormone ACTH (1-24) (Moreno and Prieto, 1993). In this case, derivatized fatty-acids with the anthroyl chromophore were used as acceptors (3-AS and 12-AS). It should be stressed that in addition to the above described spin labeled fatty-acids, this is the best characterized family of probes for this type of studies (Blatt *et al.*, 1984; Villalaín and Prieto, 1991). The Förster radius (Eq. (26) in Sub-section 12.6.2) for the Trp/3-AS and the Trp/12-AS pairs is $R_0 = 24$ Å (the same value for both pairs is due to the identical absorption spectra of both acceptors). Similar efficiencies were determined with both probes, which means that the interplanar donor-acceptor distance (Davenport *et al.* 1985) was identical in both cases, *i.e.*, Trp is located in the membrane in-between the 3- and the 12-AS probes. In this way it is not deeply buried, but neither is adsorbed near the interface.

In another work, the above described study of the inactivating peptide ShB of the potassium channel and the non-inactivating mutant ShB-L7E (Poveda *et al.*, 2003) in interaction with membranes, FRET from the Tyr residue of these peptides to *t*-PnA was used. The acceptor, *t*-PnA, is known to be internalized in the membrane (Castanho *et al.*, 1996), with the tetraene chromophore centered at a distance of 12.1 Å from the interface (de Almeida *et al.*, 2002). The Förster radii for these systems are $R_0 = 25.8 \pm 0.2$ Å (ShB) and $R_0 = 26.4 \pm 0.2$ Å (ShB-L7E). In Figure 12.7, the variation of the energy transfer efficiency *vs.* the number of acceptor per R_0^2 is shown. FRET is more efficient from ShB as compared to that from ShB-7LE, and therefore the former peptide is more deeply located in the membrane as compared to the latter. In addition, in Figure 12.7 is also presented the theoretical expectation for energy transfer in two-dimensions and a good agreement is found for the ShB/*t*-PnA pair. Since this curve was derived assuming no interplanar separation, this experiment locates precisely the depth of Tyr inside the membrane, such as could be concluded from the "parallax model". For the ShB-7LE, an interplanar separation of 12 Å describes the data, so this peptide's fluorophore is clearly near the surface (for more details in interplanar FRET see Section 12.7).

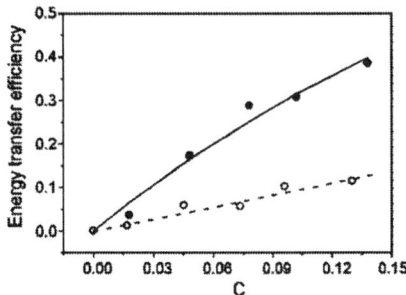

Figure 12.8. FRET efficiency between ShB peptide (closed circles) or mutant ShB-L7E peptide (open circles) and *t*-PnA as a function of C, the number of acceptor molecules per R_0^2. The solid line is the theoretical expectation for bidimensional random distribution of acceptors (in-plane FRET), and the dotted line is just a guide to the eye. Reprinted with permission from Poveda *et al.* (2003). Copyright 2003 American Chemical Society.

12.3.4. Emission Spectra of Tryptophan and Tyrosine, and Red-Edge Excitation Shift

One of the most immediate information about the fluorophore environment comes from its spectral distribution. Trp emission maximum undergoes a significant red-shift from a non-polar (~320 nm) to a solvating environment (~350 nm), and this has been used to follow, *e.g.*, protein denaturation. However, the membrane interface is a complex anisotropic medium, and from the blue-shift of Trp upon interaction with the membrane, it is not possible to obtain quantitative information about in-depth location.

Red-edge excitation shift (REES) is a very interesting phenomenon which was pioneered among others by Weber (Weber and Shinitzky, 1970; Valeur and Weber, 1977), and can be used in biophysics for obtaining topographical information (for a review see Chattopadhyay, 2003). This effect, *i.e.*, an increase of the maximum emission wavelength of fluorescence upon exciting on the tail of the absorption band, with a concomitant increase in anisotropy, is observed for fluorophores in a polar and motion-restricted environment. The effect is related to a dipole relaxation occurring on the time scale of fluorescence emission. The interfacial region of membranes is characterized by these properties, so if a strong effect is observed, it points to a surface location since the membrane core is non-polar and less rigid. We obtained for a peptide in interaction with a membrane (Santos *et al.*, 1998) an impressive REES of 18 nm which is a clear evidence of an interfacial location. However, as compared to the differential quenching methodology, no precise quantitative information can be obtained.

12.4. OBTAINING INFORMATION ON THE SECONDARY STRUCTURE OF THE PEPTIDE FROM FLUORESCENCE INTENSITY DECAYS

As previously described, Trp fluorescence emission decay kinetics is usually complex, and in addition to a sum of exponentials, continuous distributions of lifetime populations have also been used to describe the fluorescence decays (*e.g.* Vincent *et al.*, 2000). In fact, the lifetime of Trp can be influenced by a wide variety of factors that include solvent quenching, quenching by groups in the protein itself and electron transfer to the carbonyl of the peptide bond (Engelborghs, 2001). The complexity of single Trp peptides and proteins fluorescence decay kinetics can arise from two different but not mutually exclusive phenomena, namely the existence of ground state heterogeneity, usually attributed to different rotamers (*e.g.*, Willis and Szabo, 1992) and solvent relaxation processes (Lakowicz, 2000; Toptygin *et al.*, 2001). Multi-exponential decay in single Trp proteins has also been described as originating from the internal motion of Trp in the proteins, due to a coupling of internal motion and angle (and possibly distance) dependent quenching efficiency of neighboring quenchers (Tanaka *et al.*, 1994) and from reversible excited state dynamics (Engelborghs, 2001). This complexity intrinsic to Trp emission decays precludes its utilization in the recovery of direct information about, *e.g.*, secondary structure in peptides/proteins (Ladokhin, 2001). However, structural information can still be recovered from time-resolved fluorescence data of Trp residues as illustrated in the following study of the interaction of α-MSH with negatively charged vesicles, for which partition, emission shift and residual anisotropy results were presented in the previous section.

The primary structure of α-MSH is Ac-Ser-Tyr-Ser-Met-Glu-His-Phe-Arg-Trp-Gly-Lys-Pro-Val-NH$_2$. It appears that the peptide has no preferred structure in water (Biaggi *et al.*, 1997), being very flexible, whereas all its synthetic analogs with superpotent biological activity present a β-turn (or sometimes another kind of turn) stabilized within their central region comprising residues 6-9, *i.e.*, His-Phe-Arg-Trp (Sawyer *et al.*, 1980, 1982; Al-Obeidi *et al.*, 1989). This Trp-containing region of the peptide is furthermore the minimum melanotropic message sequence, essential for ligand binding and biological function (Hruby *et al.*, 1987).

The α-MSH tryptophanyl fluorescence decay in buffer is complex (described by three exponentials), and from the fits, a short component ($\tau_1 = 0.448$ ns (20°C) or 0.307 ns (37°C); $a_1 = 0.18$), an intermediate component ($\tau_2 = 2.00$ ns (20°C) or 1.36 ns (37°C); $a_2 = 0.48$) and a long component ($\tau_3 = 3.45$ ns (20°C) or 2.35 ns (37°C); $a_3 = 0.34$) are obtained. In the presence of both lipid systems at the two temperatures studied a variation of both the lifetimes and amplitudes of the components is observed. The data for the DMPC/DMPA (3:1) mixture at 20°C is depicted in Figure 12.8, and the trend of variation is similar to the one obtained for the DMPC/DMPG mixture. The short lifetime is essentially invariant while the intermediate one increases slightly and reaches a plateau and a marked increase is observed for the long one which varies from 3.45 ns in buffer up to 6.72 ns for the highest lipid concentration. Regarding the amplitudes, that of the short component shows a very slight increase; the amplitude of the intermediate component increases significantly and for the amplitude of the long one a correspondent decrease is obtained, both reaching a plateau at higher lipid concentrations. The amplitudes and the lifetimes of the components in aqueous solution are similar to those obtained by Ito *et al.* (1993).

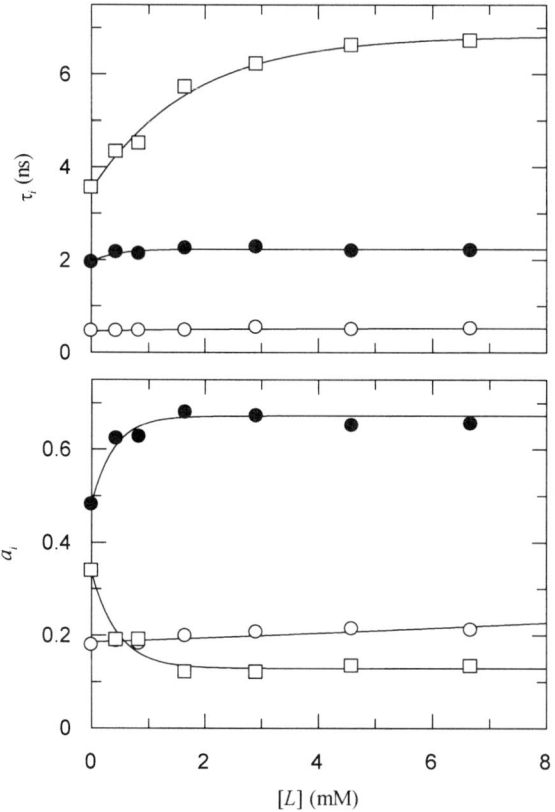

Figure 12.8. Fluorescence lifetime components τ_i and respective normalized pre-exponentials a_i for α-MSH (λ_{exc} = 295 nm, λ_{em} = 350 nm) *vs.* lipid concentration [*L*] (DMPC/DMPA (3:1) LUV) at 20°C. The solid lines are merely guides to the eye and have no physical meaning. Reprinted with permission from Contreras *et al.* (2001). Copyright 2001 Biophysical Society.

The lifetime-weighted quantum yield, $<\tau>$ (Eq. 5), of α-MSH in water at 20°C is very close to the value for Trp in aqueous solution at the same temperature ($<\tau>$ = 2.25 ns) (Lakowicz, 1999) and just slightly higher than the value for α-MSH reported in the literature ($<\tau>$ = 2.09 ns) (Ito *et al.*, 1993). If we extrapolate the values at different temperatures reported in that work, the fluorescence lifetime at 37°C ($<\tau>$ = 1.51 ns) is in very good agreement with our data.

From the partition coefficient study previously described (Eq. 6), $<\tau>_L$, the lifetime-weighted quantum yield of the peptide interacting with the membrane was determined and the values for the two membrane model systems are presented in Table 12.1. The values are longer in the DMPC/DMPG vesicles as compared to the DMPC/DMPA ones at the same temperature.

Upon interaction with the lipid system, the peptide undergoes alterations of its secondary structure and these can be appreciated from the trend of variation of both lifetime components and pre-exponentials (Figure 12.8). It is difficult to infer about the secondary structure of the peptide from the fluorescence decays in the different environments mainly because there are still very few studies made on peptides in which the Trp residue is in a position of known structure, and Dahms and Szabo (1995) point to subtleties that require caution when predicting structures. However, from the trend of variation of the time-resolved data, it can be concluded that upon membrane incorporation, α-MSH undergoes strong structural changes, which is in agreement with the IR data (Contreras et al., 2001). However, it should be stressed that this pattern of fluorescence time-resolved data only reports the peptide structure in the vicinity of the Trp residue, whereas global information about all the peptide structure is obtained from IR. Thus, there is no contradiction when the very same pattern of variation was obtained for both lipids, at variance with IR data which allowed concluding that intermolecular aggregates are the dominant structural feature in DMPG-containing vesicles, while in DMPA-containing vesicles some random structure is present.

The structural information obtained for the α-MSH peptide was possible to obtain because there was a "reference state" (the decay in buffer, where the peptide has a random structure). In fact, it is only possible to extract reliable information on the Trp vicinity of peptides and proteins when a comparison can be made (Engelborghs, 2001). The trend displayed by the lifetimes and amplitudes of the decay components can be appreciated as a function of lipid concentration, as shown here or, e.g., in the absence vs. presence of (varying concentrations of) a denaturing agent (Dahms and Szabo, 1995).

The multi-exponential behavior of single Tyr peptides and proteins has been extensively reported, and has been attributed to the existence of ground state rotamers sensing different chemical environments and interconverting slowly on the nanosecond timescale (Laws et al., 1986), and thus the fluorescence decay of those peptides and proteins should, in principle, contain some structural information.

12.5. FLUORESCENCE ANISOTROPY OF THE PEPTIDE CONTAINS STRUCTURAL AND DYNAMICAL INFORMATION

Depolarization of the fluorophore emission can occur through a variety of processes. In the absence of phenomena such as energy homotransfer (see sub-section 12.6.5), the rotational dynamics of the fluorophore is the determinant factor. This dynamics depends on parameters that include temperature, the viscosity of the surrounding medium, and the size and shape of the molecule containing the fluorophore. Consequently, important structural and dynamical information can be extracted from anisotropy measurements, both time-resolved and to a lesser extent, steady-state.

The anisotropy decays of fluorophores in biomolecules are usually complex, as several movements contribute to the depolarization process. They are usually described by a sum of exponentials,

$$r(t) = (r_0 - r_\infty)\sum_i \beta_i \exp(-t/\varphi_i) + r_\infty \tag{11}$$

where r_0 is the fundamental anisotropy (at time zero, and depends only on the fluorophore), φ_i are the rotational correlation times, β_i their respective amplitudes ($\Sigma\ \beta_i = 1$), and r_∞ the residual or limiting anisotropy. The anisotropy decay parameters are obtained using a nonlinear least-squares global analysis method by simultaneously fitting the vertically and horizontally polarized emission components of the decay to Eq. (11). The steady-state anisotropy in turn, is obtained from the steady-state intensity components (Jabłoński, 1960):

$$<r> = (I_{VV} - GI_{VH})/(I_{VV} + 2GI_{VH}) \tag{12}$$

where the different intensities I_{ij} are the steady-state vertical and horizontal components of the fluorescence emission with excitation vertical (I_{VV} and I_{VH}, respectively) and horizontal (I_{HV} and I_{HH}, respectively) to the emission axis. The latter pair of components is used to calculate the G factor ($G = I_{HV}/I_{HH}$; Chen and Bowman, 1965).

In the case of the aforementioned α-MSH, the steady-state anisotropy in water has a very low value, both at 20°C ($<r> = 0.0144 \pm 0.0003$) and at 37°C ($<r> = 0.0146 \pm 0.0007$). The time-resolved anisotropy decays to zero, with a single rotational correlation time of 0.52 ± 0.02 ns and 0.32 ± 0.01 ns, respectively. Sub-nanosecond φ values, and absence of a limiting anisotropy, justify the low values of the steady-state anisotropy. In fact, for molecules with a single rotational correlation time and complex fluorescence intensity decay, the steady-state anisotropy is given by the following composition of Perrin (1926, 1929) equations (based on the law of anisotropy addition; Weber, 1960):

$$<r> = \frac{r_0 - r_\infty}{<\tau>} \sum_i \frac{a_i}{1/\tau_i + 1/\varphi} + r_\infty \tag{13}$$

which leads to values of $<r>$ coincident with the experimental steady-state values. The rotational correlation time can alternatively be calculated assuming spherical geometry for the peptide. In this case, $\varphi - \eta V/RT$ where η is the viscosity of the solvent, and V the volume of the rotating unity. The volume of the peptide can be estimated to be 1821 $\mathring{A}^3$ (Zamyatnin, 1972) and in this case the values predicted are $\varphi = 0.45$ ns (20°C) and $\varphi = 0.29$ ns, which are only slightly shorter than the experimental ones. Thus, we can conclude that the peptide is essentially monomeric in aqueous solution for the concentrations used in fluorescence measurements (~30 μM), whereas IR spectra, carried out on more concentrated samples indicate the presence of peptide aggregates at those temperatures (Contreras et al., 2001).

When in the presence of lipid bilayers, a positive limiting anisotropy is reached, as described in Section 12.2. In this case, even in the presence of a high lipid concentration, the fraction of peptide remaining in the aqueous phase can contribute significantly to the fluorescence emission of a sample. This implies that a detailed analysis of the fluorescence anisotropy decay of the peptide at shorter times is eventually too complex, because the fitting model would have to consider at least two different peptide populations, each associated with a complex fluorescence intensity decay. However, the study of limiting anisotropies of the peptide in the membrane is not hampered by this fact, and in Section 12.2 it was shown how to obtain the value of r_∞ for the peptide interacting with the membrane. From the values give in Table 12.1, it can be seen that the limiting anisotropies obtained are in all cases very high, considering that the fluorophore is not a typical hydrophobic molecule incorporated into the membrane core. This means that the peptide is strongly adsorbed at the membrane/water and the wobbling motion of Trp is very restricted, even in the fluid phase. Irrespective of the structure adopted by the peptide in the vesicles, not only it is thermally stabilized (see previous section) but it is also very rigid. The limiting anisotropies are related to the order parameter, S, describing the equilibrium orientation distribution of the probe at times much larger than the rotational correlation time, and the fundamental anisotropy r_0, through the following relationship (Best *et al.*, 1987):

$$S^2 = \frac{r_\infty}{r_0} = \left\langle \frac{3\cos^2\theta - 1}{2} \right\rangle^2 \tag{14}$$

where θ is the angular rotation of the emission transition moment and the angle brackets indicate an average over the entire fluorophore population, assuming cylindrical symmetry. If we consider a fundamental anisotropy for Trp excitation at 295 nm of $r_0 = 0.24$ (Valeur and Weber, 1977) an angle $\theta = 25°$ is obtained for the gel phase, or $\theta = 27\text{-}31°$ for the fluid phase. In a molecular dynamics study by Pascutti *et al.* (1999), the peptide acquired, in a low dielectric constant medium, a packed conformation that remained stabilized for more than 7.0 ns of the simulation, and as already mentioned, the superpotent analogs possess a stabilized turn involving the Trp residue (Sawyer *et al.*, 1982). Based on the results presented here, it is reasonable to predict that the interaction of α-MSH with negatively-charged membranes induces and stabilizes a specific conformation, probably involving the Trp-containing message region, which could be similar to the one found for the superpotent analogs, and therefore necessary for its biological activity.

A more detailed study could be performed with the shaker B (ShB) peptide, because due to its higher partition coefficient it is possible to have a very low fraction of peptide in the aqueous phase in the presence of relatively low lipid concentrations (Poveda *et al.*, 2003). The ShB peptide comprises the 20 N-terminal amino acids (H$_2$N-MAAVAGLYGLGEDRQHRKKQ) of each subunit in the Shaker B potassium channel, the so called inactivating "ball" responsible for inducing fast, N-type inactivation in this and many other related or unrelated channels (*e.g.*, Hoshi *et al.*, 1990). Previous biophysical studies with a fluorophore-labeled ShB peptide suggested that the peptide binds to lipid vesicles with high affinity, readily adopts a strongly hydrogen-bonded intramolecular β-hairpin structure and becomes inserted into the hydrophobic bilayer in a monomeric form. In contrast, the non-inactivating mutant ShB-L7E also binds phospholipids vesicles, but is unable to either form the characteristic β-structure or get into the hydrophobic core of the bilayer. ShB peptide contains a single Tyr residue and no

Trp. Tyr displays a high intrinsic anisotropy and a fluorescence lifetime optimal to characterize nanosecond and sub-nanosecond motions in peptides and proteins (Ferreira *et al.*, 1994).

The steady-state fluorescence anisotropy measured in buffer at 22°C by exciting at 285 nm was relatively similar for both peptides ($<r>$ = 0.039 and $<r>$ = 0.051 for ShB and ShB-L7E, respectively) and much smaller than the fundamental anisotropy (r_0 = 0.29) corresponding to immobilized Tyr at that excitation wavelength (Gryczynski *et al.*, 1991).

Table 12.2. Time-resolved fluorescence parameters of ShB and ShB-L7E (rotational correlation times, φ_i, amplitudes, β_i, and residual anisotropy, r_∞) in aqueous buffer and incorporated into PA vesicles at 22°C.

Peptide	Medium	r_0	β_1	φ_1 (ns)	β_2	φ_2 (ns)	r_∞	χ^2
ShB	Buffer	0.20	0.51	0.2	0.49	0.9	0	1.0
ShB-L7E	Buffer	0.20	0.67	0.2	0.33	0.9	0	1.0
ShB	PA	0.17	0.42	0.2	0.58	4.3	0.07	1.0
ShB-L7E	PA	0.18	0.64	0.1	0.36	0.3	0.06	1.1

Reprinted with permission from Poveda *et al.* (2003). Copyright 2003 American Chemical Society.

In the time-resolved anisotropy experiments, two rotational correlation times were needed to describe the decay processes, which were similar for both peptides (Table 12.2). Since the long rotational correlation time (φ_2) was five times longer than the short one (φ_1), the total anisotropy can be interpreted as the product of two independent depolarizing processes, a first one due to fast movements of the peptide segment containing the Tyr residue [$r'(t)$], and a second one related to the global motion of the whole peptide (Lipari and Szabo, 1980):

$$r(t) = r'(t)\exp\left[\left(1 - S_2^2\right)\exp\left(-t/\varphi_{global}\right)+ S_2^2\right]$$

(15)

$$r'(t) = r(0)\left[\left(1 - S_1^2\right)\exp\left(-t/\varphi_{segmental}\right)+ S_1^2\right]$$

where S_1 and S_2 are the order parameters characterizing the internal and the whole peptide fluctuations. The anisotropy decays to zero for both peptides in buffer, thus in this case S_2 = 0. The short and long rotational correlation times obtained from the fit are related to φ_{global} and $\varphi_{segmental}$ (Table 12.3) by

$$\varphi_2 = \varphi_{segmental}$$

(16)

$$\varphi_1 = \left(\varphi_{segmental}\varphi_{global}\right)/\left(\varphi_{segmental} + \varphi_{global}\right)$$

The longest rotational correlation time, similar for both peptides, reflecting the motion of the whole peptide, can be used to estimate the equivalent hydrodynamic radius

R, assuming spherical symmetry. From $\varphi = \eta V/RT$, and as $V = 4\pi R_h^3$, a value of $R_h = 9.6$ Å is computed. By independent methods the radius is estimated to be 9.4 Å, and thus the peptide is essentially monomeric (Poveda *et al.*, 2003). The segmental rotational correlation time is restricted, as shown by the S_1 values in Table 12.3. Assuming a "wobbling in cone" model in which the molecule exists in a square-well potential beyond which further angular displacement is energetically impossible, the angular displacement can be calculated from S_1 (note that this is a different principle than the one that leads to Eq. (14)), through the relationship (Kinosita *et al.*, 1982):

$$S_i = \frac{\cos^2 \theta + \cos \theta}{2} \tag{17}$$

(or equivalently, $\cos \theta = \frac{1}{2}[(8S_i+1)^{\frac{1}{2}} -1]$). The restriction is more severe in ShB (angular displacement of 38°), as compared to ShB-L7E (47°). The steady-state anisotropy of both peptides showed a large increase upon addition of egg PA vesicles, and from a fit of Eq. (3) the partition coefficient and the steady-state anisotropy values for both peptides incorporated in the membrane were determined ($K_p = (4.5 \pm 0.5) \times 10^4$, $<r> = 0.120 \pm 0.003$ for ShB and $K_p = (2.2 \pm 0.3) \times 10^4$, $<r> = 0.093 \pm 0.003$ for ShB-L7E). The partition coefficients are much higher than those previously presented for α-MSH.

Table 12.3. Parameters from the fit of an independent two-motion model to the anisotropy decay of ShB and SHB-L7E (segmental and global correlation times, φ_i, order parameters, S_i, and cone angles, θ_{0i}) in aqueous buffer and incorporated into PA vesicles at 22°C.

Peptide	Medium	$\varphi_{segmental}$ (ns)	φ_{global} (ns)	S_1	S_2	θ_{01} (deg)	θ_{02} (deg)
ShB	Buffer	0.2	0.9	0.70	0	38	
ShB-L7E	Buffer	0.2	0.9	0.57	0	47	
ShB	PA	0.1	4.3	0.87	0.75	24	35

Reprinted with permission from Poveda *et al.* (2003). Copyright 2003 American Chemical Society.

The fluorescence anisotropy decays for the peptides ShB and ShB-L7E have very different kinetics, but in both cases a residual, time-independent value different from zero is reached (Figure 12.9), indicating a restricted motion of the Tyr side chain on the timescale of the fluorescence emission. The fitting parameters are shown in Table 12.2. Two rotational correlation times were needed to describe the decay processes. In the case of the mutated peptide, they were both very short, whereas for the ShB peptide, the shorter one was identical to the one in buffer, but the long one was much higher.

The two rotational correlation times for ShB differ by more than a order of magnitude, and again, they can be attributed to two independent molecular motions. Following the model previously used for aqueous solution, and applying Eq. (15, 16), the order parameters and the values of $\varphi_{segmental}$ and φ_{global} can be obtained (Table 12.3).

Quenching and FRET experiments show that the Tyr residue of ShB peptide is located much deeper into the membrane than that of ShB-7LE (Sub-sections 12.3.1 and 12.3.3). All the above mentioned results agree with the burying of the β-hairpin of ShB into the hydrophobic region, parallel to the phospholipids acyl chains. An additional C-terminal portion would remain lying over the membrane surface facilitating electrostatic interactions. According to this model, at least two rotational correlation times and a residual anisotropy must be extracted from the time-resolved anisotropy experiments of the ShB peptide inserted in membranes: a short correlation time which reflects internal fluctuations of the Tyr residue, a longer component corresponding to fluctuations of the whole peptide segment incorporated into the membrane, and a residual value which shows that the reorientation motion of the Tyr side chain is restricted on the timescale of the fluorescence emission due to interactions with phospholipid chains.

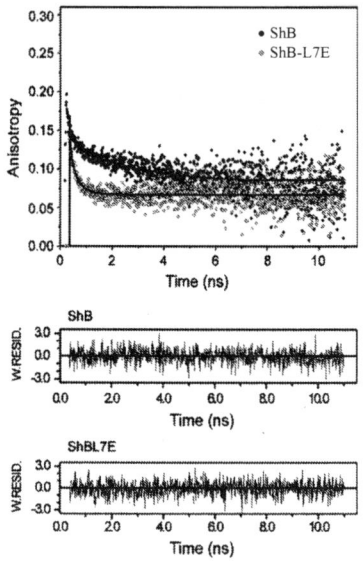

Figure 12.9. Time-resolved fluorescence anisotropy of the Tyr residue of the ShB and the mutant ShB-L7E peptides in PA LUV at 22°C. Experimental data and that from the corresponding fit (see Table 12.2 for decay parameters) are shown as points and as a continuous line, respectively. Residuals of each fit are also shown. Excitation and emission wavelengths were 285 and 320 nm, respectively. Reprinted with permission from Poveda *et al.* (2003). Copyright 2003 American Chemical Society.

Assuming that the hairpin structure behaves like a rigid body into the bilayer, we could estimate the diffusion coefficient corresponding to the rigid body fluctuations from the longest rotational correlation time as $D_\perp = \sigma/\varphi_{global}$ (Kinosita *et al.*, 1982; Vogel *et al.*, 1988) where σ is a function of the order parameter S_2 and can be computed from the polynomial approximation described by Mateo *et al.* (1996). Modeling the hairpin as a cylinder with a diameter of 8 Å and assuming a lipid viscosity of *ca.* 0.2 P (Best *et al.*, 1987) a volume of ≈ 850 Å^3 was estimated, which is in agreement with the volume of the N-terminal 1-10 amino acids in ShB, corresponding to the hydrophobic half of the peptide. The β-turn would be formed by the sequence 4-7 (VAGL) establishing four stabilizing intramolecular hydrogen bonds. The glycine residue at position 11, due to its larger degree of rotational freedom, could act as a hinge, allowing the fluctuations of the membrane embedded sequence. In this way, it is useful to calculate the amplitude of these fluctuations, which can be obtained from the order parameter S_2, assuming the wobbling in a cone model (Eq. 17). An angular displacement of 35° was obtained, which is smaller than the value reported in fluid membranes from electron paramagnetic resonance and fluorescence depolarization techniques for anchored rigid probes (for a review, see Mateo *et al.*, 1996) and helical polypeptides containing Trp (Vogel *et al.*, 1988). The small angle suggests that the restriction in the motion of the β-hairpin structure is due not only to the phospholipid chains but also to intramolecular interactions with adjacent residues and to the strong electrostatic interactions established between the C-terminal portion of the peptide and the bilayer surface. Furthermore, the short correlation time for the membrane-bound form is similar to the one obtained in buffer, but the angular displacement of this motion is smaller (24° *vs.* 38°), and indicates that the hydrogen-bonded structure adopted by the peptide in the bilayer, together with the proximity of the phospholipids acyl chains, restrict strongly the amplitude of the motion in the Tyr-containing peptide segment.

Finally, the dynamics of ShB-L7E in the membrane is very different from that of ShB, exhibiting two short rotational correlation times. These could correspond to rotational motions of short segments in a fluid medium. The limiting anisotropy value also indicates that the amplitude of these motions is restricted. If we consider that the peptide lies at the membrane surface interacting electrostatically with the phospholipid head groups, the global rotational motion of the peptide is then prevented, and only rapid albeit constrained fluctuations of the Tyr side chain can take place, in agreement with the parameters recovered from the anisotropy decays.

In sum, our results show that a wide variety of molecular details regarding the ShB *vs.* the ShBL7E peptide interaction with membranes can be obtained using the intrinsic fluorescence of their single Tyr residue.

12.6. FORMATION OF PEPTIDE-RICH PATCHES/PEPTIDE AGGREGATES *VS.* RANDOM DISTRIBUTION

The peptide fluorescence, either from Trp or Tyr residue(s), or a covalently-labeled probe, can be highly dependent on its aggregation state. For instance, if there is the possibility of self-quenching, the oligomerization of the peptide, leading to stable aggregates, can result in a static quenching process. On another hand, the formation of peptide rich domains without stable aggregates can lead to an increased dynamic self-quenching, due to the fact that the local concentration of peptide is higher than that expected on the basis of a random distribution on the whole bilayer. The situations of static and dynamic quenching can be easily distinguished if both steady-state and time resolved measurements are performed. Additionally, FRET, due to its strong distance-dependence, is also very useful addressing these questions.

12.6.1. Aggregation State and Lateral Distribution of M13 Major Coat Protein from Self-Quenching Studies

M13 major coat protein is the main protein component of the filamentous bacteriophage M13, with about 2800 copies. It contains a single hydrophobic transmembrane segment of approximately 20 amino acid residues, apart from an amphipathic N-terminal arm and a heavily basic C-terminus with a high density of lysine (Lys) residues (for reviews see Stopar *et al.*, 2003; Hemminga *et al.*, 1993). It has been known to exist in many aggregation states, depending on factors like isolation, reconstitution procedure, pH, ionic strength, and amphiphile composition (Hemminga *et al.*, 1993; Stopar *et al.*, 1997). The mechanism of phage assembly in the *Escherichia coli* membrane is not yet completely understood, but the assembly site is thought to be composed of a dynamic protein-lipid network, characterized by the absence of a preferential association between M13 coat protein and/or lipids (Hemminga *et al.*, 1993), which allows storage of monomeric coat protein at very high local concentrations, as the insertion of the protein in the assembling phage particle is only possible on the monomeric form (Russel, 1991). The type of interactions between lipids and coat protein that allows the formation of this structure is largely unknown. It has been proposed that the self-association behavior of transmembrane proteins incorporated in lipid membranes is influenced by hydrophobic matching conditions on the protein-lipid interface (Mouritsen and Bloom, 1984; Killian, 1998). The monomeric protein is expected to be stable under perfect matching conditions with the surrounding phospholipid milieu. In case of hydrophobic mismatch at the protein-lipid interface, it is possible that the boundary lipids reorganize themselves, to lower the tension created by exposure of hydrophobic acyl chains or amino-acid residues, which can be achieved by ordering/disordering of the phospholipids (Nezil and Bloom, 1992). Moreover, if the hydrophobic mismatch is too high for correction with small adjustments of bilayer hydrophobic thickness, this might result in protein aggregation to obtain minimization of the protein-lipid contacts (Ren *et al.*, 1999; Lewis and Engelman, 1983; Mobashery *et al.*, 1997; Mall *et al.*, 2001).

In order to obtain information on the influence of lipid bilayer composition on the lateral distribution and oligomerization properties of M13 coat protein in membrane model systems, several fluorescence methodologies (fluorescence self-quenching, absorption and emission spectra, and energy transfer) were applied (Fernandes *et al.*, 2003), using the protein derivatized with n-(4,4-difluoro-5,7-dimethyl-4-bora-3a,4a-diaza-s-indacene-3-yl)methyl iodoacetamide (BODIPY FL C_1-IA) or n-(iodoacetyl)aminoethyl-1-sulfonaphthylamine (IAEDANS). The fluorescence decay of BODIPY in the labeled mutant proteins incorporated in 1,2-dioleoyl-*sn*-glycero-3-phosphocholine (DOPC; C18:1 acyl chains), 1,2-dimyristoleoyl-*sn*-glycero-3-phosphocholine (DMoPC; C14:1 acyl chains)/DOPC, 1,2-dierucoyl-*sn*-glycero-3-phosphocholine (DEuPC; C22:1 acyl chains)/DOPC, DMoPC, and DEuPC bilayers was described by two components, $\tau_1 = 6.23$ ns ($a_1 = 0.90$) and $\tau_2 = 3.27$ ns, which leads to an average lifetime, calculated by (*e.g.*, Sillen and Engelborghs, 1998):

$$\bar{\tau} = \sum_i a_i \tau_i^2 \bigg/ \sum_i a_i \tau_i \tag{18}$$

of $\bar{\tau} = 6.1$ ns, as measured in samples with [BD-M13 coat protein]$_{eff}$ < 10^{-3} M (BODIPY-labeled protein effective membrane concentration). For the determination of protein effective concentration in the membranes, the lipid molar volumes were calculated from the reported lipid area (72 Å^2) and membrane thicknesses (Tristam-Nagle *et al.*, 1998; Lewis and Engelman, 1983). From the fluorescence lifetimes, the dynamical contribution for quenching can be discriminated.

The effects of the collisional contribution of self-quenching on the fluorescence lifetime for a molecule with a complex decay are described by the Stern-Volmer equation (Sillen and Engelborghs, 1998):

$$<\tau>_0/<\tau> = 1 + <k_q> \cdot \bar{\tau}_0 \, [F] \tag{19}$$

The subscript 0 indicates the absence of quencher (in our case, the values for the sample with the lowest protein concentration), $<k_q>$ is the bimolecular quenching rate constant, the brackets indicate that this is a (complex) average of the rate constants related to each component of the decay, and $[F]$ is the concentration of the fluorophore. By increasing effective protein concentration, and considering a random protein distribution in the bilayers, linear self-quenching Stern-Volmer plots are obtained (Fernandes *et al.*, 2003) and $<k_q>$ values are recovered (Table 12.4).

Table 12.4. Bimolecular diffusion rate constants, $<k_q>$, diffusion coefficients, D, and apparent sphere-of-action radii (R_S) recovered from BODIPY fluorescence emission self-quenching from BODIPY-labeled T36C and A35C mutants of M13 major coat protein.

Systems	$< k_q > / (10^9 \, M^{-1}s^{-1})$	$D / (10^{-7} \, cm^2s^{-1})$	R_s (Å)
T36C in DOPC	2.3	0.7	14
A35C in DOPC	1.7	0.4	14
A35C in DMoPC	2.6	0.8	23
A35C in DEuPC	5.0	2.6	27
T36C in DMoPC/DOPC	6.6	4.2	14
T36C in DEuPC/DOPC	20	22	14

Reprinted from Fernandes *et al.*, (2003) with permission. Copyright 2003 Biophysical Society.

In the analysis of the steady-state fluorescence intensities, the static quenching (described by the sphere-of-action model) has to be taken into account. The combined contributions of the collisional and the sphere-of-action effects on the fluorescence intensity are given by Loura *et al.* (2000) as

$$I_F = \frac{C \times [F]}{1/\tau_0 + <k_q> [F]} \exp(-V_s N_A [F]) \tag{20}$$

Here, I_F is the fluorescence intensity, C is a constant, and V_s is the sphere-of-action volume, from which the sphere-of-action radius is obtained. In the absence of protein aggregation this radius should be close to the sum of the van der Waals radii. Eq. (20) was used to analyze the fluorescence steady-state quenching profiles for BD-M13 coat protein in DOPC, DMoPC/DOPC, DEuPC/DOPC, DMoPC, and DEuPC (Figure 12.10). Because $<k_q>$ values were obtained from the lifetime quenching profile, the fluorescence quenching contribution of the sphere-of-action effect could be retrieved (Table 12.4).

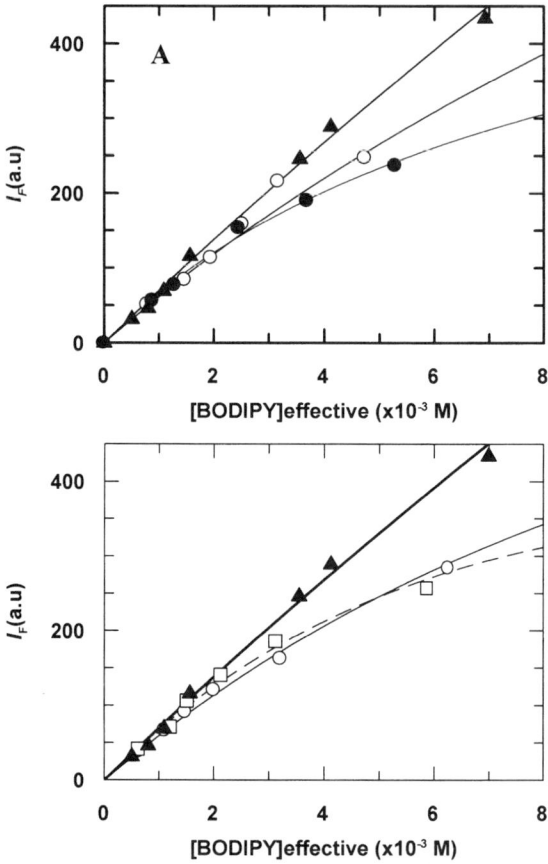

Figure 12.10. Fluorescence steady-state BODIPY self-quenching profile. A - Protein incorporated in DOPC (▲), DMoPC/DOPC (60/40 mol/mol) (○), and DEuPC/DOPC (60/40 mol/mol) (●). Eq.(20) is fitted to the data on the basis of dynamical quenching and a sphere-of-action quenching model (14.4 Å radius) (—) for the protein in all lipid systems. B – Protein incorporated in DOPC (▲), DMoPC (○), and DEuPC (□). Eq. (20) fit to data from DOPC bilayers using a sphere-of-action radius of 14 Å (—), from DEuPC bilayers with a sphere-of-action radius of 27 Å (---), and from DMoPC bilayers using a sphere-of-action radius of 23 Å (—). These higher values are evidence of aggregation. Reprinted from Fernandes *et al.* (2003) with permission. Copyright 2003 Biophysical Society.

The bimolecular quenching constants calculated from the self-quenching results for BD-M13 coat protein incorporated in DOPC, DMoPC/DOPC (60/40 mol/mol) DEuPC/DOPC (60/40 mol/mol), DEuPC, and DMoPC allow the estimation of the labeled protein molecular diffusion coefficient (*D*) using the Smoluchowski equation (Lakowicz, 1999), taking into account transient effects (Umberger and Lamer, 1945). For this calculation and the validity of the 3D approach see sub-section 12.6.3. Taking 6 Å for BODIPY collisional radius, a value of 7.0×10^{-8} cm^{2}s^{-1} is obtained for $D_{\text{BD-M13 coat protein}}$ in DOPC bilayers, which has the same order of magnitude as the values of D for the M13 coat protein incorporated in fluid bilayers reported in the literature (Smith *et al.*, 1979, 1980).

However, for the lipid mixtures in which the predominant lipid does not hydrophobically match with the hydrophobic core of the M13 coat protein, the values obtained for D are unreasonably high. For the DMoPC/DOPC bilayers, a value of 4.2×10^{-7} cm^{2}s^{-1} is obtained, whereas in DEuPC/DOPC it is even higher (2.2×10^{-6} cm^{2}s^{-1}). If $D_{\text{BD-M13 coat protein}}$ in pure vesicles of DOPC is reporting a random distribution in the bilayer, the values in these mixtures are likely to be reporting protein segregation effects in the bilayer. This is caused by the hydrophobic mismatch constraints that the protein finds when incorporated in bilayers with too-long, or too-short phospholipids in their composition, leading to formation of localized areas with increased DOPC plus protein content. In this way, the apparent effective concentration in Eq. (19), should be higher than the one assumed on the basis of a random distribution, which led to an overestimation of $<k_q>$, and consequently of D.

This rationalization is supported by D values obtained from BODIPY labeled protein in the pure mismatching lipid DMoPC and DEuPC (Table 12.4). These values are smaller than the ones obtained from the mixtures, and the diffusion coefficient in pure DMoPC is almost identical to the value in pure DOPC. For pure vesicles of DEuPC, $D_{\text{BD-M13 coat protein}}$ is larger than in DOPC, but it is still much smaller than the value obtained from the DEuPC/DOPC mixture. These dynamic self-quenching results indicate therefore that, although in pure vesicles of DEuPC there are already more collisions between BODIPY groups than would be expected from a random distribution of labeled protein in the bilayer (probably due to aggregation), when DOPC is added the probability of collision increases greatly. At least in part, this can be explained in terms of protein segregation into DOPC-enriched microdomains. In DMoPC/DOPC the effect is similar, but the bimolecular quenching constant is smaller than in DEuPC/DOPC. In Figure 12.10, the obtained steady-state quenching profiles are presented together with the theoretical expectation for a sphere-of-action quenching model (Eq. 20). For the BODIPY-labeled protein in the DOPC-containing lipid systems (DOPC, DMoPC/DOPC, and DEuPC/DOPC) the results are well described using a sphere-of-action radius of 14 Å. For the pure mismatching lipids it is necessary to use larger radii to describe the results using Eq. (21).

In case that a complex is formed, the model to describe static quenching effects should take into account its equilibrium constant. For monomer/dimer equilibrium of only one molecular species, the fluorescence intensity is given by:

$$I_F = \frac{C \times [F]}{1/\tau_0 + <k_q>[F]} \times \frac{-1 + \sqrt{1 + 8K_a[F]}}{4K_a} \tag{21}$$

where K_a is the oligomerization constant. However, for our system, in which there are two different protein species (labeled and unlabeled, where the unlabeled class includes both non-labeled mutant and wild-type protein), there will be several combinations of protein species within an aggregate. For a dimer, there will be three different combinations available—labeled protein/labeled protein, labeled protein/unlabeled protein, and unlabeled protein/unlabeled protein—but only formation of the first one induces self-quenching of BODIPY. A complexation model describing fluorescence static quenching in our system will have to account for this fact. From the knowledge of the concentration of each species, the fraction of labeled protein participating in oligomers (dimers/trimers) containing more than one BODIPY labeled protein (and as a result non-fluorescent), can be obtained for a given K_a. The resulting set of nonlinear equations is promptly solved (for a given total protein concentration, labeling efficiency, labeled protein concentration, aggregation number, and K_a) using adequate mathematical software.

In Figure 12.11, a simulation was included for a small degree of protein aggregation in DOPC, DMoPC, and DEuPC bilayers. In these simulations it was considered that, due to the small degree of self-association considered, there was no change in M13 coat protein distribution and dynamics, and that in an oligomer, the fluorescence intensity of a BODIPY group is reduced to zero by the presence of another BODIPY group in the same aggregate. For DOPC bilayers, the prediction using a low fraction of aggregation (25% for dimerization and 10% for trimerization) clearly overestimates the extent of aggregation at the high labeled protein concentration, the range where this methodology is more sensitive. In agreement, from Figure 12.10, it is clear that the data for the three DOPC-containing lipid systems are rationalized on the basis of dynamic quenching and a sphere of action, without need for assumption of aggregation. The recovered radius (R_s = 14 Å) is close to the sum of the Van der Waals radii. These results indicate that BD-M13 coat protein in the studied DOPC-containing bilayers does not oligomerize. This conclusion is further supported by the absence of BODIPY dimers in our samples, which would be revealed in the absorption/emission spectra.

Still, the results from self-quenching on pure bilayers of hydrophobic mismatching lipids (DEuPC and DMoPC) point to some aggregation, as the sphere-of-action radii recovered from the fit of Eq. (20) were very unrealistic (27 Å and 23 Å for DEuPC and DMoPC, respectively). Simulations for BODIPY emission self-quenching due to aggregation are compared with the experimental data in these lipid systems in Figure 12.11. DMoPC data could be reasonably described using a K_a = 20 for dimerization and a K_a = 1300 for trimerization (13% of aggregated protein at the protein concentration of the most concentrated data point in both simulations).

For the data obtained in DEuPC, the degree of aggregation is higher than in DMoPC (a larger sphere-of-action radius was recovered from the fit of Eq. (20)). This result is in agreement with the observations of Meijer et al. (2001), who, from ESR studies, reported that the protein appeared to exist in several orientations/conformations or in an aggregated form while incorporated in DEuPC bilayers. The larger extent of coat protein aggregation observed in the longer lipid bilayers can be explained by the fact that negative hydrophobic mismatch is considered to be energetically less favorable than positive mismatch (Killian, 1998; Mall et al., 2001).

Regarding the DEuPC/DOPC and DMoPC/DOPC lipid mixtures used in the present study, no change in conformation or orientation was found (CD spectroscopy/IAEDANS fluorescence emission spectra, not shown), and, as described above even for the protein in DEuPC/DOPC (60/40 mol/mol) no aggregation was detected (BODIPY self-quenching). These results point to stabilization of the protein by the hydrophobic matching phospholipid (DOPC), which was probably achieved by protein segregation to

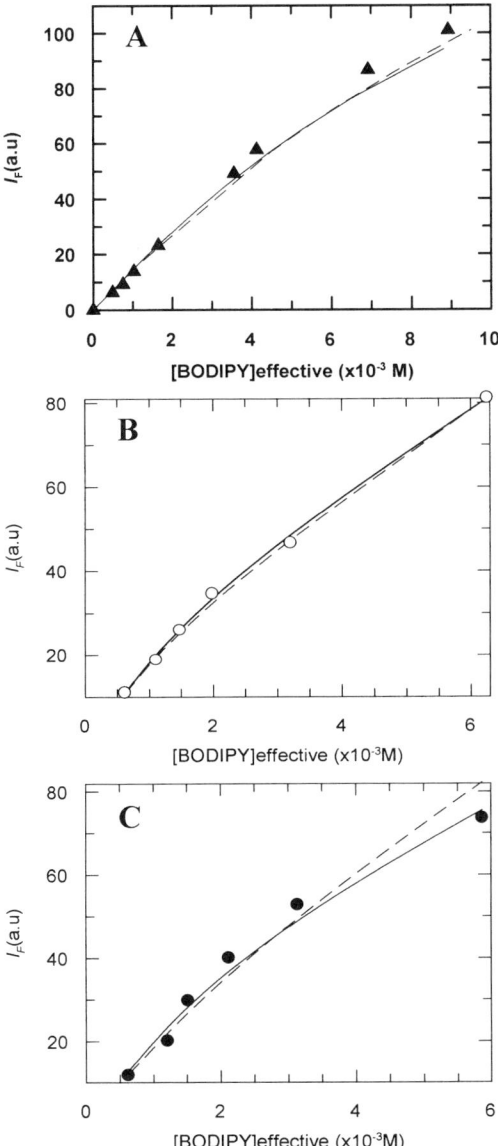

Figure 12.11. Fluorescence steady-state data for BODIPY self-quenching at different labeled protein concentrations. A- Protein incorporated in DOPC (▲) and simulations considering protein oligomerization including static quenching: (—) due to trimerization of the protein with $K_a = 1000$ ($\approx$ 10% total protein oligomerization). (---) due to dimerization of the protein with $K_a = 10$ ($\approx$ 25% total protein oligomerization). B- Protein incorporated in DMoPC (○) and simulations for dimerization of protein with $K_a = 20$ ($\approx$ 13% total major coat protein oligomerization for the most concentrated data point) (---), and trimerization of the protein with $K_a = 1300$ ($\approx$ 13% total M13 coat protein oligomerization for the most concentrated data point) (—). C - Protein incorporated in DEuPC (●) and simulations for dimerization of protein with $K_a = 30$ ($\approx$ 17% total protein oligomerization for the most concentrated data point) (---) and trimerization of the protein with $K_a = 7500$ ($\approx$ 25% total protein oligomerization for the most concentrated data point) (—). Reprinted from Fernandes *et al.* (2003) with permission. Copyright 2003 Biophysical Society.

domains enriched in that phospholipid, partly explaining the high M13 coat protein apparent molecular diffusion coefficients obtained for the protein incorporated in DEuPC/DOPC and DMoPC/DOPC bilayers.

Although an increase in BODIPY emission dynamic self-quenching was already visible for pure DEuPC bilayers, the bimolecular diffusion rate constant value in DEuPC/DOPC bilayers is much higher, and still, this can only be explained by segregation to DOPC-enriched microdomains.

12.6.2. Headgroup and Acyl Chain-Length Effects on Lateral Distribution of M13 Major Coat Protein Studied by FRET

The energy transfer study was performed using IAEDANS –labeled protein as donor and BODIPY-labeled as acceptor, due to the good overlap of the former's emission with the latter's absorption. Energy transfer studies were performed for the labeled protein incorporated in DOPC, DOPC/DOPG (80/20 mol/mol), DOPE/DOPG (70/30 mol/mol), DEuPC/DOPC (60/40 mol/mol), and DMoPC/DOPC (60/40 mol/mol). It was intended to study the influence of electrostatic interactions and hydrophobic mismatch in the aggregation and compartmentalization properties of the M13 coat protein. Due to the nonlamellar character of phosphatidylethanolamines (PE), it was necessary to include a fraction of lamellar lipids (phosphatidylglycerol, PG) in the lipid mixture for bilayer stabilization.

Experimental energy transfer efficiency is obtained from

$$E = 1 - I_{DA}/I_D \tag{22}$$

where I_D and I_{DA} are the donor fluorescence intensities in absence and presence of acceptor, respectively. To obtain topological information, this observable should be compared with theoretical expectations.

For the in-plane FRET, the decay of donor fluorescence in the presence of acceptor, assuming a radius of exclusion of acceptors (R_e) around the donor (in this case the sum of the van der Waals radii of the chromophores), and a random distribution in the plane of the membrane considered as infinite, becomes (Wolber and Hudson, 1979):

$$\rho_{cis}(t) = \exp\left\{-\pi R_0^2 n_2 \gamma\left[\frac{2}{3}, \left(\frac{R_0}{R_e}\right)^6 \left(\frac{t}{<\tau>}\right)\right]\left(\frac{t}{<\tau>}\right)^{1/3} + \pi R_e^2 n_2\left(1 - \exp\left[-\left(\frac{R_0}{R_e}\right)^6 \left(\frac{t}{<\tau>}\right)\right]\right)\right\} \tag{23}$$

where R_0 is the Förster radius (see below), n_2 is the acceptor numerical density (number of acceptors per unit area), and

$$\gamma(x,y) = \int_0^y z^{x-1} \exp(-z)dz \tag{24}$$

is the incomplete Gamma function.

From these equations and admitting that $R_0 \gg R_e$ (as verified in our study) Wolber and Hudson (1979) obtained the analytical solution for energy transfer efficiencies:

$$E = 1 - \sum_{j=0}^{\infty} \left(-\pi \Gamma(2/3) R_0^2 n_2 \right) \times \frac{\Gamma(j/3+1)}{j!} \tag{25}$$

where Γ is the complete gamma function (which is given by Eq. (24) when the upper limit of the integral is ∞).

The Förster radius is given by:

$$R_0 = 0.02108 \cdot \left[\frac{\left(\int \lambda^4 \varepsilon(\lambda) f(\lambda) d\lambda \right) \kappa^2 \Phi_D}{n^4} \right]^{1/6} \tag{26}$$

where κ^2 is the orientation factor, n is the refractive index of the medium, Φ_D is the donor quantum yield, $f(\lambda)$ is the normalized emission spectra of the donor and $\varepsilon(\lambda)$ is the absorption spectra of the acceptor. If the λ units in Eq. (26) are nm, the calculated R_0 has Å-units (Berberan-Santos and Prieto, 1987).

For our pair, using $\Phi_D = 0.64$ (determined in this study) and assuming $\kappa^2 = 2/3$ (the isotropic dynamic limit) and $n = 1.4$ (Davenport et al., 1985), we obtain $R_0 = 49$ Å. The value $\kappa^2 = 2/3$ was considered because for fluorophores in the centre of a fluid bilayer, the rotational freedom should be sufficiently high to randomize orientations. This is supported by the reasonably low steady-state anisotropy values obtained for the IAEDANS and BODIPY probes labeled on the T36C M13 coat protein mutant ($<r>_{\text{AEDANS}} = 0.14$, $<r>_{\text{BODIPY}} = 0.23$; for a detailed discussion, see Loura et al., 1996).

The results for BD-M13 coat protein in DOPC, DOPC/DOPG, DOPE/DOPG, DMoPC/DOPC, and DEuPC/DOPC bilayers are presented in Figure 12.12. The centered position of BODIPY in the bilayer allows for a simplification of the energy transfer analysis for a two-dimensional situation, as described by Eq. (25), i.e., there is no need to consider interplanar (bilayer) FRET geometry (Loura et al., 2001a; this situation is described in Section 12.7). Simulations of energy transfer for random distribution of acceptors using Eq. (25) can therefore be compared with our experimental results (Figure 12.12). The energy transfer efficiencies obtained for BD-M13 coat protein in the DMoPC/DOPC and DEuPC/DOPC bilayers support the other results discussed above for these mixtures, as they can only be explained by protein segregation in the bilayer or severe aggregation (Figure 12.12 B). However, as discussed above, the data obtained by fluorescence emission self-quenching indicate that segregation into DOPC-enriched domains (rather than aggregation) is the major phenomenon in these lipid mixtures.

Although the results can be reasonably explained on the basis of protein segregation to 60% of the total bilayer area (Figure 12.12 B), this rationalization should be considered an oversimplification, and is presented as an illustration. Indeed, the measured efficiencies are only reporting the average BD-M13 coat protein surface density that each IAEDANS-labeled protein is sensing. Probably there will be M13 coat protein interacting with the mismatching phospholipids, but the majority of the proteins will be preferentially surrounded by DOPC, and microdomains enriched in DOPC and M13 coat protein should be formed.

The lipid mixtures used in this work for hydrophobic matching studies (DMoPC/DOPC and DEuPC/DOPC) are thought to be considerably closer to ideality than the ones used in the studies of gel/fluid coexistence by Dumas et al. (1997), and of natural and pyrene-derivatized lipids by Lehtonen and Kinunnen (1997). However, protein segregation was observed in our study for both DMoPC/DOPC and DEuPC/DOPC, whereas in DOPC random distribution of protein in the bilayer was confirmed. Also interestingly, the degree of segregation appears to be similar for both mixtures (Figure 12.12 B), which was not expected due to the observed larger aggregation degree of coat protein in DEuPC, and therefore to an apparent larger packing difficulty with the longer lipid.

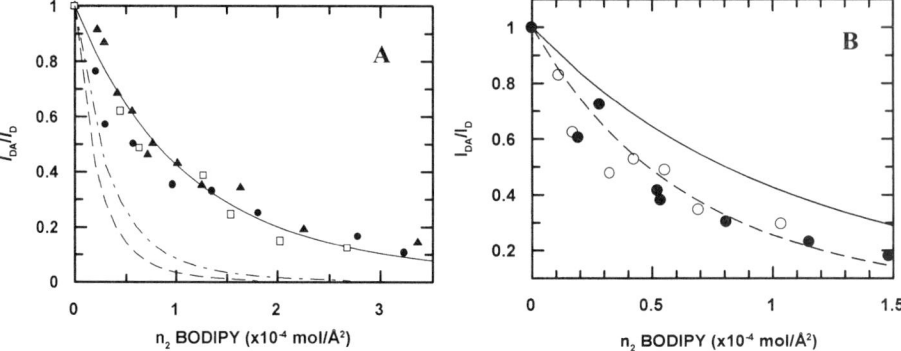

Figure 12.12. A - Donor (IAEDANS) fluorescence quenching by energy transfer acceptor (BODIPY). (▲) - Experimental data for DOPC, (●) – DOPC/DOPG (80/20 mol/mol) and (□) – DOPE/DOPG bilayers. (70/30 mol/mol). (—) - Theoretical expectation (Eq. 25) for a random distribution of acceptors. Energy transfer simulation for total co-localization of M13 coat protein considering 20% (----) and 30% (— - —) of the total surface area available. B - Donor (IAEDANS) fluorescence quenching by energy transfer acceptor (BODIPY). (—) - Theoretical expectation for a random distribution of acceptors. (---) - Simulation for a segregation of major coat protein considering 60% of the total surface area available. (●) - Experimental data for DEuPC/DOPC (60/40 mol/mol), (○) –experimental data for DMoPC/DOPC (60/40 mol/mol). I_{DA} and I_D were obtained by integration of donor decays. Reprinted with permission from Fernandes et al. (2003). Copyright 2003 Biophysical Society.

The formation of local structures within the thermodynamic fluid phase (see Introduction Section) for a mixture of two PCs with a 4-carbon difference in acyl-chain lengths (DMPC/DSPC) was theoretically predicted by Mouritsen and Jørgensen (1994). The clusters size in the Monte-Carlo configurations obtained by these authors appears to be very small (10–20 molecules at most). However, significant alterations in FRET efficiency, as measured by Lehtonen et al. (1996) in mixtures of unsaturated PCs, and by us between IAEDANS-labeled protein and BODIPY-labeled protein, require that the domain size should be of the order of magnitude of R_0, which is ≅ 5 nm. In our system, this large domain size might be a consequence of protein-induced phase separation.

In this work we also studied whether similar heterogeneities could be induced by the presence of positively charged M13 coat protein in bilayers composed of mixtures of anionic and neutral phospholipids. Due to the basic character of M13 coat protein C-terminal, it is reasonable to consider the possibility of anionic phospholipid-enriched domain induction by M13 coat protein incorporation in the bilayer. The formation of these domains could actually help explain some of the mechanisms involved in the creation of the phage assembly site.

Assuming that the hypothetical domains were composed by all the protein and DOPG content in the sample, and that the protein would be randomly distributed inside them, we can, using Eq. (25), obtain theoretical curves describing the energy transfer within these domains. These plots are compared with the experimental data points for M13 coat protein incorporated in DOPC/DOPG (80/20 mol/mol) and DOPE/DOPG (70:30 mol/mol) in Figure 12.12 A. It is concluded that the segregation of M13 coat protein to a PG-rich phase in the mixed systems, induced by electrostatic interactions between the positively charged protein and the negatively charged phospholipid, is ruled out. In fact, this process would lead to a greater local surface density of acceptor, and therefore, a very significant increase of energy transfer efficiencies should be observed, contrary to the data.

On the whole, from this study, it is clearly concluded that the M13 coat protein monomeric state is highly stable when incorporated in bilayers containing hydrophobic matching phospholipids, which had been suggested in other studies (Stopar et $al.$, 1997; Spruijt et $al.$, 1989; Sanders et $al.$, 1991). The lack of anionic phospholipids has no effect on the protein oligomerization properties at the protein concentrations used in this study. When the protein is incorporated in pure vesicles of mismatching lipid there is evidence for protein aggregation but for mixtures of lipids containing both hydrophobic-matching (DOPC) and -mismatching (DEuPC or DMoPC) phospholipids, the protein probably segregates to domains enriched in DOPC, which can explain the stability of the monomeric species of the protein in these lipid systems. This segregation effect is only observed when hydrophobic mismatching phospholipids are present, suggesting that the hydrophobic matching conditions on the protein-lipid interface are more important than electrostatic interactions between the M13 coat protein and the phospholipids, for the protein lateral distribution on the bilayer.

12.6.3. γM4 Lateral Distribution from Fluorescence Self-Quenching Studies

In the case of the peptide γM4 from the muscle AChR, the presence of a dynamic self-quenching process gave valuable information on the lateral distribution of peptide in the chol-poor ld phase $vs.$ the chol-rich lo phase (de Almeida et $al.$, 2004).

The fluorescence of the γM4 peptide incorporated in multilamellar vesicles (MLV) made of POPC/chol (60:40 mol/mol) is shown in Figure 12.13. The spectral position (not shown) and the steady-state anisotropy, $<r>$ (Table 12.5), are insensitive to the presence of chol, $i.e.$, they are the same in ld or lo phases, and the anisotropies are reasonably high and typical of a Trp-containing peptide strongly immobilized in a membrane.

The lifetime-weighted quantum yields $<\tau>$, both for the ld and lo phases, are shown in Table 12.5 for several γM4 peptide concentrations. In the ld phase, upon increasing the concentration from 0.7 mol% up to 7 mol%, the Trp lifetime-weighted quantum yield was reduced to one-half. This can be accounted for by the presence of Cys and Lys residues in the γM4 transmembrane domain, which are known to be efficient quenchers of Trp fluorescence (Chen and Barkley, 1998). Thus, at higher concentrations the proximity between γM4 peptides leads to a decrease of the Trp intrinsic fluorescence due to an intermolecular self-quenching process.

Table 12.5. Photophysical properties of the Trp^{453} residue in the $\gamma M4$ peptide incorporated into POPC/chol vesicles with low (ld phase) or high (lo phase) chol content at room temperature. $\lambda_{exc} = 288$ nm and $\lambda_{em} = 340$ nm. The concentration of $\gamma M4$ peptide is expressed in mol% relative to total lipid. See text for further details.

$\gamma M4$ (mol %)	$<\tau>_{ld}$ (ns)	$<\tau>_{lo}$ (ns)	$\overline{\tau}_{ld}$ (ns)	$\overline{\tau}_{lo}$ (ns)	$<r>_{ld}$	$<r>_{lo}$
0.7	3.04±0.05	1.47±0.10	4.29	2.84	0.089±0.006	0.091±0.012
3.0	2.11±0.08	1.44±0.11	3.53	2.69	0.078±0.007	0.076±0.006
7.0	1.43±0.08	—	2.81	-	0.080±0.008	-

Reprinted with permission from de Almeida *et al.* (2004). Copyright 2004 Biophysical Society.

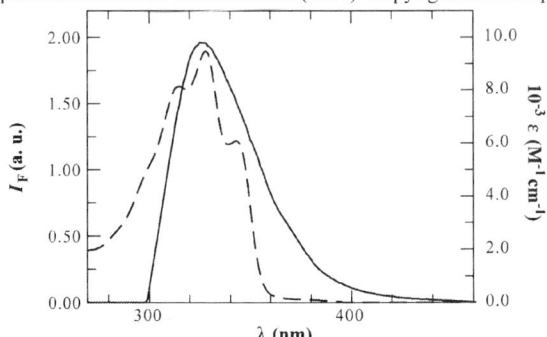

Figure 12.13. Emission spectrum of $\gamma M4$ ($\lambda_{exc} = 288$ nm) in POPC vesicles (solid line) and absorption spectrum of DHE in POPC/chol (3:2) vesicles (dotted line) at room temperature. Reprinted with permission from de Almeida *et al.* (2004). Copyright 2004 Biophysical Society.

For the lo phase, the lifetimes do not show such a strong variation, and in addition, the values for the lower concentrations (0.7% and 3 %) in lo are close to those found for the higher concentrations in ld (7 %). Also, as the dynamic self-quenching mechanism depends on the quencher concentration sensed by the fluorophore, in the case of $\gamma M4$ in chol-rich vesicles, a higher effective peptide concentration occurs, augmenting the extent of the self-quenching process (the quenching process in this case is that of Trp fluorescence by other residues of the peptide, and thus is not exactly the same as described for M13, where only the fluorophore group is involved in the quenching process).

The effect of the collisional contribution of self-quenching on the fluorescence lifetime for a molecule with a complex decay is described by the Stern-Volmer equation (Eq. 19). In this case, the calculation is performed separately for each concentration, and thus $\overline{\tau}_0$ can be replaced by $\overline{\tau}_{0q}$, which is given by (Sillen and Engelborghs, 1998):

$$\overline{\tau}_{0q} = \sum_i a_{0i} \tau_i \tau_{0i} \Big/ \sum_i a_{0i} \tau_i \tag{27}$$

The subscript 0 indicates again the absence of quencher.

The bimolecular quenching rate constant $<k_q>$ is related to the diffusion coefficient of the fluorophore (D) via the Smoluchowski equation (Lakowicz, 1999), taking into account transient effects (Umberger and Lamer, 1945):

$$< k_q >= 4\pi N_A (2R_c)(2D)\left[1 + 2R_c / (2\tau_{0q} D)^{1/2}\right] \qquad (28)$$

where N_A is the Avogadro constant and R_c is the collisional radius.

This equation assumes diffusion in an isotropic 3D medium. If the membrane were strictly bidimensional, different boundary conditions for the Smoluchowski formalism could be applied (Razi-Naqvi, 1974). The best approach to the specific situation of probe diffusion in a membrane is the one used by Owen (1975), in which the finite bilayer width (cylindrical geometry) is taken into account. Owen introduced the parameter τ_s, which defines the transition from the spherical (3D) to the cylindrical geometry, its value being $\tau_s = 16$ ns when considering the bilayer (Wiener and White, 1992) and the peptide/Trp (Zamyatnin, 1972) parameterization. This value of τ_s is much longer than the longest fluorescence lifetime of Trp[453] in the γM4 peptide (~ 5 ns), and longer than our experimental time-window (15.3 ns = 15.3 ps/channel ×1,000 channels); the 3D framework approximation is therefore essentially correct. If we consider that the lifetime-weighted quantum yield of the more diluted γM4 peptide concentration is the limiting $<\tau>_0$ (no quenching occurring) (Table 12.5), a value of $<k_q> = 2.6 \times 10^9$ mol^{-1} dm^3 s^{-1} is obtained from Eq. (19) for the 7 mol % peptide concentration. This value assumes a random peptide analytical concentration in the lipid $[F] = 0.083$ M^{-1}, determined on the basis of its 7 mol % concentration and considering a volume for the POPC molecule of 1,263 Å^3 (Small, 1986). A similar value ($<k_q> = 2.8 \times 10^9$ mol^{-1} dm^3 s^{-1}) could be obtained for the other peptide concentration, 3%.

If in Eq. (28) we assume the collisional radius for the dimer to be 10 Å, diffusion coefficients of $D = 14 \times 10^{-8}$ cm^2 s^{-1} (for 7% peptide) and $D = 12 \times 10^{-8}$ cm^2 s^{-1} (for 3% peptide) are obtained. These values have the same order of magnitude or are slightly higher than those typically found for diffusion in an ld phase ($D = 1.1 \times 10^{-8}$ cm^2 s^{-1} (Dietrich et al., 2001), $D = 1$-3×10^{-8} cm^2 s^{-1} for transmembrane proteins and $D = 9$-14×10^{-8} cm^2 s^{-1} for a fluorescent lipid derivative (Vaz et al., 1982). Although this is a comparison with values obtained from fluorescence recovery after photobleaching studies of large proteins or from phospholipid single particle tracking studies, the agreement is reasonably good within the framework of the approximations used.

It is interesting to note that in pure lo phase, for the lower peptide concentration studied, the lifetime is already quite low (Table 12.5) and similar to the value for the highest concentration in ld phase. As previously described, self-quenching was observed for a single ld phase (it is known that several amino acid side-chains are effective quenchers of Trp fluorescence), and therefore only a higher effective peptide concentration can be invoked to rationalize the lo data. This means that the peptide is not randomly distributed in the lo phase and that patches with a higher local concentration of γM4 are formed. Within the framework of a Stern-Volmer analysis (Eq. 19), this higher peptide concentration gives rise to a higher decrease in lifetime. It can therefore be concluded that the formation of γM4 peptide-rich patches is strongly induced by chol.

12.6.4. FRET from γM4 (Trp[453]) to Dehydroergosterol (DHE): Sterol Segregation in a One-Phase System *vs.* Peptide-Rich Patches

In order to obtain information on the affinity of the γM4 peptide for chol, FRET measurements were carried out using Trp[453] and the fluorescent chol analogue DHE as donor and acceptor, respectively. In this series of experiments the totally reduced monomeric species of γM4 was used, since the disulfide-bonded dimer would introduce an additional complexity in the analysis of FRET data: two Trp donors would be present in the dimer, the system topology would lose cylindrical symmetry around each donor, and a complex geometry regarding the excluded volume for transfer would have to be considered. To further reduce the complexity of the system, the monomeric γM4 species was reconstituted in small unilamellar vesicles (SUV) in order to avoid a multilayer geometry and to minimize light scattering data biasing. Another important advantage of SUV relative to larger model membranes is the absence of DHE tail-to-tail dimerization that is observed in 100 nm diameter vesicles (Loura and Prieto, 1997) and which would also render the analysis of the FRET results more difficult. DHE has no bulky fluorophores in its structure, thus lacking the disadvantages of other chol analogs, which were shown to be excluded from chol-rich phases (Loura *et al.*, 2001b). At the chol concentration at which the FRET study was carried out (40 mol %), the system is largely in the lo phase (Mateo *et al.*, 1995; de Almeida *et al.*, 2003; Figure 12.2). The complexity of studying FRET in a biphasic system (Loura *et al.*, 2001a) is therefore circumvented. The typical radius for a SUV (~25-50 nm) is much larger than 1.5 R_0, and the system can be safely assumed as being planar with respect to FRET (Eisinger *et al.*, 1981).

The Förster radius was calculated next from the experimental data in Figure 12.13 –the donor's emission (Trp[453]) and the acceptor's (DHE) absorption. Values of $\kappa^2 = 2/3$, and $n = 1.44$ (Davenport *et al.*, 1985) were considered in Eq. (26). The fluorescence quantum yield $\Phi_D = 0.067$ was estimated on the basis of the Trp lifetime-weighted quantum yield as compared to that of *N*-acetyltryptophanamide, a well-known model for this amino acid residue within a polypeptide chain (Szabo and Rayner, 1980). This procedure avoids all the errors related to the determination of quantum yields in scattering media using steady-state data, and it does not take into account effects of static quenching by sulphydryl groups. A value of $R_0 = 20$ Å was obtained.

The fluorescence decay of Trp[453] is described by a sum of three exponentials, and for the determination of FRET efficiency, E, decay integration was carried out in order to determine the lifetime-weighted quantum yield:

$$E = 1 - \left(\int_0^\infty i_{DA}(t)dt \right) / \int_0^\infty i_D(t)dt = 1 - <\tau>_{DA} / <\tau>_D \qquad (29)$$

where $i_D(t)$ and $i_{DA}(t)$ are the fluorescence decays of donor in the absence and presence of acceptor, respectively. It should be pointed out that this study could not be carried out using steady-state techniques (Eq. 22) due to the strong absorption overlap of donor and acceptor, leading to large inner filter effects, and the low steady-state intensities for the reduced species, preventing the measurement of reliable absolute intensities. The variation of E upon increasing acceptor concentration is shown in Figure 12.14. FRET efficiency was found to be lower than that expected on the basis of a random distribution.

The theoretical expectation for a random distribution of acceptors (Figure 12.14) involves consideration of the following structural features: i) The peptide helix was approximated to a cylinder with 10 Å diameter (Bowie, 1997), with the donor located at the axis; ii) the peptide was allowed to incorporate into the bilayer in either direction, *i.e.*, the Trp^{453} donor can be located close to any of the two membrane/water interfaces, and DHE can in turn be located on either leaflet. Because R_0 is not small with respect to the membrane thickness, both in-plane (Eq. 23) and out-of-plane (Eq. (36) in the next section) transfer to acceptors in the other membrane leaflet were considered (the interplanar donor-acceptor distance was fixed as $w = 22$ Å on the basis of molecular models). In the calculation of the surface density of acceptors, the condensing effect of chol on the lipid surface was taken into account ($\sim 7\text{Å}^2$ / molecule; Smaby *et al.*, 1997), together with an area/molecule of 66.4 Å^2 for POPC (Chiu *et al.*, 1999) and 37.7 Å^2 for chol (Smaby *et al.*, 1997).

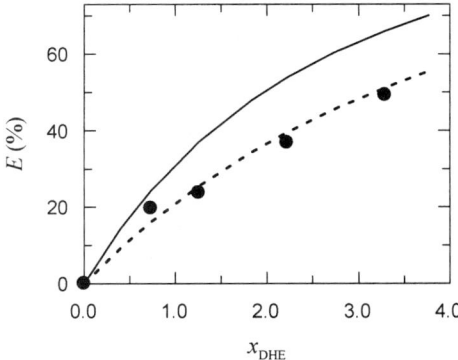

Figure 12.14. FRET efficiency from γM4 (reduced monomeric species) Trp^{453} to DHE ($R_0 = 20$ Å) *vs.* acceptor mole fraction. The circles are experimental data points. The solid line is the calculated FRET efficiency for a random distribution of acceptors in the plane of the membrane with an exclusion radius of 5 Å, to account for the area occupied by the donor containing α-helix. Both in-plane and out-of-plane (interplanar distance of 22 Å) FRET are considered. See text for further details on the calculations. The dotted line is also based on a random distribution model considering only 62% DHE molecules available as acceptors. Reprinted with permission from de Almeida *et al.* (2004). Copyright 2004 Biophysical Society.

The experimentally determined values of FRET efficiency, E, were significantly lower than expected, and the data could only be rationalized with an effective sterol concentration $\sim 38\%$ lower than the analytical expected concentration (see Figure 12.14). This does not necessarily mean that there is a tendency to exclude the sterol away from the γM4 peptide vicinity, but the area available for the dispersion of sterol is probably reduced because of the formation of peptide rich-patches. This estimated area reduction can, in turn, be compared with that derived from the above-mentioned self-quenching data. The same degree of quenching is obtained for 7 mol% peptide in ld and for all concentrations in lo. Since the diffusion coefficient in lo is 3-fold lower than in ld (*e.g.* Dietrich *et al.*, 2001), it can be assumed that the effective concentration of peptide is 3-fold higher in the patches. Using the area per lipid molecule for the lo phase (computed as described above) it can be concluded that the γM4 peptide occupies ~30% of the area in the patches in agreement with the FRET data.

12.6.5. γM4 Structure and Organization from Energy Homotransfer Studies

In the absence of a reducing agent, and because there is a Cys residue in the peptide sequence, γM4 has the ability to form dimers, although the monomeric species predominates (as concluded from polyacrylamide gel electrophoresis in denaturing conditions). The dimer has two Trp[453] residues, one in each helix, which can lead to energy homotransfer (energy migration) between these two Trp residues. This interaction can be used to obtain structural information (distances) on the peptide (de Almeida *et al.*, 2004). As a consequence of energy migration, Trp anisotropy may decrease. The expected value for the dimer anisotropy $<r>_D$ can be determined following, *e.g.*, Runnels and Scarlata (1995), as

$$<r>_D = <r>_M (1+(R_0/l)^6)/ (1+2(R_0/l)^6) + <r>_{ET} (R_0/l)^6/(1+2(R_0/l)^6) \tag{30}$$

where l is the distance between the two chromophores, R_0 is the Förster radius for energy migration, $<r>_M$ is the anisotropy of the initially excited molecule, and $<r>_{ET}$ is the anisotropy of the second molecule of the pair. This second contribution can be disregarded because it is smaller than 4% of the first one at all times (Berberan-Santos and Valeur, 1991). To obtain the inter-Trp distance, we fed the experimentally determined value of $<r>_D$ for the dimer into Eq. (30). In the energy heterotransfer experiments described in the previous sub-section, the reduced species, *i.e.*, the monomer was used. However, the monomer anisotropy $<r>_M$ could not be evaluated because of the strong fluorescence quenching produced by the -SH group in Cys[451]. This fact also rules out any biasing of steady-state anisotropy data, due to the presence of the monomeric species. The steady-state intensity was too low to yield reliable data. We therefore resorted to literature values from a systematic study of time-resolved anisotropy in which Trp were introduced in different positions along an α-helical peptide in a fluid membrane (Vogel *et al.*, 1988). Trp residues located near the surface, such as Trp-1, or at a shallow position in the hydrocarbon core, such as Trp-6 in the cited work (a condition identical to that of the γM4 peptide under study), constitute a suitable model to obtain $<r>_M$.

We calculated the steady-state anisotropy by integration of the reported time-resolved data of Vogel *et al.* (1988) according to $i(t) = \Sigma_i a_i \exp(-t/\tau_i)$ and Eq. (11). The steady-state anisotropy results from:

$$<r> = \int_0^\infty i(t)r(t)dt \bigg/ \int_0^\infty i(t)dt \tag{31}$$

In this equation, both the lifetime data of Vogel *et al.* (1988) and our own were used, leading in both cases to an identical value for the anisotropy: $<r>_M = 0.082 \pm 0.013$, which is not significantly different from that obtained for the dimeric peptide in the present study (Table 12.5). It can therefore be concluded that there is no relevant energy migration in the dimer, allowing us to set a lower boundary for the inter-Trp distance.

A Förster radius $R_0 = 10$ Å for energy migration was calculated according to Eq. (26) considering $\Phi_D = 0.13$. This value was obtained again by comparing the lifetime-weighted quantum yields of the γM4 peptide and that of *n*-acetyltryptophanamide. From Eq. (30) it is predicted that for interchromophore distances larger than $\approx 2R_0$, the degree of depolarization as a consequence of energy migration is insignificant. This corresponds to a distance of 20 Å in the present case. In fact, for two helices connected by a disulfide bond, each Trp residue would be in an approximately diametrically opposed position, because for an ideal α-helix, an arc of ~200° would be subtended between Trp[453] and Cys[451] (Figure 12.15 A).

Now that intra-peptide energy migration was discarded, the reasons why anisotropy is largely concentration-independent (Table 12.5), *i.e.*, the lack of efficient intermolecular energy migration in either the ld or lo phases, should be discussed.

Before this analysis, the decrease in fluorescence lifetime due to the intermolecular collisional self-quenching must be considered. It should be stressed that energy migration does not affect the lifetime or intensities, *i.e.*, there is no quenching (Valeur, 2001) and only anisotropy decreases. However, the decrease in lifetime by some other mechanism results in an increase in anisotropy (Eq. 13). Thus the decrease in anisotropy due to energy migration would be hidden due to a compensatory reduction in fluorescence lifetime, as described in Section 12.5. Considering Eq. (13), a value of limiting anisotropy $r_\infty = 0.066$ (experimentally determined), and $r_0 = 0.173$ for excitation at $\lambda = 288$ nm (Valeur and Weber, 1977), a global rotational correlation time $\varphi = 1$ ns is obtained.

These result facilitates the determination of the expected anisotropies for the higher concentrations (3% and 7%); they are $\langle r \rangle = 0.090$, and $\langle r \rangle = 0.094$, respectively. As can be seen, they are very close to the experimentally determined ones (Table 12.5), *i.e.*, the decrease in fluorescence lifetime does not imply a significant variation of anisotropy. This is certainly related to the large residual anisotropy.

From the invariance of anisotropy with concentration it can safely be concluded that there is no significant intermolecular energy migration. For the energy migration in a bi-dimensional system such as a membrane, Snyder and Freire (1982), from Monte-Carlo simulations, obtained a decrease in anisotropy for a random distribution of fluorophores in a membrane (and although this was derived for an isotropic rotor, it is relevant to compare these expectations with our data). Considering the Förster radius, $R_0 = 8.5$ Å, as described above (but now calculated from a lower quantum yield, reflected in an also lower lifetime-weighted quantum yield, Table 12.5), and using the data from Figure 12.9 in the work of Snyder and Freire (1982), it can be concluded that for the highest concentration of γM4 studied here (7%, which corresponds to 0.22 molecules within a circle of radius R_0), the reduction in anisotropy should be at most 15%, in full agreement with our data.

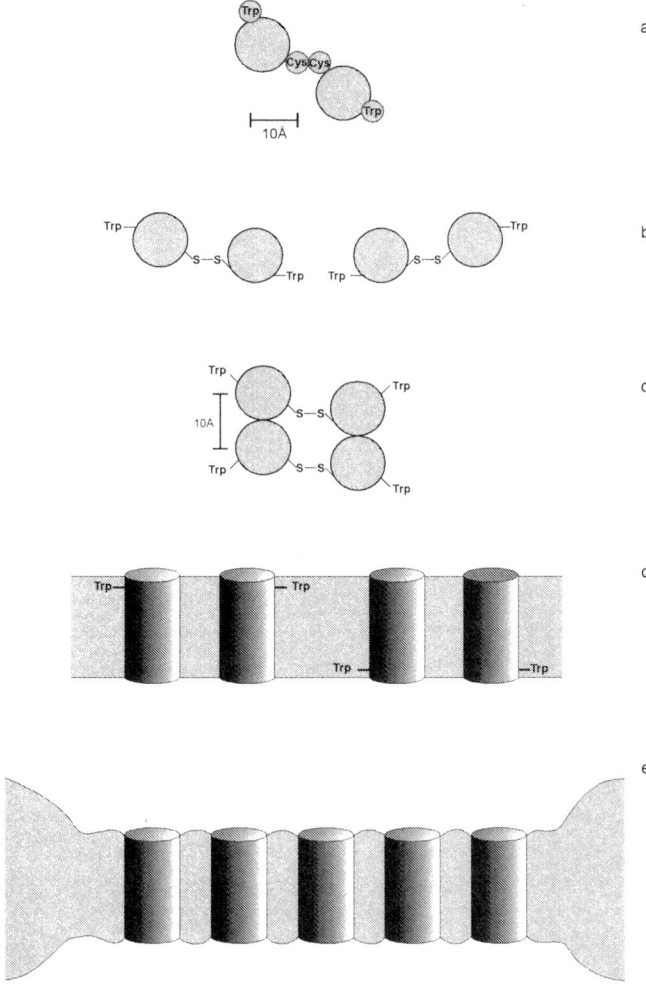

Figure 12.15. Putative organization of γM4 in chol-poor and chol-rich systems. A plausible structural framework to account for the absence of energy migration between Trp[453] residues in the γM4 peptide, and the variation of lifetime-weighted quantum yield with peptide concentration on the ld and lo phases is provided. A) Top view of the α-helical peptide showing the relative positions of Trp[453] and Cys[451], and a probable geometry for a disulfide bonded dimer. B) Possible geometry for a linear aggregate. This structure is ruled out. C) Parallel aggregate in end-on view. This structure is not ruled out by energy homotransfer data. D) Anti-parallel aggregate, lateral view. This structure is not ruled out by energy homotransfer data. E) Peptide-rich patch. The model depicts the distribution of γM4 in POPC/chol vesicles with high chol content (lo phase). The negative mismatch causes a disordering effect in the vicinity of a peptide molecule, and other peptides accommodate better in a region close to where other peptides localize. In the ld (chol-poor phase) the peptide is randomly distributed due to the very good matching between the hydrophobic thickness of the bilayer and the peptide. Reprinted with permission from de Almeida *et al.* (2004). Copyright 2004 Biophysical Society.

From the self-quenching experiments, it was concluded that peptide-rich patches occur, at least in the lo phase. Because no energy migration is operative, and due to the small Förster radius for migration, we can obtain structural information about these patches. The absence of energy migration implies that: i) there is no molecular contact of Trp residues in one dimer with those of another dimer. This indicates that no linear aggregates are formed (Figure 12.15 B). If this type of dimer-dimer contact did occur, a strong decrease in anisotropy such as, for example, observed for DHE (Loura and Prieto, 1997) would take place. ii) The formation of dimer "parallel" aggregates cannot be discarded (Figure 12.15 C). For this geometry, an inter-Trp minimum distance of 10 Å could be expected. The anisotropy would drop to 78% of its initial value (Eq. 30), and if a small enough fraction of these aggregates were formed, this would not entail high global migration efficiency. It should be stressed that as stated above for the intramolecular process, there would be no transfer for the other two Trp residues. Anyway, anti-parallel inter-helical contacts (Figure 12.15 D) would not produce significant depolarization because the Trp residues would be too far apart, *i.e.*, they would be separated by at least the hydrophobic thickness of the bilayer. The method employed in this work to obtain peptide-containing vesicles cannot discard the occurrence of both possible peptide orientations in the bilayer. It is known, in addition, that anti-parallel contacts are more probable due to stabilization by the opposite sign macrodipole moment in the anti-parallel helix (Bowie, 1997) and in lo, with a larger hydrophobic thickness, less of the helix termini should protrude to the solvent, thus decreasing solvent screening and increasing the strength of the dipole-dipole interaction relative to ld. This can allow for other types of helix-helix interactions like close packing (note that at the extreme of the hydrophobic helix opposite to the Trp^{453} side, there is a threonine (Thr^{469}) and a glycine residue (Gly^{470}), both with small side chains that should favor helix approximation, in addition to Gly^{462} and Thr^{463}, approximately in the middle of the γM4 helix).

If the γM4 peptide formed linear aggregates, there would be no significant interhelix stabilization. This effect is maximal for the case of anti-parallel dimers. However, there would be no need to invoke such a limiting situation if long-range interactions were operative. As shown in Figure 12.15 *E*, patches could be formed with no direct molecular contact, *i.e.*, a small number of lipid molecules would occur in between the peptides. The helix would induce a local disordering on the acyl chains, together with bilayer deformation, to better match with the hydrophobic portion of the peptides. Once the bilayer was deformed, with an energetic penalty (Lundbaek *et al.*, 2003), other helices would also be better accommodated nearby, justifying the formation of peptide-enriched patches. Patches with an increased local peptide concentration would, in turn, favor the occurrence of dimers.

These properties of a representative membrane-embedded segment of the AChR, obtained from different fluorescence methodologies, may bear relevance to the organization of the γ-subunit α-helical bundle motif and the AChR membrane-spanning region at large. The tendency of the hydrophobic γM4 peptide to maximize peptide-peptide interactions in the presence but not in the absence of chol may be related to the ability of this sterol to stabilize the α-helix content of the native AChR (Fong and McNamee, 1987; Butler and McNamee, 1993) and to the inability of reconstituted AChR to respond to agonist stimulation with a cation flux in the absence of chol (Rankin *et al.*, 1997).

12.7. PROTEIN/PEPTIDE-LIPID SELECTIVITY: COMPOSITION AND SIZE OF THE ANNULAR REGION

Peptides and proteins interacting with two-component (or multi-component) bilayers are frequently reported to exhibit lipid selectivity, *i.e.*, are preferentially surrounded by one lipid component relative to the other (Dumas *et al.*, 1997; Lehtonen and Kinnunen, 1997; Fahsel *et al.*, 2002). The traditional technique used to address this question has been ESR spectroscopy (*e.g.*, Marsh and Horváth, 1998). However, due to its sensitivity to probe concentration, FRET offers a large potential in this field, and both qualitative (Wang *et al.*, 1988) and quantitative though approximate models (Antollini *et al.*, 1996; Bonini *et al.*, 2002; see Fernandes *et al.* (2004) and Loura *et al.* (2001a) for critical discussions) have been published. In this section, we describe a fluorescence study of lipid selectivity concerning a transmembrane protein for which a new FRET methodology was specifically derived (Fernandes *et al.*, 2004).

As described in the previous section, M13 major coat protein has a single hydrophobic transmembrane segment. Headgroup phospholipid selectivity ESR studies had already been performed with aggregated forms of the protein (Peelen *et al.*, 1992; Wolfs *et al.*, 1989, Datema *et al.*, 1987), pointing to a moderate preference for anionic lipids PA, phosphatiylserine (PS) and PG over zwitterionic lipids PC and PE. This effect most probably results from electrostatic interaction of anionic head groups with the highly basic C-terminal domain of the protein, which contains 4 Lys residues. In the study described below, head group selectivity was studied by FRET from a covalently-bonded fluorophore in the protein (essentially monomeric) to different labeled phospholipids. In addition, selectivity for hydrophobically matching phospholipids relative to non matching counterparts was also investigated.

The FRET model assumes two populations of energy transfer acceptors, one located in a single annular shell around the protein and the other outside the shell. The donor fluorescence decay curve has FRET contributions from both populations:

$$i_{DA}(t) = i_D(t) \cdot \rho_{annular}(t) \cdot \left[\rho_{random}(t)\right]^2 \tag{32}$$

Here i_D and i_{DA} are the donor fluorescence decay in the absence and presence of acceptors respectively, and $\rho_{annular}$ and ρ_{random} are the RET contributions arising from energy transfer to annular labeled lipids and to randomly distributed labeled lipids outside the annular shell respectively.

The acceptors in the annular shell (in our study, (7-nitro-2-1,3-benzoxadiazol-4-yl)amino (NBD)-labeled phospholipids) are located at a constant distance (d) to the fluorophore (in our study, 7-diethylamino-3((4'iodoacetyl)amino)phenyl-4-methyl-coumarin (DCIA), linked at the 36th residue), which lies at the centre of the transmembrane domain (Figure 12.16), and therefore we can assume that the energy transfer to each of these acceptors is described by the rate constant

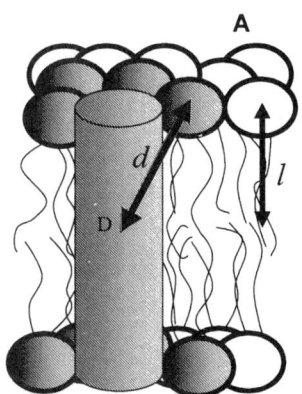

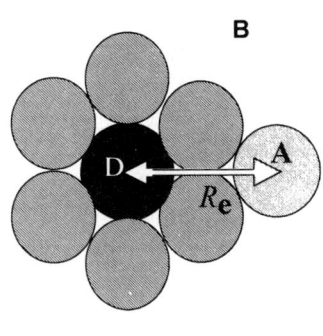

Figure 12.16. Molecular model for the FRET analysis ((A) side view; (B) top view). Protein-lipid organization presents a hexagonal geometry. Donor fluorophore from the mutant protein is located in the center of the bilayer, whereas the acceptors are distributed in the bilayer surface. Two different environments are available for the labeled lipids (acceptors), the annular shell surrounding the protein and the bulk lipid. Energy transfer to acceptors in direct contact with the protein has a rate coefficient dependent on the distance between donor and annular acceptor (Eq. 33). Energy transfer toward acceptors in the bulk lipid is given by Eq. (36) (see text for details). Reprinted with permission from Fernandes *et al.* (2004). Copyright 2004 Biophysical Society.

$$k_T = \frac{1}{\tau_D}\left(\frac{R_0}{d}\right)^6 \tag{33}$$

where τ_D is the donor lifetime (in the absence of acceptor). From spectral data, $R_0 = 39.3$ Å was measured for this pair. The NBD fluorophores are assumed to be located in the bilayer surface (Chattopadhyay and London, 1987; see also Fernandes *et al.* (2004) for details).

Considering a hexagonal type geometry for the protein-lipid arrangement (Figure 12.16 B), each protein will be surrounded by 12 annular lipids. In bilayers composed by both labeled and unlabeled phospholipids, these 12 sites will be available for both of them. The probability (μ) of one of these sites being occupied by labeled phospholipid is given by

$$\mu = K_S \frac{n_A}{n_A + n_{Lipid}} \tag{34}$$

Here, n_A is the concentration of labeled lipid, and n_{lipid} is the concentration of unlabeled lipid. K_S is the relative association constant, which reports the relative affinity of the labeled and unlabeled phospholipid. Using a binomial distribution, the probability of each occupation number (0-12 sites occupied simultaneously by labeled lipid), and finally the FRET contribution arising from energy transfer to annular lipids is computed,

$$\rho_{annular}(t) = \sum_{n-0}^{n=12} e^{-nk_T \cdot t} \cdot \binom{12}{n} \cdot \mu^n \cdot (1-\mu)^{12-n} \tag{35}$$

The FRET contribution from energy transfer to acceptors randomly distributed outside the annular region in a different plane to that of the donors is given by Davenport *et al.* (1985).

$$\rho_{trans}(t) = \exp\left\{ -2n_2\pi \cdot l^2 \cdot \int_0^{l/\sqrt{l^2+R_e}} \left[1 - \exp(-t\,b^3\,\alpha^6)\right]\alpha^{-3}\,d\alpha \right\} \tag{36}$$

where $b=(R_0/l)^2\tau_D^{-1/3}$, n_2 is the acceptor density in each leaflet, l is the distance between the plane of the donors and the planes of acceptors, and R_e is the distance between the protein axis and the second lipid bilayer (exclusion distance for bulk-located acceptors). For the purpose of this work, n_2 must be corrected for the presence of labeled lipid molecules in the annular region, which therefore are not part of the randomly distributed acceptors pool. After $i_{DA}(t)$ is calculated in Eq. (32), the theoretical energy transfer efficiency E is readily calculated by numerical integration (Eq. 29).

In one experiment, M13 major coat protein selectivity for the acceptor (1,2-dioleoyl-*sn*-glycero-3-phosphoethanolamine derivatized with NBD at the head group) was measured in bilayers of either DOPC, DEuPC or DMoPC. In a second set of measurements, several probes were used as acceptors, all studies being made in DOPC vesicles. The probes used as acceptors were phospholipids of identical acyl chains (18:1 and 12:0) and different head groups (PC, PE, PS, PG, and PA) classes, derivatized with NBD at the 12:0 chain. The complete set of experiments is described in Table 12.6.

Experimental results of FRET efficiency as a function of acceptor concentration are shown (together with the best model fits) in Figures 12.17 and 12.18, and Table 12.6 summarizes the recovered K_S values. Regarding the hydrophobic matching study, the lower value recovered in DOPC relative to those in DMoPC and DEuPC bilayers confirms the larger protein selectivity towards the hydrophobic matching unlabeled phospholipid (DOPC). On the other hand, the varying acceptor head group study, a larger selectivity for the anionic labeled phospholipids (especially the 18:1-(12:0-NBD)-PA and 18:1-(12:0-NBD)-PS probes) is inferred. The latter results confirm those of Peelen *et al.* (1992), obtained using ESR and aggregated protein, and in fact the relative association constants ratios ($K_S(PX)/K_S(PC)$) obtained in the two works are almost identical.

One important difference between the ESR and FRET techniques is that the latter is not restricted to the lipids adjacent to a given protein molecule. Not only labeled lipids in the first shell of lipids will be potential acceptors to a donor-labeled integral protein, but also the acceptors in the other lipid shells surrounding the protein will contribute to the final result. For that reason, this study also seems to confirm the hypothesis of perturbation (in terms of lipid distribution) induced by the protein to largely limit itself to an annular shell of lipids in direct contact with the protein, in the case of the M13 major coat protein (possibly related to the fact that it has a sole transmembrane segment). Moreover, as commented above, our recovered $K_S/K_S(PC)$ agree with those obtained from ESR measurements, which assume a single layer of annular lipid.

Table 12.6. Relative association constants of labeled phospholipids towards M13 major coat protein.

Labeled phospholipid	Bilayer composition	K_S	$K_S/K_S(PC)^a$
((18:1)2-PE-NBD)	DOPC (18:1)2PC	1.4	—
((18:1)2-PE-NBD)	DEuPC (22:1)2PC	2.1	—
((18:1)2-PE-NBD)	DMoPC (14:1)2PC	2.9	—
(18:1-(12:0-NBD)-PE)	DOPC	2.0	1.0
(18:1-(12:0-NBD)-PC)	DOPC	2.0	1.0
(18:1-(12:0-NBD)-PG)	DOPC	2.3	1.1
(18:1-(12:0-NBD)-PS)	DOPC	2.7	1.3
(18:1-(12:0-NBD)-PA)	DOPC	3.0	1.5

[a] K_S (PC) is the relative association constant of (18:1-(12:0-NBD)-PC).
Reprinted with permission from Fernandes *et al.* (2004). Copyright 2004 Biophysical Society.

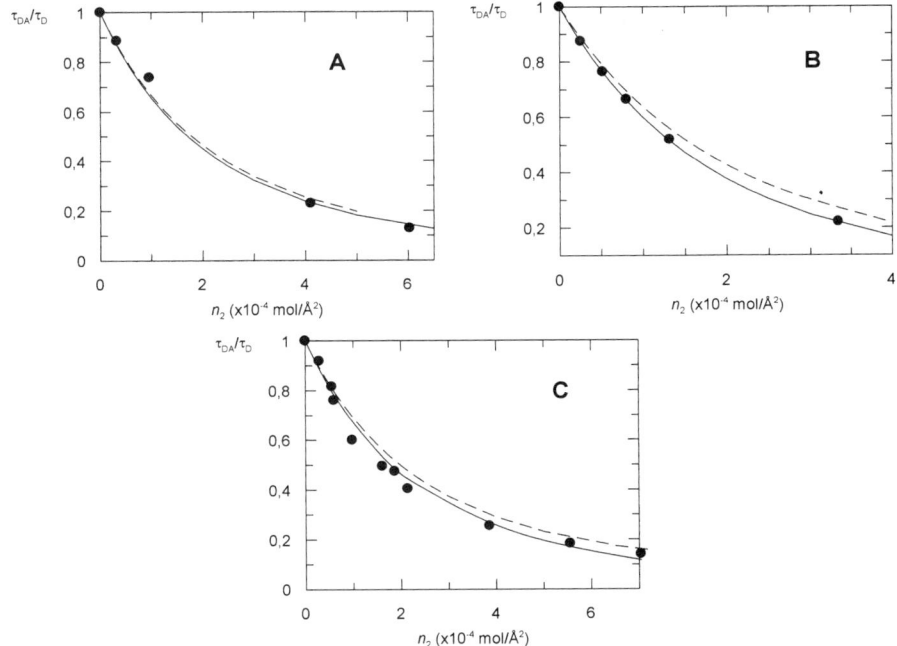

Figure 12.17. Donor (DCIA-labeled protein) fluorescence quenching by FRET acceptor ((18:1)$_2$-PE-NBD) in pure PC bilayers with different hydrophobic thicknesses. (•) - Experimental FRET efficiencies. (—) - Theoretical simulations obtained from the annular model for protein-lipid interaction using the fitted K_S. (---) - Simulations for random distribution of acceptors ($K_S = 1.0$) - A – Labeled protein incorporated in DOPC (fitted $K_S = 1.4$); B – Labeled protein incorporated in DMoPC (fitted $K_S = 2.9$); C – Labeled protein incorporated in DEuPC (fitted $K_S = 2.1$). Reprinted with permission from Fernandes *et al.* (2004). Copyright 2004 Biophysical Society.

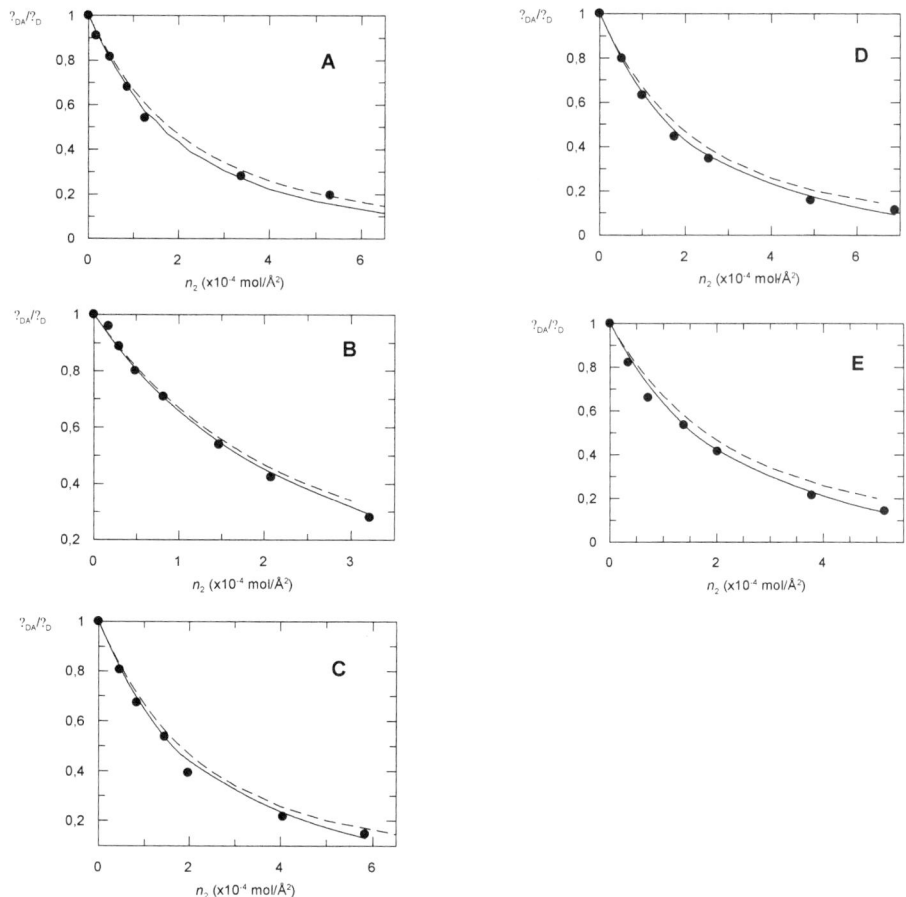

Figure 12.18. Donor (DCIA-labeled protein) fluorescence quenching by energy transfer acceptor (18:1-(12:0-NBD)-PX) (X stands for the different head group structures) in pure bilayers of DOPC. (•) - Experimental energy transfer efficiencies. (——) -Theoretical simulations obtained from the annular model for protein-lipid interaction using the fitted K_S. (---) - Simulations for random distribution of acceptors ($K_S = 1.0$) - A – PC labeled phospholipid (fitted $K_S = 2.0$); B - PE labeled phospholipid (fitted $K_S = 2.0$); C - PG labeled phospholipid (fitted $K_S = 2.3$); D - PS labeled phospholipid (fitted $K_S = 2.7$); E - PA labeled phospholipid (fitted $K_S = 3.0$). Reprinted with permission from Fernandes et al. (2004). Copyright 2004 Biophysical Society.

The FRET methodology has three interesting features. Firstly, by choosing donor-acceptors with different Förster radii it is also possible to study mainly the first-shell of lipids or also the outside shells as was the case in the present study. The joint analysis of results coming from these different donor-acceptor pairs could allow for an even more detailed description of the protein-lipid arrangement in more complex systems. In our study, the relatively large R_0 value for the used donor-acceptor pair meant that the experimental quenching curves shown in Figures 12.17 and 12.18 look similar at first sight. It is impressive that the analysis methodology is able to retrieve significant K_s

values nevertheless. Of course, this study could still be improved by the use of a donor-acceptor pair with a smaller R_0 value, closer to the distances under measurement. Second, the more economic character of fluorescence studies, which do not require the same amount of material as ESR, should be stressed. Finally, although this model leads to a somewhat complex decay law, it is actually not necessary to analyze the decay curves with this law to recover the relevant parameters. The theoretical curves are conveniently simulated and integrated in a worksheet in order to calculate the theoretical FRET efficiencies. These can be matched to experimental values by varying the K_S value (the sole unknown parameter). The experimental FRET efficiencies could be obtained from steady-state data. In our case, we obtained them from integration of donor decay curves because these are less prone to artifacts (*e.g.*, light scattering, inner filter effects, measurement of absolute intensities), which in any case, could in principle be corrected for in a steady-state experiment.

12.8. CONCLUDING REMARKS AND ACKNOWLEDGMENTS

In the present review, it was shown how relevant problems in the context of lipid-peptide interaction, such as peptide aggregation, transverse location in the membrane, dynamics, and selectivity for specific lipids can be addressed using fluorescence techniques and methodologies.

Fluorescence spectroscopy is well known in the life sciences due to its intrinsic sensitivity, and this certainly applies to the study of peptides in interaction with membranes, where most of the times minimal amounts of material are available. However, we would like to stress that, in addition to this fact, if topological modeling and state-of-the-art methodologies are used, quantified information regarding both structure and dynamics can be obtained.

Trp and Tyr are present in most membrane-interacting peptide sequences and, in addition, with current technical abilities, alterations on the sequence are trivial and these intrinsic fluorophores can be easily introduced. Although they do have a complex photophysics, such as their fluorescence decay, this can be used to obtain further information on the system. In case that no intrinsic fluorescent residue is present, other residues can be used as quenchers of fluorophores in the membrane, thus providing information on peptide-lipid interaction. A fluorophore group can also be covalently attached to the peptide.

In this moment, in the area of membrane biophysics, the existence of lipid domains, and some related issues such as the role of cholesterol, liquid-ordered phases and the existence of "rafts" are the leading areas of research. Regarding membrane domains, the simpler models are the binary lipid systems. A quantitative global treatment of the peptide photophysics taking into account phase separation is seldom carried out. This would imply, *e.g.*, the determination of the partition coefficient of the peptide between the lipid phases, and the consequences of the presence of the domains, such as different peptide aggregation states. In the present work illustrative cases both in binary and one-phase systems were described in detail.

Finally we acknowledge Fundação para a Ciência e a Tecnologia (Portugal) for financial support, namely projects and R.F.M. de A.'s grant (SFRH/BD/943/2000) under the program POCTI. We also want to thank the other co-authors of our cited work.

12.9. REFERENCES

Abrams, F. S., and London, E., 1992, Calibration of the parallax fluorescence quenching method for the determination of membrane penetration depth: refinement and comparison of quenching by spin-labelled probes, *Biochemistry* **31**:5312.

Almgren, M., 1991, Kinetics of excited state processes in micellar media, in: *Kinetics and Catalysis in Microheterogeneous Systems.* M. Gratzel and K. Kalyanasundaram, eds., Marcell Dekker, New York, pp. 63-113.

Al-Obeidi, F., Castrucci, A. M. L., Hadley, M. E., and Hruby, V. J., 1989, Potent and prolonged acting cyclic lactam analogues of α-melanotropin: design based on molecular dynamics, *J. Med. Chem.* **32**:2555.

Antollini, S. S., Soto, M. A., Bonini de Romanelli, I., Gutierrez-Merino, C., Sotomayor, P., and Barrantes, F. J., 1996, Physical state of bulk and protein-associated lipid in nicotinic acetylcholine receptor-rich membrane studied by laurdan generalized polarization and fluorescence energy transfer, *Biophys. J.* **70**:1275.

Barrantes, F. J., Antollini, S. S., Blanton, M. P., and Prieto, M., 2000, Topography of nicotinic acetylcholine receptor membrane-embedded domains, *J. Biol. Chem.* **275**:37333.

Berberan-Santos, M. N., and Prieto, M. J. E., 1987, Energy transfer in spherical geometry. Application to micelles, *J. Chem. Soc. Faraday Trans. 2* **83**:1391.

Berberan-Santos, M. N., and Valeur, B., 1991, Fluorescence depolarization by electronic energy transfer in donor-acceptor pairs of like and unlike chromophores, *J. Chem. Phys.* **95**:8048.

Best, L., John, E., and Jähnig, F., 1987, Order and fluidity of lipid membranes as determined by fluorescence anisotropy decays, *Eur. Biophys. J.* **15**:87.

Biaggi, M. H, Riske, K. A., and Lamy-Freund, M. T., 1997, Melanotropic peptides-lipid bilayer interaction. Comparison of the hormone α-MSH to a biologically more potent analog, *Biophys. Chem.* **67**:139.

Blatt, E., Chatelier, R. C., and Sawyer, W. H., 1984, The transverse location of fluorophores in lipid bilayers and micelles as determined by fluorescence techniques. *Phothochem. Photobiol.* **39**:477.

Bloom, M., and Mouritsen, O. G., 1995, The evolution of membranes, in: *Handbook of Biological Physics, vol. 1A: Structure and Dynamics of Membranes - from Cells to Vesicles.* R. Lipowsky, E. Sackmann, eds., Elsevier Science B.V., Amsterdam, pp. 65-95.

Bonini, I. C., Antollini, S. S., Gutierrez-Merino, C., and Barrantes, F. J., 2002, Sphingomyelin composition and physical asymmetries in native acetylcholine receptor-rich membranes, *Eur. Biophys. J.* **31**:417.

Bowie, J. U., 1997, Helix packing in membrane proteins, *J. Mol. Biol.* **272**:780.

Brown, D. A., and London, E., 2000, Structure and function of sphingolipid- and cholesterol-rich membrane rafts, *J. Biol. Chem.* **275**:17221.

Butler, D. H., and McNamee, M. G., 1993, FTIR analysis of nicotinic acetylcholine receptor secondary structure in reconstituted membranes, *Biochim. Biophys. Acta* **1150**:17.

Castanho, M. A. R. B., and Prieto, 1992, Fluorescence study of the macrolide pentaene antibiotic filipin in aqueous solution and in a model system of membranes, *Eur. J. Biochem.* **207**:125.

Castanho, M. A. R. B., Prieto, M., and Acuña, A. U., 1996, The tranverse location of the fluorescent probe *t*-parinaric acid in lipid bilayers, *Biochim. Biophys. Acta* **1279**:164.

Chattopadhyay, A., 2003, Exploring membrane organization and dynamics by the wavelength-selective fluorescemce approach, *Chem. Phys. Lipids* **122**:3.

Chattopadhyay, A., and London, E., 1987, Parallax method for direct measurement of membrane penetration depth utilizing fluorescence quenching by spin-labelled phospholipids, *Biochemistry* **26**:1987,39.

Chen, R., and Bowman, R. L., 1965, Fluorescence polarization: measurement with ultraviolet-polarizing filters in a spectrophotofluorometer, *Science* **147**:729.

Chen, Y., and Barkley, M. D., 1998, Toward understanding tryptophan fluorescence in proteins, *Biochemistry* **37**:9976.

Chiu, S.-W., Jakobson, E., Subramaniam, S., and Scott, H. L., 1999, Combined Monte Carlo and molecular dynamics simulation of fully hydrated dioleoyl and palmitoyl-oleyl phosphatidylcholine lipid bilayers, *Biophys. J.* **77**:2462.

Contreras, L. M., de Almeida, R. F. M., Villalaín, J., Fedorov, A., and Prieto, M., 2001, Interaction of α-melanocyte stimulating hormone with binary phospholipids membranes: structural changes and relevance of phase behaviour, *Biophys. J.* **80**:2273.

Dahms, T. E. S., and Szabo, A. G., 1995, Probing local secondary structure by fluorescence: time-resolved and circular dichroism studies of highly purified neurotoxins, *Biophys. J.* **69**:569.

Datema, K. P., Wolfs, C. J. A. M., Marsh, D., Watts, A., and Hemminga, M. A., 1987, Spin label electron spin resonance study of bacteriophage M13 coat protein incorporation into mixed lipid bilayers, *Biochemistry* **26**:7571.

Davenport, L., Dale, R. E., Bisby, R. H., and Cundall., R. B., 1985, Transverse location of the fluorescent probe 1,6-diphenyl-1,3,5-hexatriene in model lipid bilayer membrane systems by resonance energy transfer, *Biochemistry* **24**:4097.

de Almeida, R. F. M., Loura, L. M. S., Fedorov, A., and Prieto, M., 2002, Non-equilibrium phenomena in the phase separation of a two-component lipid bilayer, *Biophys. J.* **82**:823.

de Almeida, R. F. M., Fedorov, A., and Prieto, M., 2003, Sphingomyelin / phosphatidylcholine / cholesterol phase diagram: boundaries and composition of lipid rafts, *Biophys. J.* **85**:2406.

de Almeida, R. F. M., Loura, L. M. S., Prieto, M., Watts, A., Fedorov, A., and Barrantes, F. J., 2004, Cholesterol modulates the organization of the γM4 transmembrane domain of the muscle nicotinic acetylcholine receptor, *Biophys. J.* **86**:2261.

de Planque, M. R. R., Kruijtzer, J. A. W., Liskamp, R. M. J., Marsh, D., Greathouse, D. V., Koeppe II, R. E., de Kruijff, B., and Killian, J. A., 1999, Different membrane anchoring positions of tryptophan and lysine in synthetic transmembrane alpha-helical peptides, *J. Biol. Chem.* **274**:20839.

Dietrich, C., Bagatolli, L. A., Volovyk, Z. N., Thompson, N. L., Levi, M., Jacobson, K., and Gratton, E., 2001, Lipid rafts reconstituted in model membranes, *Biophys. J.* **80**:1417.

Dumas, F., Sperotto, M. M., Lebrun, M.-C., Tocanne, J.-F., and Mouritsen, O. G., 1997, Molecular sorting of lipids by bacteriorhodopsin in dilauroylphosphatidylcholine /distearoylphosphatidylcholine lipid bilayers, *Biophys. J.* **73**:1940.

Eberle, A. N., 1988, *The Melanotropins. Chemistry, Physiology and Mechanism of Action*, Karger Publishers, Basel.

Edidin, M., 1998, Defining and imaging membrane domains. *Biol. Skr. Dan. Vid. Selsk.* **49**:19.

Edidin, M., 2003, Lipids on the frontier: a century of cell-membrane bilayers, *Nat. Rev. Mol. Cell Biol.* **4**:414.

Eisinger, J., Blumberg, W. E., and Dale, R. E., 1981, Orientational effect in intra- and intermolecular long range excitation energy transfer, *Ann. N. Y. Acad. Sci.* **366**:155.

Engelborghs, Y., 2001, The analysis of time resolved protein fluorescence in multi-tryptophan proteins, *Spectrochim. Acta A Mol. Biomol. Spectrosc.* **57**:2255.

Fahsel, S., Pospiech, E. M., Zein, M., Hazlet, T. L., Gratton, E., and Winter, R., 2002, Modulation of concentration fluctuations in phase separated lipid membranes by polypeptide insertion, *Biophys. J.* **83**:334.

Fernandes, F., Loura, L. M. S., Prieto, M., Koehorst, R., Spruijt, R. B., and Hemminga, M. A., 2003, Dependence of M13 major coat protein oligomerization and lateral segregation on bilayer composition, *Biophys. J.* **85**:2430.

Fernandes, F., Loura, L. M. S., Koehorst, R., Spruijt, R. B., Hemminga, M. A., Fedorov, A., and Prieto, M., 2004, Quantification of protein-lipid selectivity using FRET: application to the M13 major coat protein, *Biophys. J.* **87**:344.

Fernandes, M. X., García de la Torre, J., and Castanho, M. A. R. B., 2002, Joint determination by Brownian dynamics and fluorescente quanching of the in-depth location profila of biomolecules in membranes, *Anal. Biochem.* **307**:1.

Ferreira, S. T., Stella, L., and Gratton, E., 1994, Conformational dynamics of bovine Cu, Zn superoxide dismutase revealed by time-resolved fluorescence spectroscopy of the single tyrosine residue, *Biophys. J.* **7**:1185.

Fong, T. M., and McNamee, M. G., 1987, Stabilization of acetylcholine receptor secondary structure by cholesterol and negatively charged phospholipids in membranes, *Biochemistry* **26**:3871.

Graham, I. G., Gagné, J., and Silvius, J. R., 1985, Kinetics and thermodynamics of calcium induced lateral phase separations in phosphatidic acid containing bilayers, *Biochemistry* **24**:7123.

Gryczynski, I., Steiner, R. F., and Lakowicz, J. R., 1991, Intensity and anisotropy decays of the tyrosine calmodulin proteolytic fragments, as studied by GHz frequency-domain fluorescence, *Biophys. Chem.* **39**:69.

Hemminga, M. A., Sanders, J. C., Wolfs, C. J. A. M., and Spruijt, R. B., 1993, Lipid-protein interactions involved in bacteriophage M13 infection, *Protein-Lipid Inter. New Compr. Biochem.* **25**:191.

Hoshi, T., Zagotta, W. N., and Aldrich, R. W., 1990, Biophysical and molecular mechanisms of Shaker potassium channel inactivation, *Science* **250**:533.

Hruby, V. J., Wilkes, B. C., Hadley, M. E., Al-Obeidi, F., Sawyer, T. K., Staples, D. G., deVaux, A. E., Dym, O., Castrucci, A. M. de L., Hintz, M. F., Riehm, J. P., and Rao, R., 1987, α-Melanotropin: the minimal active sequence in the frog skin bioassay, *J. Med. Chem.* **30**:2126-2130.

Hurley, J. H., and Meyer, T., 2001, Subcellular targeting by membrane lipids, *Curr. Opin. Cell. Biol.* **13**:146-152.

Ipsen, J. H., Karlström, G., Mouritsen, O. G., Wennerström, H., and Zuckermann, M. J., 1987, Phase equilibria in the phosphatidylcholine-cholesterol system, *Biochim. Biophys. Acta* **905**:162-172.

Ito, A. S., Castrucci, A. M. de L., Hruby, V. J., Hadley, M. E., Krajcarski, D. T., and Szabo, A. G., 1993, Structure-activity correlations of melanotropin peptides in model lipids by tryptophan fluorescence studies, *Biochemistry* **32**:12264.

Jabłoński, A., 1960, On the notion of emission anisotropy, *Bull. Acad. Pol. Sci.* **8**:259.

Johnson, J. E., and Cornell, R. B., 1999, Amphitropic proteins: regulation by reversible membrane interactions, *Mol. Membr. Biol.* **16**:217.

Killian, J. A., 1998, Hydrophobic mismatch between proteins and lipids in membranes, *Biochim. Biophys. Acta* **1376**:401.

Kinosita, K., Jr., Ikegami, A., and Kawato, S., 1982, On the wobbling-in-cone analysis of fluorescence anisotropy decay, *Biophys. J.* **37**:461.

Korlach, J., Schwille, P., Webb, W. W., and Feigenson, G. W., 1999, Characterization of lipid bilayer phases by confocal microscopy and fluorescence correlation spectroscopy, *Proc. Natl. Acad. Sci. USA* **96**:8461.

Ladokhin, A. S., 1997, Distribution analysis of depth-dependent fluorescence quenching in membranes. *Meth. Enzymol.* **278**:462.

Ladokhin, A. S., 2001, On the interpretation of decay-associated fluorescence spectra in proteins, *Biopolymers Cell* **17**:221.

Lafleur, M., Faucon, J. F., Dufourcq, J., and Pezolet, M., 1989, Perturbation of binary phospholipid mixtures by melittin: a fluorescence and Raman spectroscopy study, *Biochim. Biophys. Acta* **980**:85.

Lakowicz, J. R., 1999, *Principles of Fluorescence Spectroscopy*, 2nd Ed., Kluwer Academic/Plenum Press, New York.

Lakowicz, J.R., 2000, On spectral relaxation in proteins, *Photochem. Photobiol.* **72**:421.

Laws, W. R., Ross, J. B. A., Wyssbrod, H. R., Beechem, J. M., Brand, L., Sutherland, J. C., 1986, Time-resolved fluorescence and ^{1}H-NMR studies of tyrosine and tyrosine analogues: correlation of NMR-determined rotamer populations and fluorescence kinetics, *Biochemistry* **25**:599.

Lehrer, S. S., 1971, Solute perturbation of protein fluorescence. The quenching of the tryptophanyl fluorescence of model compounds and of lysozyme by iodide ion. *Biochemistry* **10**:3254.

Lehtonen, J. Y. A., and Kinnunen, P. K. J., 1997, Evidence for phospholipids microdomain formation in liquid crystalline liposomes reconstituted with *Escherichia coli* lactose permease, *Biophys. J.* **72**:1247.

Lehtonen, J. Y. A., Holopainen, J. M., and Kinnunen, P. K. J., 1996, Evidence for the formation of microdomains in liquid crystalline large unilamellar vesicles caused by hydrophobic mismatch of the constituent phospholipids, *Biophys. J.* **70**:1753.

Lentz, B. R., Barrow, D. A., and Hoechli, M., 1980, Cholesterol-phosphatidylcholine interactions in multilamellar vesicles, *Biochemistry* **19**:1943.

Lewis, B. A., and Engelman, D. M., 1983, Bacteriorhodopsin remains dispersed in fluid phospholipid bilayers over a wide range of bilayer thickness, *J. Mol. Biol.* **166**:203.

Lipari, G., and Szabo, A., 1980, Effect of vibrational motion on fluorescence depolarization and nuclear magnetic resonance relaxation in macromolecules and membranes, *Biophys. J.* **30**:489.

London, E., and Brown, D. A., 2000, Insolubility of lipids in triton X-100: physical origin and relationship to sphingolipid/cholesterol membrane domains (rafts), *Biochim. Biophys. Acta* **1508**:182.

Loura, L. M. S., and Prieto, M., 1997, Dehydroergosterol structural organization in aqueous medium and in a model system of membranes, *Biophys. J.* **72**:2226.

Loura, L. M. S., Fedorov, A., and Prieto, M., 1996, Resonance energy transfer in a model system of membranes: application to gel and liquid crystalline phases, *Biophys. J.* **71**:1823.

Loura, L. M. S., Fedorov, A., and Prieto, M., 2000, Membrane probe distribution heterogeneity: a resonance energy transfer study, *J. Phys. Chem. B.* **104**:6920.

Loura, L. M. S., de Almeida, R. F. M., and Prieto, M., 2001a, Detection and characterization of membrane microheterogeneity by resonance energy transfer, *J. Fluoresc.* **11**:197.

Loura, L. M. S., Fedorov, A., and Prieto, M., 2001b, Exclusion of a cholesterol analog from the cholesterol-rich phase in model membranes, *Biochim. Biophys. Acta* **1511**:236.

Loura, L. M. S., de Almeida, R. F. M., Coutinho, A., and Prieto, M., 2003, Interaction of peptides with binary phospholipid membranes: application of fluorescence methodologies, *Chem. Phys. Lipids* **122**:77.

Lundbaek, J. A., Anderson, O. S., Werge, T., and Nielsen, C., 2003, Cholesterol-induced protein sorting: an analysis of energetic feasibility, *Biophys. J.* **84**:2080.

Mabrey, S., and Sturtevant, J. M., 1976, Investigation of phase transitions of lipids and lipid mixtures by high-sensitivity differential scanning calorimetry, *Proc. Natl. Acad. Sci. USA.* **73**:3862.

Mall, S., Broadbridge, R., Sharma, R. P., East, J. M., and Lee, A. G., 2001, Self-association of model transmembrane helix is modulated by lipid structure, *Biochemistry* **40**:12379.

Marsh, D., 1990, *Handbook of Lipid Bilayers*, CRC Press, Boca Raton.

Marsh., D., and Horváth, L. I., 1998, Structure, dynamics and composition of the lipid-protein interface. Perspectives from spin-labelling, *Biochim. Biophys. Acta* **1376**: 267.

Mateo, C. R., Acuña, A. U., and Brochon, J.-C., 1995, Liquid-crystalline phases of cholesterol/lipid bilayers as revealed by the fluorescence of *trans*-parinaric acid, *Biophys. J.* **68**:978.

Mateo, C. R., Souto, A. A., Amat-Guerri, F., and Acuña, U., 1996, New fluorescent octadecapentaenoic acids as probes of lipid membranes and protein-lipid interactions *Biophys. J.* **71**:2177.

McMullen, T. P. W., and McElhaney, R. N., 1995, New aspects of the interaction of cholesterol with dipalmitoylphosphatidylcholine bilayers as revealed by high-sensitivity differential scanning calorimetry, *Biochim. Biophys. Acta* **1234**:90.

Meijer, A. B., Spruijt, R. B., Wolfs, C. J. A. M., and Hemminga, M. A. 2001, Membrane-anchoring interactions of M13 major coat protein, *Biochemistry* **40**:8815.

Milhiet, P. E., Domec, C., Giocondi, M.-C., Van Mau, N., Heitz, F., and Le Grimellec, C., 2001, Domain formation in models of the renal brush border membrane outer leaflet, *Biophys. J.* **81**:547.

Mobashery, N., Nielsen, C., and Andersen, O. S., 1997, The conformational preference of gramicidin channels is a function of lipid bilayer thickness, *FEBS Lett.* **412**:15.

Moreno, M. J., and Prieto, M., 1993, Interaction of the peptide hormone adrenocorticotropin, ACTH (1-24), with a membrane model system: a fluorescence study, *Photochem. Photobiol.* **57**:431.

Mouritsen, O. G., and Bloom, M., 1984, Mattress model of lipid-protein interactions in membranes, *Biophys. J.* **46**:141.

Mouritsen, O. G., and Jørgensen, K. 1994, Dynamical order and disorder in lipid bilayers, *Chem. Phys. Lipids* **73**:3.

Mouritsen, O. G., and Jørgensen, K., 1997, Small-scale lipid-membrane structure: simulation vs. experiment, *Curr. Opin. Struct. Biol.* **7**:518.

Needham, D., and Nunn, R. S., 1990, Cohesive properties (elastic deformation and failure) of lipid bilayer membranes containing cholesterol, *Biophys. J.* **58**, 997.

Nezil, F. A., and Bloom, M., 1992, Combined influence of cholesterol and synthetic amphiphilic peptides upon bilayer thickness in model membranes, *Biophys. J.* **61**:1176.

Owen, C. S., 1975, Two dimensional diffusion theory: cylindrical diffusion model applied to fluorescence quenching, *J. Chem. Phys.* **62**:3204.

Pascutti, P. G., El-Jaik, L. J., Bisch, P. M., Mundim, K. C., and Ito, A. S., 1999, Molecular dynamics simulation of α-melanocyte stimulating hormone in a water-membrane model interface, *Eur. Biophys. J.* **28**:499.

Peelen, S. J. C. J., Sanders, J. C., Hemminga, M. A., and Marsh, D., 1992, Stoichiometry, selectivity and exchange dynamics of lipid-protein interaction with bacteriophage M13 coat protein studied by spin label electron spin resonance. Effects of protein secondary structure, *Biochemistry* **31**:2670.

Perrin, F., 1926, Polarization de la lumière de fluorescence. Vie moyenne des molécules dans l'état excité, *J. Phys. Radium* **7**:390.

Perrin, F., 1929, La fluorescence des solutions: induction moléculaire, polarization et durée d'émission, *Ann. Phys. (Paris)* **12**:169.

Poveda, J. A., Prieto, M., Encinar, J. A., González-Ros, J. M., and Mateo, C. R., 2003, Intrinsic tyrosine fluorescence as a tool to study the interaction of the Shaker B "ball" peptide with anionic membranes, *Biochemistry* **42**:7124.

Rankin, S. E., Addona, G. H., Kloczewiak, M. A. Bugge, B., and Miller, K. W., 1997, The cholesterol dependence of activation and fast desensitization of the nicotinic acetylcholine receptor, *Biophys. J.* **73**:2446.

Razi-Naqvi, K., 1974, Diffusion-controlled reactions in two dimensional fluids: discussion of measurements of lateral diffusion of lipids in biological membranes, *Chem. Phys. Lett.* **28**:2303.

Ren, J., Lew, S., Wang, J., and London, E. 1999, Control of the transmembrane orientation and interhelical interactions within membranes by hydrophobic helix length, *Biochemistry* **38**:5905.

Runnels, L. W., and Scarlata, S. F., 1995, Theory and application of fluorescence homotransfer to melittin oligomerization, *Biophys. J.* **69**:1569.

Russel, M., 1991, Filamentous phage assembly, *Mol. Microbiol.* **5**:1607.

Sanders, J. C., van Nuland, N. A. J., Edholm, O., and Hemminga, M., 1991, Conformational and aggregation of M13 coat protein studied by molecular dynamics, *Biophys. Chem.* **41**:193-202.

Santos, N. C., Prieto, M., and Castanho, M. A. R. B., 1998, Interaction of the major epitope region of HIV protein gp41 with membrane model systems. A fluorescence spectroscopy study, *Biochemistry* **37**:8674.

Santos, N. C., Prieto, M., and Castanho, M. A. R. B., 2003, Quantifying molecular partition into model systems of biomembranes: an emphasis on optical spectroscopic methods, *Biochim. Biophys. Acta* **1612**:123.

Sawyer, T. K., Sanfilippo, P. J., Hruby, V. J., Engel, M. H., Heward, C. B., Burnett, K. B., and Hadley, M. E., 1980, 4-Norleucine, 7-D-phenylalanine-α-melanocyte-stimulating hormone: a highly potent α-melanotropin with ultralong biological activity, *Proc. Natl. Acad. Sci. USA* **77**:5754.

Sawyer, T. K., Hruby, V. J., Darman, P. S., and Hadley, M. E., 1982, [half-Cys^4,half-Cys^{20}]-α-melanocyte-stimulating hormone: a cyclic α-melanotropin exhibiting superagonist biological activity, *Proc. Natl. Acad. Sci. USA* **79**:1751.

Sillen, A., and Engelborghs, Y., 1998, The correct use of "average" fluorescence parameters, *Photochem. Photobiol.* **67**:475.

Silvius, J. R., 1992, Cholesterol modulation of lipid intermixing in phospholipid and glycosphingolipid mixtures. Evaluation using fluorescent lipid probes and brominated lipid quenchers, *Biochemistry* **31**:3398.

Simon, J. A., Williamson, I. M., East, J. M., and Lee, A. G., 2003, Interactions of anionic phospholipids and phosphatidylethanolamine with the potassium channel KcsA. *Biophys. J.* **85**:3828.

Simons, K., and Ikonen, E., 1997, Functional rafts in cell membranes, *Nature* **387**:569.

Singer, S. J., and Nicolson, G. L., 1972, The fluid mosaic model of the structure of cell membranes, *Science* **175**:720.

Smaby, J. M., Momsen, M. M., Brockman, H. L., and Brown, R. E., 1997, Phosphatidylcholine acyl unsaturation modulates the decrease in interfacial elasticity induced by cholesterol, *Biophys. J.* **73**:1492.

Small, D. M., 1986, *The Physical Chemistry of Lipids: from Alkanes to Phospholipids (Handbook of Lipid Research, Vol. 4)*. Plenum Press, New York.

Smith, L. M., Smith, B. A., and McConnell, H. M., 1979, Lateral diffusion of M-13 coat protein in model membranes, *Biochemistry* **18**:2256.

Smith, L. M., Rubenstein, J. L. R., Parce, J. W., and McConnell, H. M., 1980, Lateral diffusion of M-13 coat protein in mixtures of phosphatidylcholine and cholesterol, *Biochemistry* **19**:5907.

Snyder, B., and Freire, E., 1982, Fluorescence energy transfer in two dimensions. A numeric solution for random and non-random distributions, *Biophys. J.* **40**:137.

Spruijt, R. B., Wolfs, C. J. A. M., and Hemminga, M. A., 1989, Aggregation related conformational change of the membrane-associated coat protein of bacteriophage M13, *Biochemistry* **28**:9158.

Stopar, D., Spruijt, R. B., Wolfs, C. J. A. M., and Hemminga, M. A., 1997, In situ aggregational state of M13 bacteriophage major coat protein in sodium cholate and lipid bilayers, *Biochemistry* **36**:12268.

Stopar, D., Spruijt, R. B., Wolfs, C. J. A. M., and Hemminga, M. A., 2003, Protein-lipid interactions of bacteriophage M13 major coat protein, *Biochim. Biophys. Acta* **1611**:5.

Szabo, A. G. and Rayner, D. M., 1980, The time resolved emission spectra of peptide conformers measured by pulsed laser excitation, *Biochem. Biophys. Res. Commun.* **94**:909.

Tanaka, F., Tamai, N., Mataga, N., Tonomura, B., and Hiromi, K., 1994, Analysis of internal motion of single tryptophan in *Streptomyces* subtilisin inhibitor from its picosecond time-resolved fluorescence, *Biophys. J.* **67**: 874.

Thewalt, J. L., and Bloom, M., 1992, Phosphatidylcholine: cholesterol phase diagrams. *Biophys. J.* **63**:1176.

Toptygin, D., Savtchenko, R. S., Meadow, N. D., and Brand, L., 2001, Homogeneous spectrally- and time-resolved fluorescence emission from single-tryptophan mutants of IIA^{Glc} Protein. *J. Phys. Chem. B* **105**:2043.

Tristam-Nagle, S., Petrache, H. I., and Nagle, J. F., 1998, Structure and interactions of fully hydrated dioleoylphosphatidylcholine bilayers, *Biophys. J.* **75**:917.

Umberger, J. Q., and Lamer, V. K., 1945, The kinetics of diffusion controlled molecular and ionic reactions in solution as determined by measurements of the quenching of fluorescence, *J. Am. Chem. Soc.* **67**:1099.

Valeur, B., 2001, *Molecular Fluorescence. Principles and Applications*, Wiley-VCH, New York.

Valeur, B., and Weber, G., 1977, Anisotropic rotations in 1-naphthylamine, existence of a red-edge transition moment normal to the ring plane, *Chem. Phys. Lett.* **45**:140.

Valeur, B., and Weber, G., 1977, Resolution of the fluorescence excitation spectrum of indole into the ^1L_a and ^1L_b excitation bands, *Photochem. Photobiol.* **25**:441.

Vaz, W. L. C., Criado, M., Madeira, V. M. C., Schoellmann, G., and Amd Jovin, T. M., 1982, Size dependence of the translational diffusion of large integral membrane proteins in liquid-crystalline phase lipid bilayers. A study using fluorescence recovery after photobleaching, *Biochemistry* **21**:5608.

Villalaín, J. B., and Prieto, M., 1991, Location and interaction of n-(9-anthroyloxy)-stearic acid probes incorporated in phosphatidylcholine vesicles, *Chem. Phys. Lipids* **59**:9.

Vincent, M., Gilles, A.-M., de la Sierra, I. M. L., Briozzo, P., Bârzu, O., Gallay, J., 2000, Nanosecond fluorescence dynamic stokes shift of tryptophan in a protein matrix, *J. Phys. Chem. B* **104**:11286.

Vist, M. R., and Davis, J. H., 1990, Phase equilibria of cholesterol/dipalmitoylphosphatidylcholine mixtures: ^{2}H nuclear magnetic resonance and differential scanning calorimetry, *Biochemistry* **29**:451.

Vogel, H., Nilsson, L., Rigler, R., Vogues, K.-L., and Jung, G., 1988, Structural fluctuations of a helical polypeptide traversing a lipid bilayer, *Proc. Natl. Acad. Sci. USA* **85**:5067.

Wang, S., Martin, E., Cimino, J., Omann, G., Glaser, M., 1988, Distribution of phospholipids around gramicidin and D-beta-hydroxybutyrate dehydrogenase as measured by resonance energy transfer, *Biochemistry* **27**:2033.

Weber, G., 1960, Fluorescence-polarization spectrum and electronic electronic-energy transfer in tyrosine, tryptophan and related compounds, *Biochem. J.* **75**:335.

Weber, G., and Shinitzky, M., 1970, Failure of energy transfer between identical aromatic molecules on excitation at the long wave edge of the absorption spectrum, *Proc. Natl. Acad. Sci. USA* **65**:823.

Welti, R., and Glaser, M., 1994, Lipid domains in model and biological membranes, *Chem. Phys. Lipids* **73**:121.

Wiener, M. C., and White, S. H., 1992, Structure of a fluid dioleoylphosphatidylcholine bilayer determined by joint refinement of x-ray and neutron diffraction data. III. Complete structure, *Biophys. J.* **61**:437.

Willis, K. J., and Szabo, A. G., 1992, Conformation of parathyroid hormone: time-resolved fluorescence studies, *Biochemistry* **31**:8924.

Wimley, W. C., and White, S. H., 1996, Experimentally determined hydrophobicity scale for proteins at membrane interfaces, *Nat. Struct. Biol.* **3**:842.

Wolber, P. K., and Hudson, B. S., 1979, An analytical solution to the Förster energy transfer problem in two dimensions, *Biophys. J.* **28**:197.

Wolfs, C. L. A. M., Horváth, L. I., Marsh, D., Watts, A., Hemminga, M. A., 1989, Spin label ESR of bacteriophage M13 coat protein in mixed lipid bilayers. Characterization of molecular selectivity of charged phospholipids for the bacteriophage M13 coat protein in lipid bilayers, *Biochemistry* **28**:995.

Wu, S. H., and McConnell, H. M., 1975, Phase separations in phospholipid membranes, *Biochemistry* **14**:847.

Zamyatnin, A. A., 1972, Protein volume in solution, *Prog. Biophys. Mol. Biol.* **24**:107.

HIGH-THROUGHPUT TISSUE IMAGE CYTOMETRY

Peter T.C. So[1], Timothy Ragan[1,2], Karsten Bahlmann[1,2], Hayden Huang[3],
Ki Hean Kim[1], Hyuk-Sang Kown[1], Richard T. Lee[3]

13.1. INTRODUCTION

This review describes the development and some pilot applications of a new technology: 3D tissue image cytometry. 3D tissue image cytometry quantifies tissue morphology and biochemistry in a high throughput fashion. Promising applications range from cardiology, cancer biology, to tissue engineering. In this review, we will first examine traditional cytology, histology, and cytometry techniques that are the precursors of 3D tissue image cytometry. We will consider novel 3D optical imaging methods that form the technological basis for this new technology. Experiments for the validation and characterization of this instrument are presented. An application of this technology focusing on cardiac hypertrophy study is described in detail.

13.1.1. Cytology and Histology

Cytology is the science of examine cellular physiological and pathological states based on microscopic examination. Histology is similar but is focus on the examination of tissue structures. Cytology and histology are the current standards for the diagnosis of many diseases. Both techniques compare the microscopic structure of cells or tissues with known physiological and pathological forms. Prior to microscopic examination, specimens are first subjected to fixation and staining procedures. For histology, tissue sections are treated with a fixative such as paraformaldehyde or formalin, in order to to cross link protein structures and to prevent decomposition. The sample is subsequently

[1] Massachusetts Institute of Technology, Department of Mechanical Engineering and Division of Biological Engineering, 77 Massachusetts Avenue, Cambridge, MA 02139
[2] TissueVision, Inc., 98 Line Street Suite 2, Somerville, MA 02143
[3] TissueVision, Inc., 98 Line Street Suite 2, Somerville, MA 02143

either frozen or embedded in paraffin to facilitate slicing into thin sections using a microtome. Typical histological sections are between 1 – 100 microns thick. The sections are then labeled using stains such as H&E (hematoxylin and eosin) and trichrome. These stains increase image contrast and aids in distinguishing between different microstructures; for example, H&E stains works by preferential partitioning of hematoxylin and eosin into acidic and basic compartments respectively. Importantly, molecular specific labeling can be accomplished when necessary by using antibodies conjugated with dye molecules. Since the original tissue volume is typically at least a few millimeters in size, histological specimen preparation results in tens to thousands of tissue sections. A selected subset of these tissue sections are imaged using wide field light microscopy in a manual fashion. Cytology sample preparation procedure is similar but without the complications associated with tissue sectioning and labeling. In general, highly trained pathologists are required for accurate classification of these cellular or tissue images. In selected case, more narrowly trained technicians are involved in this manual diagnosis process, such as in the diagnosis of Papanicolaou (Pap) test specimens.

13.1.2. Quantitative Cellular Cytology and Image Cytometry

Traditionally, the final step of cytological study involves the examination of the specimen morphology by pathologists, biologists, or technicians. The training of these specialists allows them to make diagnostic decision from complex images based on the morphology of cellular structures and the staining pattern of these structures. While the diagnostic accuracy and sensitivity of these trained specialists can be very high, this step is ultimately a subjective process. The quality of the diagnosis is highly dependent on the training and the competence of the specialists who examine the specimen microscopically. An excellent example is Pap test [1-3]. Pap test is an important and very common procedure for the early detection of cervical cancer. Pap test is a cytological examination of cells scarped from the cervix. The number of Pap tests performed in U.S. is large; Pap tests are recommended for mature women at a frequency of once every three years. As a screening test, the sensitivity (the ability to eliminate false negative) and specificity (the ability to eliminate false positive) are rather modest. The sensitivity and specificity for the detection of cervical intraepithelial neoplasia-1 are only 70-80% and 95% respectively. There is a significant chance of under diagnosis. Realizing that approximately half the false negative is due to under sampling and over 30% is due to operator error. Further, recognizing that each slide contains approximately 50,000-300,000 cells, operator error is not unexpected. Therefore, additional high throughput screening procedures based on image recognition algorithms, such as neural network methods, have been developed and have shown clinical efficacy [4-6]. This is a form of high throughput image cytometry. It should be noted that automated image recognition is only used as a second level screen to rapidly examine specimens judged to be negative by manual screening. Automated screening has not been used as the primary screen is partly due to the difficulty of image classification of complex structures. Further, one may note that the use of automated, quantitative image classification algorithms outside Pap test is relatively rare in cytology. This is partially due to image recognition in general can be computationally challenging. This is further due to the image recognition algorithms developed for one specific diagnostic test, such as Pap test, often cannot be generalized for other diagnosis tasks.

13.1.3. Quantitative Image Cytometry and Flow Cytometry

Automated Pap test is probably the most important application of image cytometry. The technique of image cytometry for the study of 2D cell cultures was well developed [7-11]. In image cytometry, a population of cells is placed on a microscope coverslip or other type of solid support. Using white light or fluorescence wide field microscopy or low resolution scanning microscopy, image cytometry can study a large cell populations and to obtain multi-parameter data on individual cells. By imaging at reduced resolution and implementing automating specimen translation under the microscope objective, it allows imaging at significantly higher speed than standard microscopy. This method provides image data that contains a statistically relevant population of cells. A typical implementation of a laser scanning image cytometer is described by Kamentsky [11] that used a 25X objective and had a spot size for the excitation volume of 2.5 μm full width half maximum in the radial direction and extent of 171 μm in the axial direction. A 0.5 μm step size was employed for the raster scan and area of 4.25 mm^2 was imaged per minute[11].

Similar to cytology, image cytometry provides morphological information of the cells. Unlike standard cytology, the large number of cells imaged makes manual examination of individual cells impractical. In addition to imaging hardware, automated image analysis is an important, integral component of an image cytometry system. In white light cytometry, typical classification parameters may be based on cell shape, cell size, or dye concentration. In fluorescence mode, cytometry analysis can further attain molecular level specificity. For example, para-nucleic vs cytoplasmic distribution of proteins can be readily distinguished based on antibody labeling. Biochemical states throughout cells can be monitored by a variety of fluorescent probes targeting messenger molecules such as calcium and nitrite oxide. The distribution and morphology of specific cellular organelles can also be targeted. Biochemical signals in the cell can be further spatially localized and associated with specific cellular compartments such as the nucleus or the cytoplasm. Cellular and matrix debris can be excluded from the analysis. Importantly, image cytometry is suitable for the study of cell cultures in vivo allowing the relocalization of specific cells over time. The time course of biochemical and/or morphological changes in the cell population can be studied. A typical image cytometer can operates at about kilohertz speed and has been applied to rare event detection. Fetal nucleated red blood cell, with occurrence frequencies as low as 1 in 20 million cells, can be detected using image cytometry [12].

Flow cytometry is a related technique that allows even higher throughput assay of a population of cells. Instead of studying cells on solid support, a suspension of cells is prepared and flowed through the focus of a laser beam or other excitation source in a fine stream, such that individual cells sequentially pass through the focus. Both scattering and fluorescence parameters are studied as the cells pass through the observation volume [13]. The cells are labeled with fluorescent indicators that either monitor cellular biochemical/metabolic state or target specific surface molecules on the cellular plasma membrane. Flow cytometry with throughput over 1×10^6 to 1×10^7 cells per second has been demonstrated. Flow cytometry has been used for rare event detection down to a rate of 1 cell in 10^7. While flow cytometry has higher throughput than image cytometry, the need to put cells into suspension scarifies the cellular morphological information and the ability to monitor time dependent changes.

13.1.4. 2D and 3D Tissue Image Cytometry

Just as image and flow cytometry is an extension of traditional cytology, tissue cytometry is a high throughput extension of standard histology. Tissue image cytometry has a number of advantages. The ability to quantify cellular state and morphology inside intact tissue has significant advantages over methods that works with cultured cells or cells dissociated from their tissue matrix. Cellular morphology is drastically altered between cells inside native tissues and the same cells dissociated from the tissue matrix growing as 2D culture. Cells extracted from their tissue matrix also lack mechanical and biochemical signal inputs that are presence in their native tissue resulting in changes in cellular morphology, biochemistry, and gene expression profiles. Further, cell-cell interactions are either lost or altered when cells are grown in culture or dissociated from their native tissue environment. The physiology of pancreatic islets is an example of where cellular behavior in culture and in their native environment are very different. Bennett et al. used two-photon microscopy to monitor β cells redox activity inside in intact pancreatic islets[14, 15]. β cell redox activity is monitored by their NAD(P)H level. Previous research on β cells grown as 2D cultures shows significant cell-cell variations in NAD(P)H level change in response to external glucose levels. This 2D culture experiments result in a metabolic models of glucose metabolism based on step-wise recruitment of individual β cells. However when Bennett et al. imaged β cells within intact islets they found significantly more homogeneous glucose response suggesting that the step-wise model based on 2D culture results may not be relevant to the actual physiological insulin response in the pancreas of animal or people. This difference is clearly important for the design of pharmaceuticals for diabetes treatment.

Tissue cytometry can be applied to 2D or 3D tissue specimens. For the more common 2D tissue cytometry, thin tissue sections are prepared as in conventional histology. These tissue sections are then imaged using standard image cytometery. A number of 2D image cytometers based on this principle has been developed to study cells inside in intact tissues [16-18].

While 2D tissue cytometery is a powerful technique to extract cellular and molecular level information from tissues, the section process has a number of limitations. First, mechanical sectioning of tissue into thin sections often introduces morphological artifacts due to the stress exerted onto the tissue matrix. Second, even in the presence of sectioning artifacts, it is still possible to reconstruct the 3D tissue structure using serially generated histological sections. However, this is a difficult and time consuming process given the need to register individual tissue slices. Third, typical tissue sections have thickness on the order of 10 microns. Complex and non-local structures in tissues cannot be easily studied. For example, the morphology of cardiac myocytes is difficult to be measured from 2D tissue section since myocytes are large and extends beyond a single section. Another example is in the study of vascular structure in tissues for the quantification of angiogenesis processes during tumor development. Fourth, it is difficult to place the distribution of a selected cellular sub-population in the context of the 3D architecture of a whole organ. The ability to associate cellular distribution with organ physiological architecture is important for studying problems such as cancer metastasis.

13.1.5. Biomedical Research Opportunities using high throughput 3D tissue image cytometry

3D tissue image cytometry is a new technology. As a new technology, its potential for biomedical research is still far from fully explored. In the short term, we have identified a number of immediate applications. In cardiology, the quantification of myocyte morphology is critical for understanding disease such as cardiac hypertrophy. This important application will be discussed in depth at a later section. We are further exploring the application of 3D tissue image cytometry to study two aspects of cancer biology. The first aspect is cancer progression. The process of how cancer cells extravasate through blood vessel walls and expand to form metastatic cancer at distant organs is far from completely understood. We are exploiting 3D tissue image cytometery to explore questions such as the distribution of these cancer cells on the organ level and the spatial relationship of these metastatic cancer cells with the organ vasculature. This technology further allows us to study the time course of cancer cell clearance rate from the vascular system and provides temporal information on the growth rate of the metastatic tumors. Another aspect of cancer biology that can be studied is carcinogenesis. Novel animal models have been created by genetic engineering where fluorescent markers are activated upon mutation of a particular locus in the genome [19]. However, since the nature mutation rate is exceedingly low, about 1×10^{-5} to 1×10^{-6}, these mutated cells can be detected by flow cytometry but not by traditional histology. Therefore, important questions, such as the effect of organ site on the mutation rate or clonal expansion rate, cannot be addressed. The ability of 3D tissue image cytometery to examine a large tissue volume containing a large cell population within a reasonable time period would greatly facilitate carcinogenesis studies of this type. In addition to cardiology and cancer biology, other interesting areas of application include neural biology where we are exploring the potential application of this technology to quantify neuronal connectivity as a function of animal development, stem cell research where we are mapping stem cell distribution in organs and examining the process of adult stem cell division in animals, and tissue engineering where we are examining cellular differentiation and organ formation in situ.

In addition to these immediate applications, we believe that 3D tissue image cytometry is critical to modernize histopathology. While histology is a clinical gold standard, it is far from a quantitative science. Unlike cytology where automation and quantitative is making significant progress as in automated Pap test, there are little parallel efforts in the quantification of histological data. The lack of major attempts in the quantification of histological data is partially due the fact that image classification and recognition in complex tissue setting is significantly more challenging computationally than in cell cultures. Another problem lies in many data base driven classification algorithms requires a large quantity of tissue images data as training sets which is not readily available today. However, with the advent of 3D tissue image cytometry, a large quantity of tissue digital images can now be readily acquired. Clearly, the next step lies in developing efficient recognition and classification algorithms needed to extract diagnostic information from these images.

Finally, with the quantification of histology, one may envision that tissue physiology and pathology can be better understood through modern genomic and proteomic analysis. Combining high through 3D tissue image cytometry with the ability to map gene and protein expression profiles, physiological models may be developed based on the underlying molecular and cellular process. These physiological models may allow us to understand how tissue structure is affected genetic and protein expression variations. An early example of this type of analysis may be found in areas such as cancer development. Cancer is a disease which has a very strong spatial component to its etiology [20]. Cancer cells can invade the stroma of the surrounding tissue and recruit non-malignant cells to differentiate and support the growing tumor. The arrangement of normal tissue boundaries becomes pathogenic as the expression profiles of the surrounding cells are altered by cell signaling from the malignant cells [21-23]. The application of 3D tissue image cytometry to simultaneously map tissue morphology, gene and protein expression patterns may allow us to better understand this important pathological process.

13.2. TISSUE IMAGING TECHNIQUES

3D tissue image cytometry is a recent invention. 2D image cytometry is a significantly simpler technology. 2D image cytometry couples robotic sample positioning hardware with standard wide-field white light or fluorescence microscopy. Wide field microscopy is a well understood technique; the major recent technology improvement comes from the availability of high sensitivity and high speed area detectors such as charge couple devices (CCD) and CMOS imagers. In contrast, the development of 3D tissue image cytometry must overcome a significant optical technology hurdle. Traditional microscope cannot easily image cellular structures within optically opaque thick tissues. Standard wide field microscopy has very poor resolution and contrast when image depth is further than several microns from the surface of the tissue due to scattering, absorption and aberration. Deconvolution algorithms can be applied to z-stacks of images acquired at successive depths. However, the results of deconvolution methods are typically very poor when imaging depth is thicker than a few tens of microns. Deconvolution methods are also very expensive in terms of computational time and may be impractical to be applied to process the large tissue volumes generated from the tissue cytometer.

The difficulty of imaging into 3D tissue is overcome by the development of a number of new imaging technologies: confocal reflected light and fluorescent microscopy, reflected light optical coherence tomography, and two-photon fluorescence microscopy. These techniques have succeeded in imaging tissue down to a depth of 0.5 to 1 mm with resolution ranges from fraction of a micron to ten microns.

13.2.1. Methods for Morphological Characterization of Tissue

Reflected light techniques are very effective in quantifying the structure of tissues. Light reflection occurs at interfaces or heterogeneity where there is a difference in indexes of refraction between adjacent regions. The amount of light reflected increases with index difference. Two important tissue imaging techniques that can be applied for 3D tissue image cytometry are confocal reflected light microscopy and optical coherence tomography.

13.2.1.1. Confocal Reflected Light microscopy

In reflected light mode, confocal microscopes are very useful for the imaging of thick and highly scattering tissues. Reflected light confocal systems provide useful structural information. Biological interfaces, where there are index of refraction changes such as the plasma membrane and the nuclear envelop, can be visualized. Light from a laser is focused into a specimen through a microscope objective and is scanned across the sample via mechanical scanners. Reflected light is detected in the epi-geometry and is separated from the incident light by a dichroic mirror. The beam is "de-scanned" when it retraces its path through the mechanical scanners. A pinhole aperture is placed in a conjugate position of the object plane. Light that comes from the focal plane is focus at the aperture and is transmitted. Reflected light from out-of-focus regions is defocused at the aperture and is thus rejected. The confocal pinhole thus selects signal from the focal plane and provides depth discrimination [24-26]. The resolution of confocal reflected light imaging is typically on the order of 0.15 to 0.30 microns laterally and 0.50 to 1.0 microns axially. Confocal reflected light imaging is capable of imaging down to a depth of a few hundred microns [27-29].

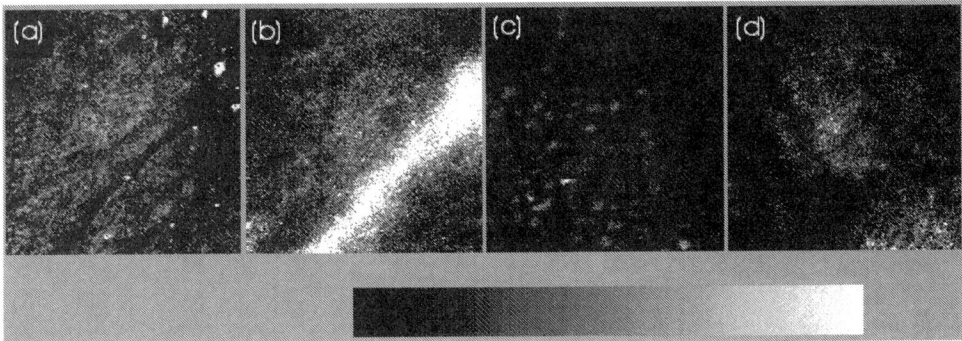

Fig. 13.1: Confocal reflected light microscopy images of ex vivo human skin. (a) stratum corneum, (b) stratum spinosum, (c) basal cells, (d) dermis.

This confocal principle has been exploited in a real-time, slit scanning, confocal microscope that is used clinically for optical biopsy and diagnostics in ophthalmology. The tandem-scanning confocal microscope, which operates in real-time, was developed by Petran and co-workers to image optical sections of thick, highly scattering biological materials[30]. Corcuff and co-workers have adapted the tandem scanning confocal microscope for in vivo skin imaging[31]. Stacks of optical sections from in vivo skin that were acquired with the tandem scanning reflected light confocal microscope can be visualized in three dimensions as orthogonal slices. Another implementation of reflected light scanning laser confocal microscope has been successful developed for video rate imaging of in vivo skin [28]. Image data demonstrating confocal reflected lighted imaging of ex vivo human skin is shown in Figure 13.1.

13.2.1.2. Optical Coherence of Tomography

Low-coherence interferometry is based on a Michelson interferometer in which the interferometer is illuminated using a low coherence source [32-34]. The interferometer measures the time delay of pulses by interfering the light that is backscattered from a sample with light that has traveled a known reference path length and thus a known time delay. A schematic of this system can be seen in Figure 13.2. The light from a source is divided into two paths by a beamsplitter. These two different paths can be referred to as the sample and reference arms. Light is retro-reflected from both the sample and the reference arms and is recombined by the same beamsplitter and directed into an optical detector. Interference is generated only if the distances that the light rays travel along the two arms are approximately the same, and their path length difference is within the coherence length of the light source. The coherence length of a light source is an inverse function of its spectral bandwidth. The coherence length of modern ultra-fast light source ranges from few microns up to tens of microns. By adjusting the length of the reference arm, signal from a specific plane in the specimen can be selected only when the optical path of the movable reference arm matches a reflecting surface in the specimen.

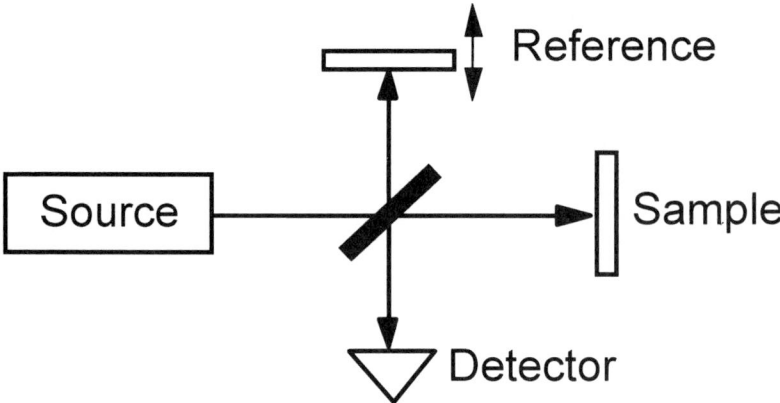

Fig. 13.2: A basic schematic of a low coherence tomography system. Light from the source is split by the dichoric mirror. The light reflected from the movable reference arm is recombined with the light reflected from the sample. Interferometric signal is detected.

Low-coherence interferometry forms the basis for an imaging technique called optical coherence tomography (OCT). OCT is an imaging technique where the reference mirror is scanned to reflecting surfaces at different depth of the specimen. The specimen is subsequently translated and a depth scan is again performed at the next location. By combining the depth scan results from consecutive locations, a sagittal image of the tissue can be produced. Optical coherence tomography typically has axial resolution of about 10 microns but can be as low as a few microns [35]. The imaging depth of this method can almost reach 1 mm in typical tissues.

13.2.2. Methods for Functional Characterization of Tissue

While reflected light imaging is very effective in obtaining morphological information from tissues. The utility of 3D tissue image cytometry is much reduced without the ability to obtain genetic, biochemical, and metabolic information from the specimen. Today, the most promising method to obtain functional information is fluorescence imaging. We discuss two modes of fluorescence imaging: confocal and two-photon microscopy. Other promising methods around the horizon that may be useful for obtaining functional information from tissues are second harmonic generation microscopy [36-44] and coherent anti-stoke Raman microscopy [45-47]. The prototype 3D tissue image cytometer developed in our laboratory is based on two-photon microscopy but other 3D imaging modality can be incorporated to provide complementary information.

13.2.2.1. Fluorescence Confocal Microscopy

Confocal microscopy can also be used in the detection of fluorescence signal. In fluorescence confocal microscopy, tissue is typically excited by UV or blue/green excitation and fluorescence emission in the visible spectral range is detected. There are three major limitations in performing fluorescence confocal microscopy in deep tissue. First, typical tissue endogenous chromophores absorb strongly in the UV and blue spectral regime. The penetration of the excitation light is very limited. Second, since much of the excitation light in the UV and blue spectral range is absorbed, there is significant photobleaching and photodamage throughout the tissue. Third, the scattering effect of tissue is an inverse function of wavelength. Therefore, the use of short wavelength excitation light results in further reduction of the penetration depth due to scattering effects. Today, fluorescence confocal microscope is not often used for deep tissue imaging. Experiments in human skin show that UV excitation has imaging depth on the order of 20-40 mm and increases to about 100 mm for green excitation [28].

13.2.2.2. Two-Photon Fluorescence Microscopy

Two-photon fluorescence microscope is typically preferred over fluorescence confocal microscopy for deep tissue imaging. Two-photon fluorescence microscopy was developed by Denk, Webb and co-workers in 1990. Today, two-photon microscopy is the method of choice for fluorescence microscopic deep tissue imaging with sub-micron resolution.

13.2.2.2a. Basic Two-photon Fluorescence Microscopy

Fluorophores can be excited by the simultaneous absorption of two photons each having half the energy needed for the excitation transition. Since the two-photon excitation probability is significantly less than the one-photon probability, two-photon excitation occurs only at appreciable rates in regions of high temporal and spatial photon concentration. The high spatial concentration of photons can be achieved by focusing the laser beam with a high numerical aperture objective to a diffraction-limited spot. The high temporal concentration of photons is made possible by the availability of high peak power mode-locked lasers. In general, two-photon excitation allows 3-D biological structures to be imaged with resolution comparable to confocal microscopes but with a number of significant advantages: (1) Conventional confocal techniques obtain 3-D resolution by using a detection pinhole to reject out of focal plane fluorescence. In contrast, two-photon excitation achieves a similar effect by limiting the excitation region to a sub-micron volume at the focal point. This capability of limiting the region of excitation instead of the region of detection is critical. Photo-damage of biological specimens is restricted to the focal point. Since out-of-plane chromophores are not excited, they are not subject to photobleaching. (2) Two-photon excitation wavelengths are typically red-shifted to about twice the one-photon excitation wavelengths. The significantly lower absorption and scattering coefficients ensure deeper tissue penetration. (3) The wide separation between the excitation and emission spectra ensures that the excitation light and the Raman scattering can be rejected without filtering out any of the fluorescence photons. This sensitivity enhancement improves the detection signal to background ratio.

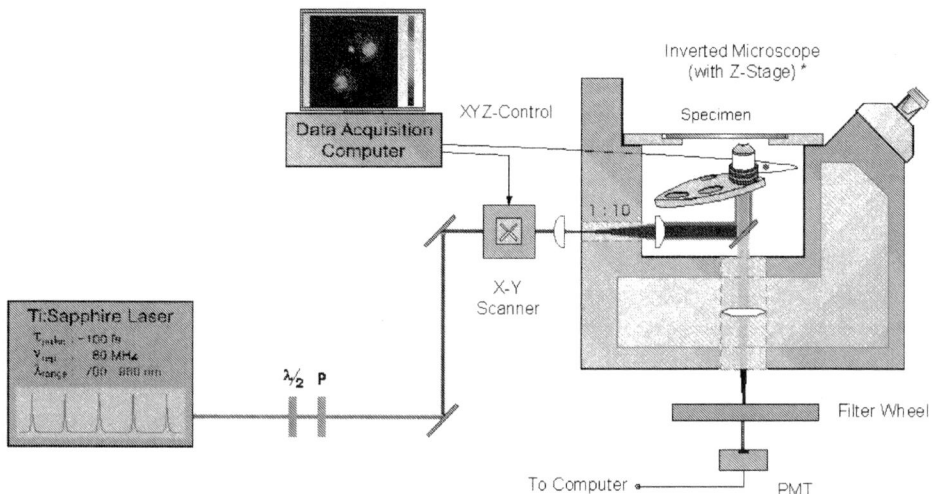

Fig. 13.3: A typical schematics of a two-photon microscope.

Depth discrimination is the most important feature of two-photon microscopy. For one-photon excitation in a spatially uniform fluorescent sample, equal fluorescence intensities are contributed from each z-section above and below the focal plane assuming negligible excitation attenuation. On the other hand, in the two-photon case over 80% of the total fluorescence intensity comes from a 1 μm thick region about the focal point for objectives with numerical aperture of 1.25. Thus, 3-D images can be constructed as in confocal microscopy, but without confocal pinholes. This depth discrimination effect of the two-photon excitation arises from the quadratic dependence of two-photon fluorescence intensity upon the excitation photon flux which decreases rapidly away from the focal plane. The spatial resolution of two-photon microscopy is comparable to one-photon methods. For the excitation of the same fluorophore, the two-photon resolution is roughly half the one-photon confocal resolution. This lower spatial resolution is due to the use longer wavelength light. For a 1.25 N.A. objective using excitation wavelength of 960 nm, the typical point spread function has FWHM of 0.3 µm in the radial direction and 0.9 µm in the axial direction. Two-photon excitation provides better suppression of higher order Airy rings.

The design of a basic two-photon excitation microscope has been described (Figure 13.3) [48]. A mode-locked infrared titanium-sapphire laser is used. The essential features are its high average power (1.5 W), its high repetition rate (80 MHz), and its short pulse width (150 fs). The high peak power optimizes two-photon excitation while minimizing scattered light contamination from the excitation source. This is particularly important in deep tissue study where excitation efficiency should be maximized but the power deposited in the tissue should be minimized. The beam expanded laser light is directed into the microscope via a galvanometer-driven x-y scanner. Images are generated by raster scanning the x-y mirrors. The excitation light enters a microscope via a modified epi-luminescence light path. The scan lens is positioned such that the x-y scanner is at its eye-point while the field aperture plane is at its focal point. Since the objectives are infinity-corrected, a tube lens is positioned to re-collimate the excitation light. The scan lens and the tube lens function together as a beam expander which over-fills the back aperture of the objective lens. Proper over-filling of the back aperture is required for the objective to achieve the manufacturer specified numerical aperture. The excitation light is reflected by the dichroic mirror to the objective. The dichroic mirrors are custom made short pass filters which maximize reflection in the infrared and transmission in the blue-green region of the spectrum. The objective used should feature high numerical aperture, water immersion, and large image area. Axial scan is achieved by driving the objective using a piezo-positioner. The fluorescence emission is collected by the same objective and transmitted through the dichroic mirror along the emission path. An additional barrier filter is needed to further attenuate the scattered excitation light because of the high excitation intensity used using suitable short pass filters such as BG39 Schott glass filter. The number of available fluorescence photons is always a limiting factor for deep tissue imaging. For deep tissue imaging, we implement a single photon counting signal detection system. The fluorescence signal at each pixel is detected by a high sensitivity photomultiplier tube (PMT). The PMT current pulses induced by individual photons are converted to digital electronic signal using a custom photon discriminator. The number of collected photons is counted using a home built interface circuit and the result is transferred to the data acquisition computer. Image is reconstructed from this digital data.

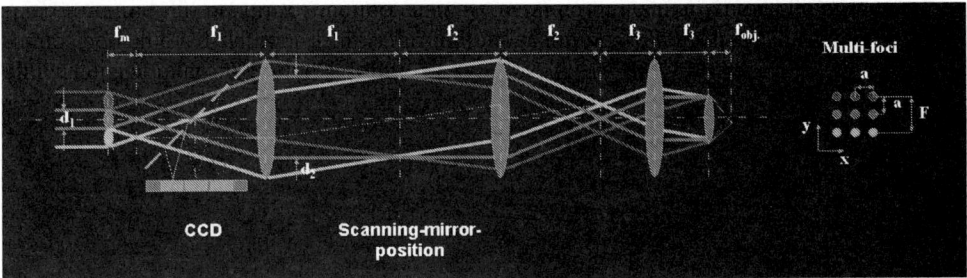

Fig. 13.4. Schematics of multi-focal multi-photon microscope using a lenslet array.

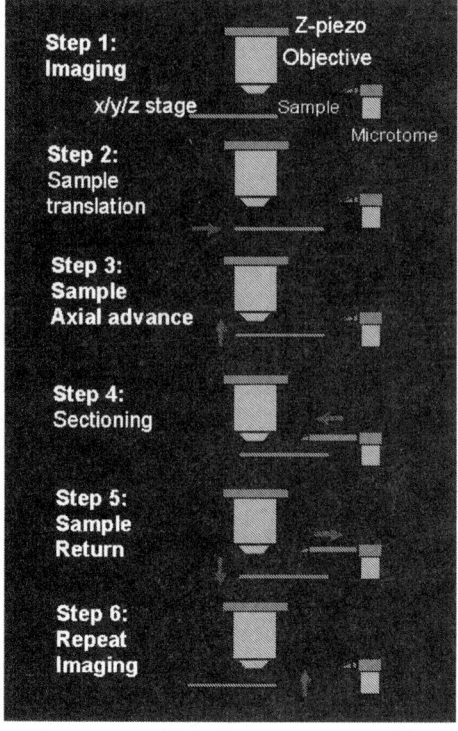

Fig. 13.5: Imaging and tissue removal steps using a microtomy system.

13.2.2.2b. High Speed Two-Photon Fluorescence Microscopy

A major limitation of using a standard two-photon microscope for 3D tissue image cytometry is its low speed which typically have frame rate of 0.5 Hz. A number of high speed two-photon microscopes have been developed to address this limitation. One approach to high speed two-photon imaging system is based on the line-scanning. Image acquisition time is reduced by covering the image plane with a line instead of a point [49]. The line focus is typically achieved by using a cylindrical element in the excitation beam path. The resulting fluorescent line image is acquired with a spatially resolved detector such as a CCD camera. The main drawback associated with line scanning is the inevitable degradation of the image resolution, especially in the axial direction. A second approach is based on high speed raster scanning of a single diffraction-limited spot by using either a high-speed resonance scanner [50] or rotating polygonal mirror [51]. For high speed, single focus scanning, a large, single-point detector, such as a photomultiplier tube or an avalanche photodiode, can be used. The spatial information is encoded by the timing design of the raster scan pattern. A third approach is multiphoton multifocal microscopy (MMM) [52, 53]. This approach is based projecting multiple two-photon excited foci in the specimen using a lenslet array element (Figure 13.4). The imaging speed is improved since multiple foci can be scanning in parallel. Further, since these foci can be made sufficiently spatially and temporally separated, this approach does not cause resolution degradation as in line scanning. A CCD camera is typically used to image the foci as they are scanned across the specimen. Multi-anode PMT can also be used as detector for MMM and is more effective than CCD system for deep tissue imaging.

13.2.3. Tissue Microtomy and Robotics Sample Handling

The penetration depth of two-photon microscope is bout 200 µm. However, for 3D tissue image cytometry, it is desirable to image deeper into specimens or even the whole organ of small animals. As an example, tumor metastasis has been studied in animal models and been shown to occur over great distances [54, 55]. Today, there is no optical solution for microscopic imaging of specimen with thickness on the order of centimeters. An effective solution for fixed or frozen specimen is the incorporation of an automated microtome into a high throughput multiphoton microscope. Specimen can be imaged up to a depth of 200-300 microns using multi-photon microscopy. For a fixed tissue block, the top layer that have been imaged can then be removed using the automatic microtome and exposes a new section for imaging. By repeating this process, it is possible to image through an entire fixed sample and obtain high-resolution 3D images of macroscopic volumes (Figure 13.5). A major limitation of the microtomy approach is that in vivo imaging is not possible. However, the ability to image tissue in vivo is often not critical for most tissue cytometry applications.

Another limitation of tissue image cytometry is the limited field of view of high numerical aperture objectives. The best commercial high numerical aperture objectives have field of view of a few hundred microns and up to about one millimeter. While this problem may be potentially addressed by developing custom large objectives with wider field of view, this solution requires significant research and development effort. A more practical alternative is the development of high speed robotic specimen stage that allows rapid and accurate translation of the specimen under the objective. Most commercial microscope specimen translation stages are based on servo or stepper motors that have settling time on the order of one second. Further, most of these stages lack feedback control and have positioning resolution on the order of microns. To address the limitation of these commercial systems, we have developed robotic specimen stages incorporating high speed linear motors, optical encoders, and digital feedback control loops. This system is capable of specimen position with settling time on the order of 5 ms and has positioning accuracy of 0.2 microns.

13.2.4. Image Visualization, Analysis and Quantification

Major stride has been made in the development of 3D tissue image cytometry hardware. As more advanced instruments are developed, the data acquisition rate of 3D tissue image cytometer significant increases. Consider the imaging of a small animal organ such as a mouse heart. The volume is this order is about 1 cm^3. If this volume is imaged at 1 micron resolution, the image data is on the order of a few Tbytes. Using MMM technology with high speed specimen stage, this volume can be imaged on the order of about five hours. This extremely high data rate presents significant computation challenge in terms of data storage and visualization. Data storage concerns can be partially alleviated by efficient data compression and the ever lowering cost of data storage medium. In contrast, data visualization of this large quantity of data is an even great challenge. Today, there is no image visualization system capable of real time rendering of this data volume. Unlike normal 3D rendering routines, the visualization of a data volume representing a whole organ will require rapid rendering of sub-volumes at different resolution hierarchy. Further, the development of dedicated hardware rendering engine is likely need for this visualization task.

While storage and visualization is difficult, they are not comparable to the challenge in pattern recognition, structural classification, data mining, and model building. Most of these pattern recognition and structural classification algorithms are computationally intensive even for small data set. The computational time required to process Tbyte size data is formidable. The challenge in data mining and model building further lies in our far from complete understanding on how molecular and cellular level processes affect the observed tissue structure and biochemistry.

However, for simple recognition and classification task, reasonable progress can be made. As an example, one of the typical tasks is to recognize, count, and map the distribution of fluorescent cells inside a tissue matrix. This task is critical in applications such as mapping metastatic cancer cells within tissues. Since the fluorescent label of these cells are known, the automatic recognition of these cells can be based on unique spectral signature of these cells. However, since tissues often have significant autofluorescence, segmentation and recognition algorithm based on simple spectral distinction is often insufficient. Instead a multi-dimensional classification algorithm is more robust. This multi-dimensional classification algorithm can be based on additional spectral characteristics, such as intensity level, or morphological information, such as cell size and shape (Figure 13.6).

13.3. EXPERIMENTAL REALIZATION OF HIGH THROUGHPUT TISSUE IMAGE CYTOMETRY

We have developed a high throughput tissue image cytometery based on coupling a multiphoton multifoci microscope with a microtome system and a high speed robotic specimen positioning stage. A number of promising proof-of-concept experiments are presented below.

13.3.1. Deep Tissue 3D Imaging

The potential of applying two-photon microscopy to study tissue physiology has been recognized since its inception [56]. Today, two-photon microscopy is particularly widely used in two areas: neurobiology [57-67] and embryology [68-71]. Imaging in neurobiology and embryology can rely on the use of extrinsic fluorescent probes. However, tissue imaging can also be performed base on intrinsic fluorescence [72-75]

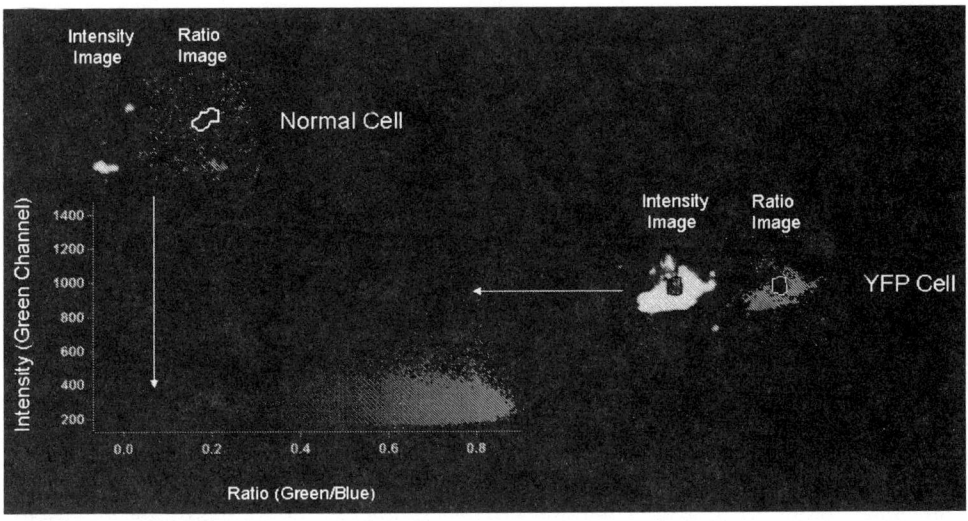

Fig. 13.6: The screen output an image classification routine for the detection of green fluorescent protein (GFP) expressing cells in mouse skin tissue. The nuclei of all the cells in the skin are further labeled with DAPI, a blue nucleic dye. GFP expressing cells can be selected based on its intensity and its spectral signature (Green/Blue ratio). GFP cells should be bright green objects whereas non-GFP expression cells should be bluer and dimmer. Each point in the plot represents an object after image segmentation. The position of the points in the plot provides information about the objects' intensity and color. For example, points at the lower left side of the plot represent dim blue objects whereas points on the upper right side of the plot present bright green objects. Therefore, the points on the upper right region of the plot are more likely to represent GFP cells. Representative images of objects corresponding to these points can be rapidly recalled from the original data and allow manual classification. Additionally, morphology based classification parameters such as cell size and cell shape can be used to as an aid in object classification

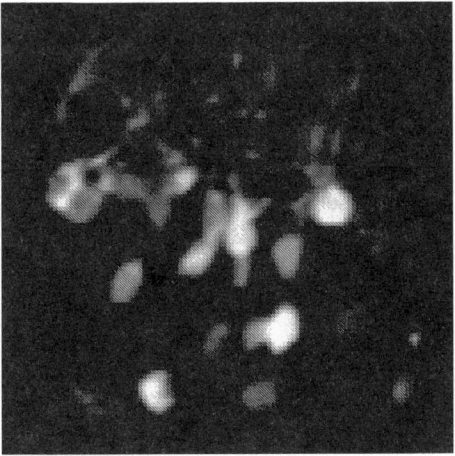

Fig. 13.7. GFP expressing neurons in a mouse brain slice were imaged at a depth exceeding 100 microns

Today, most deep tissue two-photon imaging experiments are based on single focus scanning system. We have chosen to implement our 3D tissue image cytometery based on MMM technology because of its suitability for high speed imaging. However, the applicability of MMM for deep tissue imaging has not been demonstrated. Since MMM requires the use of position resolved detectors such as CCDs or multi-anode PMTs, one expect that the scattering of the emission photons by the opaque tissue matrix would partially degrade the image projected on these position resolved detectors. Nevertheless, we have successfully demonstrated that MMM has adequate penetration depth for high speed tissue imaging. We have imaged green fluorescent protein expressing neurons in mouse brains using MMM down to a depth of over 100 microns (Fig. 7).

13.3.2. Quantitative Rare Cell Detection in 3D

We further characterize the ability of this tissue cytometery in quantify rare events in 3D environment. We prepared collagen matrix tissue samples seeded with two populations of mouse 3T3 NIH fibroblasts. One of the populations was labeled with the cytoplasm marker CellTracker Green, and the nuclei label Hoechst 33342. The other population was labeled with Hoechst 33342 only. Cell mixtures with varying relative concentrations ranging from 1:1 up to 1:10000 green cells per total cells were prepared. These cell preparations were then dispersed into a collagen matrix.

We imaged millimeter size areas of this collagen matrix seeded with cells down to a depth of 80 microns. This was accomplished by imaging individual image blocks with dimensions of 200 by 200 microns in the planer direction and 80 microns in the axial direction. The image resolution is about 1 micron laterally and 2 micron axially. An overlap of approximately 10 microns between image stacks was kept which allowed the images to be mosaic into a single large volumetric image. The emission light was separated into two channels using a dichroic mirror at 500 nm.

Figure 13.8a is a ratiometric image of the two channel data 20 μm into the surface of a 7 mm by 5.5 mm area of a collagen matrix which was seeded with 1 green cell per 1000 cells. Figure 13.8b shows an enlargement of a region within the larger image containing a single cell labeled with CellTracker green. Five samples were prepared at concentrations of 1:1, 1:10, 1:100, 1:1000, and 1:10000 green cells per total cells. The images were analyzed by computer using custom image analysis algorithms. Briefly, the images were first separated into their two component channels. A median filter was applied to the blue channel containing the cell nuclei, and then imaged was thresholded using an adaptive thresholding algorithm based on the method of Otsu[76]. A distance transform was then applied to the image in order to calculate the distance between each feature pixel and the nearest background pixel. This transformed image was again thresholded in order to identify markers for the nuclei. Finally, a labeling algorithm was applied to the image in order to count the number of disconnected components in the image and record their volumes. To identify the green cells, a median filter was applied to the second channel of the original image, and a ratiometric image was produced by ratioing the two median filtered channels. The green cells could then be identified and thresholded on the basis of their spectroscopic signature. A labeling algorithm was then used on the thresholded image to count the number of green cells in the image. Figure 13.8c shows a comparison between the measured green cell fractions with the expected mixing ratios.

13.4. AN APPLICATION IN CARDIAC HYPERTROPHY STUDY

13.4.1. Overview, Genetic Factors, and Histopathological Symptoms

The term "cardiac hypertrophy" is general and has many different manifestations[77, 78]. In this article, cardiac hypertrophy refers to an enlargement of all or part of the heart, which can arise from several different causes. Some causes of hypertrophy include genetic mutations (for example, familial hypertrophic cardiomyopathies), responses to pressure or volume overloads (such as hypertension or mitral valve prolapse), or loss of function of part of the heart (such as after myocardial infarction). Volume overload leads to longer myocytes, while pressure overload leads to thicker myocytes; pressure-overload hypertrophy is more commonly studied due to its higher morbidity and mortality.

Of recent interest are the genetic influences on the timing, severity and development of cardiac hypertrophy. For example, changes in expression patterns of desmin, lamin A/C, and α-myosin heavy chains are all known to lead to cardiomyopathies, which are typically accompanied by hypertrophy and fibrosis to varying degrees.

While overall heart enlargement can be determined using organ-scaled imaging such as x-rays, myoctye hypertrophy can be identified from histological sections derived from small heart biopsies. Some hypertrophied myocytes are thicker than those of non-hypertrophied myocytes, with the latter having average cross-sectional diameters of 5-25 microns, and some Purkinje Cells having cross-sectional diameters up to 40 microns

13.4.2. The Need for High Throughput Tissue Analysis

While molecular techniques for creating specific over- or under-expression of genes have advanced at a rapid pace, the modalities for interpreting the effects of these changes have not kept pace. Studies relying on established histological techniques are two-dimensional and used for an 'overall picture' rather than detailed quantization. Whole organ imaging techniques have low resolution (such as those obtained from x-rays or ultrasound) or limited specificity to markers of interest (magnetic resonance imaging). None of these techniques are suitable for generating three-dimensional high-resolution data necessary to detect morphological changes at the single cell level, especially if the changes are subtle.

Subtle effects that might arise from genetic alterations include changes in myocyte organization, such as the orientation of the myocytes, changes in nuclear structure and volume, and changes in cell-cell contact area. These alterations are not easily determined from two-dimensional sections, yet they require subcellular resolution. Therefore, the full benefit of the increasingly detailed molecular studies being performed cannot be achieved without a high-throughput scanning system that can image in three-dimensions at micron-leveled resolution.

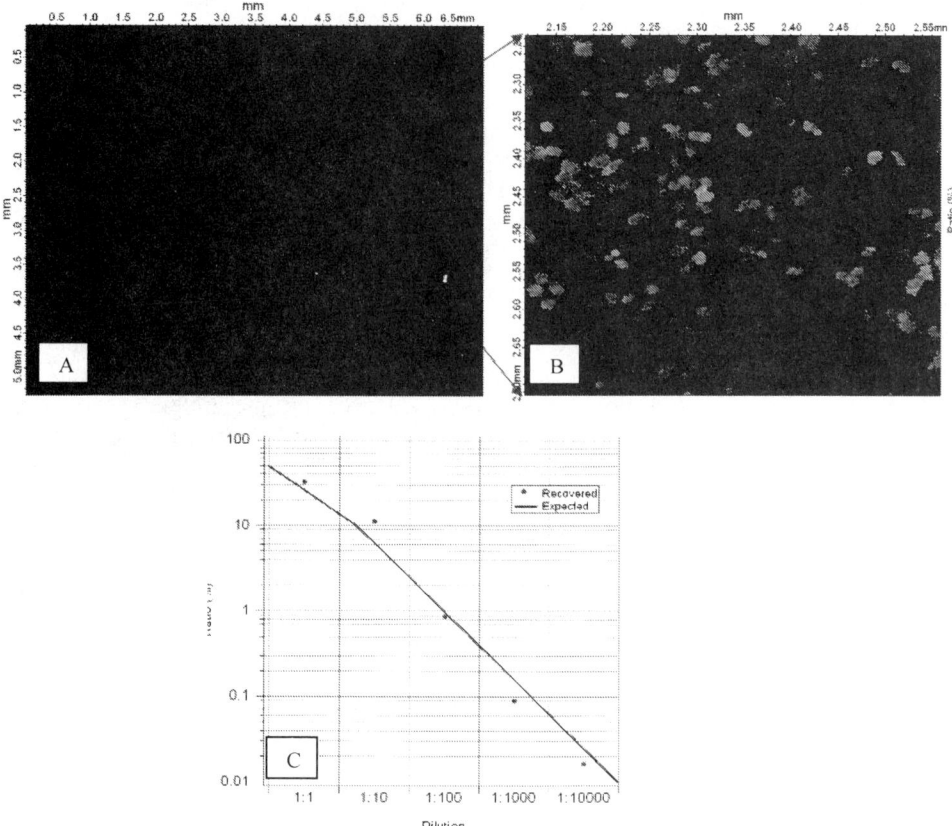

Fig. 8: Rare cell detection within a 3D collagen matrix. (a) A composite image was created by stitching together many smaller images. (b)an expanded view of a section of the original image. (c) The correspondence between the expected cell count and the measured cell count

13.4.3. Specimen Preparation

Mice were anesthetized and labeled using a tail-vein injection. For nuclear labeling, the cell-permeant Hoechst stain was used. For endothelial labeling, a fluorescently-tagged lectin was injected, and for the matrix between intact cells, a fluorescently-tagged maleimide was used. For two-channel scanning, only two of these labels were used. The fluorescent tags for lectin and maleimide were chosen to be spectrally different to minimize the amount of cross-labeling that would appear. After a suitable incubation period, the animals were sacrificed and the organs were harvested, fixed and embedded in paraffin. Standard histological slices were taken to expose the organ's region of interest and to assess the strength of the label.

Fig. 9: a) Mouse Heart tissue which has been intravital labeled with Hoechst 33342 for the nuclei and Texas Red labeled lectin for the blood vessel wall b) Mouse heart tissue which has been intravitally labeled with a Texas-Red allowing clear delineation of myocyte morphology.

13.4.4. High Throughput Imaging

We have applied 3D tissue image cytometer to accurate quantify myocyte morphological states in the mouse heart. Figure 13.9 is a 3D image of a section of a mouse heart that has been fixed with 4% paraformaldehyde. The heart has imaged at a resolution of 0.7 microns in the lateral direction and 2 microns in the axial direction (Figure 13.9a). The vasculature and the nuclei which have been stained via intravital labeling before the mouse was sacrificed using the nuclear stain DAPI and a lectin-Alexa 488 linker which stains the endothelial walls of the blood vessels. A small region outlined in red of the cross section is zoomed in to show the vasculature and nuclei. We are currently applying this technique to study the 3D morphology of individual myocytes. Myocytes have dimensions on the order of 100 microns which makes accurate measurements of their morphology extremely difficult or impossible using traditional 2D histopathology. Maleimide labeling are used to locate the boundaries of individual myocytes. Figure 13.9b is a cross sectional image of heart tissue stained with this protocol.

13.5. CONCLUSION AND OUTLOOK

3D tissue image cytometry has the potential to greatly enhance traditional histological practice. Major progress has been made in the imaging instrumentation. Whole organ of small animals can be imaged at micron resolution on the order of one day. Future development clearly will clearly further increase the throughput of this type of instrument. Today, the greater challenges of tissue cytometry lie in software development in visualization, image recognition, image classification, and data mining. With the availability of this class of instrument, we can now address interesting biomedical problems in areas spanning cardiology, cancer biology, and tissue engineering that require a quantification of the population statistics from cells inside their native tissue environment. The future lies in using this high throughput technology for understanding tissue physiology and pathology based on the underlying genetic and molecular expression profiles.

13.6. ACKNOWLEGEMENT

PTS, TR, KB, KHK, HSK acknowledge NIH supports: R21/R33 CA84740-01, P01 HL64858-01A1, R33 CA091354-01A1. TR and KB acknowledge additional NIH support:

13.7. REFERENCES

1. Nanda, K., D.C. McCrory, E.R. Myers, L.A. Bastian, V. Hasselblad, J.D. Hickey, and D.B. Matchar, Accuracy of the Papanicolaou test in screening for and follow-up of cervical cytologic abnormalities: a systematic review, *Ann Intern Med*, **132**(10), 810-9 (2000).
2. Sherman, M.E., M. Schiffman, R. Herrero, D. Kelly, C. Bratti, L.J. Mango, M. Alfaro, M.L. Hutchinson, F. Mena, A. Hildesheim, J. Morales, M.D. Greenberg, I. Balmaceda, and A.T. Lorincz, Performance of a semiautomated Papanicolaou smear screening system: results of a population-based study conducted in Guanacaste, Costa Rica, *Cancer*, **84**(5), 273-80 (1998).
3. Tumer, K., N. Ramanujam, J. Ghosh, and R. Richards-Kortum, Ensembles of radial basis function networks for spectroscopic detection of cervical precancer, *IEEE Trans Biomed Eng*, **45**(8), 953-61 (1998).
4. Kok, M.R. and M.E. Boon, Consequences of neural network technology for cervical screening: increase in diagnostic consistency and positive scores, *Cancer*, **78**(1), 112-7 (1996).
5. Koss, L.G., Cervical (Pap) smear. New directions, *Cancer*, **71**(4 Suppl), 1406-12 (1993).
6. Rennie, J., Cancer catcher. Neural net catches errors that slip through Pap tests, *Sci Am*, **262**(5), 84 (1990).
7. Darzynkiewicz, Z., E. Bedner, X. Li, W. Gorczyca, and M.R. Melamed, Laser-scanning cytometry: A new instrumentation with many applications, *Exp Cell Res*, **249**(1), 1-12 (1999).
8. Kamentsky, L.A. and L.D. Kamentsky, Microscope-based multiparameter laser scanning cytometer yielding data comparable to flow cytometry data, *Cytometry*, **12**(5), 381-7 (1991).
9. Kamentsky, L.A., D.E. Burger, R.J. Gershman, L.D. Kamentsky, and E. Luther, Slide-based laser scanning cytometry, *Acta Cytol*, **41**(1), 123-43 (1997).
10. Kamentsky, L.A., L.D. Kamentsky, J.A. Fletcher, A. Kurose, and K. Sasaki, Methods for automatic multiparameter analysis of fluorescence in situ hybridized specimens with a laser scanning cytometer, *Cytometry*, **27**(2), 117-25 (1997).
11. Kamentsky, L.A., Laser scanning cytometry, *Methods Cell Biol*, **63**, 51-87 (2001).
12. Bajaj, S., J.B. Welsh, R.C. Leif, and J.H. Price, Ultra-rare-event detection performance of a custom scanning cytometer on a model preparation of fetal nRBCs, *Cytometry*, **39**(4), 285-94 (2000).
13. Rieseberg, M., C. Kasper, K.F. Reardon, and T. Scheper, Flow cytometry in biotechnology, *Appl Microbiol Biotechnol*, **56**(3-4), 350-60 (2001).
14. Bennett, B.D., T.L. Jetton, G. Ying, M.A. Magnuson, and D.W. Piston, Quantitative subcellular imaging of glucose metabolism within intact pancreatic islets, *J. Biol. Chem.*, **271**(7), 3647-51 (1996).
15. Piston, D.W., S.M. Knobel, C. Postic, K.D. Shelton, and M.A. Magnuson, Adenovirus-mediated knockout of a conditional glucokinase gene in isolated pancreatic islets reveals an essential role for proximal metabolic coupling events in glucose-stimulated insulin secretion, *J. Biol. Chem.*, **274**(2), 1000-4 (1999).
16. Gorczyca, W., Z. Darzynkiewicz, and M.R. Melamed, Laser scanning cytometry in pathology of solid tumors. A review, *Acta Cytol*, **41**(1), 98-108 (1997).
17. Gorczyca, W., V. Sarode, G. Juan, M.R. Melamed, and Z. Darzynkiewicz, Laser scanning cytometric analysis of cyclin B1 in primary human malignancies, *Mod Pathol*, **10**(5), 457-62 (1997).
18. Hendricks, J.B., Quantitative histology by laser scanning cytometry, *J. Histotechnology*, **24**, 59-62 (2001).

19. Hendricks, C.A., K.H. Almeida, M.S. Stitt, V.S. Jonnalagadda, R.E. Rugo, G.F. Kerrison, and B.P. Engelward, Spontaneous mitotic homologous recombination at an enhanced yellow fluorescent protein (EYFP) cDNA direct repeat in transgenic mice, *Proc Natl Acad Sci U S A*, **100**(11), 6325-30 (2003).

20. Liotta, L.A. and E.C. Kohn, The microenvironment of the tumour-host interface, *Nature*, **411**(6835), 375-9 (2001).

21. Leethanakul, C., V. Knezevic, V. Patel, P. Amornphimoltham, J. Gillespie, E.J. Shillitoe, P. Emko, M.H. Park, M.R. Emmert-Buck, R.L. Strausberg, D.B. Krizman, and J.S. Gutkind, Gene discovery in oral squamous cell carcinoma through the Head and Neck Cancer Genome Anatomy Project: confirmation by microarray analysis, *Oral Oncol*, **39**(3), 248-58 (2003).

22. Knezevic, V., C. Leethanakul, V.E. Bichsel, J.M. Worth, V.V. Prabhu, J.S. Gutkind, L.A. Liotta, P.J. Munson, E.F. Petricoin, 3rd, and D.B. Krizman, Proteomic profiling of the cancer microenvironment by antibody arrays, *Proteomics*, **1**(10), 1271-8 (2001).

23. Gohongi, T., D. Fukumura, Y. Boucher, C.O. Yun, G.A. Soff, C. Compton, T. Todoroki, and R.K. Jain, Tumor-host interactions in the gallbladder suppress distal angiogenesis and tumor growth: involvement of transforming growth factor beta1, *Nat Med*, **5**(10), 1203-8 (1999).

24. Wilson, T., *Confocal Microscopy*. 1990, London: Academic Press.

25. Wilson, T. and C.J.R. Sheppard, *Theory and Practice of Scanning Optical Microscopy*. 1984, New York: Academic Press.

26. Masters, B.R., *Selected papers on confocal microscopy*. 1996, Bellingham: SPIE.

27. Rajadhyaksha, M., M. Grossman, D. Esterowitz, R.H. Webb, and R.R. Anderson, In vivo confocal scanning laser microscopy of human skin: melanin provides strong contrast, *J Invest Dermatol*, **104**(6), 946-52 (1995).

28. Masters, B.R., Three-dimensional confocal microscopy of human skin in vivo: autofluorescence of normal skin, *Bioimages*, **4**, 13-19 (1996).

29. Masters, B.R., Three-dimensional confocal microscopy of the living in situ rabbit cornea, *Optics Express*, **3**(9) (1998).

30. Petran, M., M. Hadravsky, M.D. Egger, and R. Galambos, Tandem scanning reflected light microscope, *J. Opt. Soc. Am.*, **58**, 661-664 (1958).

31. Corcuff, P., C. Bertrand, and J.L. Leveque, Morphometry of human epidermis in vivo by real-time confocal microscopy, *Arch Dermatol Res*, **285**(8), 475-81 (1993).

32. Fujimoto, J.G., M.E. Brezinski, G.J. Tearney, S.A. Boppart, B. Bouma, M.R. Hee, J.F. Southern, and E.A. Swanson, Optical biopsy and imaging using optical coherence tomography, *Nat Med*, **1**(9), 970-2 (1995).

33. Hee, M.R., C.A. Puliafito, C. Wong, J.S. Duker, E. Reichel, J.S. Schuman, E.A. Swanson, and J.G. Fujimoto, Optical coherence tomography of macular holes, *Ophthalmology*, **102**(5), 748-56 (1995).

34. Schuman, J.S., M.R. Hee, A.V. Arya, T. Pedut-Kloizman, C.A. Puliafito, J.G. Fujimoto, and E.A. Swanson, Optical coherence tomography: a new tool for glaucoma diagnosis, *Curr Opin Ophthalmol*, **6**(2), 89-95 (1995).

35. Fujimoto, J.G., Optical coherence tomography for ultrahigh resolution in vivo imaging, *Nat Biotechnol*, **21**(11), 1361-7 (2003).

36. Pons, T., L. Moreaux, O. Mongin, M. Blanchard-Desce, and J. Mertz, Mechanisms of membrane potential sensing with second-harmonic generation microscopy, *J Biomed Opt*, **8**(3), 428-31 (2003).

37. Moreaux, L., T. Pons, V. Dambrin, M. Blanchard-Desce, and J. Mertz, Electro-optic response of second-harmonic generation membrane potential sensors, *Opt Lett*, **28**(8), 625-7 (2003).

38. Yang, C. and J. Mertz, Transmission confocal laser scanning microscopy with a virtual pinhole based on nonlinear detection, *Opt Lett*, **28**(4), 224-6 (2003).

39. Brown, R.M., Jr., A.C. Millard, and P.J. Campagnola, Macromolecular structure of cellulose studied by second-harmonic generation imaging microscopy, *Opt Lett*, **28**(22), 2207-9 (2003).

40. Campagnola, P.J. and L.M. Loew, Second-harmonic imaging microscopy for visualizing biomolecular arrays in cells, tissues and organisms, *Nat Biotechnol*, **21**(11), 1356-60 (2003).

41. Mohler, W., A.C. Millard, and P.J. Campagnola, Second harmonic generation imaging of endogenous structural proteins, *Methods*, **29**(1), 97-109 (2003).

42. Campagnola, P.J., A.C. Millard, M. Terasaki, P.E. Hoppe, C.J. Malone, and W.A. Mohler, Three-dimensional high-resolution second-harmonic generation imaging of endogenous structural proteins in biological tissues, *Biophys J*, **82**(1 Pt 1), 493-508 (2002).

43. Campagnola, P.J., H.A. Clark, W.A. Mohler, A. Lewis, and L.M. Loew, Second-harmonic imaging microscopy of living cells, *J Biomed Opt*, **6**(3), 277-86 (2001).

44. Campagnola, P.J., M.D. Wei, A. Lewis, and L.M. Loew, High-resolution nonlinear optical imaging of live cells by second harmonic generation, *Biophys J*, **77**(6), 3341-9 (1999).

45.   Nan, X., J.X. Cheng, and X.S. Xie, Vibrational imaging of lipid droplets in live fibroblast cells with coherent anti-Stokes Raman scattering microscopy, *J Lipid Res*, **44**(11), 2202-8 (2003).

46.   Cheng, J.X., S. Pautot, D.A. Weitz, and X.S. Xie, Ordering of water molecules between phospholipid bilayers visualized by coherent anti-Stokes Raman scattering microscopy, *Proc Natl Acad Sci U S A*, **100**(17), 9826-30 (2003).

47.   Cheng, J.X., Y.K. Jia, G. Zheng, and X.S. Xie, Laser-scanning coherent anti-Stokes Raman scattering microscopy and applications to cell biology, *Biophys J*, **83**(1), 502-9 (2002).

48.   So, P.T., C.Y. Dong, B.R. Masters, and K.M. Berland, Two-photon excitation fluorescence microscopy, *Annu Rev Biomed Eng*, **2**, 399-429 (2000).

49.   Guild, J.B. and W.W. Webb, Line scanning microscopy with two-photon fluorescence excitation, *Biophys J*, **68**, 290a (1995).

50.   Fan, G.Y., H. Fujisaki, A. Miyawaki, R.K. Tsay, R.Y. Tsien, and M.H. Ellisman, Video-rate scanning two-photon excitation fluorescence microscopy and ratio imaging with cameleons, *Biophys. J.*, **76**(5), 2412-20 (1999).

51.   Kim, K.H., C. Buehler, and P.T.C. So, High-speed, two-photon scanning microscope, *Appl. Opt.*, **38**(28), 6004-9 (1999).

52.   Bewersdorf, J., R. Pick, and S.W. Hell, Mulitfocal multiphoton microscopy, *Opt. Lett.*, **23**, 655-657 (1998).

53.   Brakenhoff, G.J., J. Squier, T. Norris, A.C. Bliton, M.H. Wade, and B. Athey, Real-time two-photon confocal microscopy using a femtosecond, amplified Ti:sapphire system, *J Microsc*, **181**(Pt 3), 253-9 (1996).

54.   Hoffman, R., Green fluorescent protein imaging of tumour growth, metastasis, and angiogenesis in mouse models, *Lancet Oncol*, **3**(9), 546-56 (2002).

55.   Hoffman, R.M., Visualization of GFP-expressing tumors and metastasis in vivo, *Biotechniques*, **30**(5), 1016-22, 1024-6 (2001).

56.   Denk, W., J.H. Strickler, and W.W. Webb, Two-photon laser scanning fluorescence microscopy, *Science*, **248**(4951), 73-6 (1990).

57.   Denk, W., K.R. Delaney, A. Gelperin, D. Kleinfeld, B.W. Strowbridge, D.W. Tank, and R. Yuste, Anatomical and functional imaging of neurons using 2-photon laser scanning microscopy, *J. Neurosci. Methods*, **54**(2), 151-62 (1994).

58.   Fetcho, J.R. and D.M. O'Malley, Imaging neuronal networks in behaving animals, *Curr. Opin. Neurobiol.*, **7**(6), 832-8 (1997).

59.   Yuste, R. and W. Denk, Dendritic spines as basic functional units of neuronal integration, *Nature*, **375**(6533), 682-4 (1995).

60.   Yuste, R., A. Majewska, S.S. Cash, and W. Denk, Mechanisms of calcium influx into hippocampal spines: heterogeneity among spines, coincidence detection by NMDA receptors, and optical quantal analysis, *J Neurosci.*, **19**(6), 1976-87 (1999).

61.   Denk, W., M. Sugimori, and R. Llinas, Two types of calcium response limited to single spines in cerebellar Purkinje cells, *Proc. Natl. Acad. Sci. U S A*, **92**(18), 8279-82 (1995).

62.   Svoboda, K., W. Denk, D. Kleinfeld, and D.W. Tank, In vivo dendritic calcium dynamics in neocortical pyramidal neurons, *Nature*, **385**(6612), 161-5 (1997).

63.   Svoboda, K., F. Helmchen, W. Denk, and D.W. Tank, Spread of dendritic excitation in layer 2/3 pyramidal neurons in rat barrel cortex in vivo, *Nat. Neurosci.*, **2**(1), 65-73 (1999).

64.   Helmchen, F., K. Svoboda, W. Denk, and D.W. Tank, In vivo dendritic calcium dynamics in deep-layer cortical pyramidal neurons, *Nat. Neurosci.*, **2**(11), 989-96 (1999).

65.   Shi, S.H., Y. Hayashi, R.S. Petralia, S.H. Zaman, R.J. Wenthold, K. Svoboda, and R. Malinow, Rapid spine delivery and redistribution of AMPA receptors after synaptic NMDA receptor activation, *Science*, **284**(5421), 1811-6 (1999).

66.   Mainen, Z.F., R. Malinow, and K. Svoboda, Synaptic calcium transients in single spines indicate that NMDA receptors are not saturated, *Nature*, **399**(6732), 151-5 (1999).

67.   Maletic-Savatic, M., R. Malinow, and K. Svoboda, Rapid dendritic morphogenesis in CA1 hippocampal dendrites induced by synaptic activity, *Science*, **283**(5409), 1923-7 (1999).

68.   Summers, R.G., D.W. Piston, K.M. Harris, and J.B. Morrill, The orientation of first cleavage in the sea urchin embryo, Lytechinus variegatus, does not specify the axes of bilateral symmetry, *Dev. Biol.*, **175**(1), 177-83 (1996).

69.   Mohler, W.A. and J.G. White, Stereo-4-D reconstruction and animation from living fluorescent specimens, *Biotechniques*, **24**(6), 1006-10, 1012 (1998).

70.   Mohler, W.A., J.S. Simske, E.M. Williams-Masson, J.D. Hardin, and J.G. White, Dynamics and ultrastructure of developmental cell fusions in the Caenorhabditis elegans hypodermis, *Curr. Biol.*, **8**(19), 1087-90 (1998).

71. Squirrell, J.M., D.L. Wokosin, J.G. White, and B.D. Bavister, Long-term two-photon fluorescence imaging of mammalian embryos without compromising viability, *Nat. Biotechnol.*, **17**(8), 763-7 (1999).

72. Piston, D.W., B.R. Masters, and W.W. Webb, Three-dimensionally resolved NAD(P)H cellular metabolic redox imaging of the in situ cornea with two-photon excitation laser scanning microscopy, *J. Microsc.*, **178**(Pt 1), 20-7 (1995).

73. So, P.T.C., H. Kim, and I.E. Kochevar, Two-photon deep tissue ex vivo imaging of mouse dermal and subcutaneous structures, *Opt. Exp.*, **3**(9), 339-50 (1998).

74. Masters, B.R., P.T. So, and E. Gratton, Multiphoton excitation fluorescence microscopy and spectroscopy of in vivo human skin, *Biophys. J.*, **72**(6), 2405-12 (1997).

75. Kim, K.H., P.T.C. So, I.E. Kochevar, B.R. Masters, and E. Gratton, Two-photon fluorescence and confocal reflected light imaging of thick tissue structures, *SPIE Proc.*, **3260**, 46-57 (1998).

76. Otsu, N., Threshold Selection Method from Gray-Level Histograms, *IEEE Trans Systems Man and Cybernetics*, **9**, 62-66 (1979).

77. Dorn, G.W., 2nd, J. Robbins, and P.H. Sugden, Phenotyping hypertrophy: eschew obfuscation, *Circ Res*, **92**(11), 1171-5 (2003).

78. Dorn, G.W., 2nd and H.S. Hahn, Genetic factors in cardiac hypertrophy, *Ann N Y Acad Sci*, **1015**, 225-37 (2004).

WHITHER FLUORESCENCE BIOSENSORS?[1]

Richard Thompson[2]

14.1. INTRODUCTION

Over the past couple of decades, fluorescence-based biosensors have emerged as a powerful means for performing chemical analyses. From our standpoint, fluorescence-based biosensors use a biological (or biomimetic) molecule to transduce the presence or level of an analyte as a change in some fluorescence observable such as intensity ratio or lifetime. In general we use the term "sensor" to mean an indicator system that does not consume a reagent (or the analyte) like an enzyme substrate, and thus in principle can measure the analyte continuously over a period of time (Thompson and Walt 1994). The majority of such sensors bind the analyte reversibly, and at equilibrium the fractional occupancy of the binding site is a simple function of the analyte activity. The purpose of this chapter is to identify issues and opportunities for the development of fluorescence biosensors, both as research tools and for practical applications, especially in the clinical realm. It is not intended as a comprehensive review: several recent and forthcoming articles and books (including this one) serve this purpose admirably (Lakowicz 1999; Wolfbeis 1991) (Thompson 2004). Thus we will highlight selected developments in recognition chemistry, transduction mechanisms, and fluorescent labels; as well as new target analytes, techniques for *in vivo* measurements, new imaging approaches, and new opportunities in clinical analysis.

14.2. STATE OF THE ART

It is our contention that fluorescence sensors based on biomolecule recognition represent the best hope for analysis of many targets, particularly in complex matrices such as the interiors of cells. While the most popular biosensor is the electrochemical glucose sensor based on glucose oxidase, the capability of coupling chemical information with spatial (anatomical) information offered by fluorescence imaging techniques (like microscopy) confers an overwhelming advantage over other transduction modes for research applications.

---

[1] With apologies to the shade of Gregorio Weber and his "Whither Biophysics?" in the *Annual Reviews of Biophysics and Biophysical Chemistry*, 1990, which served as an inspiration.

[2] University of Maryland School of Medicine, Baltimore, Maryland 21201.

While many powerful techniques exist for chemical analysis in various media such as ICP-MS, HPLC, and electrophoresis, nearly all are "single pixel" and offer spatial resolution only with difficulty, if at all.

While many fluorescence sensors based on non-biomimetic recognition have been described and are being developed (Fernandez-Gutierrez and Munoz de la Pena 1985; Haugland 1996; White and Argauer 1970) , sensors based on biological molecules offer advantages difficult to secure otherwise; we will generally refer to the former as "small molecule" sensors to distinguish them. There are three main reasons for this. First is selectivity: biological molecules such as antibodies and enzymes offer (and have demonstrated) unmatched ability to selectively recognize and bind analytes of all kinds in the presence of myriad potential interferents. For instance, carbonic anhydrase can quantitate subpicomolar levels of free Cu(II) in a matrix as complex as sea water, which contains potentially interfering cations such as Ca or Mg at billion-fold higher levels (Zeng et al. 2003). While some techniques listed above exhibit this level of sensitivity, they typically require significant sample pretreatment to work in complex media. Molecular beacons and other fluorescent nucleic acid probes routinely identify target sequences of nucleic acids in the presence of literally billions of nucleotides of background. Of course, excellent small molecule fluorescent indicators are known for many analytes such as Ca (Grynkiewicz et al. 1985), glucose (DiCesare and Lakowicz 2002), pH (Haugland 1996), and oxygen , but these tend to be analytes that are present at relatively high to medium levels (millimolar to micromolar, with Ca down to nanomolar), whereas there are no small molecule indicators at all for many analytes of interest (for example leukotrienes, most saccharides, most nucleotides), especially those likely to be present at trace levels. The second is adaptability: biological recognition molecules can often be readily modified to suit the user's requirements for a particular application, while this may be difficult or infeasible altogether for small molecules. For instance, several groups have used combinatorial synthesis approaches with proteins or RNA aptamers to modify or improve sensitivity and selectivity (Hunt and Fierke 1997) By comparison (although combinatorial approaches are well known for small molecules (Thompson and Ellman 1996)), it would appear difficult to construct usefully large libraries of compounds (for instance) closely related to Fura-2, which are tightly constrained in their structure by the need to maintain selectivity in cation binding as well as fluorescence transduction. Indeed, Fierke and her colleagues showed that mutagenic approaches (both directed and combinatorial) could improve the kinetics of analyte binding to a protein transducer up to ten thousand-fold (Huang et al. 1996), which for small molecule recognition chemistry such as cyclophanes for metals would appear very difficult to implement synthetically. A third advantage is flexibility in transduction: in small molecule indicators the fluorophore is usually intimately coupled with the recognition moiety to generate a useful fluorescence change; a good portion of the art of fluorescent indicator design and synthesis goes into assuring this coupling. By comparison, with macromolecule systems it is often straightforward to engineer transduction by resonance energy transfer (Forster 1948)to produce substantial changes in fluorescence intensity, (Miyawaki et al. 1997) , intensity ratio (Thompson et al. 2002), anisotropy (Thompson et al. 1998), or lifetime (Ozinskas et al. 1993; Thompson et al. 1996b). An additional key advantage for biologically-based sensors was demonstrated by Tsien and his colleagues with a protein-based biosensor which could be expressed within a living cell (Miyawaki et al. 1997). They fused genes for blue and green-emitting variants of the green fluorescent protein (GFP) to the N– and C-termini of calmodulin, which undergoes a large conformational change upon binding calcium ion; the conformational change changes the inter-fluorophore distance and consequently the energy transfer efficiency (**see Figure 14.1**). Other attempts to implement this basic approach have been less successful (Pearce et al. 2000), but the possibilities of this method are striking.

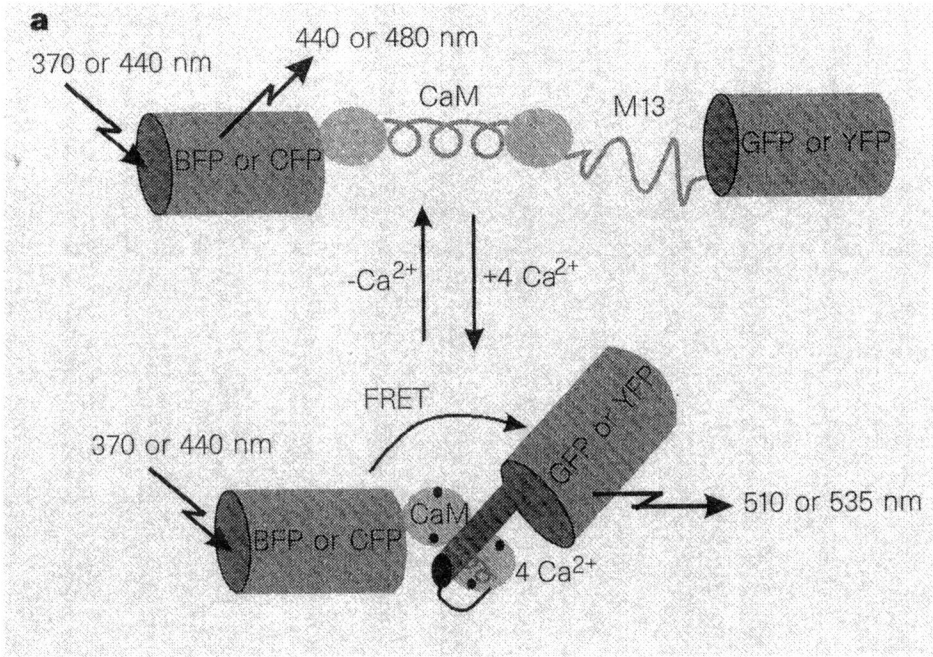

Figure 14.1. FRET-based Ca sensor of Miyawaki, et al.

14.3. NEW RECOGNITION CHEMISTRY

As the palette of known protein and other macromolecule structures expands, we have a correspondingly larger choice of potential binding sites for a variety of potential analytes. Added to this is the power of new software for QSAR (Quantitative Structure-Activity Relationships), which enable medicinal chemists to predict the affinity of different pharmacophores for particular receptors (of known structure). For the analyst the problem is inverted, in that one has a small molecule analyte for which one must find or devise a receptor. While in principle one can always raise an antibody to such a hapten or antigen, antibodies (particularly monoclonals) may not have high affinity; the target is often toxic itself, metabolizable, or otherwise unsuited for raising antibodies; and one is still left with the problem of transducing the binding event as a change in fluorescence. Hellinga's group has gone a long way to finding a fairly general approach to the problem of generating recognition molecules to essentially arbitrary target molecules (Looger et al. 2003); for this effort they made receptors (based on the maltose binding protein platform (Li and Cass 1991)) which recognized analytes as diverse as trinitrotoluene (an explosive), lactate, and serotonin. While the affinities for the analytes were for the most part modest (micromolar range), the basic computational approach provides an entry into a receptor (particularly for non-natural analytes). Furthermore, there is the prospect of improving the receptor by designed changes as well as combinatorial (random or quasi-random) mutation; such sensor proteins are also excellent candidates for molecular evolution, particularly for unnatural targets not previously "seen" by the proteome.

Another intriguing new result in recognition chemistry is the Suslick group's development of porphyrin derivatives with unusual affinity and selectivity for small neutral species (Rakow and Suslick 2000). Such analytes (being small) do not make many Van der Waals contacts and may only make a single hydrogen bond, so they don't bind most proteins with much affinity,... Consequently, typical polypeptides bind these ligands with poor affinity, if at all. Rakow and Suslick recognized that some of these targets bind as ligands to porphyrin derivatives with good affinity and accompanying color changes, which led them to develop a colorimetric array detector. While some of these porphyrin derivatives may also be fluorescent and thus directly applicable to a fluorescence sensor, of greater interest is the new accessibility and prospective transduction of analytes for which binding proteins are unlikely to be available, and for which even the approach of Looger, et al., is unlikely to provide high affinity and selectivity receptors.

14.4. NEW TRANSDUCTION MECHANISMS

From our perspective, one of the areas in fluorescence biosensing that has required and will continue to require substantial effort is transduction, which is the means by which the presence or level of the analyte is transduced as a change in fluorescence. As described above, finding or making a biomolecule that recognizes and perhaps binds to a target molecule can be difficult, but at least some potential approach to isolating such macromolecules is usually apparent, often because the target must be recognized by some biomolecule to be of concern for toxicity reasons. By comparison, transduction is often an issue. Of course, if the analyte is itself fluorescent transduction is usually straightforward (Tromberg et al. 1987), but such cases are the exception. Many means of transducing recognition have been described in the literature, but some are indirect, require consumption of a reagent, and/or produce only a change in intensity which compromises accuracy and complicates calibration (Thompson and Walt 1994).

For instance, the well-known fluorescence polarization immunoassays (**Figure 14.2**) are competitive assays requiring a known amount of fluorescent-labeled antigen be present to compete with the unknown amount in the sample. The fact that both the antigen from the sample and fluorescent-labeled antigen must be able to diffuse to and from the antibody makes constructing a continuously responding sensor difficult, since the labeled antigen can diffuse away by the same pathway the unknown antigen diffused in on. Similarly, enzyme-based fluorescence assays which essentially measure the altered fluorescence of a product of a reaction and relate it to the level of the analyte (which may itself be the enzyme, a substrate, or a modulator of the enzyme activity (Spencer et al. 1973)) also rely on a separate reagent, which is likely to be consumed. This reliance on a second chemical entity is a general problem, and approaches which eschew a second, diffusing component (particularly if consumed) are termed "reagentless"and are preferred.

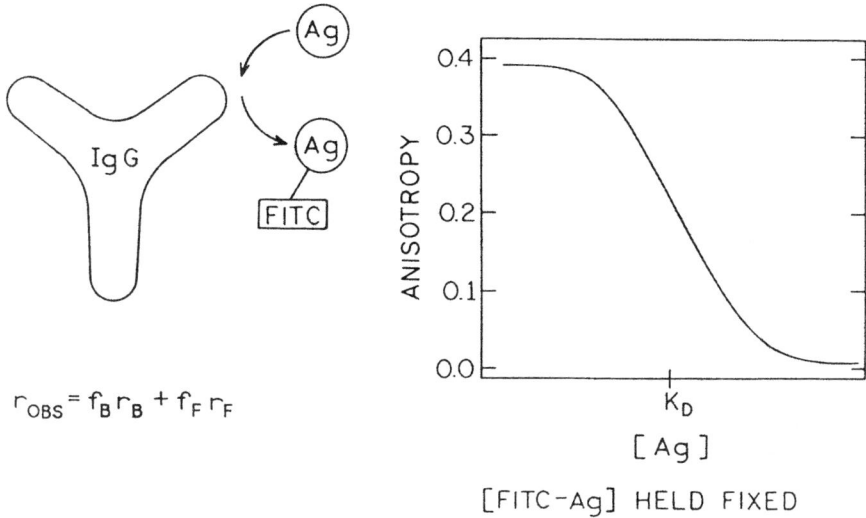

$$r_{OBS} = f_B r_B + f_F r_F$$

Figure 14.2. Fluorescence polarization immunoassay. FITC-labeled antigen competes with unlabeled antigen for IgG binding sites; as the proportion of free FITC-Ag increases, the anisotropy (polarization) drops.

In many cases, binding of the analyte (or its mere presence, in the case of oxygen (Demas and DeGraff 2001)) causes quenching., which is related to the analyte concentration trivially by the Stern-Volmer relation, and is measured as a change in intensity or lifetime. Of course, the binding of ligands to proteins often quenches the intrinsic tryptophanyl fluorescence, but this is of little practical use for sensing due to the requirement for deep UV excitation, interfering fluorescence in natural specimens, and the typically modest changes observed since the ligand binding affects only the minority of tryptophans in the quencher's immediate vicinity. In this static quenching case, if binding of the analyte causes complete quenching the intensity declines but there is no lifetime change. Attaching extrinsic labels to recognition molecules has proven more successful, because the labels are much better fluorophores than tryptophan. In these cases the quenching process may be resonance energy transfer (Ullman and Schwarzberg 1981), which has the important attribute that quenching efficiency (and thus response) can be engineered by selecting the spectral overlap and spatial positioning of donor and acceptor; with protein-based recognition in particular this is easily done by molecular biology techniques (Thompson et al. 1996a). Energy transfer has the additional advantages that it permits flexibility in choice of labels (so long as they have the right overlap), results in a change in lifetime as well as intensity, and in the case of a colored analyte may be reagentless (Thompson et al. 1996b). A change in overlap integral caused by a change in absorbance spectrum of the analyte (or recognition molecule) can result in a change in energy transfer efficiency (Thompson et al. 1997), but to generate large fluorescence changes requires large changes in spectra: a change in absorbance that is readily apparent visually may perturb the overlap integral (and thus the transfer efficiency) only slightly. In a few cases, binding of the analyte promotes the binding of a second, fluorescent ligand whose emission changes dramatically, but again, these approaches are seldom reagentless (Thompson and Jones 1993).

Among the more popular reagentless approaches has been the use of polarity - sensitive (or relaxation) fluorescent labels on recognition molecules, whose emission spectra exhibit changes upon analyte binding (Dattelbaum and Lakowicz 2001; Li and Cass 1991; Marvin and Hellinga 2001; Thompson et al. 1999). Such spectral changes, if substantial enough, lend themselves to ratiometric measurements at two different wavelengths, which have several advantages. Most of the available labels which exhibit the largest effects are related to the aminonaphthalene sulfonate derivatives which appeared in the 1950's(Weber 1953). In the form of reactive labels (dansyl chloride, acrylodan, IAEDANS) these are widely used; unfortunately, they often require ultraviolet excitation and have poor extinction coefficients. More troublesome is the inability to predict or control their response in the presence (or absence) of analyte. Basically, one attaches it to the protein or other recognition molecule and hopes for the best; even precise placement near known binding sites using molecular biology techniques does not assure a useful result. While more solvent sensitive dyes are known, they appear to have poor quantum yields in aqueous media. Despite substantial efforts by Russian scientists (Mazurenko and Bakhshiev 1970)and others (MacGregor and Weber 1981), there seems to be little theory for predicting the response of such a label on the surface of a protein. If a useful theory were available, together with a visibly excited, high extinction analog of Acrylodan, it might be possible to design label-recognition molecule conjugates with predictable responses; by comparison, the current "cut and try" is tedious and offers no guarantee of success.

14.5. NEW FLUOROPHORES AND FLUORESCING STATES

Probably the most exciting development in fluorophores has not been in the molecules themselves at all, but in the perturbation of their emission by proximity to conducting particles or surfaces. Several groups have been exploring this new field, and the fascinating results so far suggest that it will potentially be of the first importance (Lakowicz et al. 2003). In particular, workers have demonstrated that emission from fluorophores near metal surfaces, instead of being distributed more or less uniformly in all directions, can be made to emit in a specified direction describing a cone. Moreover, proximity to metal surfaces can greatly enhance emissive rates, such that fluorophores once described as mediocre due to effective competing non-radiative pathways (such as nucleotides) can be induced to emit with high quantum yields. Finally, proximity to metal surfaces can also enhance energy transfer efficiency. Breaking so many of the "rules" of fluorescence emission is intrinsically appealing, and suggests this technology will be fruitful for many applications. Key issues in this technology are making the metal particles or surface itself, and controlling the proximity of the fluorophore to it, at Angstrom levels. By now there are several techniques for doing this, but whether they will emerge from the laboratory for practical applications remains to be seen.

14.6. NEW ANALYTES

Even despite decades of effort, only a modest number of potential analytes have been the object of fluorescence sensor development, and of these only a minority permit quantitation of the analyte under physiologically relevant conditions. For even a simple organism like *E. coli,* which is comprised of hundreds of types of small molecules and a few thousand different macromolecules, measuring a significant fraction of these molecules is infeasible. It is also probably unnecessary, since a large fraction of small molecules are biosynthetic intermediates present at low and stable concentrations, such as Krebs cycle intermediates. A middle ground would appear feasible, however, where some dozens of analytes could be measured. For many of these molecules assays exist, such as chromatographic analyses or Western blots, but these either do not permit worthwhile imaging, quantitation, and/or a continuous response.

So what would be interesting analytes? First and foremost would be signaling molecules, such as neurotransmitters and hormones. Dialysis probes have been used to measure neurotransmitters in the brain, but their spatial and temporal resolution is modest, and they permit no imaging. Also of interest would be macromolecules in an altered state, such as phosphorylation. Antibodies exist for phosphorylated proteins such as glycogen phosphorylase, but one would wish to observe the kinetics of phosphorylation (and dephosphorylation) in real time in response to insulin and glucagon. Detecting the triggering of enzymatic cascades such as clotting and apoptosis would be of interest, and particularly its spatial evolution following the insult. We note that the new array techniques permit the expression of thousands of genes to be monitored simultaneously, but even this exciting new technique has shortcomings. In particular, not every physiological response would appear to require production of new proteins; for instance, the net activity of HMG CoA reductase (the enzyme catalyzing the first committed step in cholesterol biosynthesis, and the target of the "statin" drugs) is modulated by allosteric effectors, covalent modification of the enzyme, and accelerated breakdown of the enzyme as well as synthesis of new enzyme molecules. Additional shortcomings of the array techniques include their modest quantitation and low time resolution of a rather rapid process. Similarly, one would like to observe the liganded state and degree of self-association of many receptors at least as much as the levels of receptors themselves. In some cases just being able to observe the presence of a particular molecule *in vivo* would be immensely useful, as in the case of the Alzheimer's disease peptide Aβ 1-42. For a fermentation or other applied biotechnology process, it might be enough to monitor the product level and a few key analytes that reflected the health of the organism such as NADH (Armiger et al. 1986), pyruvate, carbon dioxide, and succinate. Obviously there is no shortage of potential targets, but as of yet there is no approach akin to array methods which permits analyzing many analytes at once.

Indeed, it may be of really central importance to follow a number of analytes simultaneously to understand the functions of signaling molecules (like the insulin/glucagon pair) that have multiple sites of action. From a control systems theory standpoint cellular metabolism is many-fold more complex than a typical chemical manufacturing plant, in that many more agencies can influence the "output" of a particular process in the cell than a manufacturing process, and the product concentration typically modulates other pathways.

For instance, compare an industrial process like the production of ammonia for fertilizer by the Haber process from nitrogen and hydrogen gases, to biosynthesis of glucose from pyruvate, water, GTP, ATP, and reducing equivalents (gluconeogenesis). Control of ammonia synthesis is done by a computer, which modulates the synthetic reaction by changing temperature, pressure, and gas flows through the catalytic bed based on real time knowledge of these parameters, and perhaps some simple feedback loop designed using control systems theory (Kuo 1975). By comparison, the eleven steps of gluconeogenesis are modulated automatically by levels of substrates and products (AMP and acetyl CoA) as well as a separate regulator (fructose 2,6-bisphosphate) and phosphorylation of two enzymes under the control of glucagon (Pilkis et al. 1988). All of this is to say that while one might need monitor only a few parameters to understand aprocess like the Haber process without understanding it *ab initio*, to understand the control of gluconeogenesis one might wish to monitor half a dozen or more analytes simultaneously.

14.7. IN *VIVO* MEASUREMENTS

Of particular interest are measurements *in vivo*, a regime where biosensors offer great promise. The near-IR images taken of fluorophores in tissue through the skin of nude mice (Fisher et al. 1996) indicate the potential power of this approach. The attenuation due to absorbance (mostly due to hemoglobin) and scattering drops significantly with wavelength , suggesting that matters may improve at wavelengths up to perhaps 1500 nm, beyond which water absorbance becomes prohibitive (Thompson 1994b). Transdermal analyses are of particular interest in the realm of clinical diagnostics, because they offer the possibility of analysis without collecting a specimen; however, the general need to administer a fluorescent-labeled antibody or other probe into the bloodstream would appear to negate this advantage. Nevertheless, there has been substantial work on finding tumors based on spectral features in tissue fluorescence emission spectra, including clinical trials (Chang et al. 2004). For externally accessible tissues like the skin, oral cavity, and cervix such a potentially rapid, inexpensive diagnostic tool is very attractive. Intrinsic tissue fluorescence is mainly attributable to molecules such as NADH, flavins, and connective tissue proteins, none of which absorb or emit at the long wavelengths transmitted more readily by tissue. A potential solution is multiphoton excitation, wherein focused beams of infrared light from modelocked lasers excite fluorescence from fluorophores within the tissue; however, the penetration depth is still modest (cm), and would appear infeasible for looking at deeper tissues and organs. Enough prevalent cancers (breast, cervix, skin, prostate) might potentially be identified by such techniques that this remains an active and promising area of research.

14.8. NEW OPPORTUNITIES IN CLINICAL ANALYSIS?

The intersection between the rising development cost of clinical diagnostic technology and the shrinking margins arising from poor reimbursement by insurance companies has made developing clinical diagnostic technology a much less attractive business than in years past; certainly the financial risk of doing so has increased. Much current effort has gone into reducing costs of determining analytes known to be of interest, as well as adapting technology for use in the doctor's office or at home. These trends mitigate against fluorescence approaches, in that colorimetric approaches, electrochemical methods, or visual tests like the hCG pregnancy test not requiring instruments will nearly always be cheaper than fluorescence. Recognizing this, there have been a few efforts to develop instrument-less fluorescence assays (Gryczynski et al. 1999). Similarly, it is axiomatic that a clinical diagnostic provide information to the clinician of a type and timeliness to be useful in intervening in the patient's illness. Fluorescence assays and sensors have much to offer here because of their inherent speed. A third issue is the desirability of minimizing interventions with diagnostic procedures; where previously a needle stick was viewed as a modest intervention to acquire a sample, the trend for marketability as well as (notionally) safety is to use externally accessible samples where possible. Thus there is substantial interest in analysis of specimens like saliva and tears, which don't require a needle stick. All of this is not to say that new assays or sensors will not be developed, it is just that altered business and regulatory conditions have changed the objectives and raised the bar a little higher.

Probably the brightest spot in fluorescence-based clinical diagnostics has been the widespread effort to identify the presence of deleterious mutations in genes (e.g., SNPs: single nucleotide polymorphisms) using approaches such as molecular beacons. The amenability of such nucleic acid hybridization approaches to configuration as an array has made possible the screening of individuals for thousands of genes simultaneously, at greatly reduced cost. Beyond the formidable technical issues involved in doing this (which seem close to solution), there are substantial political issues of privacy and security of such personal data. In particular, knowledge of the presence of mutations (especially in oncogenes or other genes associated with various maladies) is of some interest to the patient and his/her physician, but is of overwhelming value to insurance companies because it permits them to more precisely quantify their risks (and therefore costs), particularly for enormously expensive chronic diseases like diabetes and atherosclerosis. There may be legislative remedies sought to assure maintenance of the security of such data, but taken together these factors are likely to limit the implementation of such testing, which otherwise would be valuable to the health of patients.

14.9. NEW CHALLENGES

A key issue with all equilibrium-binding assays is their reversibility and related to that, their speed of response. The equilibrium constant K_D is usually equal to the ratio of association and dissociation rate constants (kon and koff). In homogenous aqueous media near room temperature the diffusion rates of ions and small molecules (to which the on rate is limited) are typically 10^8 cm^2 sec^{-1}, with molecular oxygen up to 100-fold faster. Thus for a reversible ligand binding reaction that is tight ($K_D \sim 10^{-12}$ molar, for instance), the zero-order off rate will be 10^{-4} sec^{-1}, implying that once the ligand is bound, it will be several tens of minutes before it will come to equilibrium, and thus accurately reflect the ligand concentration. For most antibodies binding their cognate antigens the rates are slower, and thus tight binding is effectively irreversible (Levison et al. 1971). For analysis of a discrete sample (or a sensor used as an alarm) this is seldom an issue, but it comes into play in sensing applications, where we wish to know the concentration of the analyte on a continuous or quasi-continuous basis. The binding kinetics would thus appear to impose a strict limitation in the form of a tradeoff of sensitivity and speed. From a clinical standpoint the reactions which proceed *in vivo* endure the same limitation: in general, the free concentration of a tightly bound ligand would appear to change only slowly, and thus not require rapid response. However, clinically speed is always of value, and in environmental monitoring a rapid response (perhaps in a harbor survey using a towed sensor array) is very desirable indeed.

Generally, it would appear difficult to change either the off or on rates; nevertheless, some progress has been made. In the first instance, we and others have used catalysts to more rapidly equilibrate ligand binding to a receptor. Human carbonic anhydrase II (CA) binds zinc ion with very slow on- and off-rates (Henkens and Sturtevant 1968), limiting its utility as a sensor. However, it was discovered that certain carboxylic acids formed a ternary complex with CA and the Zn(II), which dissociated much faster (Pocker and Fong 1983). Because the acid could be present at high concentrations, the rate of equilibration was appreciably increased. In the second instance, the actual binding kinetics themselves could be improved by mutagenizing the CA molecule itself . An example of this is shown in **Figure 14.3**, where a variant of apocarbonic anhydrase (E117A) exhibits a much faster response than the wild type. One can imagine other approaches to speed dissociation which might be implemented, but it may be more difficult to improve the diffusion rate of the analyte. Other approaches involving array sensors said to mimic natural sensors for trace analytes such as pheromones have been described but not implemented to our knowledge. A general approach to this problem would be, we feel, a valuable contribution.

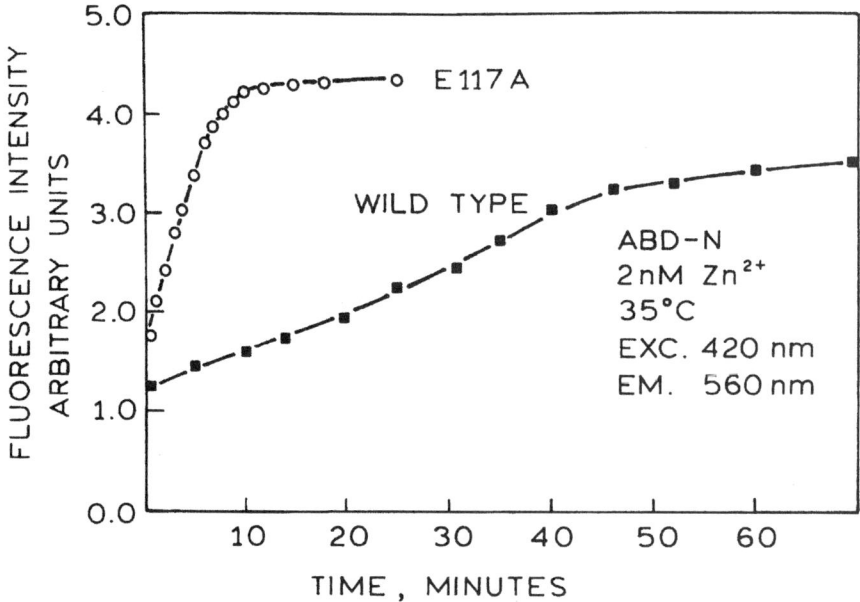

Figure 14.3. Time response of wild type (filled circles) and E117A variant apocarbonic anhydrases to 2 nM free Zn, visualized by ABD-N.

A second challenge is the development of very long wavelength fluorescent labels. Many of the transduction approaches mentioned above can be adapted in principle to use at any wavelength given a suitable label, and several workers have emphasized the value of fluorophores emitting at longer wavelengths (Thompson 1994b; Wolfbeis 1991), particularly in minimizing background fluorescence and permitting the use of desirable excitation sources such as laser diodes. From the standpoint of fiber optic sensors (Thompson 1991), particularly those designed for truly remote sensing over kilometer distances, it would be useful to work within the telecommunications bands (*circa* 1330 and 1550 nm) where fiber attenuation is very low, and sources and detectors are readily available from the telecom industry. Existing multiplexing technology would then permit the sensors to be integrated into existing fiber optic networks (Thompson 1994a). In view of the large size, instability, and insolubility of aromatic or heterocyclic infrared fluorophores at 800 nm and above, it would appear difficult to synthesize usable labels in this spectral regime; the longest commercially available labels emit a little past 800 nm.. More fruitful might be metal ion complexes, perhaps of the metal to ligand charge transfer type which have proved so useful for sensing oxygen and time-resolved fluorescence assays, or semiconductor particles of nanometer size ("quantum dots"). The advent of InGaAsP/InP photomultiplier tubes (e.g., Hamamatsu R5509) with photon counting sensitivity to 1700 nm offers the prospect of good sensitivity with such labels. Whether chemists with the necessary skills will view synthesis of such labels as a worthwhile endeavor is another question entirely.

14.10. CONCLUSION

In view of the exciting recent achievements listed above (along with many others), biosensors based on fluorescence will be of central importance in both understanding biology at the molecular, supramolecular, and cellular levels, as well as for clinical applications. Compared with other analytical modalities, fluorescence offers a combination of sensitivity, spatial resolution, flexibility, and time resolution which is unmatched. When coupled with the unrivalled selectivity and affinity (and, one might say, "evolvability") of biological or biomimetic recognition, one has a collection of tools of unsurpassed power.

14.11. ACKNOWLEDGMENTS

The author wishes to acknowledge his coworkers and collaborators for their indispensable contributions to our work, and for many fruitful discussions. He also wishes to acknowledge the Office of Naval Research, the National Institute for Biomedical Imaging and Bioengineering, and the National Science Foundation for their support and encouragement.

14.12. REFERENCES

Armiger, W. B., Forro, J. F., Montalve, L. M., Lee, J. F., and Zabriskie, P. W. (1986). "The interpretation of on-line measurements of intracellular NADH in fermentation processes." *Chemical Engineering Communication*, 45, 197 - 206.
Chang, S. K., Arifler, D., Drezek, R., Follen, M., and Richards-Kortum, R. (2004). "Analytical model to describe fluorescence spectra of normal and preneoplastic epithelial tissue: Comparison with Monte Carlo

simulations and clinical measurements." *Journal of Biomedical Optics*, 9(3), 511 - 522.

Dattelbaum, J. D., and Lakowicz, J. R. (2001). "Optical determination of glutamine using a genetically engineered protein." *Analytical Biochemistry*, 291, 89 - 95.

Demas, J. N., and DeGraff, B. A. (2001). "Applications of luminescent transition platinum group metal complexes to sensor technology and molecular probes." *Coordination Chemistry Reviews*, 211, 317 - 351.

DiCesare, N., and Lakowicz, J. R. (2002). "CHarge transfer fluorescent probes using boronic acids for monosaccharide signaling." *Journal of Biomedical Optics*, 7(4), 538 - 545.

Fernandez-Gutierrez, A., and Munoz de la Pena, A. (1985). "Determinations of inorganic substances by luminescence methods." Molecular Luminescence Spectroscopy, Part I: Methods and Applications, S. G. Schulman, ed., Wiley-Interscience, New York, 371-546.

Fisher, G., Ballou, B., Srivastava, M., and Farkas, D. L. (1996). "Far-red fluorescence-based high specificity tumor imaging in vivo." *Biophysical Journal*, 70(2), 212A.

Forster, T. (1948). "Intermolecular energy migration and fluorescence (Ger.)." *Annalen der Physik*, 2, 55 - 75.

Gryczynski, I., Gryczynski, Z., and Lakowicz, J. R. (1999). "Polarization sensing with visual detection." *Analytical Chemistry*, 71(7), 1241 - 1251.

Grynkiewicz, G., Poenie, M., and Tsien, R. Y. (1985). "A new generation of calcium indicators with greatly improved fluorescence properties." *Journal of Biological Chemistry*, 260(6), 3440-3450.

Haugland, R. P. (1996). "Handbook of Fluorescent Probes and Research Chemicals." Molecular Probes, Inc., Eugene, Oregon, 679.

Henkens, R. W., and Sturtevant, J. M. (1968). "The kinetics of the binding of Zn(II) by apocarbonic anhydrase." *Journal of the American Chemical Society*, 90, 2669 - 2676.

Huang, C.-c., Lesburg, C. A., Kiefer, L. L., Fierke, C. A., and Christianson, D. W. (1996). "Reversal of the hydrogen bond to zinc ligand histidine-119 dramatically diminishes catalysis and enhances metal equilibration kinetics in carbonic anhydrase II." *Biochemistry*, 35(11), 3439-3446.

Hunt, J. A., and Fierke, C. A. (1997). "Selection of carbonic anhydrase variants displayed on phage: aromatic residues in zinc binding site enhance metal affinity and equilibration kinetics." *Journal of Biological Chemistry*, 272(33), 20364-20372.

Kuo, B. C. (1975). *Automatic Control Systems*, Prentice-Hall, Englewood Cliffs, NJ.

Lakowicz, J. R. (1999). *Principles of Fluorescence Spectroscopy*, Kluwer Academic / Plenum Publishers, New York.

Lakowicz, J. R., Malicka, J., Gryczynski, I., Gryczynski, Z., and Geddes, C. D. (2003). "Radiative decay engineering: the role of photonic mode density in biotechnology." *Journal of Physics D: Applied Physics*, 36, R240 - R249.

Levison, S. A., Portmann, A. J., Kierszenbaum, F., and Dandliker, W. B. (1971). "Kinetic behavior of anti-hapten antibody of restricted heterogeneity by stopped flow fluorescence polarization kinetics." *Biochemical and Biophysical Research Communications*, 43, 258 - 266.

Li, Q. Z., and Cass, A. E. G. (1991). "Periplasmic binding-protein based biosensors. 1. Preliminary study of maltose binding-protein as sensing element for maltose." *Biosensors and Bioelectronics*, 6(5), 445-450.

Looger, L. L., Dwyer, M. A., Smith, J. J., and Hellinga, H. W. (2003). "Computational design of receptor and sensor proteins with novel functions." *Nature*, 423, 185 - 190.

MacGregor, R. B., and Weber, G. (1981). "Fluorophores in polar media: Spectral effects of the Langevin distribution of electrostatic interactions." *Annals of the New York Academy of Sciences*, 366, 140 - 154.

Marvin, J. S., and Hellinga, H. W. (2001). "Conversion of a maltose receptor into a zinc biosensor by computational design." *Proceedings of the National Academy of Sciences*, 98(9), 4955 - 4960.

Mazurenko, Y. T., and Bakhshiev, N. G. (1970). "Effect of orientation dipole relaxation on spectral, time, and polarization characteristics of the luminescence of solutions." *Optics and Spectroscopy (Russian)*, 28, 490 - 494.

Miyawaki, A., Llopis, J., Heim, R., McCaffery, J. M., Adams, J. A., Ikura, M., and Tsien, R. Y. (1997). "Fluorescent indicators for Ca2+ based on green fluorescent proteins and calmodulin." *Nature*, 388, 882-887.

Ozinskas, A., Malak, H., Joshi, J., Szmacinski, H., Britz, J., Thompson, R. B., Koen, P., and Lakowicz, J. R. (1993). "Homogenous model immunoassay of thyroxine by phase modulation fluorescence spectroscopy." *Analytical Biochemistry*, 213, 264-270.

Pearce, L. L., Gandley, R. E., Han, W., Wasserloos, K., Stitt, M., Kanai, A. J., McLaughlin, M. K., Pitt, B. R., and Levitan, E. S. (2000). "Role of metallothionein in nitric oxide signaling as revealed by a green fluorescent fusion protein." *Proceedings of the National Academy of Sciences*, 97(1), 477 - 482.

Pilkis, S. J., El-Maghrabi, M. R., and Claus, T. H. (1988). "Hormonal control of hepatic gluconeogenesis and glycolysis." *Annual Review of Biochemistry*, 57, 755 - 783.

Pocker, Y., and Fong, C. T. O. (1983). "Inactivation of bovine carbonic anhydrase by dipicolinate: Kinetic studies and mechanistic implications." *Biochemistry*, 22, 813 - 818.

Rakow, N. A., and Suslick, K. S. (2000). "A colorimetric sensor array for odour visualization." *Nature*, 406, 710 -

713.

Spencer, R. D., Toledo, F. B., Williams, B. T., and Yoss, N. L. (1973). "[Fluorescence polarization assay of protease]." *Clinical Chemistry*, 19, 838 - 844.

Thompson, L. A., and Ellman, J. A. (1996). "Synthesis and applications of small molecule libraries." *Chemical Reviews*, 96, 555-600.

Thompson, R. B. (1991). "Fluorescence-based fiber optic sensors." Topics in Fluorescence Spectroscopy Vol. 2: Principles, J. R. Lakowicz, ed., Plenum Press, New York, 345-365.

Thompson, R. B. (1994a). "Fiber optic chemical sensors." *IEEE Proceedings on Circuits and Devices*, CD-10(3), 14-21.

Thompson, R. B. (1994b). "Red and near-infrared fluorometry." Topics in Fluorescence Spectroscopy Vol. 4: Probe Design and Chemical Sensing, J. R. Lakowicz, ed., Plenum Press, New York, 151-181.

Thompson, R. B. (2004). "Fluorescence Sensors and Biosensors." CRC Press, Boca Raton, FL.

Thompson, R. B., Cramer, M. L., Bozym, R., and Fierke, C. A. (2002). "Excitation ratiometric fluorescent biosensor for zinc ion at picomolar levels." *Journal of Biomedical Optics*, 7(4), 555 - 560.

Thompson, R. B., Ge, Z., Patchan, M. W., and Fierke, C. A. (1996a). "Performance enhancement of fluorescence energy transfer-based biosensors by site-directed mutagenesis of the transducer." *Journal of Biomedical Optics*, 1(1), 131-137.

Thompson, R. B., Ge, Z., Patchan, M. W., Huang, C.-c., and Fierke, C. A. (1996b). "Fiber optic biosensor for Co(II) and Cu(II) based on fluorescence energy transfer with an enzyme transducer." *Biosensors and Bioelectronics*, 11(6), 557-564.

Thompson, R. B., and Jones, E. R. (1993). "Enzyme-based fiber optic zinc biosensor." *Analytical Chemistry*, 65, 730-734.

Thompson, R. B., Lin, H.-J., Ge, Z., Johnson, K., and Fierke, C. "Fluorescence lifetime-based determination of anions using a site-directed mutant enzyme transducer." *Advances in Fluorescence Sensing Technology III*, San Jose, California, 247 - 257.

Thompson, R. B., Maliwal, B. P., Feliccia, V. L., Fierke, C. A., and McCall, K. (1998). "Determination of picomolar concentrations of metal ions using fluorescence anisotropy: biosensing with a "reagentless" enzyme transducer." *Analytical Chemistry*, 70(22), 4717-4723.

Thompson, R. B., Maliwal, B. P., and Fierke, C. A. (1999). "Selectivity and sensitivity of fluorescence lifetime-based metal ion biosensing using a carbonic anhydrase transducer." *Analytical Biochemistry*, 267, 185-195.

Thompson, R. B., and Walt, D. R. (1994). "Emerging strategies for molecular biosensors." *Naval Research Reviews*, 46(3), 19 - 29.

Tromberg, B. J., Sepaniak, M. J., Vo-Dinh, T., and Griffin, G. D. (1987). "Fiber optic chemical sensors for competitive binding fluoroimmunoassay." *Analytical Chemistry*, 59, 1226 - 1230.

Ullman, E. F., and Schwarzberg, M. (1981). "Fluorescence quenching with immunological pairs in immunoassays." Syva Company, U.S.

Weber, G. (1953). "Rotational Brownian motion and polarization of the fluorescence of solutions." *Advances in Protein Chemistry*, 8, 415-459.

White, C. E., and Argauer, R. J. (1970). *Fluorescence Analysis: A Practical Approach*, Marcel Dekker, Inc., New York.

Wolfbeis, O. S. (1991). "Fiber Optic Chemical Sensors and Biosensors." CRC Press, Boca Raton.

Zeng, H. H., Thompson, R. B., Maliwal, B. P., Fones, G. R., Moffett, J. W., and Fierke, C. A. (2003). "Real-time determination of picomolar free Cu(II) in seawater using a fluorescence-based fiber optic biosensor." *Analytical Chemistry*, 75(24), 6807 - 6812.

OPHTHALMIC GLUCOSE MONITORING USING DISPOSABLE CONTACT LENSES

Ramachandram Badugu[1], Joseph R. Lakowicz[1,*], and Chris D. Geddes[1,2,*]

15.1. INTRODUCTION

As a common medical condition that produces excessive thirst, continuous urination and severe weight loss, Diabetes[#] has interested medical researchers for over three millennia. Unfortunately it wasn't until the early 20th century that the prognosis for this condition became any better than it was 3000 years ago. Today, approximately 150 million people worldwide are affected by diabetes. With its prevalence still rising, diabetes still continues to fascinate, practitioners and researchers alike, by its elusive cause and its many manifestations.

While the *Ebers Papyrus*, which was written around 1500 BC, excavated in 1862 AD from an ancient grave in Thebes, Egypt, described the first reference to Diabetes Mellitus, it was physicians in India at around the same time that developed the first crude test for diabetes. They observed that the urine from people with diabetes attracted ants and flies. They subsequently named the condition "madhumeha" or "honey urine" [1].

The important elements of our current understanding of diabetes can be traced to the early to mid 19th century. In 1815 Eugene Chevreul in Paris concluded that the sugar in urine was indeed Glucose, the first quantitative test for glucose in urine being developed by Von Fehling some years later in 1848 [2].

Over one-hundred and fifty years later, significant attention is still given to the development of physiological glucose monitoring [3-36]. This is because one important aspect for diabetes management, involves the tight control of blood glucose levels, so as to manage food intake and the dosage and timing of insulin injection. Tests for determining serum glucose concentration typically require blood collection by some invasive technique, usually a needle or other device causing arterial or venous puncture.

[1]-Center for Fluorescence Spectroscopy, Department of Biochemistry and Molecular Biology, [2]-Institute of Fluorescence, University of Maryland Biotechnology Institute, 725 W. Lombard St., Baltimore, MD, 21201. USA, * Corresponding authors, Geddes@umbi.umd.edu   Lakowicz@cfs.umbi.umd.edu

The term "Diabetes" was first used around 230 BC by Apollonius of Mephis, which in Greek means "to pass through" (*Dia* – through, *betes* – to go) [1].

Currently, millions of diabetics are left with very few alternatives, except to invasively draw blood many times daily to determine their blood sugar levels. To this end, many technologies have been developed over the past 20 years in an attempt to provide a technology, which promises both non-invasive and continuous physiological glucose monitoring. These include near infrared spectroscopy [3,4], optical rotation [5,6], colorimetric [7,8] and fluorescence detection [9-13], to name but just a very few.

Recently we have seen the launch of the new GlucoWatch, which approved by the FDA in 2001, is the first step towards both the continuous and non-invasive monitoring of physiological glucose. However, in addition to wearing this wrist watch based glucose sensor, it is also recommended that glucose monitoring by another blood sampling technique be additionally used from time to time. Other technologies, which are presently emerging, include, glucose monitoring skin patches, implantable glucose sensors coupled insulin pumps, and laser blood drawing, which is deemed less painful than finger pricking with a lancet or needle. Yet with all these emerging technologies, there is still a need for new technologies, which are truly non-invasive and continuous. To this end our laboratories have been developing glucose sensing contact lenses which when worn by diabetics, who often require vision correction in any case, can potentially monitor tear glucose levels, which are known to directly relate and track blood glucose levels [14-17].

In this review chapter, we employ the notion of elevated tear glucose levels during hyperglycemia to investigate, *for the first time*, the possibility of monitoring tear glucose and therefore blood glucose, using a disposable, off-the-shelf, contact lens. By incorporating new monosaccharide fluorescent signaling boronic acid containing probes (BAFs) within such a lens, we can indeed make progress towards this non-invasive and continuous approach for glucose monitoring, Figure 15.1.

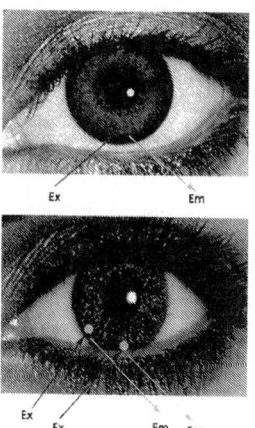 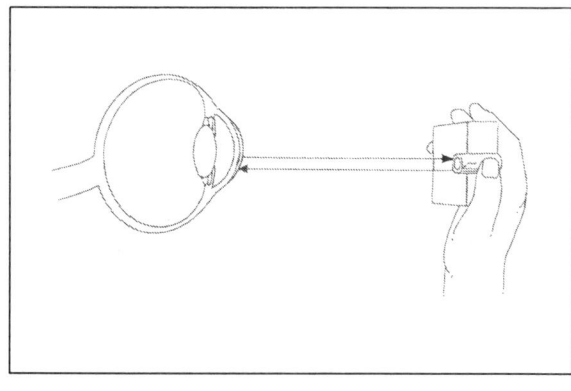

Figure 15.1. Potential methods for non-invasive continuous tear glucose monitoring. (Top left) BAF doped contact lens as described here, and (Bottom left) Sensor spots on the surface of the lens to additionally monitor other analytes in addition to glucose, such as drugs, biological markers, Ca^{2+}, K^+, Na^+, O_2 and Cl^-. Sensor regions may also allow for ratiometric, lifetime or polarization based fluorescence glucose sensing. (Right) Schematic representation of the possible tear glucose sensing device.

As with any sensors, there are several issues that have to be addressed. The first is to identify suitable transduction elements, which in the presence of glucose, can report / produce suitable signals. The second is the design of the matrix to incorporate the transduction elements. For this, we have chosen an *off-the-shelf* disposable plastic contact lens, primarily because its physiological compatibility has already been assessed, and finally, the optimization of the sensor, with regard to sensitivity, response time, reversibility and shelf-life etc. The later two issues will be discussed throughout much of this review and indeed in past papers by the authors [14-17]. For the identification of suitable transduction elements, boronic acid has been known to have high affinity for diol-containing compounds such as carbohydrates [18-20], Scheme 15.1, where the strong complexation has been used for the construction of carbohydrate sensors [21-28], transporters [29] and chromatographic materials [30]. Boronic acid compounds have also been used for the synthesis of glucose sensors [31-36], where we note the work of Shinkai [31,32] and Lakowicz [33-36] to name but just a few workers in this field. In this review chapter, we discuss how one can develop a suitable transduction mechanism for continuous glucose monitoring, using a disposable, colorless and daily use contact lens.

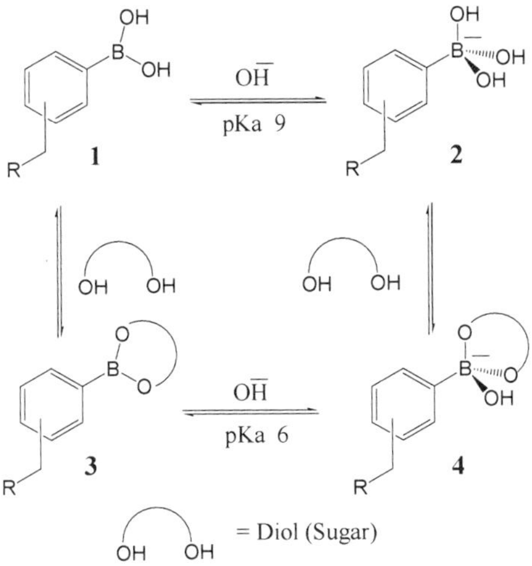

Scheme 15.1. Equilibrium between the various forms of the boronic acid in solution, and diol (sugar) interaction.

15.2. GLUCOSE SENSING USING BORONIC ACID PROBES IN SOLUTION

The potential applicability of the boronic acid probes to glucose sensing is evident from the vast amount of data published on this topic. In this regard we are interested to provide an overview on the glucose sensing using boronic acid probes in solution, which may help the reader to understand the subsequent sections of this chapter. Boronic acid forms reversible covalent bonds with vicinal dihydroxy containing compounds such as saccharides, and during this, the electronic properties and geometry at the boron atom are altered (Scheme 15.1). Subsequently, the changes in electronic properties at the boron atom are being transferred to the appended fluorophore, leading to the spectral shift and/or intensity changes observed, depending on the transduction mechanism. The equilibrium involved with the boronic acid and diol interaction is illustrated in Scheme 15.1 and can be described as follows. Boronic acids are weak Lewis Acids composed of an electron deficient boron atom and two hydroxyl groups, (**1** in Scheme 15.1), with trigonal planar conformation, and which can interact with strong bases like OH⁻ to from the anionic boronate ester form (**2** in Scheme 15.1), that having the tetrahedral geometry at the boron atom. The boronic acid group (**1**) typically shows high pK_a, around 9 [37,38]. Boronic acids couple with diols to form a boronic acid diester group (**3** in Scheme 15.1). In comparison to the boronic acid group, the boronic acid diester group is relatively more acidic ($pK_a \approx 6$) due to a higher electrophilic boron atom.

The affinity of the boronic acid is directly proportional to the number of available vicinal hydroxyl groups. Subsequently, the monophenylboronic acid shows higher affinity towards *D*-fructose over *D*-glucose because of stronger tridentate and weaker bidentate type binding with *D*-fructose and *D*-glucose respectively, with binding constants of ≈ 0.5 and 10 mM respectively [38]. The affinity of the boronic acid towards sugars is tunable by adjusting the geometry and substituents on the fluorophore moiety. A geometrically placed diboronic acid containing fluorophore shows higher affinity towards *D*-glucose over *D*-fructose. Hence glucose sensitive probes can be made with a variety of affinities, in the mM range for blood glucose [34-36], and in the μM range for tear glucose [39,40]. Also the pK_a of the boronic acid can be modulated according to the medium of interest. For example, as we can see in this section, the sugar response of most of the probes reported in the literature has been conducted in slightly elevated pH solutions such as pH 8. In contrast to this, also which is thought be one of the reasons of incompatibility of these probes within the acidic contact lens, the new probes developed based on the quinolinium moiety show an excellent sugar response in physiological pH solutions but not in slightly acidic media such as in a contact lens (see section 15.5.3). This is due to the reduced pK_a of the new quinolinium probes. An increase and decrease in boronic acid pK_a can be observed with substituents with electron donating and accepting nature, respectively, on the phenyl ring. Here we have used strategic design logic by using a electron withdrawing quaternary nitrogen center within the interacting space of boronic acid to reduce the boronic acid pK_a. Also, in contrast to the other probes, because of the positively charged quinolinium moiety, these probes are readily water soluble and are therefore potentially useful for physiological applications.

A wide range of signaling mechanisms have been employed in the design of boronic acid containing fluorophores including intramolecular charge transfer (ICT), and the photoinduced electron transfer mechanism (PET). The extent of transduction during the sugar binding results in the spectral modifications, which in-turn, can be quantified to construct the calibration curves for accurate sugar detection. In the following section we have described the response of a few representative boronic acid probes using each mechanism towards monosaccharides in buffered solutions. The molecular structures of the probes are shown in Chart 15.1.

15.2.1 Probes employing the intramolecular charge transfer mechanism (ICT).

Compounds with suitably placed electron donor and acceptor groups can show intramolecular charge transfer (ICT) features in their spectra. The extent of ICT in a system can be easily altered by changing the external parameters such as polarity of the medium; an increase in ICT is a common observation with increasing solvent polarity. An increase in ICT would result in red shift of the spectrum. An opposite observation is true when the polarity of the medium is reduced. It is also commonly observed that most probes show reduced and increased fluorescence quantum efficiencies with an increase and decrease in ICT efficiency of the system, respectively.

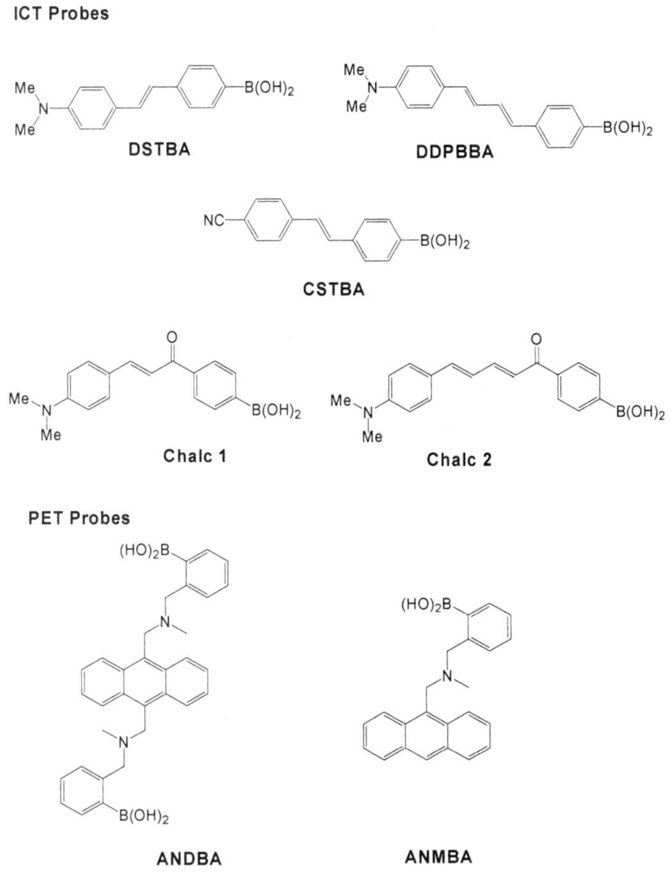

ICT Probes

PET Probes

Chart 15.1. Molecular structures of the ICT and PET probes studied in the contact lens. **DSTBA** - 4'-Dimethylaminostilbene-4-boronic acid; **CSTBA** - 4'-Cyanostilbene-4-boronic acid; **DDPBBA** - 1-(*p*-Boronophenyl)-4-(p-dimethylaminophenyl)buta-1,4-deine; **Chalc 1** – 3-[4'(Dimethylamino)phenyl]-1-(4''-boronophenyl)-prop-2-en-1-one; **Chalc 2** – 5-[4'-(Dimethylamino)phenyl]-1-(4'-boronophenyl)-pent-2,4-dien-1-one; **ANDBA** – 9,10-bis-[[*N*-methyl-*N*-(*o*-boronobenzyl)amino]methyl]anthracene; and **ANMBA** – 9-[[*N*-methyl-*N*-(*o*-boronobenzyl)amino]methyl]anthracene.

In addition, modulations in the donating and accepting capabilities of the donor and acceptor, result in spectral shifts as can be seen by the external spectral parameters. As mentioned in the last paragraph, the electronic properties of boronic acid can be altered during the binding with saccharides; boronic acid is an electron acceptor and the corresponding boronate diester is no longer an acceptor and acts as a donor when it is coupled with a suitable acceptor group on the fluorophore moiety. As a result, boronic acid probes employing the ICT mechanism show spectral shifts with changes in quantum yield upon binding with saccharides, providing unique ratiometric probes for saccharide sensing. Here we can demonstrate these features with a few examples. The molecular structures of the probes are shown in Chart 15.1. As shown, two stilbene derivatives (DSTBA and CSTBA) a polyene derivative DDPBBA and two chalcone derivatives (Chalc 1 and Chalc 2) are considered. DSTBA ad DDPBBA combines the electron-donating dimethylamino group with the electron withdrawing boronic acid group, and CSTBA combines the electron withdrawing cyano group with the boronic acid, in essence these two sets of probes show both decreased and increased ICT respectively, upon binding with glucose.

Chalcone derivatives, Chalc 1 and Chalc 2, unlike the stilbenes have the advantage of much longer wavelength emission. This is particularly attractive as longer wavelength emission potentially reduces the detection of any lens or eye autofluorescence, as well as scatter (λ^{-4} dependence), and also allows the use of cheaper and longer wavelength laser or light emitting diode excitation sources, reducing the need for UV excitation in the eye.

In chalcone probes the boronic acid group does not produce resonance forms with the electron donating amino group. The CT occurs between the dimethylamino group (electron donating group) and the carbonyl group (electron withdrawing group) (Scheme 15.2). Upon sugar binding to the boronic acid group, then a change in the electronic properties of the boron group, both when free and when complexed with sugar, leads to a change in the electronic density of the acetophenone moiety and subsequently the CT properties of the excited state of the fluorophore, noting that boronic acid group is in resonance with the carbonyl group.

Figure 15.2 shows the effect of sugar on the emission properties of DSTBA in pH 8.0 buffer-methanol (2:1, v/v) solution. The emission spectrum shows a hypsochromic shift of about 30 nm and an increase in fluorescence intensity as the concentration of fructose increases (Figure 15.2 – Top left). These dramatic and useful changes can simply be explained by the loss of the electron withdrawing property of the boronic acid group following the formation of the anionic form as shown in Scheme 15.1. A similar response has been observed with the polyene derivative DDPBBA (Figure 15.2 – Top right).

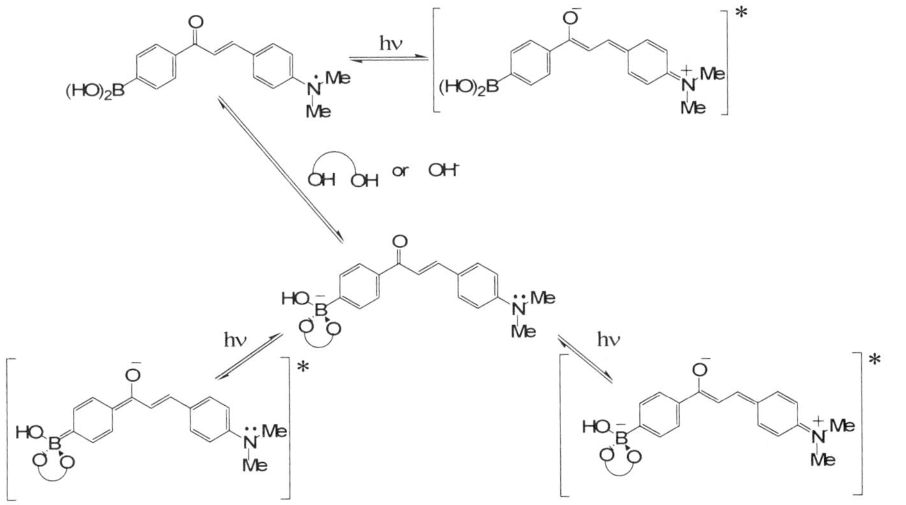

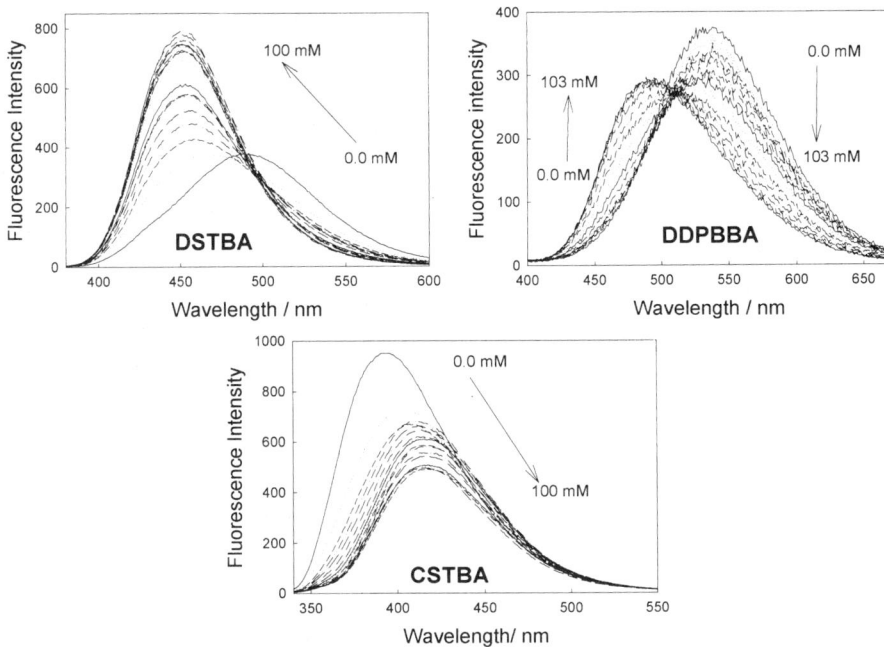

Scheme 15.2. Ground and excited state electronic distributions involved in the neutral and anionic forms of the boronic acid group of Chalc 1. For the case of interaction with OH$^-$, the diol should be replaced by two OH$^-$ groups.

Figure 15.2. Emission spectra of DSTBA (Top left), DDPBBA (Top right) and CSTBA (Bottom) in pH 8.0 buffer/methanol (2:1) with increasing concentrations of fructose, λ_{ex} for DSTBA and DDPBBA is 340 and for CSTBA, 320 nm.

The other stilbene derivative CSTBA possesses two electron withdrawing groups, boronic acid and cyano groups. In the presence of sugar as shown in Figure 15.2 – Bottom, we can observe a bathochromic shift of about 25 nm, and a decrease in the intensity at pH 8, which is opposite to that observed for DSTBA. This change has been attributed to an excited CT state present for the anionic form of CSTBA, where no CT state has been observed for the neutral form of the boronic acid group [41], suggesting that the anionic form of the boronic acid group can act as an electron donor group. Similarly Chalc 1 and Chalc 2 show an excellent response to sugar and pH, resulting from the reduced ICT in the systems upon sugar binding [39,41].

A few representative titration curves for DSTBA and CSTBA are shown in Figure 15.3, the dissociation constants for DSTBA, DDPBBA, CSTBA, Chalk 1 and Chalc 2, are shown in Table 15.1. As previously mentioned, monoboronic acid derivatives show higher affinities for *D*-Fructose, and the affinity decreases for *D*-Glucose. As for the data provided in this section, one quickly realizes that the probes based on the ICT mechanism are potential candidates for physiological glucose monitoring.

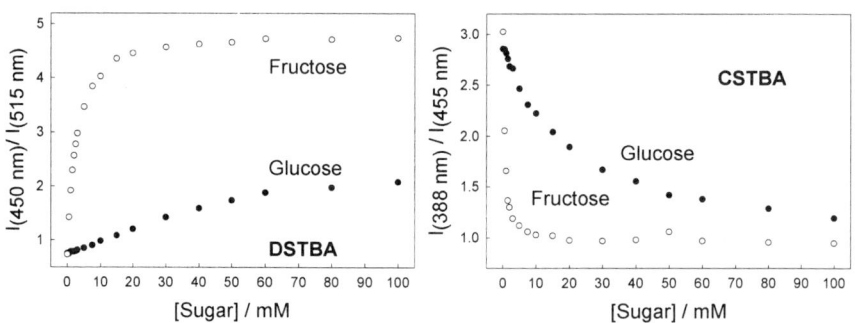

Figure 15.3. Ratiometric response plot for DSTBA (Left) and CSTBA (Right) in pH 8.0 buffer / methanol (2:1) with glucose and fructose.

Table 15.1. Dissociation constants (K_D, mM) of the boronic acid probes for glucose and fructose.

	DSTBA	DDPBBA	CSTBA	Chalc 1	Chalc 2	ANDBA	ANMBA
Glucose	98	17	18	34	30	0.51	21.3
Fructose	2.5	1.1	0.65	2.5	2.1	---	---

[a]The dissociation constant (K_D) values of DSTBA, DDPBBA and CSTBA were measured at pH 8.0 for that of Chalc 1 and Chalc 2 were obtained in pH 6.5 buffer.

15.2.2 Probes employing the photoinduced intramolecular electron transfer mechanism, (PET).

The anthracenediboronic acid derivative, ANDBA, is the one of the well-known boronic acid probes specific for glucose detection, developed by Shinkai *et al.* in 1995. The corresponding monophenylboronic acid probe ANMBA (Chart 15.1), is nonspecific for any saccharide, and shows high affinity towards fructose, as is the case with any monophenyl boronic acid based probe. The glucose specificity for ANDBA is due to the geometrically suitably appended two boronic acid moieties: the two boronic acid groups bind either side of the plane of the glucose ring (Figure 15.4). This novel approach of designing glucose specific probes has triggered vast amounts of interest in the research community over the last 10 years.

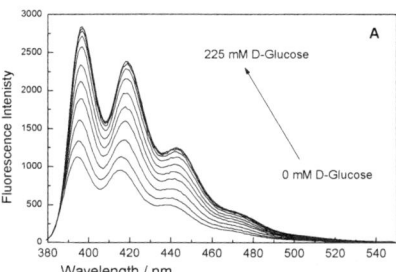

○ = Glucose   ⃫ = Anthracene

Figure 15.4. A Schematic representation of glucose binding with the diboronic acid derivative, ANDBA.

The fluorescence spectra of ANDBA and ANMBA in pH 8.0 buffer-methanol (2:1, v/v) with glucose are shown in Figure 15.5. The fluorescence intensity of these probes increases with increasing concentrations of the monosaccharides. This is because; originally in these systems a photoinduced electron transfer from the donor amine to the acceptor anthracene results in fluorescence quenching. Subsequent binding with monosaccharides, shows increased acidity and thus reduces the PET interaction in the system. The suppressed PET interaction results in the emission enhancement of that probe. Subsequently the probes show fluorescence enhancement in the presence of glucose. In contrast the monophenylboronic acid derivative ANMBA shows relatively weaker affinity towards glucose over the diboronic acid derivative ANDBA. The measured dissociation constants (K_D) are 21.3 and 0.51 mM for ANMBA and ANDBA, respectively, with glucose (Table 15.1).

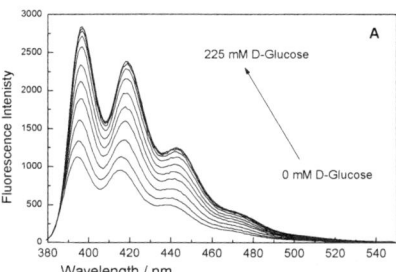
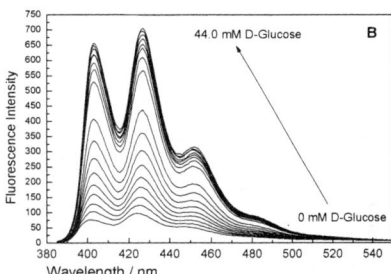

Figure 15.5. Fluorescence spectral changes of ANMBA (A) and ANDBA (B) in pH 8.0 buffer / methanol (1:2) with increasing concentrations of glucose. λ_{ex} = 365 and 380 nm for ANMBDA and ANDBA, respectively.

Figure 15.6 shows frequency domain decay profiles of ANMBA and ANDBA with glucose. Similar to the wavelength ratiometric sensing of analytes, lifetime based sensing has intrinsic advantages for biomedical fluorescence sensing and is thus considered to provide for a more reliable analytical method. As seen from figure 15.6, these probes show increases in lifetime with increasing concentrations of glucose. A fluorescence lifetime change from 9.8 to 12.4 and 5.7 to 11.8 ns for ANMBA and ANDBA respectively, has been observed. Although these probes are insoluble in water or indeed in physiological fluids, based on the results from steady-state and lifetime measurements, they serve to demonstrate that these probes, especially ANDBA, are potential PET based boronic acid probes for physiological glucose sensing applications.

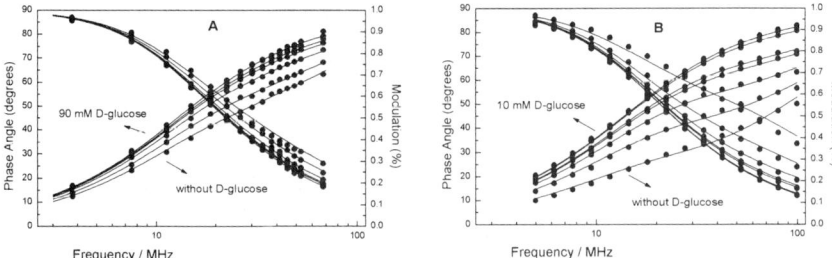

Figure 15.6. Effect of glucose on the frequency-domain decay profiles for ANMBA (A) and ANDBA (B) in pH 8.0 buffer / methanol (2:1) mixture. The lines represented the global fits with two lifetimes.

15.3. LENS FEASIBITY STUDY

As briefly mentioned in the introduction, the continuous monitoring of glucose may be possible using a contact lens embedded with a suitable glucose sensing probe, whose glucose sensing response is retained within the contact lens environment. In this regard we are interested to see the applicability of the probes discussed in section 15.2 towards glucose monitoring in a disposable, off-the-shelf, contact lens.

15.3.1 Lens doping and contact lens holder

The contact lenses were washed several times with Millipore water at 20°C to remove the salts and other preservatives in the contact lens. The contact lens is a polyvinyl alcohol type photocured polymer which swells slightly in water. Its hydrophilic character readily allows for the diffusion of the analytes in tears. The probe doping was conducted by incubating the lenses in a high concentration of the respective BAFs solution for 24 hrs before being rinsed with Millipore water. Lenses were used directly after being prepared. Doped contact lenses, which were allowed to leach excess dye for 1 hr, were inserted in the contact lens holder, Figure 15.7. The quartz lens holder, which was used in all the lens studies, has dimensions of 4*2.5*0.8 cm, all 4 sides being of optical quality. The contact lens is mounted onto a stainless steel mount of dimensions 4*2*0.4 cm, which fits tightly within the quartz outer holder. A circular hole in the center of the mount with a 1.5 cm ID, has a raised quartz lip, which enables the lens to be mounted. The mount and holder readily allow for ≈ 1.5 cm^3 of solution to be in contact with the front and back sides of the lens for the sugar sensing experiments. Buffered solutions of sugars were then added to the lens. Fluorescence spectra were typically taken 15 mins after each sugar addition to allow the lens to reach equilibrium. Excitation and emission was performed using a Varian Fluorometer, where the geometry shown in Figure 15.7 - right, was employed to reduce any scattering of the excitation light, the concave edge of the lens facing towards the excitation source. We additionally tested the lens excited from the convex edge, just as would be used in the eye, and encouragingly found identical results.

15.3.2 Response of ICT probes with in the contact lens

15.3.2.1 Stilbene Derivatives

Figure 15.8 shows the emission spectra and titration curves for a DSTBA doped contact lens towards both glucose and fructose. As expected the magnitude of the response towards fructose is greater, reflecting the higher affinity of mono boronic acids for fructose [41]. Comparing the response of DSTBA in both solution and lens (Figure 15.3 left and Figure 15.8 Bottom), we can see that an opposite response is observed in the lens, where the emission spectra similarly shows a blue shift, accompanied by a decrease in intensity as the fructose concentration is increased. In addition the sugar affinity is decreased slightly in the lens.

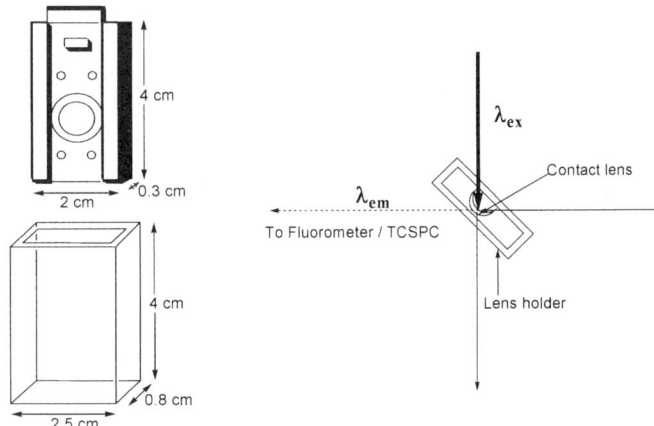

Figure 15.7. Contact lens mount and quartz holder (Left top and Bottom, respectively) and experimental geometry used for contact lens glucose sensing (Right).

Similarly Figure 15.9 shows the response of CSTBA in the lens for both glucose and fructose. While a similar reduction in intensity is observed as compared to solution, no red shift in the emission is observed, indicative of a reduction in the electron donating capability of the anionic sugar bound form.

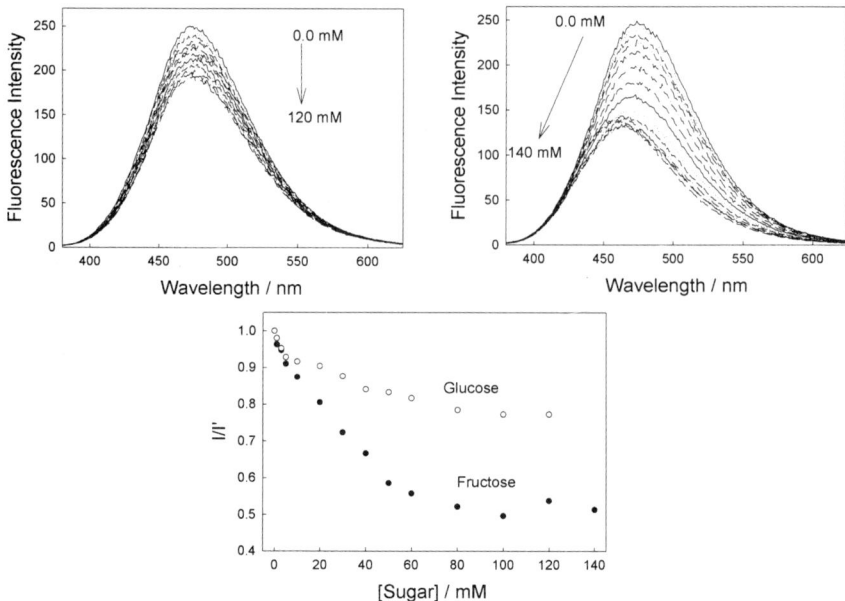

Figure 15.8. Emission spectra of the DSTBA doped contact lens, pH 8.0 buffer / methanol (2:1), with increasing concentrations of glucose (Top left) and corresponding spectra with increasing concentrations of fructose (Top right). λ_{ex} = 340 nm. Intensity ratio plot for the DSTBA doped contact lens towards both glucose and fructose (Bottom), where I and I' are the intensities in the presence and absence of sugar respectively at λ_{em} max.

The lack of suitable spectral shifts in the presence of sugar eliminates, at this stage, the possibility of wavelength ratiometric sensing as shown for the solution based measurements in Figures 15.3. Subsequently, Figure 15.8-bottom and Figure 15.9-bottom compares the responses of the stilbene probes, DSTBA and CSTBA, based on a simple intensity ratio measurement. It is interesting to see the much greater response for fructose for CSTBA in the lens as compared DSTBA, where notable changes in intensity occur at < 20 mM [fructose]. However the glucose response of DSTBA in the contact lens appears more promising for [glucose] < 10 mM, where a 10 % fluorescence intensity change is observed for ≈ 10 mM glucose at pH 8.0. The greater response of CSTBA within the contact lens may arise from the relatively lower pK_a value for CSTBA over the DSTBA probe. The apparent pK_a values for CSTBA and DSTBA are 8.17 and 9.14, respectively.

3.2.2 Polyene Derivative

The spectral response of DDPBBA in the contact lens is also different to that observed in solution, c.f. Figure 15.2-top right and Figure 15.10, where a decrease in intensity is typically observed for increasing sugar concentration, and a slight blue shift is evident for fructose binding. This is in contrast to solution-based responses which show both a blue shifted and increased emission, Figure 15.2-top right. While the general spectral changes observed for both DSTBA (Figure 15.8) and DDPBBA (Figure 15.10), are similar, a greater dynamic response to sugar is observed for DSTBA as compared to DDPBBA, c.f. figure 15.8 – bottom and Figure 15.10 – bottom. In addition the response of DDPBBA towards both glucose and fructose are similar over the sugar concentration range studied, Figure 15.10 – bottom, as compared to the significantly different responses observed for both sugars for DSTBA and CSTBA, Figure 15.8- bottom and Figure 15.9- bottom.

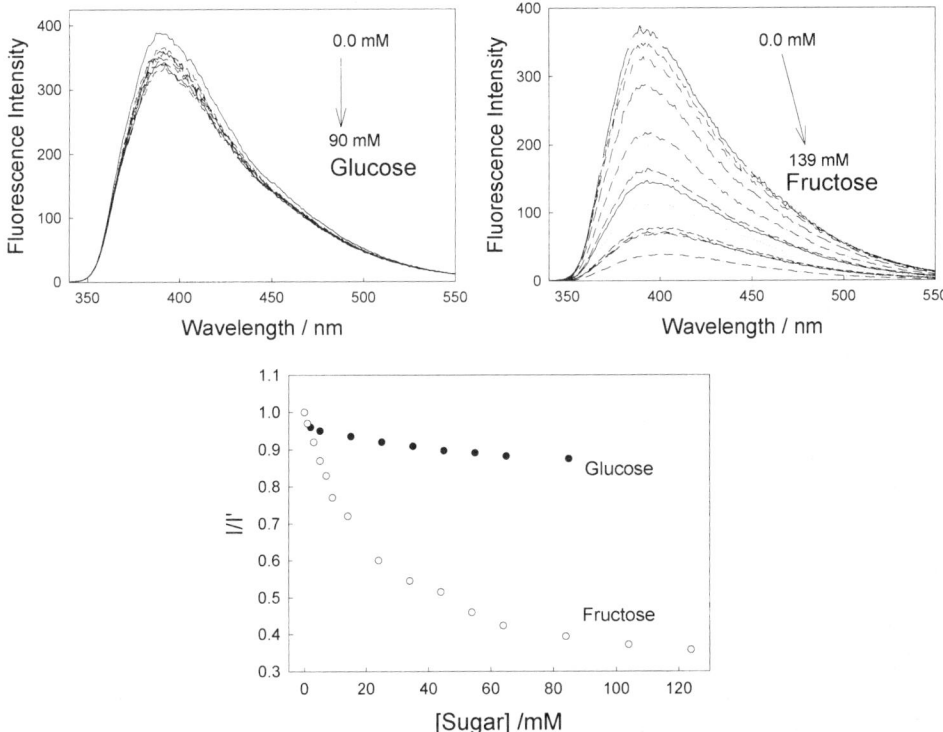

Figure 15.9. Emission spectra of the CSTBA doped contact lens, pH 8.0 buffer / methanol (2:1), with increasing concentrations of glucose (top left) and corresponding spectra with increasing concentrations of fructose (top right). λ_{ex} = 320 nm. Intensity ratio plot for CSTBA doped contact lens towards both glucose and fructose (bottom), where I and I' are the intensities in the presence and absence of sugar respectively at λ_{em} max.

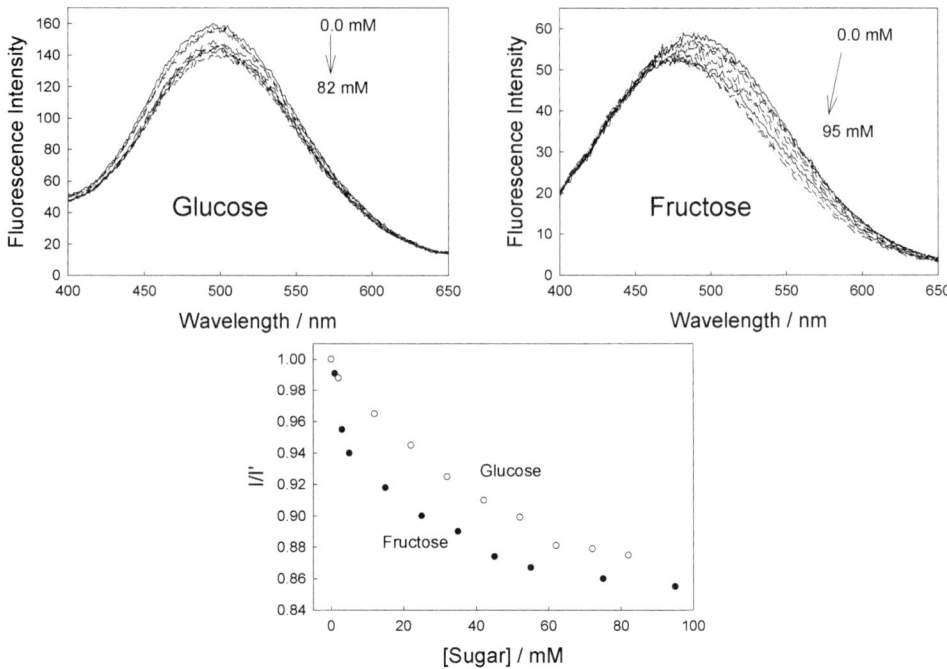

Figure 15.10. Emission spectra of the DDPBBA doped contact lens, pH 8.0 buffer / methanol (2:1), with increasing concentrations of glucose (Top left) and corresponding spectra with increasing concentrations of fructose (Top right). λ_{ex} = 320 nm. Intensity ratio plot for DDPBBA doped contact lens towards both glucose and fructose (Bottom), where *I* and *I'* are the intensities in the presence and absence of sugar respectively at λ_{em} max.

3.2.3 Chalcone Derivatives

The response of Chalc 1 and Chalc 2 doped contact lenses towards sugar are shown in Figures 15.11 and 15.12 respectively. Both chalcone doped lenses display similar responses to sugar, Figures 15.11 and 15.12 – bottom respectively, only their respective emission wavelengths differ. Chalc 1 shows an emission centered around 560 nm in the lens as compared to 580 nm in solution (not shown), while Chalc 2 shows an emission centered at ⬚ 630 nm as compared to 665 nm in solution. In contrast to the responses observed in solution, a reduction in fluorescence intensity is observed for both Chalc 1 and 2 doped contact lenses. Interestingly, the solution response for Chalc 2 towards 100 mM fructose at pH 8.0 produces a ≈ 3 fold increase in fluorescence emission, as compared to the ≈ 2.6 fold reduction for the same fructose concentration in the contact lens.

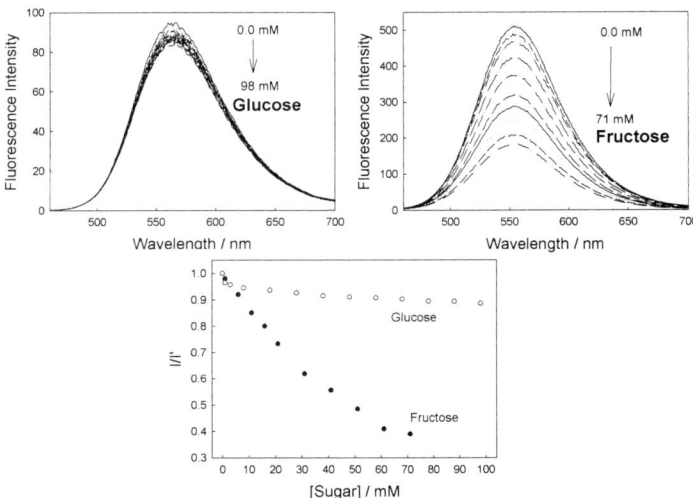

Figure 15.11. Emission spectra of Chalc 1 doped contact lens, pH 8.0 buffer / methanol (2:1), with increasing concentrations of glucose (Top left) and corresponding spectra with increasing concentrations of fructose (Top right). λ_{ex} = 430 nm. Intensity ratio plot for Chalc 1 doped contact lens towards both glucose and fructose (bottom), where I and I' are the intensities in the presence and absence of sugar respectively at λ_{em} max.

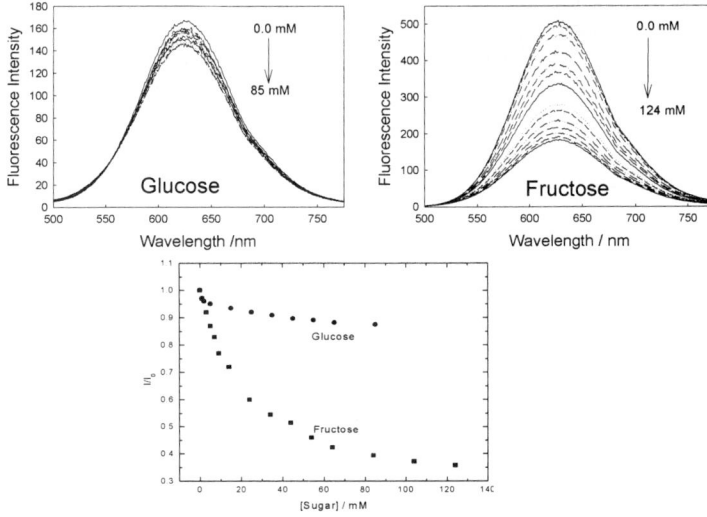

Figure 15.12. Emission spectra of Chalc 2 doped contact lens, pH 8.0 buffer / methanol (2:1), with increasing concentrations of glucose (Top left) and corresponding spectra with increasing concentrations of fructose (Top right). λ_{ex} = 460 nm. Intensity ratio plot for Chalc 2 doped contact lens towards both glucose and fructose (Bottom), where I and I' are the intensities in the presence and absence of sugar respectively at λ_{em} max.

15.3.3. Response of the PET probes within the contact lens

The emission spectra of ANDBA in the lens with glucose are presented in Figure 15.13. The response of the probe towards glucose and fructose in the lens is compared in figure 15.13 -right. It is evident from the figure that ANDBA in a contact lens has lost its ability to respond to glucose, an insignificant spectral change is observed within the contact lens. Similar to ANDBA, the monophenylboronic acid derivative ANMBA shows non responsive spectral behavior towards glucose and fructose within the contact lens. Still it is of interest to note here that the diboronic acid derivative ANDBA still shows the glucose selectivity over the fructose, Figure 15.13-right.

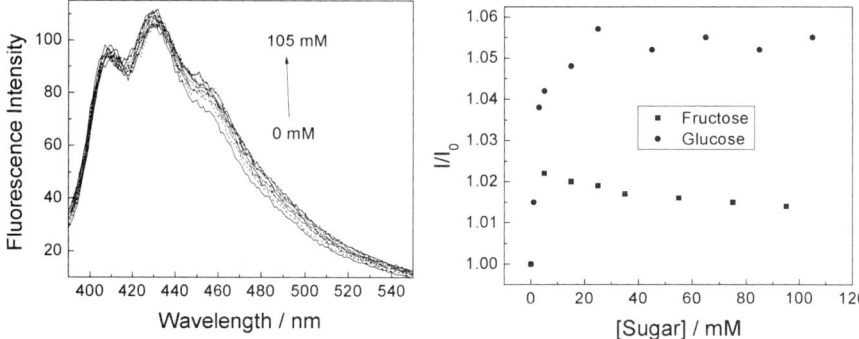

Figure 15.13. Emission spectra of ANDBA doped contact lens, pH 8.0 buffer / methanol (2:1), with increasing concentrations of glucose (Left). $\lambda_{ex} = 365$ nm. Intensity ratio plot for ANDBA doped contact lens towards both glucose and fructose (Right), where I and I_0 are the intensities in the presence and absence of sugar respectively at λ_{em} max.

In most cases the response towards glucose in the lenses was significantly different, with drastically reduced glucose responses, as compared to the response typically observed at physiological pH. To understand the different response of these probes within the contact lens, a compared to that in buffer we have estimated the local pH and polarity of the contact lens and effect of the pH on the signaling behavior of the probes. Figure 15.14 shows the emission spectra of DDPBBA in different pH solutions in the presence of increasing concentrations of glucose. As with all the other probes studied, we observed that the response towards glucose in the contact lens was similar to the response observed in ≈ pH 6.0 bulk solution, c.f. Figure 15.10-top right and 15.14 bottom right. We subsequently doped the well-known pH sensitive probe *fluorescein* [41] within the lenses, and determined that the lenses had an unbufferable pH of ≈ 6.1, Figure 15.15-left. However, the influence of an acidic lens pH does not solely explain the spectral shifts observed in Figures 15.10 and 15.14, with DDPBBA additionally showing a 50 nm hyposchromic shift in the contact lens. An investigation of the polarity within the contact lens, using the

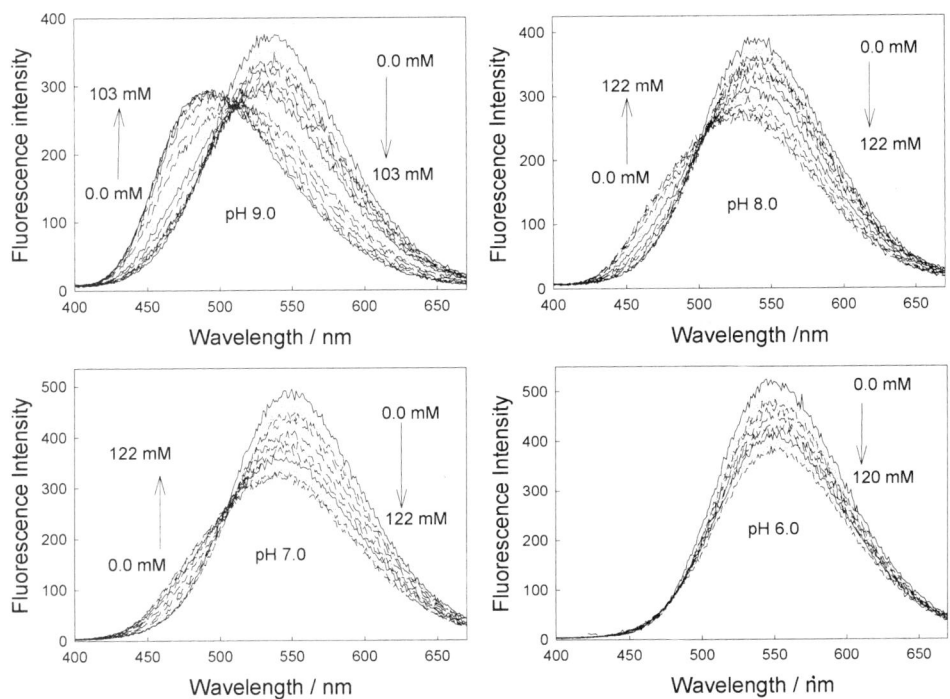

Figure 15.14. Emission spectra of DDPBBA in different pH media (buffer/methanol, 2:1) with increasing glucose concentrations. λ_{ex} = 340 nm.

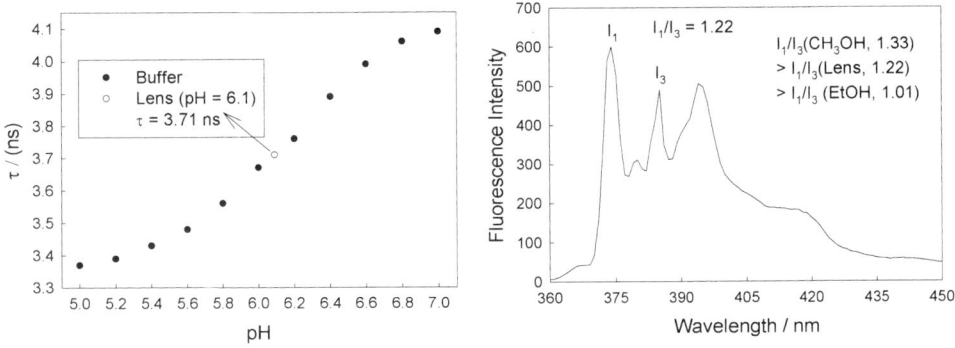

Figure 15.15. Fluorescein lifetime versus pH of the medium, and the lifetime of a fluorescein doped contact lens, (Left). Fluorescence spectra of pyrene doped contact lens to asses the polarity inside the contact lens (Right). The obtained I_1/I_3 data is close to that of methanol.

pyrene I_1 and I_3 band ratio's (I_1/I_3) [14,42,43], Figure 15.15-right, revealed a lens polarity similar to that of methanol, which in hindsight was not too surprising, given the nature of the polymer, i.e. it is PVA based. Thus the difference in the polarity and pH may cause the difference in the response of the probes within the contact lens when compared to that in solution.

15.4. RATIONALE FOR THE DESIGHN OF NEW GLUCOSE SENSING PROBES

Feasibility studies of doped lenses using the above mentioned boronic acid probes produced poor glucose responses, rationaled as due to the mildly acidic pH and methanol like polarity within the contact lens, and the subsequent effect on the transduction mechanisms, ICT or PET. This was also not surprising given the fact that these probes were designed for sensing at a physiological pH of ≈ 7.4, the probes typically having pK_a around 9. Hence to obtain a notable glucose response in the contact lens polymer, it was deemed necessary to design new probes with significantly reduced sugar-bound pK_a values. In addition to the environmental parameters and constraints of pH and polarity, the probes also have to be sensitive to the very low concentrations of tear glucose, ≈ 500 μM for a healthy person, increasing up to several mM for diabetics, recalling that the *blood glucose* levels for a healthy person are ≈ 10-fold higher [14-17].

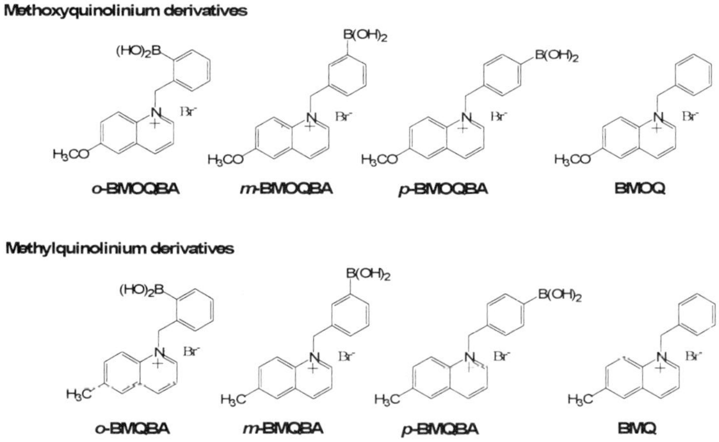

Chart 15.2. Molecular structure of the boronic acid probes based on quinolinium nucleus as the fluorophore. *o-*, *m-*, and *p-* BMOQBA – *N*-(2, 3, 4-boronobenzyl)-6-methoxylquinolinium bromide, and the control compound (BMOQ–*N*-benzyl-6-methoxyquinolinium bromide); *o-*, *m-*, and *p-*BMQBA – *N*-(2, 3, 4, -boronobenzyl)-6-methylquinolinium bromide, and the corresponding control compound (BMQ–*N*-benzyl-6-methylquinolinium bromide)

As mentioned in section 15.2, the pK_a of phenyl boronic acid is known to be tunable with the appropriate substituents [44], for example, an electron withdrawing group reduces the pK_a of the sugar bound form, while an electron donating group increases it. We therefore considered the interaction between the quaternary nitrogen of the 6-methyl- and methoxyquinolinium moieties, and the boronic acid group, which reduces the pK_a of the probe. In this regard we have synthesized 2 new classes of isomeric boronic acid containing probes (8-probes in total), Chart 15.2, where the spacing between the interacting moieties, quaternary nitrogen of the 6-methyl- or methoxyquinolinium and boronic acid groups, enables both an understanding of the sensing mechanism to be realized, and the selection of the most suitable isomer based on its glucose binding affinity. In addition, control compounds (BMQ and BMOQ), which do not contain the boronic acid moiety, and are therefore insensitive towards sugar, were synthesized to understand the spectral properties and responses of the probes, Chart 15.2.

15.5. GLUCOSE SENSING PROBES BASED ON THE QUINILOINIUM MOIETY

The new quinolinium based boronic acid containing probes (o-BMQBA – N-(2-boronobenzyl)-6-methylquinolinium bromide, m-BMQBA–N-(3-boronobenzyl)-6-methylquinolinium bromide, p-BMQBA–N-(4-boronobenzyl)-6-methylquinolinium bromide) and the control compound (BMQ–N-benzyl-6-methylquinolinium bromide), were conveniently prepared using a generic one step synthetic procedure described elsewhere by the authors. Similarly, the corresponding 6-methoxyquinoline nucleus; o-, m- and p-BMOQBA and a control compound BMOQ, were prepared in an analogous manner to the methyl quinolinium probes. These probes, because of the quaternized nitrogen center, are readily water soluble, alleviating the need to use methanol to solubalize the probes as discussed in section 15.2.

15.5.1. Photophysical Characterization of the Quinolinium probes

Figure 15.16 shows representative absorption and emission spectra for the *ortho*-isomers of the BMQBA and BMOQBA probes. Photophysical data of the probes is shown in Table 15.2. Typical absorption and emission band maximum of the probes can be seen at 318 and 450 nm, and 319 and 427 nm for the BMOQBA and BMQBA probes respectively. The additional absorption band at ≈ 350 nm for BMOQBA is attributed to the n→π* transition of the oxygen of 6-methoxy group [45]. The excitation independent emission band at ≈ 450 nm indicates only one ground-state species is present for BMOQ and BMOQBAs probes. The large Stokes-shifted fluorescence emission band of ≈ 100 nm is ideal for fluorescence sensing, allowing easy discrimination of the excitation wavelengths [41,45]. All probes were found to be readily water soluble, a function of the quaternized structure.

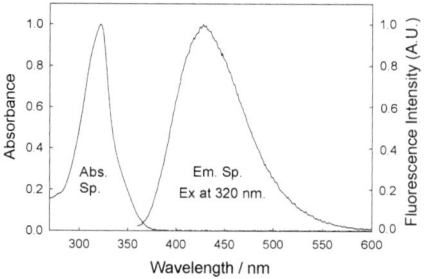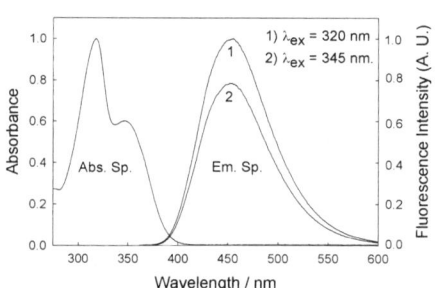

Figure 15.16. Absorption and emission spectra of *o*-BMQBA (Left) and *o*-BMOQBA (Right) in water. The spectra are representative of the respective isomeric phenylboronic acid containing fluorophores and the control compounds.

Table 15.2 shows the quantum yield values for the probes in water, obtained from a spectral comparison with N-(3-sulfopropyl)-6-methoxyquinolinium [(SPQ) (Φ_f = 0.53 in water [46])], where we can see that the BMOQBA probes have significantly higher quantum yields as compared to the BMQBA probes. Another reference compound, N-methyl-6-methylquinolinium bromide (MMQ) previously published by the authors [47] exhibits very similar spectral properties to that of BMQ and BMQBA probes, except for a noticeable quantum yield and mean lifetime difference, approximately 10-fold higher than its methyl quinolinium counterpart, namely BMQBA. This indicates an interaction between the phenyl ring and methylquinolinium moiety of the BMQBA and BMQ probes, Table 2, which is not present, or present to a much lesser extent, for the BMOQBA probes. We have attributed the relatively shorter lifetime and quantum yields of the new BMQBA probes and control compound to a photo-induced electron transfer mechanism, where the phenyl ring of BMQBA is the *donor*, and the methylquinolinium moiety is the *acceptor*. In the case of BMOQ and BMOQBA probes because of relatively more electron donating methoxy group, methoxyquinolinium is a weak acceptor and hence the PET mechanism is insignificant in these systems. Subsequently, the BMOQBA probes were found to have monoexponential lifetimes ($\approx$ 24.9 $\rightarrow$ 26.7 ns) as compared to the BMQBA probes which were biexponential in water, with significantly reduced lifetimes of 2.18 (46 % amplitude) and 4.74 ns (54 % amplitude), for the *ortho*-isomer. Both control compounds were found to have monoexponential lifetimes in water, with the BMOQ over ten times longer, 27.3 ns, as compared to BMQ, 2.59 ns. These lifetime changes further support the electron transfer hypothesis, the B$^-$(OH)$_3$ present at neutral pH further reducing the lifetime of the boronic acid probes, i.e. **2** in Scheme 15.1.

Table 15.2 Photophysical data of the quinolinium probes in water at room temperature.

Property	o-BMOQBA	m-BMOQBA	p-BMOQBA	BMOQ
λ_{abs} (max) / nm	318, 346	318, 347	318, 346	318, 347
λ_{em} (max) / nm	450	450	451	453
Φ_f	0.46	0.51	0.49	0.54
τ_f / ns	26.7	25.9	24.9	27.3
	o-BMQBA	m-BMQBA	p-BMQBA	BMQ
λ_{abs} (max) / nm	319	322	322	322
λ_{em} (max) / nm	427	427	427	427
Φ_f	0.043	0.025	0.023	0.045
τ_f / ns	4.01[a]	3.72[a]	2.10[a]	2.59

[a] Mean fluorescence lifetime.

In addition to the quantum yield and fluorescence lifetime differences between the phenyl ring containing BMQBA and BMQ probes, and the MMQ probe, we can also see lifetime differences between the BMQBA isomers themselves. We have attributed these changes due to the changes in electron donating ability of the different phenyl isomers, and additionally to their different through-space / through-bond interactions [48,49] with the methylquinolinium moiety, noting again that some $[B^-(OH)_3]$ is likely to be present at neutral pH.

The quantum yield values of the BMOQBA isomers are slightly lower than that of the control compound (BMOQ), the quantum yield values increasing in the order *ortho-*, *para-* and *meta-*. In contrast the monoexponential fluorescence lifetimes of the isomers increased in the order *para-*, *meta-* and then *ortho-*, which was slightly surprising as the quantum yields and lifetimes usually change in unison. Similar to having the highest quantum yield, the control compound also had as expected, the longest lifetime of 27.3 ns. One explanation for these differences between the isomers, lies in the interaction between the boronate ester form ($B^-(OH)_3$, **2** in Scheme 15.1) present in solution and the positively charged nitrogen center at neutral pH, the extent of which being determined by the isomer spacing.

Figure 15.17 shows the emission spectra of *o*-BMOQBA and *o*-BMQBA in buffer media whose pH is increased from pH 3 to 11. The emission spectra of the new boronic acid containing probes typically show a steady decrease in fluorescence intensity with increase in pH from 3 to 11. In contrast, the control compounds, (BMOQ and BMQ) having no boronic acid group, show no change in fluorescence intensity. Subsequently the corresponding titration curves in the absence and in the presence of 100 mM glucose and fructose, obtained by plotting the normalized intensities at band maximum versus pH, are shown in Figure 15.18.

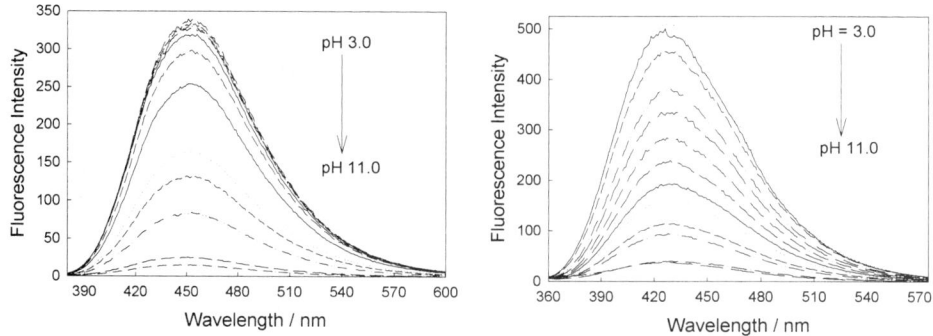

Figure 15.17. Emission spectra of *o*-BMOQBA (Left) and *o*-BMQBA (Right) in buffer media. λ_{ex} for BMOQBA and BMQBA was 345 and 320 nm, respectively.

The apparent pK_a values obtained from the titration curves shown in Figure 15.18 are shown in Table 15.3. We can see considerably reduced pK_a values for the new phenylboronic acid containing fluorophores in buffered media, as compared to the typical boronic acid probes reported in the literature [31-36], which are in the range 8-9. For comparison the pK_a values for the probes mentioned in Section 15.3 are depicted in Table 15.3. The quaternary nitrogen of the quinolinium nucleus not only reduces the pK_a of the probes, but also serves to stabilize the boronatediester, formed upon sugar complexation. This in turn increases the affinity of the probes for sugar as shown in Table 3. Hence the reduced sugar bound pK_a of these new probes, coupled with their increased glucose affinity, is most attractive for our glucose sensing contact lens application, noting our previous findings of a lens pH around 6.1 [14].

To understand this new fluorescence signaling based mechanism and the response of the probes towards both pH and monosaccharides, it is informative to consider the schematic representation shown in Figure 15.19. The boronic acid group is an electron-deficient Lewis acid having an sp²-hybridized boron atom with a trigonal planar conformation. The anionic form of the boronic acid, formed in high pH solutions, is characterized by a more electron rich sp³-hybridized boron atom with a tetrahedral geometry. The change in the electronic properties and the geometry at the boron atom induces the fluorescence spectral changes of the probes. It is well-known that the quinine/quinoline compounds exhibit high quantum yields in acidic media, from the corresponding quaternized salt [46,47]. Similarly here, the boronic acid probes are more fluorescent in acidic solutions. However, when the pH of the medium is increased the electron density on the boron atom is increased, facilitating the partial neutralization of the positively charged quaternary nitrogen of the quinolinium moiety. We have termed this interaction as a *charge neutralization-stabilization mechanism*, and a schematic representation of this mechanism with regard to glucose binding / sensing is illustrated in Figure 15.19. In contrast, the control compounds BMQ and BMOQ are unperturbed in the presence of pH or monosaccharides.

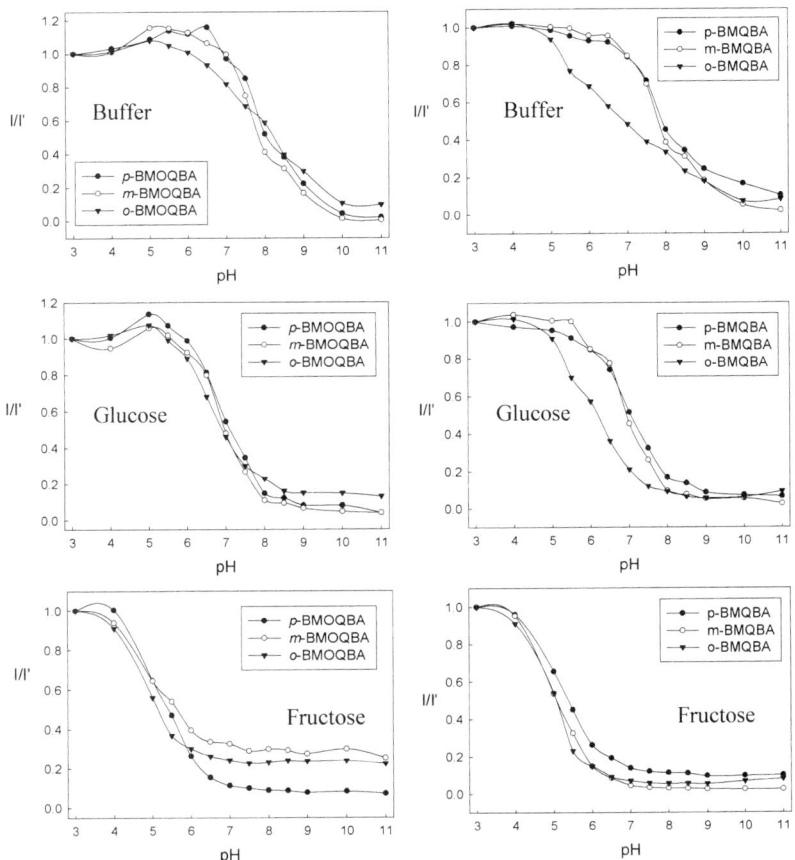

Figure 15.18. The ratio of the emission intensities at band maximum as a function of pH for BMOQBAs (Top left) and BMQBAs (Top Right) and with 100 mM Glucose (Middle left and Middle right) and 100 mM Fructose (Bottom left and Bottom right) respectively.

Table 15.3 Apparent pK_a values for boronic acid probes in buffer and effect of 100 mM sugars.

Probe	In buffer	+100 mM Glucose	+100 mM Fructose
o-BMOQBA	7.90	6.62	4.80
m-BMOQBA	7.70	6.90	5.00
p-BMOQBA	7.90	6.90	5.45
o-BMQBA	6.70	6.10	5.00
m-BMQBA	7.75	6.85	5.05
p-BMQBA	7.80	6.95	5.45
DSTBA	9.14[a]	8.34	6.61
CSTBA	8.17	7.30	5.84
DDPBBA	8.90	6.97	6.20
Chalc 1	7.50	---	5.40
Chalc 2	7.50	---	5.20

[a]The pK_a values for the probes DSTBA, CSTBA, DDPBBA, Chalc 1 and Chalc 2 are from Ref. xx.

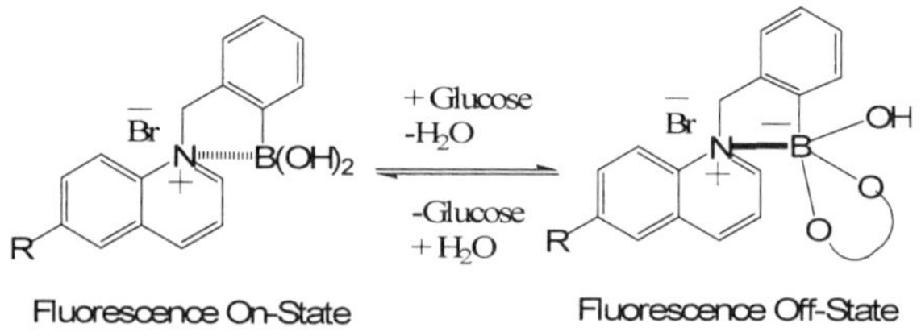

Figure 15.19. A schematic representation of the charge neutralization-stabilization mechanism with regard to glucose sensing. The bold-line between the N^+ and boron atom in the structure shown in the right side of the equation indicates the increased interaction between them, and is not intended to show covalent bond formation between the two atoms.

15.5.2 Sugar response of the Quinolinium probes in solution

The monosaccharide induced spectral changes of the probes are shown in Figure 15.20. In an analogous manner to that described for increasing pH above, we observed a systematic decrease in fluorescence intensity of the boronic acid containing probes in pH 7.5 phosphate buffer, for increasing glucose concentrations. The corresponding titration curves obtained by plotting I' divided by I, where I' and I are fluorescence intensities at 427 nm for BMQBA and 450 nm for BMOQBA, in the absence and presence of sugar respectively, versus glucose concentration, are also shown in Figure 15.21. The right-hand column of Figure 15.21 shows the response of the probes in the tear glucose concentration range.

For the BMOQBA probes we typically see a greater response for the *para-* isomer, with a 2.4-fold change in signal for 50 mM glucose. Interestingly, a ≈ 13 % change in signal is observed up to 2 mM glucose, noting that glucose levels in tears can change from ≈ 500 μM for a healthy person, and up to 5 mM for diabetics. For *ortho*-BMQBA, then a ≈16 % change in fluorescence signal can be observed over a similar glucose range. Interestingly, for both classes of probes, the *para*-isomers show the weakest response towards glucose, Figure 15.21. Given that the through-bond mechanism is expected to be prevalent for the *para*-isomers, and the *ortho*-isomer is expected to show both through bound and through space interactions, then a greater response was indeed expected, as is observed, for the *ortho*-isomer. However, the much weaker response of the *para*-isomer suggests that steric hindrance, with regard to glucose binding, is not present here.

The dissociation constants of the probes with both glucose and fructose in pH 7.5 phosphate buffer are presented in Table 15.4. As expected, a higher affinity for fructose is observed (lower K_D value), which is a general observation for monophenyl boronic acid derivatives [31-36], but it should be noted that the concentration of fructose in tears is substantially lower than for glucose [14-17]. A comparison of the trends in glucose response observed in Figure 15.21, and the recovered K_D values in Table 15.4, show some differences, which we have attributed to the difficulties encountered during data fitting. While beyond the scope of this text, these fitting difficulties clearly reflect the need for a new kinetic sugar binding function with our new probes. Further studies are underway in this regard. Based on the data presented here, these probes having lower pK_a values, may be useful for the continuous monitoring of glucose using disposable contact lenses.

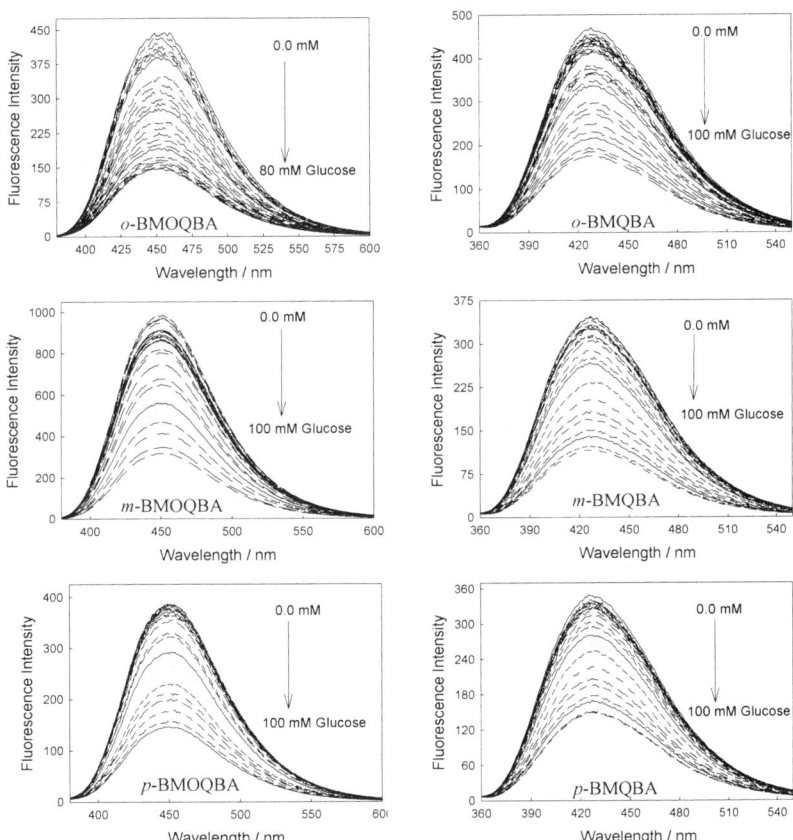

Figure 15.20. Emission spectra of BMOQBAs (Left column) and BMQBAs (Right column) in pH 7.5 phosphate buffer with increasing glucose concentration. The λ_{ex} for BMOQBAs was 345 nm and for BMQBAs 320 nm.

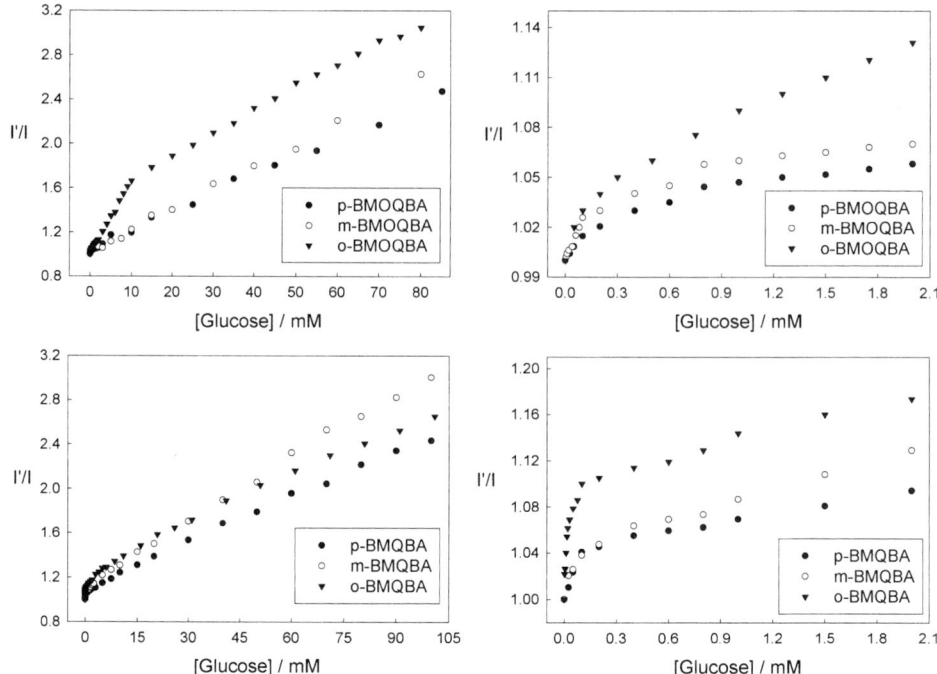

Figure 15.21. Respective intensity ratio for all three isomers of BMOQBAs in the absence, I', and in the presence, I, of glucose, (Top left), and in the *tear glucose concentration* range (Top right). The corresponding plots for BMQBAs are also shown in the blood glucose concentration range and tear glucose concentration range, (Bottom right and Bottom left, respectively).

Table 15.4 Dissociation constants, K_D (mM), of the probes with glucose and fructose in buffer and in the contact lens.

| Probe | Glucose | | Fructose | |
	Buffer	Lens	Buffer	Lens
o-BMOQBA	49.5	322.6	0.65	84.7
m-BMOQBA	1000	54.6	1.8	4.9
p-BMOQBA	430.0	111.1	9.1	34.7
o-BMQBA	100	17.9	4.7	34.8
m-BMQBA	476	58.1	13.2	21.6
p-BMQBA	370	128.2	13.8	12.9

15.5.3 Response of the new Glucose Signaling Probes in the Contact Lens

Doped contact lenses, which were previously washed and allowed to leach dye for 1 hour were tested with both glucose and fructose. Buffered solutions of sugars were added to the lens, pH 7.5 phosphate buffer, in an analogous manner to ocular conditions. Fluorescence spectra were typically taken 15 mins after each sugar addition to allow the lens to reach equilibrium. The 90 % response time, the time for the fluorescence signal to change by 90 % of the initial value, was $\approx$ 10 minutes.

Figure 15.22 shows the response of *o*-BMOQBA and *o*-BMQBA, Top and Bottom left respectively, for increasing concentrations of glucose injected into the 1.5 cm³ contact lens volume. Similar to the solution based measurements, the probes show a decrease in fluorescence intensity, which we attribute to the complexation of glucose with boronic acid and the subsequent charge neutralization mechanism described earlier. We were again able to construct the *I'/I* plots, where *I'* is the intensity in the absence of sugar, Figure 15.22 Right. As was observed in solution, Fructose had a greater response, reflecting the greater affinity of mono phenyl boronic acid derivatives for fructose. However, in the low sugar concentration ranges, < 2 mM sugar, the response towards both sugars was comparable [15-17]. Differences in the response towards glucose for the isomers could also be observed in the lenses, Figure 15.23, where *m*-BMOQBA was found to have the greatest response amongst this class of probes. From Figure 15.23 we can clearly see a greater response in the lens towards sugars than in our solution based studies at pH 7.5, with *p*-BMQBA showing a greater than 20 % fluorescence signal change with as little as 2 mM glucose. This wasn't unexpected, and is simply explained by the pK_a of the probes being < 7, the probes being compatible with the mildly acidic lens environment. The dissociation constants (K_D) obtained for all six probes with glucose and fructose within the contact lens are shown in Table 15.4. As seen from the table, the affinity of the probes towards sugar is relatively less in the contact lens, and that may be due to the nonbufferable pH as mentioned earlier in Section 15.3.

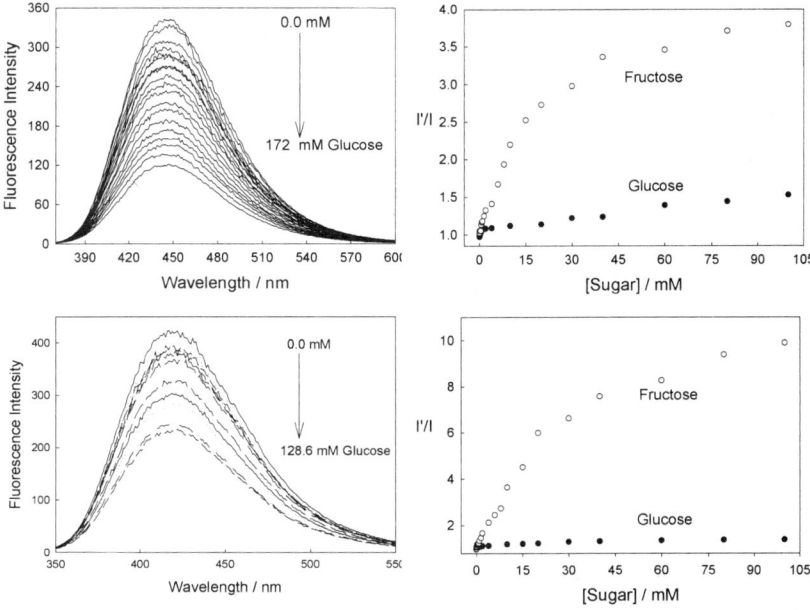

Figure 15.22. The emission spectra of an *o*-BMOQBA and *o*-BMQBA doped contact lens in the presence of increasing glucose concentrations (Top left and Bottom Left, respectively). λ_{ex} = 345 nm for *o*-BMOQBA and 320 nm for *o*-BMQBA. The corresponding emission intensity ratio at band maximum for *o*-BMOQBA and *o*-BMQBA doped contact lens in the absence, *I'*, and presence, *I*, of both Glucose and Fructose (Top right and Bottom right, respectively).

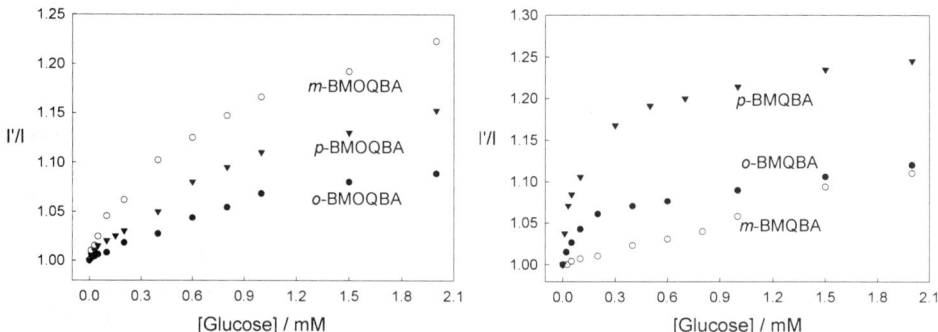

Figure 15.23. The response of BMOQBAs (Left) and BMQBAs (Right) in the contact lens and in the *tear glucose concentration range.*

Figure 15.24 directly compares the response of the probes in both the contact lens and buffer, where we can see a comparable if not better response towards glucose in the lens in the low concentration range of sugar. However, in the high concentration range, the probe doped contact lens shows a smaller response towards sugar, and we are uncertain at this time but we speculate that the binding efficiency of the probes or the local environment such as pH and polarity of the lens still play a role on the binding interaction of the probes. Also, the probe location in the contact lens, glucose diffusion, and probe leaching from the lens may well complicate the sensing response of the probes from within the contact lens as evident from the I' / I Vs [sugar] plots. In this regard, the response of doped lenses also shows complex behavior towards mM fructose concentrations, with much simpler kinetics observed in the tear glucose concentration range (data not shown).

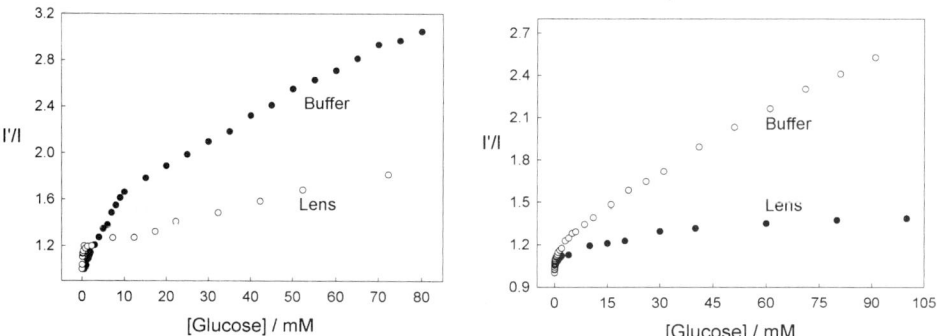

Figure 15.24. A comparison of the emission intensity ratio for the *o*-BMOQBA doped contact lens with that obtained in pH 7.5 phosphate buffer, in the absence, *I'*, and presence, *I*, of Glucose (Left) and the corresponding plot for *o*-BMQBA (Right).

For all our glucose contact lens sensing studies we repeated these doped lens experiments several times, and in all cases the trends were reproducible. It is difficult to assess the effect of the PVA hydroxyl groups of the contact lens polymer on the response of boronic acid to sugar, but our studies with solutions of glycerol indicated that sugar had much higher binding affinities than glycerol hydroxyl groups. We therefore speculate that sugars will preferentially bind boronic acid groups in the PVA lens polymer. In any event the boronic acid probes function well (reversibly) towards sugars in the lens, in what is likely to be an environment saturated with PVA hydroxyl groups.

15.6. PROBE LEACHING, INTERFERENTS AND SHELF LIFE

Leaching studies of the probes from the contact lens polymer were undertaken using the lens holder shown in Figure 15.7, which contained $\approx$ 1.5 cm^3 buffer at 20^0 C. A Varian fluorometer measured the intensity change as a function of time to determine the percentage signal change, corresponding to dye leaching. It should be noted that with no sample present, no intensity fluctuations or drifts were observed, indicating stability of the fluorometer Xenon-arc source.

We were able to observe up to about 8 % change in fluorescence intensity, attributed due to dye leaching, for the BMOQBA class of probes, with very little change after about 25 minutes, Figure 15.25. In contrast, the BMQBA probes show a much greater extent of leaching over the same time period and under identical conditions. Given that the BMQBA probes typically showed a greater response towards glucose in the lens, then this suggests that the BMQBA probes may be more accessible in the lens, than the BMOQBA probes. In addition, similar results were obtained at 38^{0}C but with a different leaching rate.

In all our lens response studies described here, lenses were pre-leached to a steady-state fluorescence intensity before use. After glucose measurements were undertaken the outer lens fluid volume surrounding the contact lens was found to be non-fluorescent indicating that dye had not leached from the lens during actual glucose sensing measurements. It should be noted that while chemistries are available to covalently label our probes within the contact lens polymer, which would eliminate any leaching, it is an important design concern for our approach that the lenses remain unmodified, so that their physiological characteristics and compatibilities remain unchanged. In fact our approach is targeted at reducing future lens redesign costs for industry, by using simple probe doping.

As with all sensors it is important to consider the effects of potential interferents and sensor shelf-life on the working response of the device. Throughout much of this paper we have shown the response of the probes towards fructose, primarily because of its well-known greater affinity for the boronic acid moiety [31-36]. However, the concentration of fructose in blood is $\approx$ 10 times lower than glucose [14], a relationship which is also thought to occur in tears [14]. Hence fructose is not thought to be a major interferent in tears, simply shown here to place the binding trends in context. However, tears are a complex mixture of proteins and other analtyes, such as sodium (120-170 mM), potassium (6-26 mM) and chloride (100 mM) [50]. We subsequently tested these new contact lens probes with various aqueous halides, given that Na^+ and K^+ are unlikely to perturb our fluorophores, Table 15.5. As expected, the BMOQBA probes are modestly quenched by chloride, with steady-state Stern-Volmer constants in the range 170-182 M^{-1}, the BMQBA probes having significantly smaller quenching constants in the range 17–44 M^{-1}. This result is simply explained by the shorter lifetime of the BMQBA probes as compared to the BMOQBA probes, and the probability of an excited-state chloride ion encounter [41,45]. Encouragingly, the BMQBA probes typically showed a greater response towards glucose, as well as being the least perturbed by aqueous chloride. In any event, simple corrections in the fluorescence signal can readily account for chloride interference on the glucose response. This is shown in Figure 15.1. Sensors spots on the surface of the lens could contain a reference chloride compound or indeed another probe sensitive to both glucose and chloride. By employing the extended Stern-Volmer equation for multiple analytes [45], one can easily correct for background interferences.

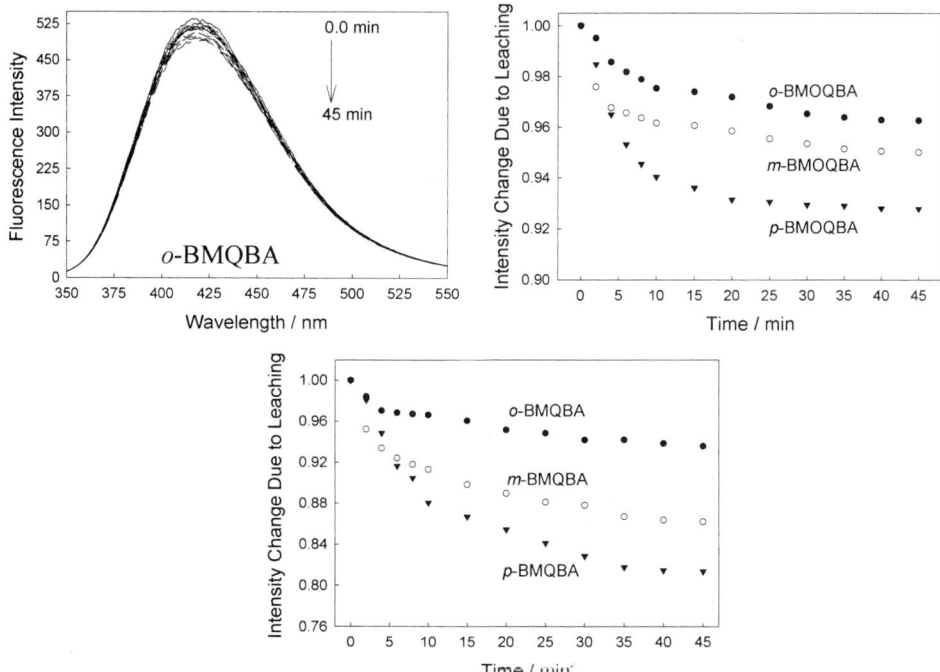

Figure 15.25. Emission spectra of *o*-BMQBA doped contact lens immersed in pH 7.5 buffer with time (Top left). λ_{ex} = 320 nm. Normalized intensity change at band maximum of BMOQBA with time due to leaching (Top right), and that for the BMQBAs (Bottom).

It is also informative to consider the pH of tears as a potential interferent, given the response of these probes to pH as shown in Figure 15.8. It is known that unstimulated tear pH levels can vary in the range 7.14 – 7.82 measured from healthy subjects at different times of the day, with a typical mean value around pH 7.45 [50]. However, a more acidic pH of less than 7.3 is found following prolonged lid closure, e.g. after sleep, which is thought to result from carbon dioxide produced by the cornea and trapped in the tear pool under the eye lids. While solutions of these new probes would be susceptible to these changes in pH, we have found that the doped lenses we studied were indeed unbufferable, hence external changes in pH are most unlikely to affect the glucose response

Table 15.5. Stern-Volmer (K_{SV}, M^{-1}) constants of the probes with halides in water.

Probe	Cl⁻	Br⁻	I⁻
o-BMOQBA	170	332[a]	471[a]
m-BMOQBA	182	413	540
p-BMOQBA	177	370	595
BMOQ	222	384	520
o-BMQBA	44.0	55.0	97.0
m-BMQBA	20.0	32.0	48.0
p-BMQBA	17.0	26.5	42.0
BMQ	35.0	55.0	71.0

[a]The concentrations of Br⁻ and I⁻ in tears is extremely low and is unlikely to be an interferent in our glucose measurements.

With regard to the glucose-sensing contact lens shelf-life, lenses that had been doped, leached and stored for several months both wet and dry, gave identical sugar sensing results, indicating no lens polymer – fluorophore interactions over this time period, or indeed probe degradation.

15.7. FUTURE DEVELOPMENTS BASED ON THIS TECHNOLOGY

15.7.1 Continuous and Non-invasive Glucose Monitoring

In this chapter, we have shown that fluorescent probes can be fabricated to be compatible with in the commercially available daily use disposable contact lenses, which have already been assessed and optimized with regard to vision correction and oxygen permeability. This has enabled the first prototype based on this new approach to be realized. With regard to glucose monitoring by this approach, we speculate on several future improvements to this technology:

- *Clear or Colored Contact Lenses*

Many boronic containing fluorophores have a visible absorption [31-36], which apart from their lack of glucose sensitivity in the lens as discussed earlier [14], would introduce color into a doped lens. While colored lenses are attractive to a few people as sports or even fashion accessories, the majority of contact lenses worn today are clear, hence our colorless lenses described here are ideal in this regard. One disadvantage however with our lenses is the requirement for an excitation and detection device as shown in Figure 15.1. One improvement to our technology could be the use of colored contact lenses, which change color due to the concentration of tear and therefore blood glucose. This can be achieved by the ground-state binding of glucose to boronic acid and the subsequent changes in fluorophore absorption spectrum. A patient wearing the lenses could simply look in the mirror, or the color even determined by an on-looker, and compared to a precalibrated color strip, one could assess the extent of hyperglycemia. This technology would be most attractive to parents of young diabetic children or for care workers of the elderly. Work in currently underway in our laboratories in this regard.

- *Sensor Spots or Doped Lenses*

As briefly mentioned earlier and shown in Figure 15.1, sensor spots on the surface of contact lenses could correct signal responses for interferents such as aqueous chloride. Indeed the spots could either be visible to the wearer (self-readout) or readable by an external monitoring device.

- *Detection Methods*

While simple colorimetric methods are likely to be the most simple to introduce to the market place, other fluorescence sensing methodologies, such as polarization, lifetime and ratiometric sensing, offer many spectroscopic sensing advantages over the simple intensity measurements described for our lenses [41]. For example, fluorescence lifetime and ratiometric measurements are independent on total light intensity or indeed fluctuations in ambient room light.

15.7.2 Clinical Condition and Diagnosis from Tears

Given that tear glucose can be continuously and non-invasively monitored using our lens approach, then it may be possible for both clinical condition assessment and disease diagnosis using the contact lens sensing platform, and suitably designed fluorophores. In fact, as compared to saliva, tears represent a more stable body fluid of low protein concentration and with only modest variations in pH.

For example, for clinical condition assessment, Na^+, K^+, Ca^{2+}, Mg^{2+}, pH, Histamine, Urea, Lactate, Cholesterol and Glucose in tears are known to directly track or relate to the serum levels [50]. Indeed, one could potentially even track body core temperature using thermochromic type probes [51], embedded within contact lenses or even sensor spots.

For disease diagnosis, the possibility of diagnosing; Glaucoma, Sjogrens disease, Lysosomal storage diseases, corneal ulceration and bacterial infections could be realized by designing lenses to detect; Catecholamines, Lysozyme, Lysosomal enzymes, Collagenase and α-Antitrypsin respectively. It is possible that many other clinical conditions and diseases could also be either monitored or diagnosed via this approach, although relatively little tear biochemistry is known [50].

15.7.3 Drug Testing, Compliance and Screening

Saliva is and has been used for therapeutic drug monitoring, by predicting the free fraction of drugs in blood from that determined in saliva. Changes in the free drug levels can have important clinical consequences, since either toxic or sub-therapeutic levels may exist, even when the total drug concentration is in the normal range. Coupled with the fact that saliva concentrations varying greatly in both pH and composition, obviates the need for a novel clinical sensing platform for drug testing, compliance and screening. As mentioned earlier, tear fluid has a relatively lower protein concentration with only slight changes in pH, where the passage of drugs from plasma to tears takes place by diffusion of the non-protein bound fraction [50]. Subsequently, tears have already been used to assess the concentration of antibiotics such as Ampicillin and the Anticonvulsants, Phenobarbital and Carbamazepine [50]. However, these drug concentration measurements are inherently difficult due to the 5 -10 μL total tear volume to be sampled [50]. Indeed, while tear glucose levels have been known to be elevated during hyperglycemia for nearly 70 years, it is the difficulties associated with tear collection, which has limited the practical use of tears for diabetes mellitus assessment. However, similar to our glucose sensing contact lenses, it may be possible to develop lenses for drug testing, based on either colorimetric or other fluorescence spectroscopic based methodologies.

15.8. CONCLUDING REMARKS

We have developed a range of new glucose sensing contact lenses, by doping strategically designed fluorescent probes into commercially available contact lenses. The probes are completely compatible with the new lenses and can readily detect glucose changes up to several mM glucose, appropriate for the tear glucose concentration range for diabetics, i.e. ≈ 500 μM $\rightarrow 5$ mM [14].

The lenses have a 90 % response time of about 10 minutes, allowing the continuous and noninvasive monitoring of ocular glucose. This is a significant improvement over enzymatic methods based on blood sampling by finger pricking, with many diabetics begrudgingly testing between 4 and 6 times daily.

With diabetes being widely recognized as one of the leading causes of death and disability in the western world, we believe our boronic acid doped contact lens approach and findings, are a notable step forward towards the continuous and non-invasive monitoring of physiological glucose.

15.9. ACKNOWLEDGEMENTS

The authors would like to thank the University of Maryland Biotechnology Institute and the NIH, National Center for Research Resources, RR-08119, for financial support.

15.10. REFERENCES

1. Principles of Diabetes Mellitus, edited by Leonid Poretsky, Kluwer Academic Plenum Publishers, Norwell Massachusetts, USA, 2002.
2. V. C. Medvei, The 18th Century and the beginning of the 19th Century, In: The history of clinical endocrinology: a comprehensive account of endocrinology from earliest times to present day, Parthenon Publishing, New York, 1993.
3. M. R. Robinson, R. P. Eaton, D. M. Haaland, G. W. Koepp, E. V. Thomas, B. R. Stallard and P. L. Robinson (1992). Non-invasive glucose monitoring in diabetic patients: A preliminary evaluation, *Clin. Chem.* 38, 1618-1622.
4. H. M. Heise, R. Marbach, T. H. Koschinsky, and F. A. Gries (1994). Non-invasive blood glucose sensors based on near-infrared spectroscopy, *Ann. Occup. Hyg.,* 18, 439-447.

5.   W. F. March, B. Rabinovitch, R. Adams, J. R. Wise and M. Melton (1982). Ocular Glucose sensor, *Trans. Am. Soc. Artif. Intern. Organs*, 28, 232-235.

6.   B. Rabinovitch, W. F. March and R. L. Adams (1982). Non-invasive glucose monitoring of the aqueous humor of the eye, Part 1, Measurement of very small optical rotations, *Diabetes Care*, 5, 254-258.

7.   G. M. Schier, R. G. Moses, I. E. T. Gan, and S. C. Blair (1988). An evaluation and comparison of reflolux Iiand Glucometer II, two new portable reflectance meters for capillary blood glucose determination, *Diabetes Res. Clin. Pract.*, 4,177-181.

8.   W. Clarke, D. J. Becker, D. Cox, J. V. Santiago, N. H. White, J. Betschart, K. Eckenrode, L. A. Levandoski, E. A. Prusinki, L. M. Simineiro, A. L. Snyder, A. M. Tideman and T. Yaegar (1988). Evaluation of a new system for self blood glucose monitoring, *Diabetes Res. Clin. Pract.*, 4, 209-214.

9.   W. Trettnak and O. S. Wolfbeis (1989). Fully reversible fiber-optic glucose biosensor based on the intrinsic fluorescence of glucose-oxidase, *Anal. Chim. Acta*, 221,195-203.

10. D. Meadows and J. S. Schultz (1988). Fiber optic biosensor based on fluorescence energy transfer, *Talanta*, 35, 145-150.

11. L. Tolosa, H. Malak, G. Rao, and J. R. Lakowicz (1997). Optical assay for glucose based on the luminescence decay time of the long wavelength dye Cy5, *Sensors Actuators B.*, 45, 93-99.

12. L. Tolosa, I. Gryczynski, L. R. Eichorn, J. D. Dattelbaum, F. N. Castellano, G. Rao and J. R. Lakowicz (1999). Glucose sensors for low cost lifetime-based sensing using a genetically engineered protein, *Anal. Biochem.*, 267,114-120.

13. S. D'Auria, N. Dicesare, Z. Gryczynski, I. Gryczynski, M. Rossi and J. R. Lakowicz (2000). A thermophilic apoglucose dehydrogenase as a nonconsuming glucose sensor, *Biochem. Biophys Res. Commun.*, 274, 727-731.

14. R. Badugu, J. R. Lakowicz, and C. D. Geddes (2004). The non-invasive continuous monitoring of physiological glucose using a novel monosaccharide-sensing contact lens, *Anal. Chem.*, 76, 610-618.

15. R. Badugu, J. R. Lakowicz, and C. D. Geddes (2003). A Glucose Sensing Contact Lens: A Non-Invasive Technique for Continuous Physiological Glucose Monitoring, *J. Fluorescence*, 13, 371-374.

16. C. D. Geddes R. Badugu, and J. R. Lakowicz, (2004). Contact lenses may provide window to blood glucose, *Biophotoincs international*, February (2), 50-53.

17. R. Badugu, J. R. Lakowicz, and C. D. Geddes (2004). Ophthalmic glucose sensing: A novel monosaccharide sensing disposable and colorless contact lens, *The Analyst*, 129, 516-521

18. J. M. Sugihara and C. M. Bowman (1958). Cyclic Benzeneboronate Esters, *J. Am. Chem. Soc.*, 80(10), 2443-2446.

19. J. P. Lorand and J. O. Edwards (1959). Polyol Complexes and Structure of the Benzeneboronate Ion, *J. Org. Chem.*, 24(6), 769-774.

20. G. Springsteen and B. Wang (2002). A detailed examination of boronic acid–diol complexation *Tetrahedron*, 58(26), 5291-5300.

21. T. D. James, K. R. A. S. Sandanayake and S. Shinkai, (1995). Chiral discrimination of monosaccharides using a fluorescent molecular sensor, *Nature*, 374, 345.

22. J. C. Norrild and H. Eggert (1995). Evidence for monodentate and bidentate boronate complexes of glucose in the furanose form – application of (1)J(C-C)-coupling-constants as a structural probe, *J. Am. Chem. Soc.*, 117(5), 1479-1484.

23. H. Eggert, J. Frederiksen, C. Morin and J. C. Norrild (1999). A new glucose-selective fluorescent bisboronic acid. First report of strong alpha-furanose complexation in aqueous solution at physiological pH, *J. Org. Chem.*, 64(11), 3846-3852.

24. W. Yang, H. He, D. G. Drueckhammer (2001). Computer-guided design in molecular recognition: Design and synthesis of a glucopyranose receptor, *Angew. Chem. Int. Ed.*, 40(9), 1714-1718.

25. W. Wang, S. Gao, and B. Wang (1999). Building Fluorescent Sensors by Template Polymerization: The Preparation of a Fluorescent Sensor for D-Fructose, Org. Letts, 1(8) 1209-1212.

26. S. Gao, W. Wang and B. Wang (2001). Building Fluorescent Sensors for Carbohydrates Using Template-Directed Polymerizations, *Bioorg. Chem.*, 29, 308-320.

27. J. J. Lavigne, E. V. Anslyn (1999). Teaching Old Indicators New Tricks: A Colorimetric Chemosensing Ensemble for Tartrate/Malate in Beverages, *Angew. Chem. Int. Ed.*, 38(24), 3666-3669.

28. J. Yoon and A. W. Czarnik (1992). Fluorescent chemosensors of carbohydrates. A means of chemically communicating the binding of polyols in water based on chelation-enhanced quenching, *J. Am. Chem. Soc.*, 114, 5874-5875.

29. B. D. Smith, S. J. Gardiner, T. A. Munro, M. F. Paugam and J. A. Riggs (1998). Facilitated transport of carbohydrates, catecholamines, and amino acids through liquid and plasticized organic membranes, *J. Incl. Phenom. Mol. Recogn. Chem.* 32, 121-131.

30. S. Soundararajan, M. Badawi, C. M. Kohlrust, J. H. Hagerman (1989).Boronic acids for affinity – chromatography – spectral methods for determinations of ionization and diol-binding constants, *Anal. Biochem.*, 178, 125-134.

31. T. D. James, K. R. A. S., and S. Shinkai (1994). A glucose-selective molecular fluorescence sensor, *Angew, Chem. Int. Ed.*, 33(21), 2207-2209.

32. T. D. James, K. R. A. S. Sandanayake, R. Iguchi, and S. Shinkai (1995). Novel saccharide-photoinduced electron-transfer sensors based on the interaction of boronic acid and amine, *J. Am. Chem. Soc.*, 117(35), 8982-8987.

33. N. Dicesare and J. R. Lakowicz (2001). Evaluation of two synthetic glucose probes for fluorescence-lifetime based sensing, *Anal. Biochem.*, 294, 154-160.

34. N. Dicesare and J. R. Lakowicz (2001). Wavelength-ratiometric probes for saccharides based on donor-acceptor diphenylpolyenes, *J. Photochem. Photobiol. A: Chem.*, 143, 39-47.

35. N. Dicesare and J. R. Lakowicz (2001). New color chemosensors for monosaccharides based on Azo dyes, *Org. Lett.*, 3(24), 3891-3893.

36. N. Dicesare and J. R. Lakowicz (2002). Chalcone-analogue fluorescent probes for saccharides signaling using the boronic acid group, *Tet. Lett.*, 43, 2615-2618.

37. V. V. Karnati, X. Gao, S. Gao, W. Yang, W. Ni, S. Sankar and B. Wang (2002). A glucose-selective fluorescence sensor based on boronic acid-diol recognition, *Bioorg. Med. Chem. Lett.*, 12, 3373-3377.

38. N. Dicesare and J. R. Lakowicz (2002). Charge transfer fluorescent probes using boronic acids for monosaccharide signaling, *J. Biomedical Optics,* 7(4), 538-545.

39. R. Badugu, J. R. Lakowicz, C. D. Geddes (2004). Fluorescence Sensors for Monosaccharides Based on the 6-Methylquinolinium Nucleus and Boronic Acid Moiety: Application to Ophthalmic Diagnostics., *Talanta*, - In press.

40. R. Badugu, J. R. Lakowicz, C. D. Geddes (2004). Boronic acid fluorescent sensors for monosaccharide signaling based on the 6-methoxyquinolinium heterocyclic nucleus: Progress towards noninvasive and continuous glucose monitoring, *Bioorg. Med. Chem.* Manuscript Submitted.

41. J. R. Lakowicz, Principles of Fluorescence Spectroscopy, 2nd Edition, Kluwer/Academic Plenum Publishers, New York, 1997.

42. N. J. Turro, B. H. Baretz and P. I. Kuo (1984). Photoluminescence probes for the investigation of interactions between sodium dodecylsulfate and water-soluble polymers, *Macromolecules*, 17(7), 1321-1324.

43. K. Kalyanasundaram and J. K. Thomas (1977). Environmental effects on vibronic band intensities in pyrene monomer fluorescence and their application in studies of micellar systems, *J. Am. Chem. Soc.*, 99(7), 2039-2044.

44. N. Dicesare, J. R. Lakowicz (2001). Spectral properties of fluorophores combining the boronic acid group with electron donor or withdrawing groups. Implication in the development of fluorescence probes for saccharides, *J. Phys. Chem. A*, 105(28), 6834-6840.

45. C. D. Geddes (2001). Optical halide sensing using fluorescence quenching: theory, simulations and applications-a review, *Meas. Sci. and Tech.*,12(9), R53.

46. O. S. Wolfbeis, E. Urbano (1982). *J. Heterocyclic Chem.*, 19, 841-843.

47. C. D. Geddes, K. Apperson, J. Karolin, D. J. S. Birch (2001). Chloride sensitive probes for biological applications, *Dyes & Pigments*, 48, 227-231.

48. M. A. Fox, M. Chanon, Eds. *Photoinduced Electron Transfer;* Elsevier: New York, 1998, Parts A-D.

49. G. J. Kavarnos, *Fundamentals of Photoinduced Electron Transfer;* VCH: New York, 1993.

50. N. J. Van Haeringen (1981). Clinical Biochemistry in Tears, *Survey of Ophthalmology*, 26 (2), 84-96,

51. N. Chandrasekharan and L. Kelly, Progress towards fluorescent molecular thermometers, in *Reviews in Fluorescence 2003*, edited by C. D. Geddes and J. R. Lakowicz, Kluwer Academic Plenum Publishers, New York, 2004.

PROGRESS IN LANTHANIDES AS LUMINESCENT PROBES

Jeff G. Reifernberger, Pinghau Ge, Paul R. Selvin[1]

16.1. ABSTRACT

Using luminescent lanthanides, instead of conventional fluorophores, as donor molecules in resonance energy transfer measurements offers many technical advantages and opens up a wide-range of new applications. Advantages include farther measurable distances (~100 Å) with greater accuracy, and insensitivity to incomplete labeling. We have also generated new luminescent lanthanide compounds with various advantages over more conventional probes. Applications highlighted include the study of ion channels in living cells and measuring *in vitro* conformation changes within smooth muscle myosin.

16.2. INTRODUCTION TO LANTHANIDE PROBES:

Luminescent lanthanide chelate complexes have unusual spectral characteristics when compared to typical organic fluorophores. These include millisecond lifetimes, sharply spiked emission spectra, high quantum yield, unpolarized emission, and a broad range of emission energies extending from the blue to the red part of the spectrum. Lanthanides are therefore an alternative to organic fluorophores particularly where there are problems of background autofluorescence [1, 2] and as donors in fluorescence (luminescence) resonance energy transfer to measure nanometer conformational changes and binding events within proteins [3-5].

First we briefly discuss the luminescent and photophysical characteristics of lanthanides as well as new lanthanide compounds that are more soluble in water, followed by a brief review of FRET theory and measurement, highlighting those areas

---

[1] Loomis Laboratory of Physics, University of Illinois, 1110 W. Green St., Urbana, IL 61801, (tel.) 217-244-3371; (fax) 217-244-7187, selvin@uiuc.edu

where lanthanides differ from conventional probes. We then show a number of applications where LRET has enabled new types of systems to be studied: ion channels in living cells and the molecular motor myosin in vitro.

16.3. STRUCTURAL AND PHOTOPHYSICAL CHARACTERISTICS OF LANTHANIDE PROBES:

Figure 16.1a shows a typical polyaminocarboxylate chelate, DTPA-cs124, in which a Tb^{3+} atom is bound. It contains an organic chromophore (carbostyril 124), which serves as an antenna or sensitizer, absorbing the excitation light and transferring the energy to the lanthanide ion. An antenna is necessary because of the extremely weak absorbance of the lanthanide (1 $M^{-1}cm^{-1}$, or 10^4-10^5 times smaller than conventional organic fluorophores). The complexes also contain a chelate (DTPA) which serves several purposes, including binding the lanthanide tightly, shielding the lanthanide ion from the quenching effects of water, and acting as a scaffold for attachment of the antenna and a reactive group (either amine or thiol), the latter for coupling the chelate complex to biomolecules. Once bound to a biomolecule, the probe can be used in resonance energy transfer application (if an acceptor dye is chosen to be around) [3, 5-17].

In general, there are two types of lanthanide complexes. In the first type of complex, the chelate backbone serves as both the coordination site for the lanthanide ion and the sensitizer, to absorb photons and transfer energy to the bound lanthanide. Chelates used in this type of complex include β-diketone, pyridine derivatives, cryptands, terphenyl-based compounds (see figure 16.1b for an example) [18, 19]. In the second type of lanthanide complexes, which will be the focus of this chapter, the complexes have relatively distinct components such that the chelate, antenna, and reactive group are separate components. An example of this second type of structure was shown figure 1a. This second approach has the advantage of allowing for the optimization of both the structure of the coordination site and the sensitizer separately.

Fig. 2a shows the emission spectra and Fig. 2b shows the excited state lifetime characteristics of the DTPA-cs124 bound to either terbium or europium. These two are, by far, the most useful lanthanides. Dy and Sm are the only other two lanthanides that emit in the visible, but with much weaker intensity [20]. Excitation of the antenna is in the ultraviolet, typically utilizing a pulsed nitrogen laser (337 nm), although flash lamps can also be used. Emission is in the green for Tb^{3+} and in the red for Eu^{3+}. This large Stokes shift enables easy discrimination against excitation light. As seen in figure 2a and 2b, Eu^{3+} and Tb^{3+} emission are sharply spiked in wavelength, with long (millisecond) excited state lifetimes. These attributes are important for resonance energy transfer applications (see below). The sharply-spiked spectra occur because emission is atomic-like and the chelate shields the atom from broadening effects of the solvent.

The sharply-spiked spectra occur because emission arises from high-spin—high-spin transition within the lanthanide atom. Tb^{3+} emission occurs between a $^5D_4$ excited state to a ground state of ^7F_J (J = 0 - 6) while Eu^{3+} emission is from a $^5D_0$ excited state to a ^7F_J (J = 0 - 6) ground state [21]. The long lifetimes for both lanthanides occur because the ground state and excited state involve electrons in the 4f shell and hence the transitions are formally parity forbidden [21]. Nevertheless, a small admixture of 5d electrons makes the transitions possible.

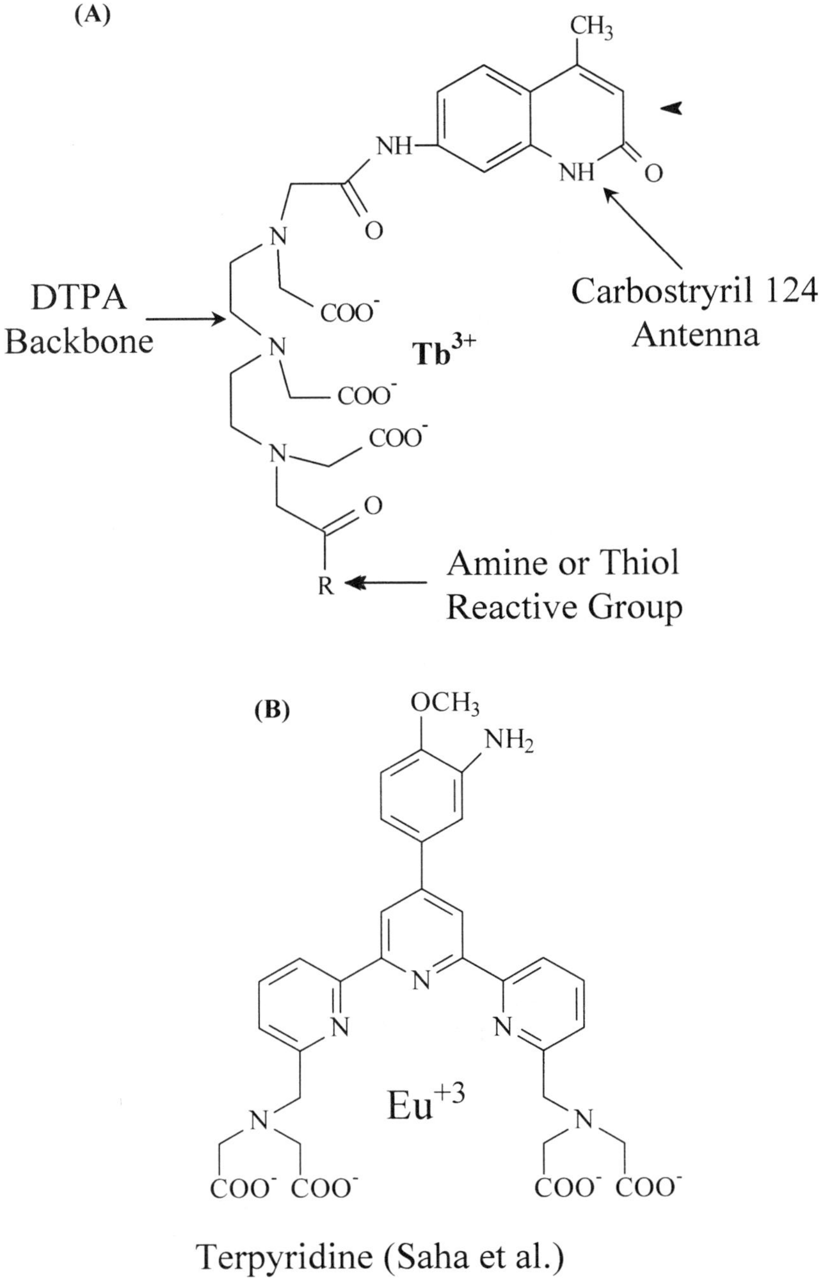

Figure 16.1. Structure of representative chelates: a) figure of Tb-DTPA-cs124 b) Terpyridine with Eu[3+] bound.

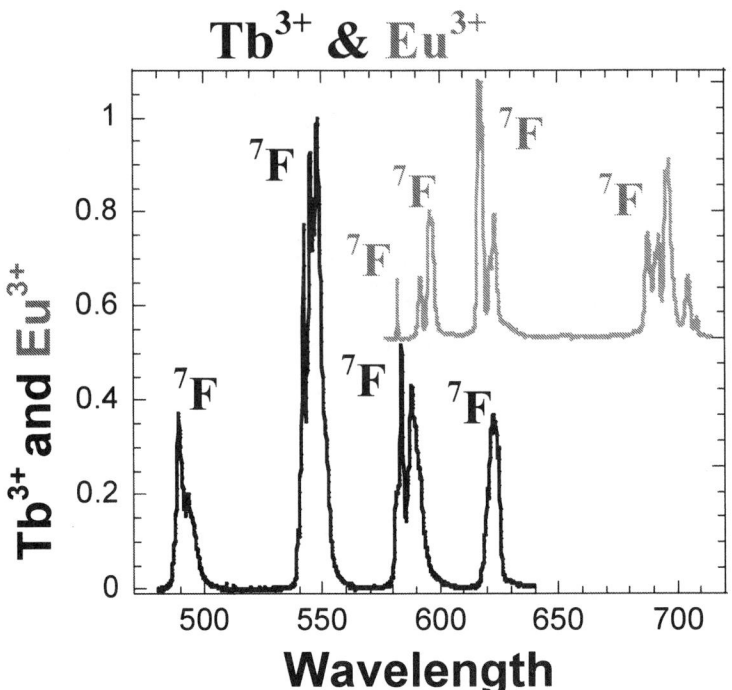

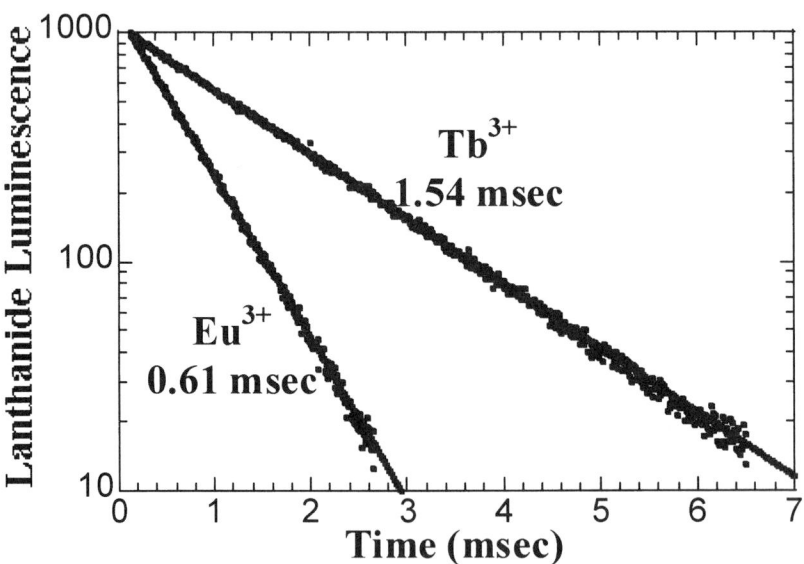

Figure 16.2. a) Emission spectra and b) lifetime of Tb^{3+}- and Eu^{3+} – DTPA-cs124.

Despite the unusual nature of the atomic states, emission within the lanthanides primarily arises from electric dipole transitions with the exception being the $^5D_0$ to $^7F_1$ transition in Eu^{3+}, which arises from a magnetic dipole [21-23]. This is important because electric dipole transitions are the same mechanism used by organic fluorophores. Hence the electric field produced by a lanthanide donor (regardless of whether it is from an electric or magnetic dipole) and by an organic donor have the same distance dependence, i.e. they both decrease as $1/R^3$ for distances $<<$ wavelength of light. Ultimately, this leads to the same distance dependence, R^{-6}, for resonance energy transfer measurements using either lanthanides or organic donors.

Another unique spectral feature of lanthanides when they are bound to a chelate-antenna complex is that the emitted light is unpolarized [23]. This is important when lanthanide complexes are used in energy transfer as donors. Energy transfer not only has a R^{-6} distance dependence between the donor and the acceptor, but there is also an orientational dependence as well. For example if the donor and acceptor are rigid and perpendicular to each other, then there will be no energy transfer from the donor to the acceptor no matter how close the two dyes may be to each other. Until recently, it had always been assumed that the emission of the Tb^{3+} and Eu^{3+} within the chelate-antenna would be mostly unpolarized due to the nature of the initial and final quantum states of the lanthanides. Reifenberger et al. froze lanthanide-chelate-antenna complex in a glycerol-water mixture and were able to measure the anisotropy of their emission (an anisotropy near zero corresponds to unpolarized emission). They found that the anisotropy of Eu-DTPA-cs124 was very small and actually zero in some of the transitions. Tb-DTPA-cs124's emission was also unpolarized. Figure 16.3 and figure 16.4 shows the anisotropy of Eu^{3+} and Tb^{3+} when bound to DTPA-cs124. (The anisotropy was essentially the same for both Eu^{3+} and Tb^{3+} bound to another chelate-antenna complex, TTHA-cs124 [23].

The emission quantum yield for Terbium or Europium in the chelates is also quite high [24]. This is important because the efficiency of energy transfer is proportional to the donor quantum yield (Equations 3 and 5 below). By lanthanide quantum yield, Q_{Ln}, here we mean the probability that the lanthanide will emit a photon given that the *lanthanide* is excited. This definition is very similar to that used with conventional fluorophores although there is a subtlety. Lanthanide excitation is a two step process: the antenna absorbs a photon, and then passes this energy onto the lanthanide with some finite probability ($\equiv Q_{transfer} \leq 1$) (Fig. 3). The lanthanide then emits with some probability – i.e. the quantum yield mentioned above, Q_{Ln}. The overall probability that the lanthanide will emit a photon (Q_{total}), given that an excitation photon was *absorbed* by the complex (antenna), is:

$$Q_{total} = Q_{Ln} \times Q_{transfer} \qquad (1)$$

Figure 16.5 is a diagram of the process described in equation 1. For organic fluorophores $Q_{transfer} \equiv 1$ and hence $Q_{total} = Q_{Ln}$. For Tb^{3+} and Eu^{3+} in polyaminocarboxylate chelates such as in Fig 1a, $Q_{transfer} = 0.4 - 0.75$ and $Q_{total} = 0.1-0.4$ [24]. In any case, the efficiency of energy transfer (related to R_0, the distance at which half the donor's energy is transferred to the acceptor, see Equation 5) is proportional to Q_{Ln}, and Q_{total} is only relevant in that it affects the total brightness of the sample.

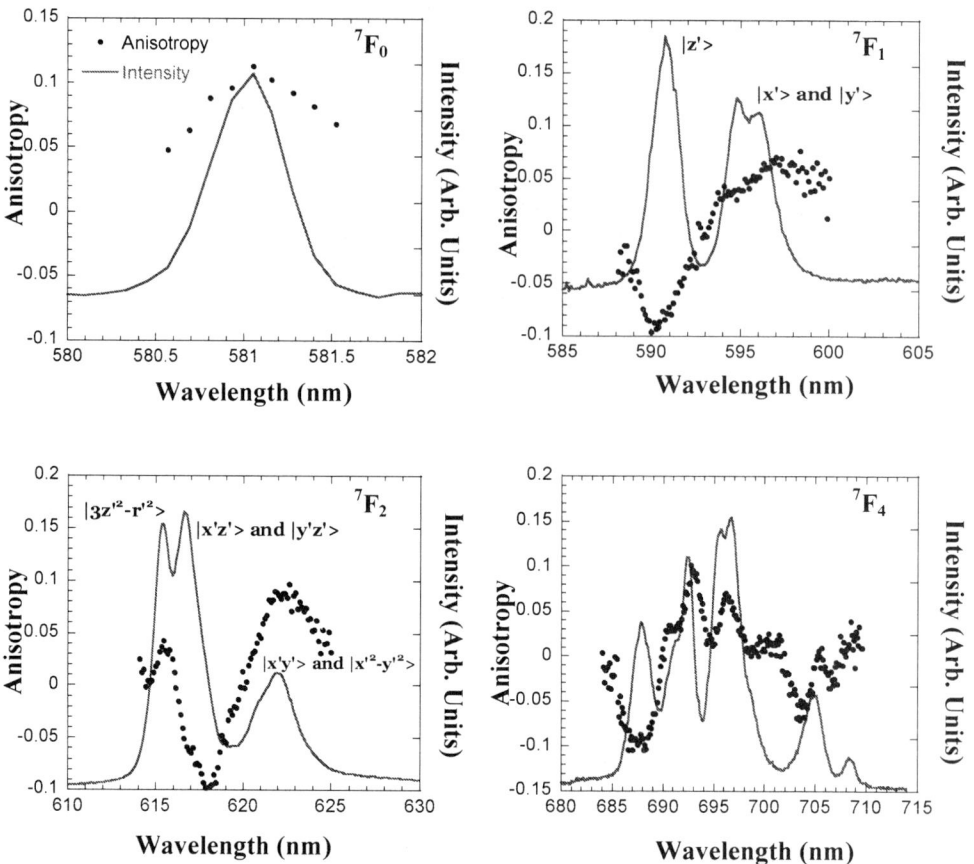

Figure 16.3: Anisotropy of Eu-DTPA-cs124 for the four major transitions within Eu^{3+}

In order for the lanthanides to be useful in bioassays, the chelates must have a reactive group for attachment to biomolecules. Fortunately, the standard reactive groups can be coupled to the chelates: amine reactive groups such as isothiocyanates [25] and thiol-reactive groups such as maleimides, bromoacetamides, pyridyl dithio groups [26] have been made for the polyaminocarboxylate chelates. The reactive groups, can, however, lead to more complicated photophysics in that they can interact with the antenna molecules or adopt multiple conformations, leading to multi-exponential lanthanide decays, particularly with terbium [26].

Several amine-reactive or thiol-reactive lanthanide chelates have been synthesized (figure 16.6 and figure 16.7) in our lab. Due to the fact that DTPA dianhydride is commercially available and not expensive, DTPA dianhydride is widely used as the backbone of the structure in which both an antenna molecule and reactive group are attached. Antenna molecules containing a free amino group are reacted with one of the two dianhydride bonds to attach the antenna molecule to the backbone. The remaining anhydride group is either converted into an amine-reactive or a thiol-reactive functional group via either an alkyl diamine or a hydrazine linker.

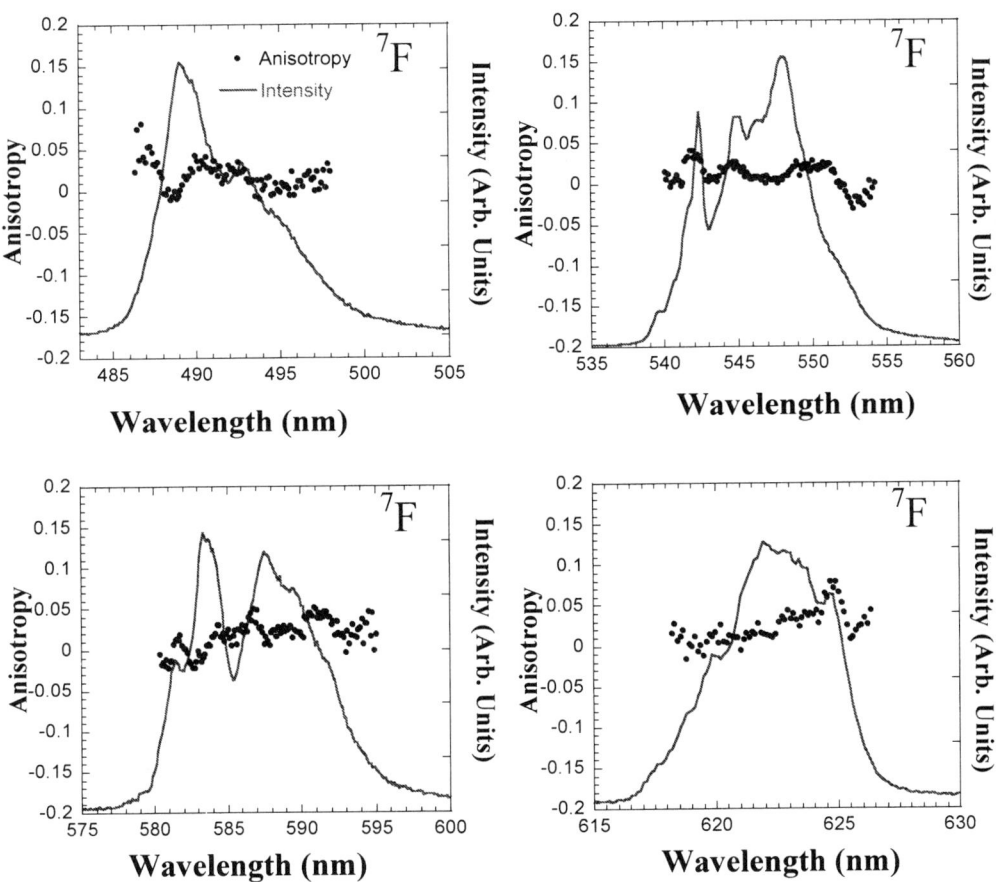

Figure 16.4: Anisotropy of Tb-DTPA-cs124 for the four major transitions within Tb^{3+}

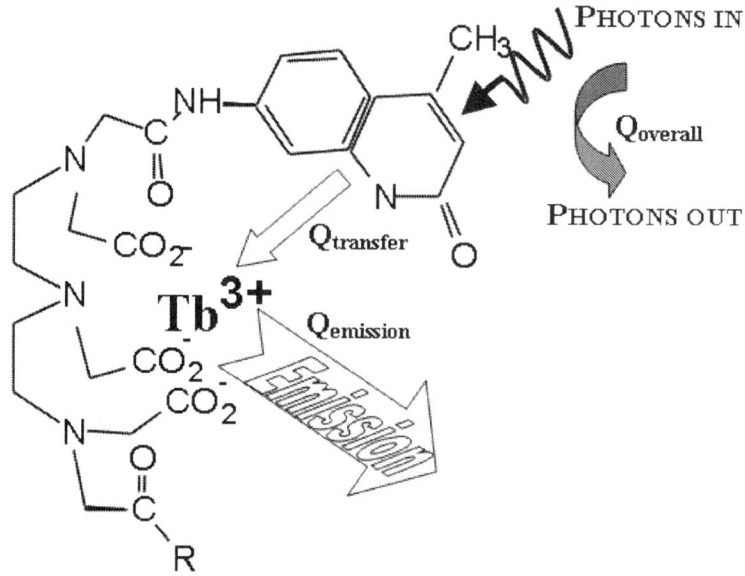

Figure 16.5: Definition of quantum yields.

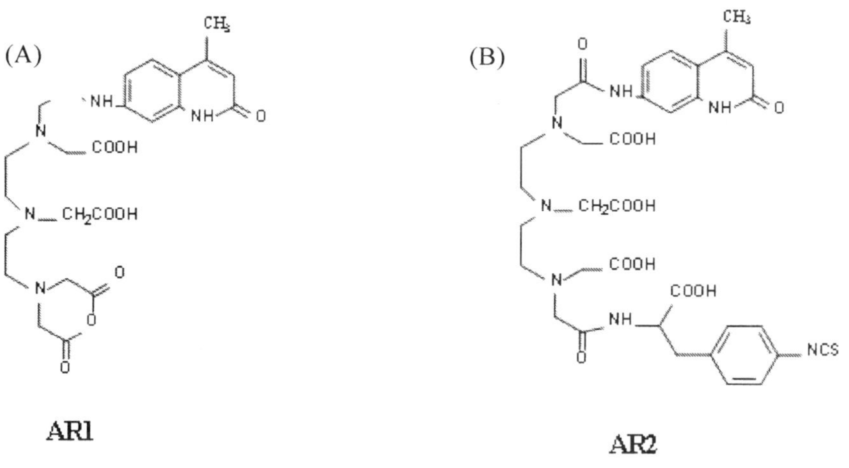

(A)   (B)

ARl   AR2

Figure 16.6: Amine reactive chelate complexes.

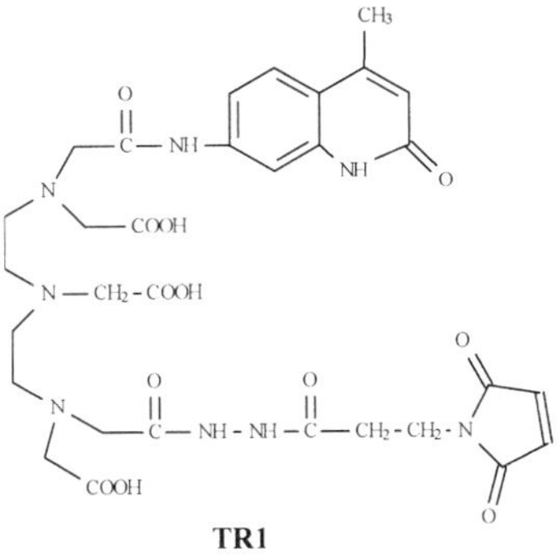

TR1

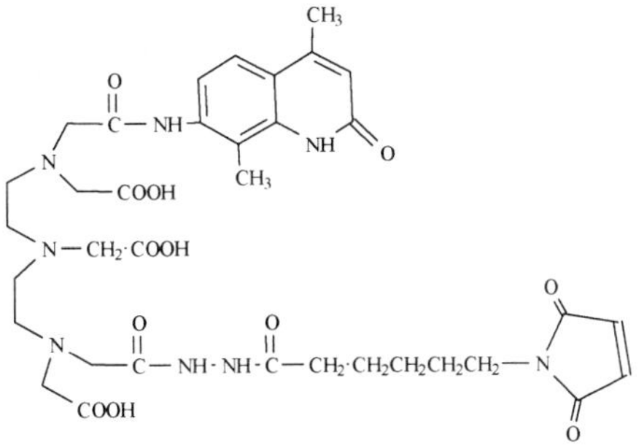

TR2

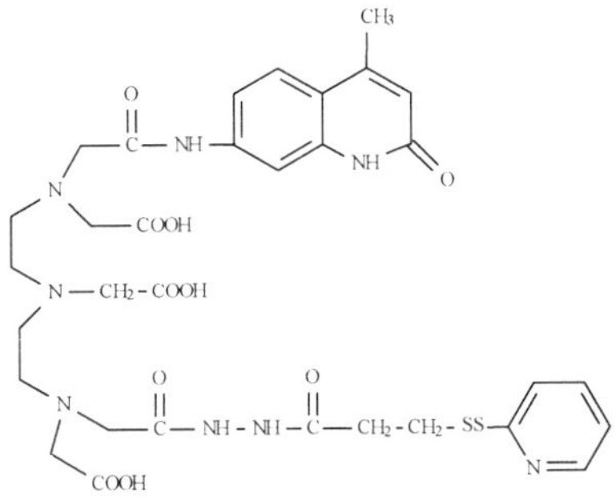

TR3

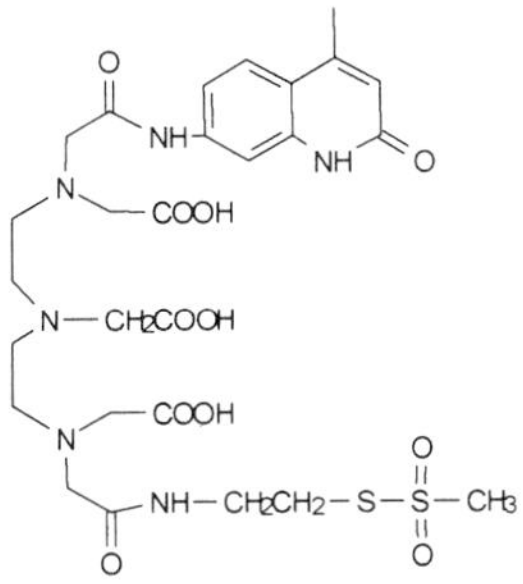

TR4

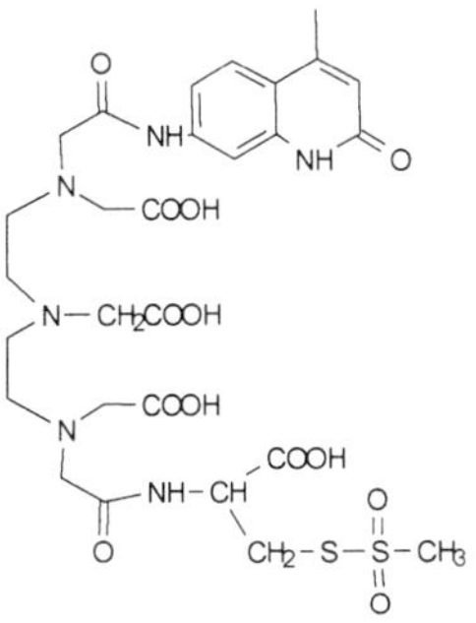

TR5

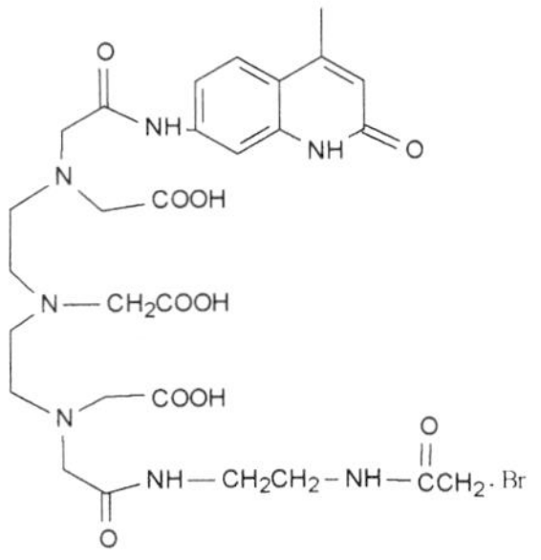

TR6

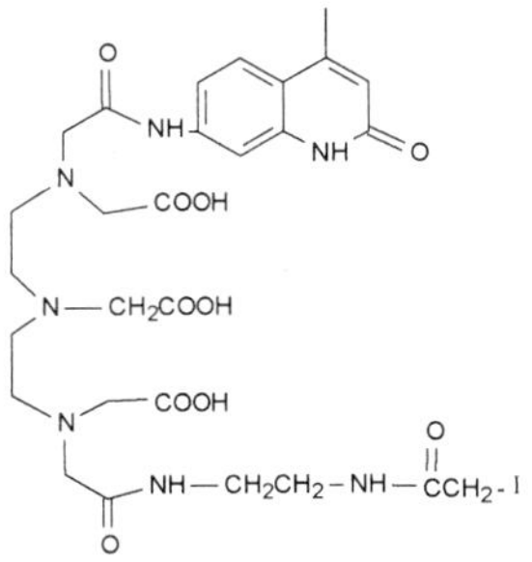

TR7

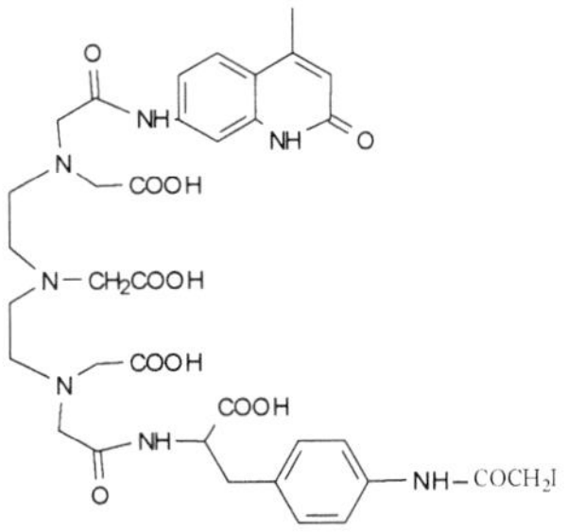

TR8

Figure 16.7: Thiol reactive chelate complexes.

We have synthesized two different amine reactive complexes in our lab. The first, **AR1** shown in figure 16.6a, is an in situ formed intermediate formed during synthesis and cannot be separated [25]. The second, **AR2** shown in figure 16.6b can rapidly react with a free amine group at pH ~9. Figure 16.7 shows the main thiol-reactive forms of chelate-antenna complexes, which include maleimide (**TR1, TR2**), pyridyldithiol (**TR3**), MTS (methanethiosulfonate, **TR4, TR5**), and haloacetyl amide (**TR6, TR7, TR8**). The maleimide form has the advantage of a quick, efficient, and irreversible reaction with thiols. **TR1** has been used as a donor in LRET experiments in the studies of muscle proteins and ion channels [15, 17, 27]. MTS is a labeling reagent frequently used in ion channel studies [28]. While having two lifetime components, the Tb^{3+} and Eu^{3+} complexes of purified MTS chelate **TR5**, have shown a predominant long lifetime component with an amplitude greater than 90%. The long lifetime component remains even after labeling to proteins and ion channels [29]. Two different haloacetyl amide chelates have been synthesized in our lab as well, bromoacetamide and iodoacetamide. The bromoacetamide chelate, **TR6**, has exhibited single exponential decay in its emission measurement with a lifetime of 1.51 ms. Upon reaction with a reduced form of glutathione, the adduct still exhibits a single exponential decay in its lifetime [30]. However, the disadvantage of this form of the chelate is that it is not very reactive towards thiols and requires relatively high pH, which can also denature the proteins that are being labeled [31]. The iodoacetamide chelates, **TR7** and **TR8**, have increased reactivity and labeling can occur at a neutral pH. However their adducts to biomolecules often show bi- or tri- exponential decay [29].

16.3.1. Modified Chelates and Antennas and Their Effect on Lanthanide luminescence:

Chelates

The most commonly used chelate is DTPA (diethylenetriaminepentaacetic acid) attached to an antenna molecule, cs124. Other polyaminocarboxylates have also been used (see figure 16.8) to attach to cs124 [32]. These chelates have shown sensitization to both terbium and europium except TETA-cs124 (Table 16.1). Tb^{3+} and Eu^{3+} can normally take 9 coordination atoms in their inner sphere [21]. DTPA-cs124 and DOTA-cs124 are 8-dentate chelates and therefore their lanthanide complexes have ~1 coordinated water molecule to the ions. Because of the solvent quenching effect of water, their lanthanide complexes, particularly for Eu^{3+}, have shortened lifetimes compared to the theoretical maximums. In contrast, TTHA-cs124 is a 10-dentate chelate, which can provide better protection to lanthanide ions from solvent attack. As the result, the lanthanide complexes of TTHA-cs124 have significantly longer lifetimes compared to that of DTPA-cs124 and DOTA-cs124 with calculated number of water bound in the inner sphere at ~0.2.

Table 16.1. Lifetimes and the Number of Waters Coordinated in the Inner Sphere of Lanthanide Chelates

Complexes Tb(Eu)	τ_{H2O} (ms)	τ_{D2O} (ms)	τ_{H2O}/τ_{D2O}	**No. of waters**	Relative intensity
DTPA-cs124	1.55 (0.62)	2.63 (2.42)	0.59 (0.26)	1.1 (1.26)	1 (1)
TTHA-cs124	2.10 (1.19)	2.37 (1.79)	0.89 (0.66)	0.2 (0.3)	1.1 (2.67)
DOTA-cs124	1.54 (0.62)	2.61 (2.25)	0.59 (0.27)	1.1 (1.23)	1.1 (0.66)

Table 16.2. Brightness and lifetimes of Tb^{3+} and Eu^{3+} complexes of cs124 derivative chelates

Metal	Chelate	Relative Brightness (%)	Lifetimes (ms)
Tb^{3+}	DTPA-cs124 ("Benchmark")	100	1.53
	DTPA-**d4**	62	1.58
	DTPA-**d4**-EMCH	72	1.86 76%; 0.44 24%
	DTPA-**d10**	80	1.74
	DTPA-**d10**-EMCH	58	2.16 82%; 0.92 18%
	DTPA-**d10**-EDA-Br	64	1.68
	DTPA-**d5**	79	0.93 45%; 0.65 55%
	DTPA-**d5**-EMPH	37	0.82 85%; 0.37 15%
	DTPA-**d3**	36	0.82 24%; 0.50 76%
	DTPA-**d9**	87	1.63
	DTPA-**d9**-EMCH	4	1.85 59%; 0.25 41%
	DTPA-**d1**	80	1.38
	DTPA-**d6**	100	1.09
	DTPA-**d7**	N/A	N/A
	DTPA-**d2**	77	1.29
	DTPA-**d2**-EMCH	20	1.43 74%; 0.41 26%
Eu^{3+}	DTPA-cs124 ("Benchmark")	100	0.61
	DTPA-**d4**	77	0.605
	DTPA-**d4**-EMCH	22	0.42 76%; 0.14 24%
	DTPA-**d10**	100	0.603
	DTPA-**d10**-EMCH	26	0.57 60%; 0.12 40%
	DTPA-**d10**-EDA-Br	43	0.60
	DTPA-**d5**	38	0.64 68%; 0.41 32%
	DTPA-**d5**-EMPH	34	0.55 75%; 0.11 25%
	DTPA-**d3**	75	0.60
	DTPA-**d9**	60	0.605
	DTPA-**d9**-EMCH	N/A	N/A
	DTPA-**d1**	52	0.60
	DTPA-**d6**	173	0.60
	DTPA-**d7**	152	0.53
	DTPA-**d2**	100	0.60
	DTPA-**d2**-EMCH	10	0.44 68%; 0.11 32%
	DTPA-**d8**	300	0.62
	TTHA-**d8**	170	1.19

Antenna

Substituted cs124 derivatives have been used as antenna molecules to lanthanide ions (Fig. 9) [30, 33]. Among these cs124 derivatives, compounds **d1-d5** introduced an additional hydrophilic group on the aromatic ring structure, making them more soluble in aqueous media, especially for **d4**. The light absorption and energy transfer to the lanthanide of an antenna molecule is a complex process and the mechanism has not been fully characterized [34-37]. It is therefore difficult to predict how the antenna's structure will affect the photophysics of the corresponding lanthanide-chelate complex.

Table 16.2 lists the lifetimes and relative intensity of the Tb^{3+} and Eu^{3+} complexes of chelates made from various cs124 derivatives. Generally, the Tb^{3+} and Eu^{3+} complexes of these chelates have comparable emission brightness to the benchmark chelate, DTPA-cs124. By introducing hydrophilic groups to the sensitizers, the chelates derived from **d1, d4, d3, d5, d2** possess increased solubility in water, especially in the case of **d4**. Although they show comparable emission brightness to the reference complex, DTPA-cs124, they often have shorter lifetimes. Contrary to this trend, Tb^{3+}-DTPA-**d4** shows slightly longer lifetime, and its brightness is about 62% that of Tb^{3+}-DTPA-cs124. Clearly, the introduction of 6-SO_3H on cs124 has no dramatic effect on the emission properties of its lanthanide complex.

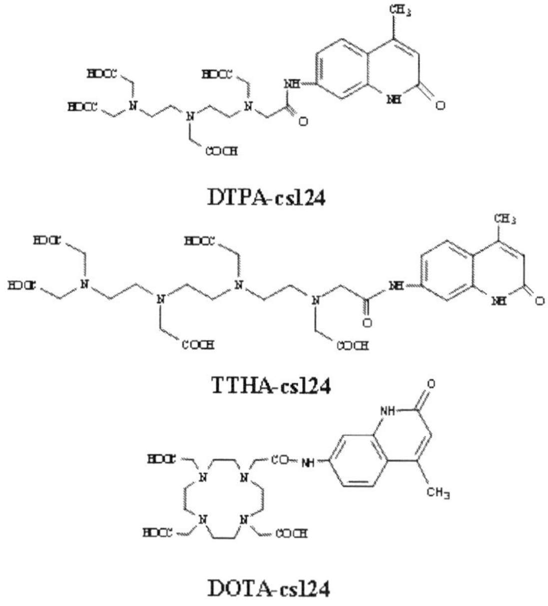

DTPA-cs124

TTHA-cs124

DOTA-cs124

Figure 16.8: Different chelates used in the lanthanide antenna complexes.

In some cases the cs124 derivatives are only luminescent with Eu^{3+}. Both the DTPA and TTHA chelates of **d8** do not cause Tb^{3+} to luminesce [33]. However, their Eu^{3+} complexes exhibit 3 and 1.17 times luminescent intensity than the corresponding

DTPA-cs124 complexes. It is also observed that with a substitute on the 3-postion of cs124 ring, its lanthanide chelate tends to be silent to Tb^{3+} and active to Eu^{3+}[33].

The Tb^{3+} complexes with the chelates of 6- or 8-CH_3 cs124 derivatives, especially the 8-CH_3 derivative, have longer lifetimes than DTPA-cs124 (1.74 ms vs. 1.53 ms). This suggests that higher lanthanide quantum yield was achieved in these complexes. Increased lanthanide quantum yield could result from either a decrease in solvent quenching or a decrease in energy back transfer to the antenna. The number of solvent molecules coordinated to the lanthanide ion in Tb^{3+}-DTPA-**d4** was measured using the method of Horrocks and Sudnick [38] in both H_2O and D_2O. The calculated number of water for both the chelate and the reference complex, Tb^{3+}-DTPA-cs124 is all 1.1, indicating that there is no difference in solvent quenching. Clearly, the higher quantum yield is achieved from a decrease of energy back-transfer from the excited Tb^{3+} to **d4**. In contrast, no energy back-transfer was observed for unsubstituted cs124 chelates [32].

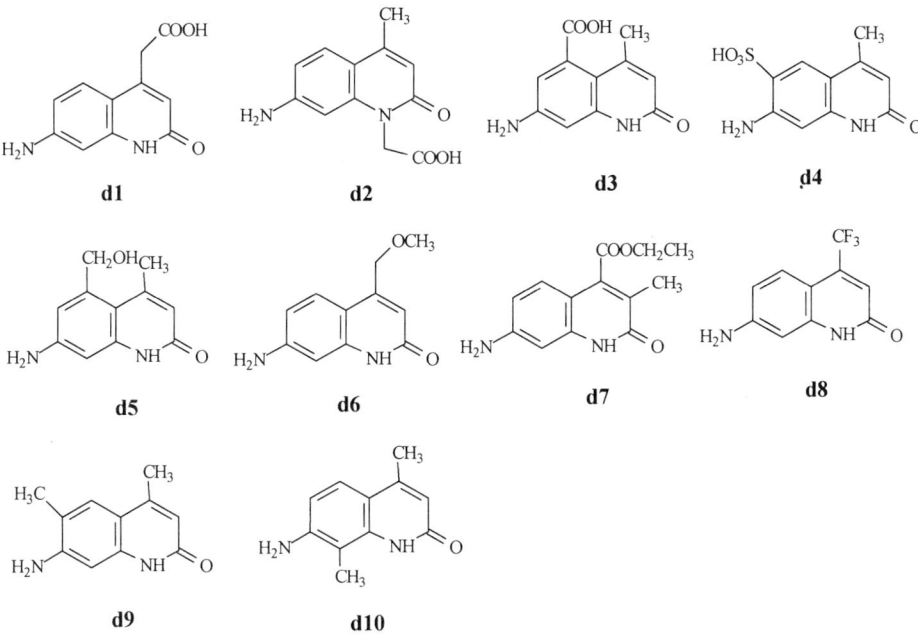

Figure 16.9: Carbostryil Derivatives.

16.3.2. Lanthanide-based Resonance Energy Transfer:

Theory of LRET

Lanthanides were first used in resonance energy transfer experiments by Horrocks and co-workers in the 1970s [39, 40]. A lanthanide was bound to a calcium-binding protein and used as a donor to transfer energy to a freely diffusing or specifically bound metal (e.g. Co). The lanthanide was excited either directly, using a powerful laser, or through a tryptophan near the lanthanide that acted as an antenna molecule. Because of the weak absorption of the metal acceptor, the donor-acceptor dipole coupling and hence measurable distances, were small (a few Angstroms). However, the work first demonstrated the idea of using lanthanides for resonance energy transfer and was useful for measuring small distances (~10 Å). Meares and co-workers in the early 1980s then used lanthanides bound to artificial chelates as donors to transfer energy to organic acceptors, in which either the donor or acceptor was free to diffuse [3]. The use of organic acceptors, which have strong absorbance where the donor emits, dramatically increased the measurable distance range (discussed further below). The primary goal in these experiments was to measure how close two probes could approach each other as one or both diffused.

LRET (luminescence, or lanthanide-based resonance energy transfer), as used in my laboratory and elsewhere, relies on the same fundamental mechanism as FRET, but instead of using two organic-based probes, uses a lanthanide donor and an organic acceptor [41], both specifically bound to a biomolecule. Both LRET and FRET utilize visible light (roughly 500 nm wavelength), yet achieve sub-nanometer resolution. In both techniques, a luminescent (fluorescent) probe, called the donor, transfers energy via a dipole-dipole interaction to a second structurally-different probe, called the acceptor [42-45]. At distances less than the wavelength of light, λ, the electric field predominantly drops off as R^{-3}. (For $R >> \lambda$, the electric field is proportional to R^{-1}). An acceptor, if nearby and containing energy levels corresponding to the frequencies of the oscillating electric field, can interact with this field, absorb the energy, and become excited. The probability of the acceptor being excited depends on the square of the electric field strength and hence decays as R^{-6} for $R << \lambda$, the relevant distance scale in FRET/LRET ($\lambda \approx 500$ nm). Energy transfer also depends on how well the acceptor energy levels match the frequencies of the donor (the so-called spectral overlap term – Equation 6, below). Finally, energy transfer may also depend on the orientation of the donor and acceptor (the κ^2 term – Equations 5 and 7) because the electric field of the donor may be polarized and anisotropic. FRET and LRET can measure distances between the probes over a range of 20-100 Å. This high spatial resolution is possible, even with optical photons, because the amount (or more precisely, the efficiency) of energy transfer (E) is a strong function of distance between the donor and acceptor fluorophores.

The efficiency of energy transfer, E, is defined as the probability that an excited donor will return to the ground state by giving its energy to an acceptor. This can be written as:

$$E = \frac{k_{et}}{k_{et} + k_{nd}} = \frac{1}{1 + \dfrac{k_{nd}}{k_{et}}} = \frac{1}{1 + \dfrac{1}{k_{et}\tau_D}} \tag{2}$$

where k_{et} is the rate of energy transfer and is distant-dependent, and k_{nd} is the rate of all other donor decay processes, such as radiative and non-radiative rates of donor decay. These latter processes clearly do not depend on the donor-acceptor distances. The donor lifetime in absence of acceptor is τ_D. Note that E depends on the ratio k_{et} to the other processes, but does not depend on the absolute donor lifetime. In FRET, donor rates (or lifetimes) and energy transfer rates are in the nanosecond range whereas in LRET they are in the millisecond range. As a side point, if the distance between the probes changes slowly on the FRET time-scale, but quickly on the LRET timescale, the two techniques can give dramatically different energy transfer efficiencies. Indeed, one signature of such dynamics is if LRET gives a much higher E than FRET [46].

Because the rate of energy transfer depends on R^{-6} distance between donor and acceptor, Equation 2 can be rewritten as:

$$E = \frac{1}{1 + \left(\dfrac{R}{R_o}\right)^6} \tag{3}$$

where R_o is the distance at which half of the energy is transferred and is generally 20-60 Å and typically 40-70 Å for LRET. The steep distance dependence, R^{-6}, arises because induced-dipole induced-dipole interactions depend on R^{-6}. (In quantum mechanical terms we talk instead of transition dipole moments, which also leads an R^{-6} dependence.) By knowing or calculating R_o, and measuring E, the distance between the probes can be found by the rearranging the terms in equation 3.

$$R = Ro\left(\frac{1}{E} - 1\right)^{\frac{1}{6}} \tag{4}$$

An important limitation of equation 4 is that R_o is often not known precisely, which limits the ability of FRET to measure absolute distances. R_o is typically calculated based on the spectral properties of the donor and acceptor, but also depends on the orientation of the donor and acceptor, which is often not precisely known. (For details, see [41, 47]). Consequently, if R_o can be determined or calculated and E measured spectroscopically, then FRET/LRET can be used as a spectroscopic ruler to determine distances [48].

R_o is usually calculated from the spectral properties of donor and acceptor [49]:

$$Ro = 0.21\left(Jq_D n^{-4} \kappa^2\right)^{\frac{1}{6}} \quad \text{(in Angstroms)} \tag{5}$$

$$J = \frac{\int \varepsilon_A(\lambda) f_D(\lambda) \lambda^4 d\lambda}{\int f_D(\lambda) d\lambda} \quad \text{in } \underline{M}^{-1}\text{cm}^{-1}\text{nm}^4 \tag{6}$$

where J is the normalized spectral overlap of the donor emission (f_D) and acceptor absorption (ε_A in units of $\underline{M}^{-1}\text{cm}^{-1}$ where $\underline{M}$ is units of Moles/liter), q_D is the quantum efficiency (or quantum yield) for donor emission in the absence of acceptor (q_D = number of photons emitted divided by number of photons absorbed), n is the index of refraction (1.33 for water; 1.29 for many organic molecules) and κ^2 is a geometric factor related to the relative orientation of the transition dipoles of the donor and acceptor and their relative orientation in space. Note that for LRET, q_D in Equation 5 is Q_{Ln}, and not $Q_{overall}$ (Equation 1). This is because Q_{Ln} determined the strength of the donor's electric field, not $Q_{overall}$.

The orientation term, κ^2, in R_o, is often a source of uncertainty in FRET measurements. It is defined as

$$\kappa^2 = \left(\cos\theta_{DA} - 3\cos\theta_D \cos\theta_A\right)^2 \qquad (7)$$

where θ_{DA} is the angle between the donor and acceptor transition dipole moments, θ_D (θ_A) is the angle between the donor (acceptor) transition dipole moment and the R vector joining the two dyes. By measuring the polarization of donor and acceptor emission, constraints on these angles can often be imposed, reducing --though usually not completely eliminating-- the uncertainty in κ^2. κ^2 ranges from 0 if all angles are 90 degrees, to 4 if all angles are zero degrees, and equals 2/3 if the donor and acceptor rapidly and completely rotate during the donor excited state lifetime [47]. If the donor emission is unpolarized as is the case for terbium, and mostly true for europium [23], and the acceptor is completely rigid and either parallel ($\kappa^2 = 2/3$) or perpendicular ($\kappa^2 = 1/3$) to the radius vector, then $1/3 < \kappa^2 < 2/3$. This limits the worst case error in R_o to -11% $+12\%$ if one simply assumes $\kappa^2 = 2/3$. Furthermore since the lanthanides have millisecond lifetimes, the acceptor will very likely rotate during this time, making κ^2 very close to 2/3. Hence, the error in distances measured via LRET due to the orientation factor is essentially negligible. This in turn makes the distance determination via LRET generally more accurate than FRET since the orientation factor in FRET is often poorly known.

Finally, R_o is also proportional to J, the spectral overlap. The lanthanides have highly spiked emission spectra in regions where several excellent dyes absorb (Fig. 2a) – e.g. the Tb^{3+} 490 nm emission peak overlaps well with Fluorescein, Green Fluorescent Protein, and Alexa 488 absorption; the Tb^{3+} 546 nm peak overlaps with Cy3, Tetramethylrhodamine, Alexa 546, and R-Phycoerythrin absorption. The Eu^{3+} 617 nm emission peak overlaps with Cy5, Alexa 633, and Allophycocyanin absorption. Consequently, J for LRET can be unusually large. When combined with a high Q_{Ln}, the R_o in LRET can also be quite large (Table 16.3).

Measuring E

In Fig. 10, an energy transfer experiment between a terbium labeled DNA and rhodamine-labeled DNA complement is shown [50]. This example highlights various ways of measuring energy transfer. In FRET and LRET, there are several ways of measuring E: a reduction in donor intensity in the presence of acceptor (due to some of the donor's energy going to the acceptor instead of into donor emission); by a decrease in donor excited state lifetime (energy transfer to the acceptor is an additional relaxation pathway of the donor's excited state); or by an increase in acceptor fluorescence (the acceptor is receiving energy from the donor and converting this energy into acceptor fluorescence). In LRET, E can also be measured via the sensitized emission lifetime (see below). In FRET, but not LRET, E can also be measured by an increase in the photostability of the donor in the presence of the acceptor because energy transfer to the acceptor decreases the donor's excited-state lifetime and photobleaching is generally proportional to the amount of time the dye spends in its excited state. Finally FRET and potentially LRET can be measured by an increase in donor intensity following photodestruction of the acceptor [51].

Table 16.3: J-values and R_o for Lanthanide Chelates and Organic Dyes

Donor-Acceptor Pairs*	J-Value (M^{-1} cm^{-1} nm^4)	R_o (Å)
Terbium to Fluorescein (bound to DNA) (ε_{max} = 75k @ 492 nm)	9.23×10^{14}	45.0
Terbium to eGFP (free) (ε_{max} = 55k @ 488 nm)	7.14×10^{14}	43.1
Terbium to TMR (bound to DNA) (ε_{max} = 100k @ 557nm)	3.80×10^{15}	57.0
Terbium to Cy3 (free) (ε_{max} = 150k @ 552nm)	5.82×10^{15}	61.2
Terbium to R phycoerythrin pH 7.5 (free) (ε_{max} = 1,960k @ 566nm)	9.60×10^{16}	97.5
Europium to Cy5 (bound to myosin) (ε_{max} = 249k @ 650nm)	8.89×10^{15}	55.2
Europium to Allophycocyanin pH 7.5 (free) (ε_{max} = 700k @ 652nm)	4.01×10^{16}	71.0

* J and Ro calculated for Terbium and Europium using corrected emission spectra and quantum yields for lanthanide bound to DTPA-cs124 in aqueous solutions (q_{Tb} = 0.48; q_{Eu} = 0.17). J and R_o in D_2O and in other chelates with same emission spectra can be determined by multiplying by the appropriate quantum yields, found in [24]. Other constants: n = 1.33; κ^2= 2/3. The emission spectra of Tb^{3+} and Eu^{3+} are insensitive to attachment to biomolecules although the absorption spectra of the acceptor dye can be somewhat sensitive to attachment. Absorption spectra of R phycoerythrin and Allophycocyanin are from Molecular Probes Inc., and Cy-3 from Amersham.

(A)

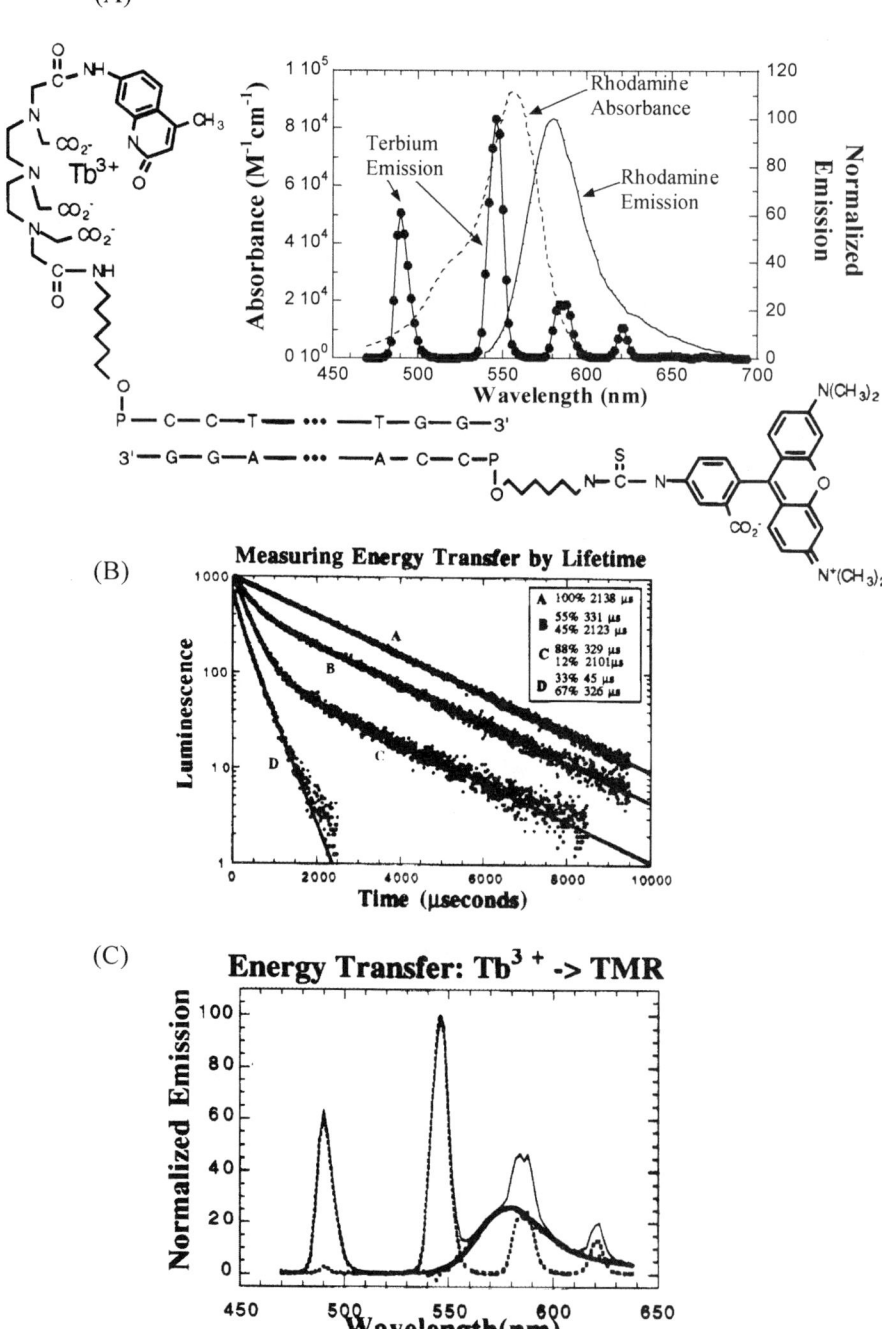

(B)

(C)

Figure 16.10: a) DNA Hybridization and model system for LRET. **b)** Lifetime data. **c)** Spectral data. (Figure adapted from [50]).

The efficiency of energy transfer (E) is then:

$$E = 1 - \frac{I_{D_A}}{I_D} = 1 - \frac{\tau_{D_A}}{\tau_D} = 1 - \frac{\tau_{A_D}}{\tau_D} = 1 - \frac{\tau_D^{bl}}{\tau_{D_A}^{bl}} \qquad (8)$$

where I_{D_A}, τ_{D_A}, $\tau_{D_A}^{bl}$ are the donor's intensity, excited state lifetime, and photobleaching time constant in the presence of acceptor, and I_D, τ_D, and τ_D^{bl} are the same parameters in the absence of acceptor. τ_{A_D} is the lifetime of the sensitized emission of acceptor and is discussed further below.

Although using absolute intensities I_{D_A} and I_D is conceptually easy, it involves matching concentrations of two different samples and hence is prone to titration errors. Lifetime measurements avoid this problem and also are able to resolve multiple species with different energy transfer efficiencies. Fig. 10c shows a single exponential donor-only lifetime, which is reduced upon hybridization with a DNA strand containing an acceptor. Starting with a single-exponential donor-only lifetime is not essential but significantly simplifies the analysis of complex donor-acceptor mixtures. Titrating in with sub-stoichiometric amounts of acceptor strand leads to two populations and hence a bi-exponential donor decay: a donor-only unhybridized strand (τ_D = 2.1 msec) and donor-acceptor double-stranded DNA (τ_{D_A} = 330 μsec). The amount of energy in the donor-acceptor pair can be calculated in equation 8 using the 330 μsec lifetime and the donor-only lifetime of a terbium-DNA hybridized to an unlabeled complementary DNA, which is 2.8 msec (data not shown). The relative populations of the two species can be determined by their pre-exponential amplitudes. Titrating in more acceptor strand increases the amplitude of the short time component but leaves its lifetime unchanged, as expected.

In LRET, E can also be measured by measuring the lifetime of the sensitized emission of acceptor – Fig 10c, curve D. The donor is excited by a pulse of light, the direct acceptor emission decays in nanoseconds, and any acceptor emission after this initial delay is therefore due only to energy transfer received by the acceptor from the long-lived donor. Its lifetime, τ_{A_D}, will follow the donor's lifetime, τ_{D_A}. Importantly, τ_{A_D} can be measured without contaminating background from either direct acceptor fluorescence (via temporal discrimination), or from donor emission (via spectral discrimination). The latter is possible because the donor is sharply spiked in emission spectra, including regions where the donor is dark, yet where the acceptor fluoresces. For example, Terbium is dark around 520 nm and 570 nm, where fluorescein and tetramethylrhodamine emit, respectively. Consequently, the temporal decay of the acceptor's sensitized emission can be measured with no background from either donor leakage or direct acceptor leakage. This sensitized emission lifetime is a very powerful advantage of LRET because it only arises from donor-acceptor pairs. In Fig. 10b, curve D, the sensitized emission lifetime is seen to closely match the short component of the donor lifetime yet does not have "contamination" from the donor-only DNA strands. The pre-exponential amplitudes of the sensitized emission decay correspond to the population of *excited* acceptors. Therefore in a multi-exponential decay, corresponding to a distribution of donor-acceptor pairs, the pre-exponential terms are the product of the individual energy transfer efficiencies and their populations [52]. This is in contrast to the donor decay in which the amplitudes are just proportional to populations.

Energy transfer also increases the acceptor's emitted intensity. In the donor-acceptor labeled sample, E can therefore be measured by comparing the residual donor fluorescence intensity, I_{D_A}, to the acceptors emission due to energy transfer I_{A_D}, and normalizing by their quantum yields (q_i's):

$$E = \frac{\dfrac{I_{A_D}}{q_A}}{\dfrac{I_{D_A}}{q_D} + \dfrac{I_{A_D}}{q_A}} \tag{9}$$

The numerator is the number of acceptor excitations, and the denominator is the total number of excitations, i.e. the number of donor excitations that lead to donor emission (first term, denominator) and the number of acceptor excitations (second term, denominator). It is necessary to normalize the emission intensities by the quantum yields because E is measured in terms of excitations, i.e. E is the fraction of the donor excitations that get converted into acceptor excitations.

Fig. 10c shows the *time-delayed* emission spectra of the donor and donor-acceptor complex (corresponding to curve C in Fig. 10b), which can be used to determine the two intensities in equation 9. The donor-acceptor sample is excited using a short excitation pulse and emission is detected after a few tens of microsecond delay. This procedure eliminates all prompt fluorescence of the acceptor. It also eliminates any contribution from acceptor-only species, if present, as well as any direct fluorescence from the antenna, both of which have nanosecond lifetime. The donor-acceptor spectra are then fit to the sum of a donor- and acceptor- spectra, with I_{D_A} being the area due to donor emission and I_{A_D} equal to the area under the acceptor emission. Note that the absolute concentrations of the donor-only species, the acceptor-only species, and the donor-acceptor species are irrelevant. In practice, the curve-fitting is done as follows: The donor-only spectra and donor-acceptor spectra are normalized at the 490 nm peak – or any point where there is no acceptor fluorescence. The donor-only curve is then subtracted from the donor-acceptor spectra and the difference is the sensitized emission curve, with area I_{A_D}. This should have the same shape as an acceptor-only emission spectra. I_{D_A} is simply the area under the donor-curve. Although we always take a donor-only spectrum as a control, we have found that the spectral shape of Tb-DTPA-cs124 does not change under any condition tested, and hence, a donor-only spectra taken once, is very likely to remain unchanged.

There are two additional points needed to properly use equation 9. First, the emission spectra must be corrected for wavelength sensitivity of the detector. This is done via conventional means using an emission source (standard lamp or a dye) whose emission spectra is known [43]. Second, the donor and acceptor quantum yields must be measured. Fortunately, we have recently determined the quantum yield of Tb^{3+} and Eu^{3+} in free polyaminocarboxylate chelates [24]. The quantum yield of lanthanide chelates bound to biomolecules can then simply be determined by comparing lifetimes to the free chelates. Acceptor quantum yields can be measured by conventional means: intensity or lifetime comparison to a standards such as Fluorescein (QY = 0.93 in 1 N NaOH [53]), Tetramethylrhodamine (QY = 0.58 in 10 mM Na-phosphate buffer, pH 7.46, 80 mM NaCl, room temperature [54]), or SulfoRhodamine 101 (QY = 1, lifetime = 4.36 nsec, in methanol [55]).

The importance of equation 9 is that it allows accurate measurement of relatively small amounts of energy transfer (distances > R_0). It is also interesting to note, though not widely appreciated, that by combining equations 4, 5, and 9, the calculated distance depends only on the acceptor quantum yield, and *not* on the donor quantum yield:

$$R = C \left(\frac{I_{D_A} q_A}{I_{A_D}} \right) \qquad (10)$$

where C is simply all the constants in R_0 except q_D. Finally, Equations 9 and 10 can also be used in conventional FRET, but here the direct excitation of acceptor must first be subtracted off [56].

Advantages of LRET

The primary drawbacks of FRET, i.e. using conventional dyes, are several-fold. a) FRET operates over a limited distance range, typically < 75 Å, because of the relatively small size of R_o and limited signal/background at larger distances (although see [57] and Alexa dyes from Molecular Probes Inc., www.probes.com). (b) E generally depends on the orientation of dyes, as well as their relative distance. (c) The finite size of the probes and uncertainty of dye position with respect to biomolecule attachment site, causes the measured dye-dye distances different than the protein-protein distances. (d) Incomplete labeling of biomolecules with dyes can make it difficult to extract distances.

While relying on the same fundamental dipole-dipole mechanism, LRET has many technical advantages over FRET. These include greater: distance accuracy and range; the ability to resolve multiple donor-acceptor distances; a greater ability to isolate signal from biomolecules labeled with both donor and acceptor, even in the presence of biomolecules labeled only with donor or only with acceptor; and less sensitivity of energy transfer to orientation of dyes. Like FRET, LRET shares the problem of sizable probes. (The chelate's atomic structure has also been determined [58], and it is roughly the same size as conventional fluorescence dyes). The linker in LRET is often somewhat shorter than for FRET probes [26].

The fundamental advantages of LRET arise because the donor emission is long-lived (Fig 2b; msec compared to nsec of acceptor or conventional dyes), sharply-spiked emission (Fig. 2a; peaks of a few nanometer width), has a high quantum yield, and is unpolarized.

1. *An order of magnitude greater accuracy in distance-determination can be achieved with LRET* because the energy transfer process is dominated by the distance between the donor and acceptor, and their relative orientations play only a minor role in determining energy transfer efficiency. A worst case scenario is 12% uncertainty in distance determination due to orientation effect. This advantage is because terbium, and usually europium emission is unpolarized [23]. This contrasts to FRET where the errors due to orientation effects can be unbounded. We have shown that angstrom changes due to protein conformational changes can readily be measured with LRET [15, 17].

2. *A 50-100-fold improvement in signal to background (S/B) can be achieved with LRET.* Specifically, energy transfer can be measured with essentially no contaminating background, a stark-contrast to many FRET dye pairs. By temporal and spectral discrimination, donor emission and acceptor emission – both intensity and lifetime — can be independently measured. This leads to dramatically improved signal to background compared to most FRET pairs. Specifically, in LRET the acceptor emission due *only* to energy transfer —the sensitized emission—can be measured with no background. Contaminating background in FRET when trying to measure energy transfer via an increase in acceptor fluorescence, arises from two sources: direct excitation of the acceptor by the excitation light (called direct acceptor fluorescence) and donor emission at wavelengths where one looks for acceptor emission. In LRET both sources are eliminated. For example, by choosing an acceptor such as fluorescein and looking around 520 nm, donor emission is dark (Fig. 2a). By using pulsed excitation and collecting light at 520 nm only after a few tens of microseconds, all the direct acceptor emission (which has nanosecond lifetime) has decayed away. Any acceptor photons emitted after a few microseconds following the excitation pulse, is therefore due only to energy transfer. Small backgrounds mean small signals, corresponding to relatively large distances (10 nm), can be measured.

3. *Samples that contain donor-only or acceptor-only can be spectrally and temporally discriminated against.* Often when labeling proteins, particularly in living cells, one gets an unknown distribution of donor-donor, donor-acceptor, acceptor-acceptor mixture. In FRET this makes distance-determination difficult. In LRET, sensitized emission from acceptor arises only from donor-acceptor labeled complex (see preceding paragraph). Energy transfer of this donor-acceptor labeled complex can then be determined by comparing the lifetime of sensitized emission (τ_{A_D}), which decays with micro- to millisecond lifetime of donor that is transferring energy to the acceptor, with the donor-only lifetime. The ability to measure energy transfer even in complex labeling mixtures is essential for the LRET studies on ion channels presented below.

16.4. INSTRUMENTATION AND APPLICATIONS:

The instrumentation to perform LRET is relatively simple, although slightly more complex than conventional steady-state fluorimeters. The general requirements are a pulsed UV excitation source and time-resolved detection. The pulsed excitation source is usually a Nitrogen laser (337nm, 5 nsec pulse-width typical, 20-50 Hz repetition rate).

For lifetime measurements, a photomultiplier tube with suitable color filters and counting electronics is used. For time-delayed spectra, a spectrometer, typically utilizing diffraction gratings, and either a time-gated photomultiplier tube or preferably a CCD, gated either electronically or with a mechanical chopper, is used. A schematic of the instrument built in our laboratory is shown in Figure 16.11 and details are given elsewhere [24, 59].

The technical advantages of LRET open up many applications. Here we highlight two representative examples that use LRET *in vivo* and *in vitro*.

Ion channels:

We have used LRET to measure conformational changes in the Shaker potassium ion channel, a voltage-gated channel involved in nerve impulses. In many ways this is an extremely demanding use of LRET. The measurement is on a living cell (Xenopus oocytes) and hence purification of completely labeled donor-acceptor species is not possible. A heterogeneous mixture of labeled proteins exists, all in the presence of non-specific labeling to other membrane components. Furthermore, two distances are expected to exist (see below) and the distance changes (as a function of voltage – see below) are quite small – a few angstroms. The technical advantages of LRET help overcome these difficulties.

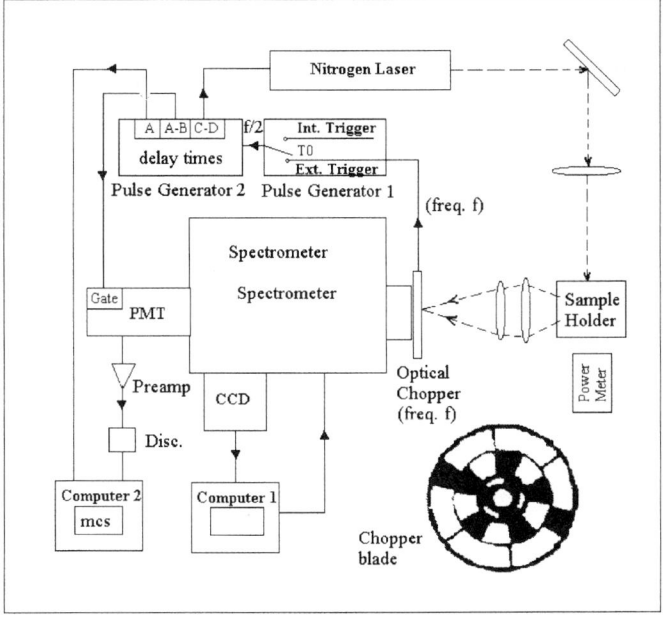

Figure 16.11: LRET Instrumentation. A pulsed nitrogen laser excites the lanthanide sample, and emission is collected by a mechanically-chopped spectrometer and CCD for time-delayed spectral measurements, or a spectrometer and electronically-gated PMT for excited-state lifetime measurements. (Figure from [59].)

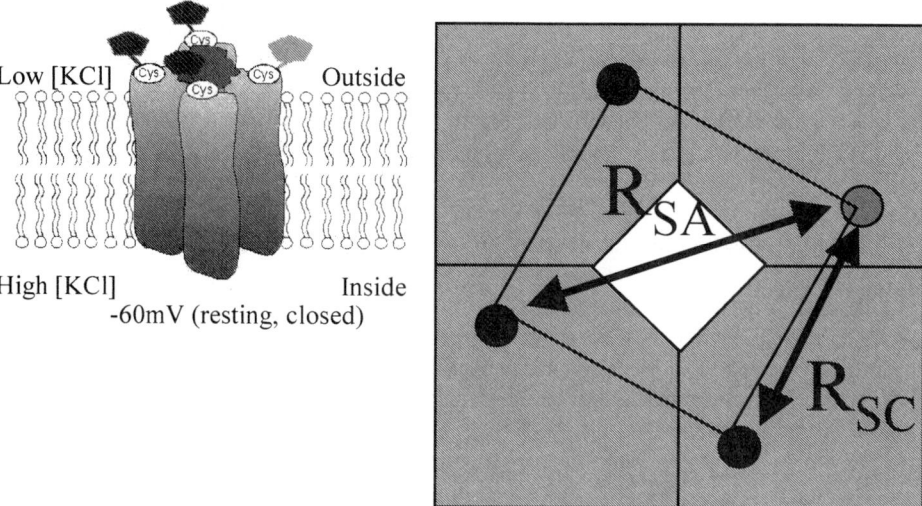

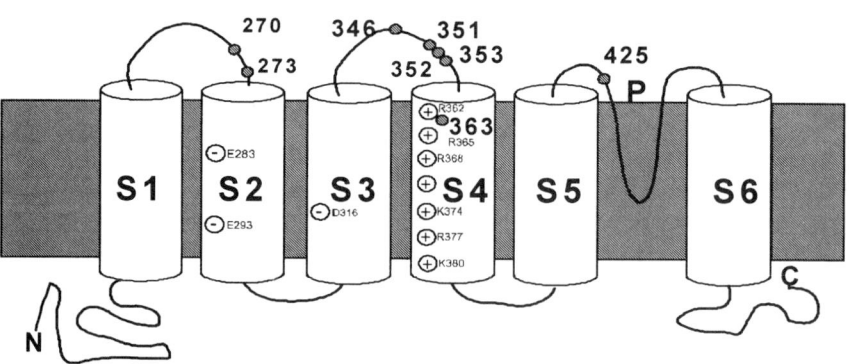

Figure 16.12 (Please see color inserts section) : Structure of Shaker potassium ion channel and labeling scheme. **a.** Side view, **b.** top view, **c.** sub-structure. The channel consists of a central pore (red in 6a) surrounded by four identical subunits. Each subunit consists of six transmembrane domains (6c) and is labeled with either a donor (blue, 6a,b), or acceptor (green, 6a,b). Labeling is done such that there are 3 donors (blue) and only one acceptor (green) per channel. Specific labeling is achieved by introducing a unique cysteine in the S3-S4 linker, near S4, which is the voltage sensor.

The channel is a transmembrane protein, consisting of four identical subunits (Fig. 12a, b) with 4-fold symmetry. Each subunit contains six transmembrane spanning segments, S1-S6 (Fig. 12c). A pore, or channel, is formed at the intersection of the four subunits, which is opened or closed, i.e. "gated", depending on the voltage across the cell membrane. At the resting transmembrane potential of approximately –60 mV the pore is closed. Upon depolarization to approximately 0 mV, the protein undergoes a conformational change that ultimately leads to an opening of the pore allowing potassium ions to flow from the inside to outside of the cell down its electrochemical gradient. This flow of ions is the current, along with sodium ions flowing through analogous sodium channels, which forms nerve impulses.

One of the transmembrane segments, S4, is known as the "voltage sensor" and contains seven positively charged amino acids. These charges feel a force due to the transmembrane potential and hence are likely to move in response to changes in the potential. A second segment, S2, also contains some positively charged residues and likely plays a secondary role in voltage sensing as well. Fundamental questions remain regarding how the channel senses and responds to voltage in the membrane.

For labeling the channel, a single engineered cysteine was introduced at various positions in the S3-S4 linker, near the top of the S4. Each channel therefore contains 4 cysteines, one on each subunit. Conveniently, the *Shaker* channel does not contain native cysteines that are reactive to extracellularly applied probes. Channels were expressed in Xenopus oocytes and labeled with a mixture of donor and acceptor probes, the donor in excess to ensure that most channels contain at most only one acceptor. Under this condition, two different donor-acceptor distances are expected (Fig 12b). A donor sees an acceptor on a contiguous subunit (distance R_{SC}) or on a subunit across the pore (distance R_{SA}). To measure these distances, we focused on measuring the sensitized emission lifetime. This has the great advantage that those channels containing all donors -- the majority of channels -- do not contribute signal and can be ignored. (Those containing all acceptors can also be ignored, although this is a very small fraction of the channels.) We therefore expect the sensitized emission lifetime to be bi-exponential, with the shorter lifetime corresponding to the greater E and shorter distance. Fig. 13a shows this behavior for a probe labeled at position 346. The two distances are in excellent agreement with the expected Pythagorean relationship. By placing probes at various positions, ranging from 363 near the top of S4, to 346 near the middle of the S3-S4 linker, we found that the intersubunit distances decreased. This implies that S3-S4, (and perhaps S4), is tilted towards the pore as one moves in the extracellular direction.

In FRET, and to a certain extent in LRET, absolute distances are always more difficult to measure than relative distances. However, to check whether our absolute distances were reasonable, we measured distances between residues 425. This residue is found in the crystal structure of the KcsA channel, a (non-voltage-gated) prokaryotic analog of Shaker containing two transmembranes per subunit, analogous to S5 and S6. We found $R_{SA} = 30$ Å, in excellent agreement with the $C_{\alpha} - C_{\alpha}$ 29 Å distance in the crystal structure [60]. Furthermore, after publication of our LRET results, other workers measured distances using "tethered linkers" and found excellent agreement in absolute distances to our results [61]. This is in sharp contrast to the FRET results [62], which yielded much larger absolute distances. The latter probably occurred because of uncertainties in donor quantum yields, and possibly because of the κ^2 factor.

Next we measured intersubunit distances as a function of voltage. Changes in lifetime, and hence distances between site 346 near S4 are shown in Fig. 13b. Fig. 13c shows a plot of R_{SC} vs. voltage, superimposed on gating charge movement. Strikingly, the changes in distance at 346 strongly mirror gating charge movement, implying that the distances we measure at 346 are related to the charge movement in S4 – and functioning of the channel. By modeling the distance vs. voltage curve, we concluded that a large transmembrane motion did not occur [17]. Furthermore, small but statistically significant changes in distance were found at positions 350, 351, 352, where 351 moved farther apart, 350 remained unchanged, and 352 moved closer together (Fig. 14a). The simplest model to account for this non-monotonic behavior is to postulate that the S3-S4 linker is helical, and undergoes a rotation about its long axis (Fig. 14b).

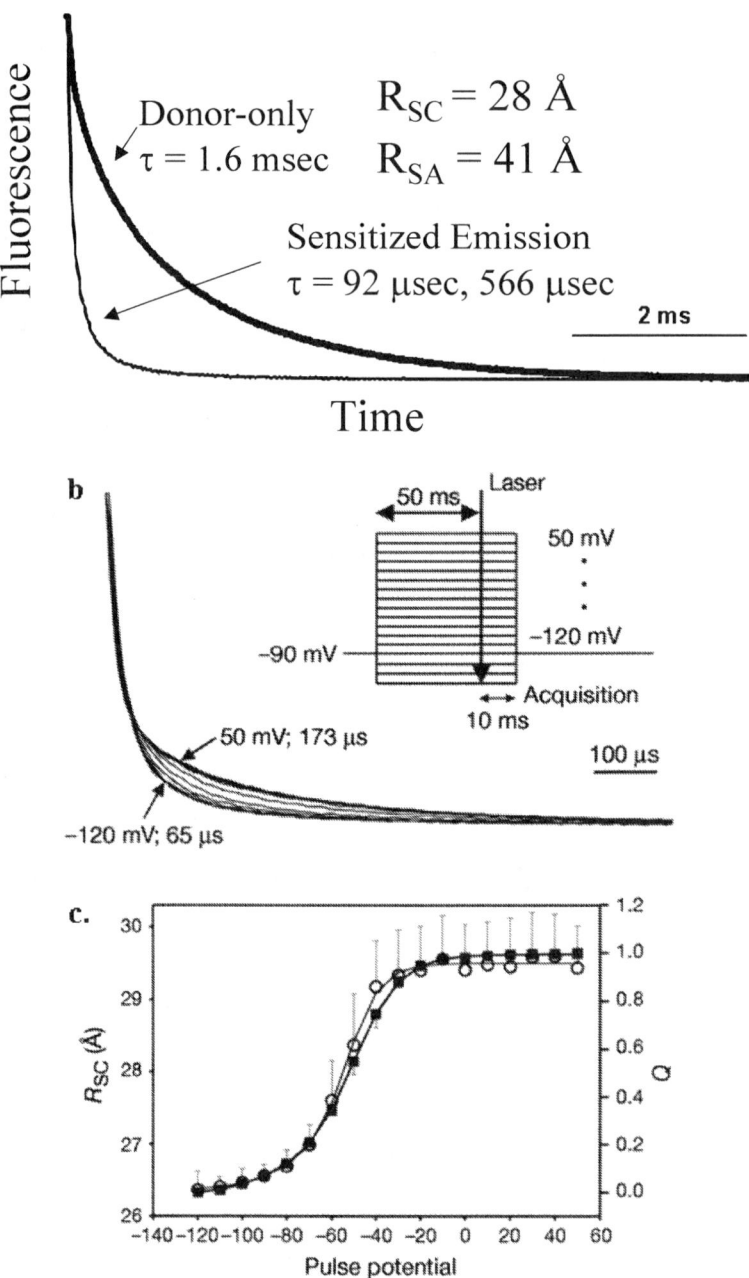

Figure 16.13: a. Biexponential sensitized emission, corresponding to two donor-acceptor distances, corresponding to distances between subunits across the channel and neighboring subunits. **b.** Voltage-dependent changes in sensitized emission arising from movement of S346C in the voltage-sensing region of Shaker potassium channel. **c.** Changes in distance between S346 and amount of charge in S4 moved across membrane potential. The changes in distance closely mirror the charge movement in S4. (Figures from [17].)

Since the S3-S4 linker distance changes are coupled closely to the charge movement of S4, S4 may also undergo a rotation (Fig. 14c) in response to voltage. That such small distance changes can be measured is a tribute to the power of LRET, although interpretation of such small distance changes must be made with caution. Interestingly, a rotation in ligand-gated ion channels [63, 64] and a transporter [65], has recently been measured, suggesting that helix rotation may be a general feature of membrane channels.

Recent work by MacKinnon et al. [66-68] have suggested that the potassium voltage sensing channels involves a large paddle consisting of S4 and part of S3. While this does not match with LRET results mentioned above, future experiments using LRET will likely help settle any discrepancies between the two models.

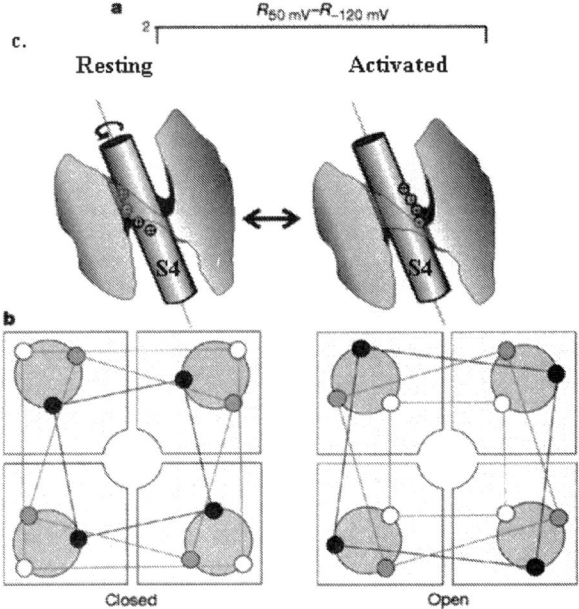

Figure 16.14: Changes in distance between sites 351-353. These data can be explained by a rotation of a helical segment of the ion channel (b), leading to a model where the voltage sensor, S4, may undergo a rotation in response to voltage (c). (Adapted from [17, 62].

Smooth Muscle ADP
Swing:

We have also used LRET to study the conformational changes that occur in smooth muscle myosin for different nucleotide/actin conditions [27]. LRET provides an excellent way to measure the absolute distances of the conformational changes of myosin as it moves through its nucleotide cycle. We can alter myosin so that only one cysteine is exposed for labeling with our lanthanide probes, and then exchange on a light chain to the myosin that is labeled with an acceptor, which in this case was TMR (or Alexa-546) [27].

Muscle contractions occur due to two proteins, actin and myosin. Nucleotide-induced conformational changes within myosin cause a relative movement of myosin with respect to actin, hence converting the chemical energy in ATP into mechanical work [69-71]. However, it is unclear how, or if, these nucleotide-induced changes depend on the presence of actin. This is critical because crystal structures have been an important tool used in determining the conformation changes within myosin during its cycle, but myosin is only crystallized in the absence of actin. If there are actin-dependent conformational changes, then the current crystal structures may represent only a small subset of the actual conformational changes that occur during the catalytic cycle of myosin.

The available myosin crystal structures, as well as spectroscopic studies, of the myosin head, indicate that the myosin powerstroke arises from a relative rotation of the light chain domain of myosin with respect to the catalytic domain when the myosin undergoes a transition from an ADP-P_i state to an ADP state. However, most myosins, including smooth and cardiac muscle myosin II, all non-muscle myosin IIs, myosin V, myosin VI, and and brush border myosin I, also display an additional rotation of the light chain domain upon release of ADP when bound to actin [72, 73]. For smooth muscle myosin, this additional rotation may be necessary before ADP can be released, thereby slowing the release of ADP. Physiologically, this may be associated with the "latch" state, i.e., smooth muscle's ability to generate high tension with minimal ATP turnover.

By measuring LRET with donor lifetimes, acceptor lifetime, and donor-acceptor intensity, we were able to explore the different myosin conformations in the presence and absence of actin [27]. Figure 16.15 shows TMR's (the acceptor) intensity for the nucleotide/actin conditions of myosin. The D-A (donor-acceptor) +ADP+AlF$_4$ (or BeF$_3$) state (also referred to as the trapped state) shows the most energy transfer and corresponds the pre-power stroke state of the myosin before it undergoes a power stroke. AlF$_4$ and BeF$_3$ mimic a free P_i. The D-A state shows the post-power stroke state (also referred to as the rigor state) following the hydrolysis of ATP and the release of all nucleotides. As is shown in figure 16.15, myosin in the presence ADP or actin are essentially the same as the rigor state. However, upon the addition of both ADP and actin, we see an increase in energy transfer. This corresponds to a state of myosin that is actin dependent. Figure 16.16 also shows the acceptor lifetime (or sensitized lifetime) for the same system. Once again the trapped states show the shortest lifetime and hence the most energy transfer, while the rigor, ADP, and actin state show have the longest lifetime and correspond to the least energy transfer. Upon the addition of actin and ADP, there is a slight increase in energy transfer corresponding to the ADP state of smooth muscle myosin. Donor lifetimes show the same result, but are not shown here.

16.5. FINAL REMARKS

New biophysical techniques invariable open up new applications. The development of new probes is leading to a dramatic expansion of the use of fluorescence in general, and FRET- (LRET)-based techniques in particular. The most pressing issue is the ability to site-specifically label probes. Temporal and spectral discrimination when using lanthanides in energy transfer measurements help decrease the sensitivity to non-specific labeling. However, particularly for cellular work, more selective means of attachments for both donors and acceptors are needed. Two different methods, one for donor, and one for acceptor, would be ideal. Genetically encoded dyes such as Green Fluorescent Proteins is one method of selective attachment [74]; dyes such as "FLASH", which bind to a highly unusual six amino acid motive via an aresenic moiety, is another [75]; dyes modified to contain Ni, which can then coordinate to a hexahistidine group engineered into a protein is yet another [76]. Using the power of these new labeling methods with the power of lanthanides will likely shed new light on biophysical systems in the near future.

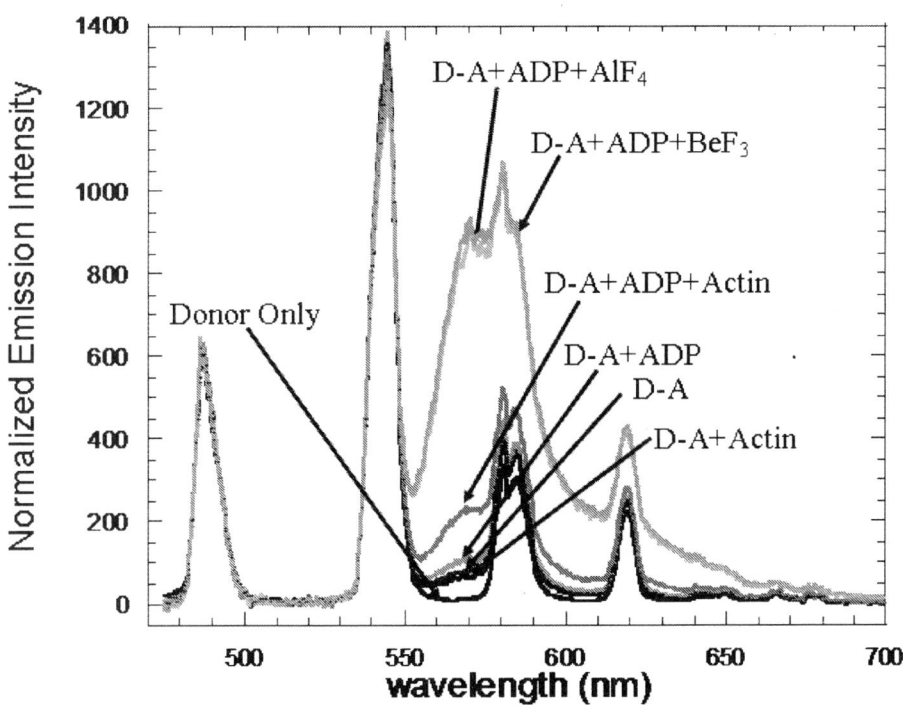

Figure 16.15: CCD (or spectra) measurements of LRET for smooth muscle myosin II.

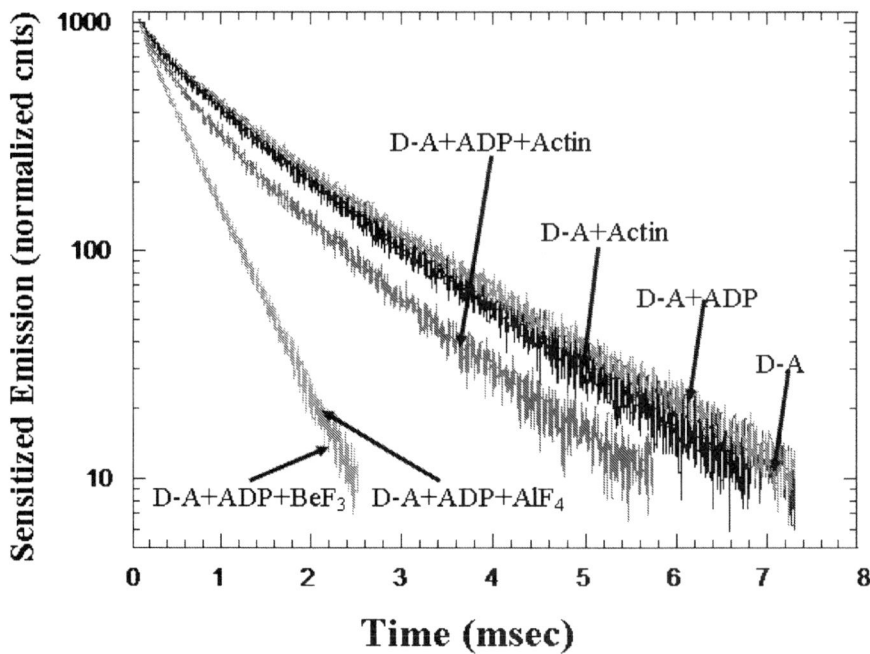

Figure 16.16: Sensitized lifetime for smooth muscle myosin II.

16.6. ACKNOWLEDGMENTS

This work was supported by NIH AR44420, NSF 9984841, and through the Material Research Laboratory, Univ. of Illinois, DOE grant DEFG 02-91ER45439.

16.7. REFERENCES

1.   Seveus, L., et al., *Use of Fluorescent Europium Chelates as Labels in Microscopy Allows Glutaraldehyde Fixation and Permanent Mounting and Leads to Reduced Autofluorescence and Good Long-Term Stability.* Microscopy Res. and Technique, 1994. **28**: p. 149-154.
2.   Marriott, G., et al., *Time-resolved delayed luminescence image microscopy using an europium ion chelate complex.* Biophysical Journal, 1994. **67**: p. 957-965.
3.   Stryer, L., D.D. Thomas, and C.F. Meares, *Diffusion-Enhanced Fluorescence Energy Transfer*, in *Ann. Rev. of Biophys. Bioeng.*, L.J. Mullins, Editor. 1982, Annual Reviews, Inc.: Palo Alto, CA. p. 203-222.
4.   Selvin, P.R., *Lanthanide-based resonance energy transfer.* IEEE J. of Selected Topics in Quantum Electronics: Lasers in Biology, 1996. **2**(4): p. 1077-1087.

5.   Mathis, G., *Probing molecular interactions with homogeneous techniques based on rare earth cryptates and fluorescence energy transfer.* Clinical Chem., 1995. **41**(9): p. 1391-1397.

6.   Mathis, G., *Rare earth cryptates and homogeneous fluoroimmunoassays with human sera.* Clinical Chem., 1993. **39**(9): p. 1953-1959.

7.   Mathis, G., et al., *Homogeneous immunoassays using rare earth cryptates and time resolved fluorescence: principles and specific advantages for tumor markers.* Anticancer Res, 1997. **17**(4B): p. 3011-4.

8.   Kolb, A.J., J.W. Burke, and G. Mathis, *A homogeneous, time-resolved fluorescence method for drug discovery,* in *High Throughput Screening: The Discovery of Bioactive Substances,* J.P. Devlin, Editor. 1997, Marcel Dekker Inc. p. 345-360.

9.   Stenroos, K., et al., *Homogeneous time-resolved IL-2-IL-2R alpha assay using fluorescence resonance energy transfer.* Cytokine, 1998. **10**(7): p. 495-9.

10.   Farrar, S.J., et al., *Stoichiometry of a Ligand-gated Ion Channel Determined by Fluorescence Energy Transfer.* J. Biol. Chem., 1999. **274**(15): p. 10100-10104.

11.   Blomberg, K., P. Hurskainen, and I. Hemmila, *Terbium and Rhodamine as Labels in a Homogeneous Time-resolved Fluorometric Energy Transfer Assay of the B Subunit of Human Chorionic Gonadotropin in Serum.* Clinical Chemistry, 1999. **45**(6): p. 855-861.

12.   Jones, S.G., et al., *Improvements in the Sensitivity of Time Resolved Fluorescence Energy Transfer Assays.* J. Fluorescence, 2001. **11**(1): p. 13-21.

13.   Heyduk, E., et al., *Conformational changes of DNA induced by binding of chironomus high mobility group protein 1a (cHMG1a).* J. Biol. Chem., 1997. **272**(32): p. 19763-19770.

14.   Heyduk, E. and T. Heyduk, *Architecture of a complex between the sigma70 subunit of Escherichia coli RNA polymerase and the nontemplate strand oligonucleotide. Luminescence resonance energy transfer study.* J Biol Chem, 1999. **274**(6): p. 3315-22.

15.   Xiao, M., et al., *Conformational changes between the active-site and regulatory light chain of myosin as determined by luminescence resonance energy transfer: The effect of nucleotides and actin.* Proc. Nat'l. Acad. Sci., USA, 1998. **95**: p. 15309-15314.

16.   Chen, J. and P.R. Selvin, *Lifetime and color-tailored fluorophores in the micro- to milli-second time regime.* J. Am. Chem. Soc., 2000. **122**(4): p. 657-660.

17.   Cha, A., et al., *Atomic scale movement of the voltage sensing region in a potassium channel measured via spectroscopy.* Nature, 1999. **402**: p. 809-813.

18.   Lis, S., et al., *Energy Transfer in Solution of Lanthanide Complexes.* Journal of Photochemistry & Photobiology A: Chemistry, 2002. **150**: p. 223-247.

19.   Saha, A.K., et al., *Time-Resolved Fluorescence of a New Europium Chelate Complex: Demonstration of Highly Sensitive Detection of Protein and DNA Samples.* J. Am. Chem. Soc., 1993. **115**: p. 11032-11033.

20.   Xu, Y.Y., et al., *Simultaneous quadruple-label fluorometric immunoassay of thyroid- stimulating hormone, 17 alpha-hydroxyprogesterone, immunoreactive trypsin, and creatine kinase MM isoenzyme in dried blood spots.* Clin Chem, 1992. **38**(10): p. 2038-43.

21.   Bunzli, J.-C.G., *Luminescent Probes,* in *Lanthanide Probes in Life, Chemical and Earth Sciences, Theory and Practice,* J.-C.G. Bunzli and G.R. Choppin, Editors. 1989, Elsevier: New York. p. 219-293.

22.   Drexhage, K.H., *Monomolecular Layers and Light.* Sci. Amer., 1970. **222**(3): p. 108-119.

23.   Reifenberger, J., et al., *Emission Polarization Properties of Europium and Terbium Chelates.* J. Phys. Chem B, 2003. **107**: p. 12862-12873.

24.   Xiao, M. and P.R. Selvin, *Quantum Yields of Luminescent Lanthanide Chelates and Far-Red Dyes Measured by Resonance Energy Transfer.* J. Am. Chem. Soc., 2001. **123**: p. 7067-7073.

25.   Li, M. and P.R. Selvin, *Amine-reactive forms of a luminescent DTPA chelate of terbium and europium: Attachment to DNA and energy transfer measurements.* Bioconjugate Chem., 1997. **8**(2): p. 127-132.

26.   Chen, J. and P.R. Selvin, *Thiol-reactive luminescent lanthanide chelates.* Bioconjugate Chem., 1999. **10**(2): p. 311-315.

27.   Xiao, M., et al., *An actin-dependent conformational change in myosin.* Nat Struct Biol, 2003. **10**(5): p. 402-8.

28.   Akabas, M.H., et al., *Acetylcholine receptor channel structure probed in cysteine-substitution mutants.* Science, 1992. **258**: p. 307-310.

29.   Ge, P. and P.R. Selvin, *Thiol-reactive Lanthanide Chelates, II.* Bioconjugate Chemistry, 2003. **14**: p. 870-876.

30.   Ge, P. and P.R. Selvin, *Carbostyril Derivatives as Antenna Molecules for Luminescent Lanthanide Chelates.* Bioconjugate Chemistry, 2004.

31. Schelte, P., et al., *Differential reactivity of maleimide and bromoacetyl functions with thiols: application to the preparation of liposomal diepitope constructs.* Bioconjug Chem, 2000. **11**(1): p. 118-23.

32. Li, M. and P.R. Selvin, *Luminescent lanthanide polyaminocarboxylate chelates: the effect of chelate structure.* J. Am. Chem. Soc., 1995. **117**: p. 8132-8138.

33. Chen, J. and P.R. Selvin, *Synthesis of 7-Amino-4-trifluoromethyl-2-(1H)-quinolinone and its use as an antenna molecule for luminescent europium polyaminocarboxylate chelates.* J. Photochem. Photobio. A:Chemistry, 2000. **5522**: p. 1-6.

34. Crosby, G.A., R.E. Whan, and R.M. Alire, *Intramolecular energy transfer in rare earth chelates: the role of the triplet state.* J. Chem. Phys., 1961. **34**: p. 743.

35. Abusaleh, A. and C. Meares, *Excitation and De-Excitation Processes in Lanthanide Chelates Bearing Aromatic Sidechains.* Photochemistry and Photobiology, 1984. **39**(6): p. 763-769.

36. Kirk, W.R., W.S. Wessels, and F.G. Prendergast, *Lanthanide-Dependent Perturbations of Luminescence in Indolylethylenediaminetetraacetic Acid-Lanthanide Chelate.* J. Phys. Chem., 1993. **97**: p. 10326-10340.

37. Alpha, B., et al., *Antenna Effect in Luminescent Lanthanide Cryptates: A Photophysical Study.* Photochemistry and Photobiology, 1990. **52**(2): p. 299-306.

38. Horrocks, W.D., Jr. and D.R. Sudnick, *Lanthanide Ion Probes of Structure in Biology. Laser-Induced Luminescence Decay Constants Provide a Direct Measure of the Number of Metal-Coordinated Water Molecules.* J. Am. Chem. Soc., 1979. **101**(2): p. 334-350.

39. Horrocks, W.D., Jr., B. Holmquist, and B.L. Vallee, *Energy transfer between Terbium(III) and Cobalt(II) in thermolysins: A new class of metal-metal distance probes.* Proc. Nat. Acad. Sci. USA, 1975. **72**(12): p. 4764-4768.

40. Horrocks, W.D., Jr. and D.R. Sudnick, *Lanthanide Ion Luminescence Probes of the Structure of Biological Macromolecules.* Accounts of Chemical Research, 1981. **14**: p. 384-392.

41. Selvin, P.R., *Principles and Biophysical Applications of Luminescent Lanthanide Probes. Annual Review of Biophysics and Biomolecular Structure*, 2002. **31**: p. 275-302.

42. Selvin, P.R., *The Renaissance in Fluorescence Resonance Energy Transfer.* Nature Structural Biology, 2000. **7**(9): p. 730-734.

43. Lakowicz, J.R., *Principles of Fluorescence.* 2 ed. 1999, New York: Kluwer Academic.

44. Selvin, P.R., *Fluorescence Resonance Energy Transfer*, in *Methods in Enzymology*, K. Sauer, Editor. 1995, Academic Press: Orlando. p. 300-334.

45. Förster, T., *Experimental and Theoretical Investigation of the Intermolecular Transfer of Electronic Excitation Energy.* Z. Naturforsch A, 1949. **4**: p. 321-327.

46. Chakrabarty, T., et al., *Holding two heads together: stability of the myosin II rod measured by resonance energy transfer between the heads.* Proc Natl Acad Sci U S A, 2002. **99**(9): p. 6011-6.

47. Dale, R.E., J. Eisinger, and W.E. Blumberg, *The orientational freedom of molecular probes.* Biophys. J., 1979. **26**: p. 161-194.

48. Stryer, L. and R.P. Haugland, *Energy Transfer: A Spectroscopic Ruler.* Proc. Natl. Acad. Sci., USA, 1967. **58**: p. 719-726.

49. Cantor, C.R. and P.R. Schimmel, *Biophysical Chemistry.* Vol. 2. 1980, San Francisco: W. H. Freeman and Co.

50. Selvin, P.R. and J.E. Hearst, *Luminescence energy transfer using a terbium chelate: Improvements on fluorescence energy transfer.* Proc. Natl. Acad. Sci, USA, 1994. **91**(21): p. 10024-10028.

51. Jovin, T.M. and D.J. Arndt-Jovin, *FRET microscopy: digital imaging of fluorescence resonance energy transfer. Applications in cell biology*, in *Microspectrofluorimetry of Single Living Cells*, E. Kohen, J.S. Ploem, and J.G. Hirschberg, Editors. 1989, Academic Press: Orlando. p. 99-117.

52. Heyduk, T. and E. Heyduk, *Luminescence energy transfer with lanthanide chelates: interpretation of sensitized acceptor decay amplitudes.* Anal Biochem, 2001. **289**(1): p. 60-7.

53. Weber, G. and F.W.J. Teale, *Determination of the Absolute Quantum Yield of Fluorescent Solutions.* Trans. Faraday Soc., 1957. **53**: p. 646-655.

54. Vamosi, G., C. Gohlke, and R. Clegg, *Fluorescence characteristics of 5-carboxytetramethylrhodamine linked covalently to the 5′ end of oligonucleotides: multiple conformers of single-stranded and double-stranded dye-DNA complexes.* Biophys J, 1996. **71**(2): p. 972-994.

55. Karstens, T. and K. Kobs, *Rhodamine B and rhodamine 101 as reference substances for fluorescence quantum yield measurements.* J. Phys. Chem., 1980. **84**: p. 1871-1872.

56. Clegg, R.M., et al., *Observing the Helical Geometry of Double-Stranded DNA in Solution by Fluorescence Resonance Energy Transfer.* Proc. Natl. Acad. Sci. USA, 1993. **90**(7): p. 2994-2998.

57.   Schobel, U., et al., *New Donor-Acceptor Pair for Fluorescent Immunoassays by Energy Transfer.* Bioconjugate Chem., 1999. **10**(6): p. 1107-1114.

58.   Selvin, P.R., et al., *Crystal structure and spectroscopic characterization of a luminescent europium chelate.* Inorganic Chemistry, 1996. **35**: p. 700-705.

59.   Xiao, M. and P.R. Selvin, *An Improved instrument for measuring time-resolved lanthanide emission and resonance energy transfer.* Rev. Sci. Inst., 1999. **70**(10): p. 3877-3881.

60.   Doyle, D.A., et al., *The Structure of the Potassium Channel: Molecular Basis of K+ Conduction and Selectivity.* Science, 1998. **280**: p. 69-77.

61.   Blaustein, R.O., et al., *Tethered blockers as molecular 'tape measures' for a voltage-gated K+ channel.* Nat Struct Biol, 2000. **7**(4): p. 309-11.

62.   Glauner, K.S., et al., *Spectroscopic mapping of voltage sensor movement in the Shaker potassium channel.* Nature, 1999. **402**(6763): p. 813-817.

63.   Horenstein, J., et al., *Protein mobility and GABA-induced conformational changes in GABA(A) receptor pore-lining M2 segment.* Nat Neurosci, 2001. **4**(5): p. 477-85.

64.   Johnson, J.P., Jr. and W.N. Zagotta, *Rotational movement during cyclic nucleotide-gated channel opening.* Nature, 2001. **412**(6850): p. 917-21.

65.   Loo, T.W. and D.M. Clarke, *Cross-linking of human multidrug resistance p-glycoprotein by the substrate, tris-(2-maleimidoethyl)amine, is altered by atp hydrolysis. Evidence for rotation of a transmembrane helix.* J Biol Chem, 2001. **276**(34): p. 31800-5.

66.   Lee, S.Y. and R. MacKinnon, *A membrane-access mechanism of ion channel inhibition by voltage sensor toxins from spider venom.* Nature, 2004. **430**(6996): p. 232-5.

67.   Jiang, Y., et al., *The principle of gating charge movement in a voltage-dependent K+ channel.* Nature, 2003. **423**(6935): p. 42-8.

68.   Jiang, Y., et al., *X-ray structure of a voltage-dependent K+ channel.* Nature, 2003. **423**(6935): p. 33-41.

69.   Rayment, I., et al., *Structure of the actin-myosin complex and its implications for muscle contraction.* Science, 1993. **261**: p. 58-65.

70.   Dominguez, R., et al., *Crystal Structure of a Vertebrate Smooth Muscle Myosin Motor Domain and Its Complex with the Essential Light Chain: Visualization of the Pre-Power Stroke State.* Cell, 1998. **94**: p. 559-571.

71.   Houdusse, A., et al., *Atomic Structure of Scallop Myosin Subfragment S1 Complexed with MgADP: A Novel Conformation of the Myosin Head.* Cell, 1999. **97**: p. 459-470.

72.   Whittaker, M., et al., *A 35-Angstrom Movement of Smooth Muscle Myosin on ADP Release.* Nature, 1995. **378**(6558): p. 748-751.

73.   Gollub, J., C.R. Cremo, and R. Cooke, *ADP release produces a rotation of the neck region of smooth myosin but not skeletal myosin.* Nature Structural Biology, 1996. **3**(9): p. 796-802.

74.   Tsien, R.Y., *The green fluorescent protein.* Annu. Rev. Biochem., 1998. **67**: p. 509-44.

75.   Griffin, B.A., S.R. Adams, and R.Y. Tsien, *Specific Covalent Labeling of Recombinant Protein Molecules Inside Live Cells.* Science, 1998. **281**: p. 269-272.

76.   Kapanidis, A.N., Y.W. Ebright, and R.H. Ebright, *Site-specific incorporation of fluorescent probes into protein: hexahistidine-tag-mediated fluorescent labeling with (Ni(2+):nitrilotriacetic Acid (n)-fluorochrome conjugates.* J Am Chem Soc, 2001. **123**(48): p. 12123-5.

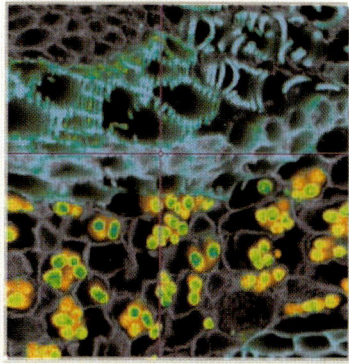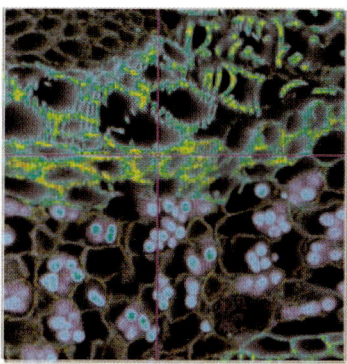

Figure 4.12 (Becker *et al,*. page 93): TCSPC lifetime image recorded in a confocal laser scanning microscope, 512 x 512 pixel scan. Left: Colour represents the mean lifetime of the double exponential decay, blue to red = 200 ps to 2 ns. Right: Colour represents the ratio of the intensity coefficients of the fast and slow decay component, a_{fast} / a_{slow}. Blue to red = 1 to 10.

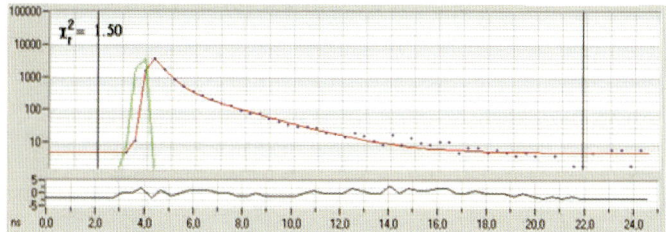

Figure 4.13 (Becker *et al,*. page 93): Fluorescence decay function in a selected pixel of the image, double exponential Levenberg-Marquardt fit and residuals of the fit. Data points are shown blue, the fitted curve red, and the instrument response function green. The pixel is marked in Figure 4.1.

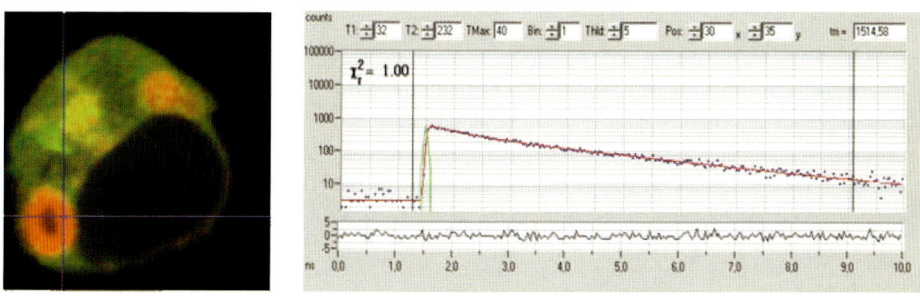

Figure 4.14 (Becker *et al,*. page 93), left: Lifetime image of an HEK cell expressing two interacting proteins labelled with CFP and YFP (left). Colour represents weighted mean lifetime, red to blue = 1500 to 2300 ps. Right: Fluorescence decay function in a selected spot. The lifetime components are 660 ps and 2.3 ns.

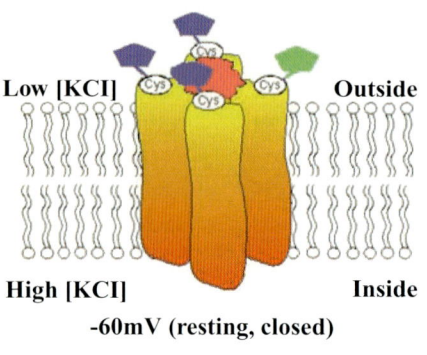

Low [KCl] Outside

High [KCl] Inside

-60mV (resting, closed)

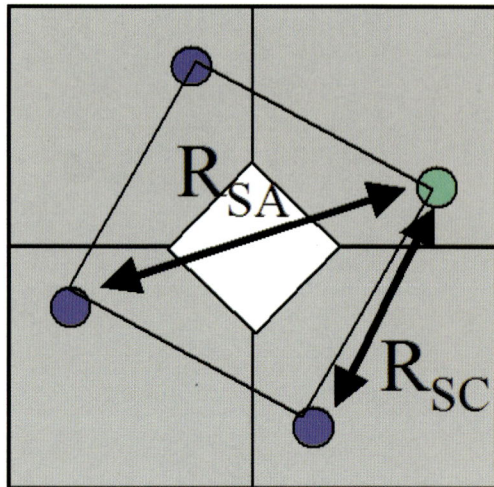

R_{SA}

R_{SC}

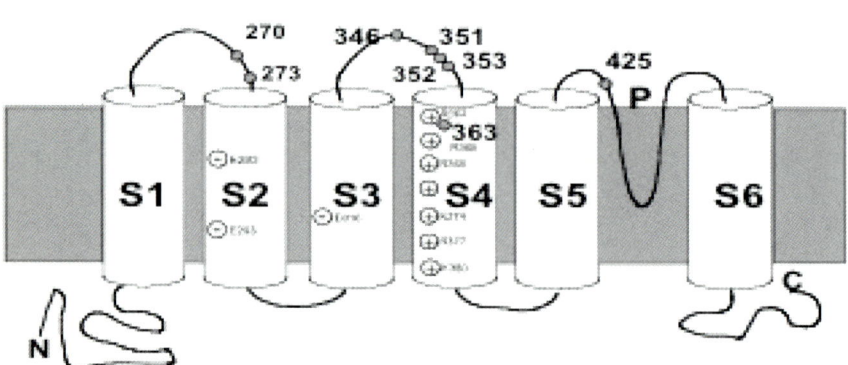

Figure 16.12 (Reifernberger *et al.,* page 422) : Structure of Shaker potassium ion channel and labeling scheme. **a.** Side view, **b.** top view, **c.** sub-structure. The channel consists of a central pore (red in 6a) surrounded by four identical subunits. Each subunit consists of six transmembrane domains (6c) and is labeled with either a donor (blue, 6a,b), or acceptor (green, 6a,b). Labeling is done such that there are 3 donors (blue) and only one acceptor (green) per channel. Specific labeling is achieved by introducing a unique cysteine in the S3-S4 linker, near S4, which is the voltage sensor.

INDEX